Hybrid Energy Systems

Other related titles:

You may also like

- PBPO159 | Oliver Probst, Sergio Castellanos, and Rodrigo Palacios | Transforming the Grid Towards Fully Renewable Energy | 2020
- PBPO115 | DC Distribution Systems and Microgrids | Tomislav Dragičević, Pat Wheeler, and Frede Blaabjerg | 2018
- PBPO095 | S.M. Muyeen and Saifur Rahman | Communication, Control and Security Challenges for the Smart Grid | 2017
- PBPO076 | Abdelhay A. Sallam and Om P. Malik | Power System Stability | 2015

We also publish a wide range of books on the following topics:
Computing and Networks
Control, Robotics and Sensors
Electrical Regulations
Electromagnetics and Radar
Energy Engineering
Healthcare Technologies
History and Management of Technology
IET Codes and Guidance
Materials, Circuits and Devices
Model Forms
Nanomaterials and Nanotechnologies
Optics, Photonics and Lasers
Production, Design and Manufacturing
Security
Telecommunications
Transportation

All books are available in print via https://shop.theiet.org or as eBooks via our Digital Library https://digital-library.theiet.org.

IET ENERGY ENGINEERING SERIES 262

Hybrid Energy Systems

Planning, operation, and optimization

Edited by
Huu Hieu Nguyen, Van Ga Bui, Thanh Viet Dinh,
Minh Quan Duong, and Tran T. Hoang

The Institution of Engineering and Technology

About the IET

This book is published by the Institution of Engineering and Technology (The IET).

We inspire, inform and influence the global engineering community to engineer a better world. As a diverse home across engineering and technology, we share knowledge that helps make better sense of the world, to accelerate innovation and solve the global challenges that matter.

The IET is a not-for-profit organisation. The surplus we make from our books is used to support activities and products for the engineering community and promote the positive role of science, engineering and technology in the world. This includes education resources and outreach, scholarships and awards, events and courses, publications, professional development and mentoring, and advocacy to governments.

To discover more about the IET, please visit https://www.theiet.org/.

About IET books

The IET publishes books across many engineering and technology disciplines. Our authors and editors offer fresh perspectives from universities and industry. Within our subject areas, we have several book series steered by editorial boards made up of leading subject experts.

We peer review each book at the proposal stage to ensure the quality and relevance of our publications.

Get involved

If you are interested in becoming an author, editor, series advisor, or peer reviewer please visit https://www.theiet.org/publishing/publishing-with-iet-books/ or contact author_support@theiet.org.

Discovering our electronic content

All of our books are available online via the IET's Digital Library. Our Digital Library is the home of technical documents, eBooks, conference publications, real-life case studies and journal articles. To find out more, please visit https://digital-library.theiet.org.

In collaboration with the United Nations and the International Publishers Association, the IET is a Signatory member of the SDG Publishers Compact. The Compact aims to accelerate progress to achieve the Sustainable Development Goals (SDGs) by 2030. Signatories aspire to develop sustainable practices and act as champions of the SDGs during the Decade of Action (2020–30), publishing books and journals that will help inform, develop, and inspire action in that direction.

In line with our sustainable goals, our UK printing partner has FSC accreditation, which is reducing our environmental impact to the planet. We use a print-on-demand model to further reduce our carbon footprint.

Published by The Institution of Engineering and Technology, London, United Kingdom

The Institution of Engineering and Technology (the "**Publisher**") is registered as a Charity in England & Wales (no. 211014) and Scotland (no. SC038698).

First published 2025

The Institution of Engineering and Technology
Futures Place,
Kings Way, Stevenage,
Herts, SG1 2UA,
United Kingdom

www.theiet.org

British Library Cataloguing in Publication Data

A catalogue record for this product is available from the British Library

ISBN 978-1-83953-977-0 (hardback)
ISBN 978-1-83953-978-7 (PDF)
ISBN 978-1-80705-009-2 (EPUB3)

Typeset in India by MPS Limited

Cover image: Tan Dao Duy/Moment via Getty Images

Contents

Acknowledgments

This book is the product of close collaboration between Vietnamese and international scholars. We extend our profound gratitude to all who so generously shared their time, expertise and insight. We are likewise indebted to the reviewers and editors of the IET, whose meticulous and professional stewardship ensured the highest standards of quality throughout. Finally, we wish to acknowledge The University of Danang – University of Science and Technology, whose unwavering support and provision of essential facilities enabled our research teams to bring this book to fruition.

Huu Hieu Nguyen et al.

About the editors

Huu Hieu Nguyen is an associate professor and rector of The University of Danang – University of Science and Technology, Vietnam. Prior to that, he was doctoral/postdoctoral researcher at Grenoble Electrical Research Center of Institut National Polytechnique de Grenoble, France. He was involved with national-level projects including research on optimizing the distribution grid using renewable energy sources, and manufacturing of surveillance robots for automating unmanned substations.

Van Ga Bui is a professor at the Faculty Mechanical Engineering of Transport, The University of Danang – University of Science and Technology, Vietnam. Prior positions include Rector of The University of Danang – University of Science and Technology, President of the University of Danang, Vice Minister of Education and Training, Vietnam. He was doctoral/postdoctoral researcher at Ecole Centrale de Lyon (France). He is an Editor-in-chief, Journal of Science and Technology, The University of Danang. His research focuses on fluid mechanics, combustion, renewable energy and air pollution.

Thanh Viet Dinh is an Associate Professor and Vice Rector at Dong A University, Vietnam. He received his PhD degree from Vinnytsia National Technical University (Ukraine). He had been working at the University of Danang for 27 years before joining Dong A University. He has also served as a visiting scholar at Washington State University (USA), the Frankfurt Institute for Advanced Studies (Germany), and Grenoble Institute of Technology (France). His research interests include power system analysis, transmission and distribution networks, renewable energy, and smart grid technologies, with a strong focus on advancing sustainable and resilient energy systems.

Minh Quan Duong is an associate professor at the Faculty of Electrical Engineering, The University of Danang - University of Science and Technology, Vietnam. He received his Ph.D. from Politecnico di Milano, Italy, and has been a visiting scholar at Arizona State University and Indiana University in the United States, as well as the Frankfurt Institute for Advanced Studies in Germany. He is a member of the IEEE Task Force on Fuzzy Systems in Renewable Energy and Smart Grids. His research interests include renewable energy in power systems, power system planning, and smart grids.

Tran T. Hoang is currently a Senior Planning Engineer – Future Systems at Powerco, New Zealand. He completed his PhD in Electrical Engineering at G2Elab, Grenoble Institute of Technology, France, in 2020. He subsequently worked at the National Institute of Solar Energy, French Alternative Energies and Atomic Energy Commission (2020–mid 2022), as a Research Fellow at the University of Auckland, New Zealand (2022–2023), and as a Principal Power System Consultant at Mott MacDonald, New Zealand (2023–March 2025). His research interests focus on renewable energy integration, grid-forming and grid-following inverters, and Advanced Distribution Management Systems (ADMS).

Chapter 1

Introduction: hybrid energy systems: planning, operation, and optimization

Huu Hieu Nguyen[1], Kim Hung Le[1] and Minh Quan Duong[1]

Global demographic forecasts estimate that the world's population will swell to roughly 9.7 billion by 2050 [1]. This unprecedented growth, combined with accelerating urbanization, rising living standards, and rapid industrialization, is poised to add formidable pressure to the planet's energy systems. In megacities across Asia, Africa, and Latin America where skyscrapers, data centers, and mass transit networks are springing up at record speed–energy consumption is set to climb not merely for basic residential needs, but also for climate-resilient infrastructure such as advanced cooling systems, desalination plants, and flood-control mechanisms. These trends mean that electricity demand in cities alone could soon eclipse that of entire countries today. Emerging economies are projected to supply more than 60% of this additional consumption, reflecting both their demographic dynamism and their aspirations for higher per capita income. Industrial corridors in India and Southeast Asia, for example, are expanding steel, cement, and chemical output—all heavily energy-intensive sectors—while simultaneously electrifying transport fleets to curb local air pollution. According to the International Energy Agency (IEA), overall global energy demand is likely to stand about 45% above its 2019 baseline by mid-century [2], even after accounting for significant gains from efficiency measures and the rising share of renewables. For policymakers, this surge presents a complex trilemma: energy must remain secure to avoid supply disruptions, affordable to underpin economic competitiveness, and sustainable to honor climate-stabilization goals. Achieving these objectives will require unprecedented investment in grid modernization, large-scale energy storage, green hydrogen, and carbon-capture technologies, alongside robust regulatory frameworks that promote innovation and equitable access. Businesses, too, must adapt—by decarbonizing their value chains, embracing circular economy practices, and deploying digital tools that squeeze more productivity from every kilowatt. The scale and urgency of the challenge are clear; so too are the opportunities for those ready to lead the next energy transition.

Conventional, centrally dispatched generation fleets that rely on coal, oil, and natural-gas turbines—together with legacy, radial transmission networks—are increasingly incapable of safeguarding frequency, voltage, and resource adequacy in the face of surging global electricity demand. Escalating concerns over

[1] Department of Electrical Engineering, The University of Danang – University of Science and Technology, Vietnam

greenhouse-gas emissions, ambient-air contaminants, and the volatility of fossil-fuel supply chains are catalyzing a systemic pivot toward cleaner, more sustainable technologies, with renewable-based power conversion platforms occupying center stage. This paradigm shift has driven rapid innovation and cost compression across the renewable spectrum, underpinning stricter "green" and "sustainability" benchmarks that now frame investment and policy decisions [3]. Modern renewable portfolios embrace an extensive array of non-deployable primary energy resources, including photovoltaic (PV) and concentrated solar-thermal power, onshore and offshore wind turbines, conventional and run-of-river hydropower, enhanced geothermal systems, sustainably harvested biomass and advanced biofuels, tidal-range and tidal-stream generators, as well as wave-energy converters. By virtue of harnessing naturally replenished fluxes, these resources avert the irreversible depletion of terrestrial or marine materials and significantly mitigate life-cycle carbon footprints. High penetration levels of variable renewable energy (VRE), however, introduce new operational and planning challenges. Fluctuating irradiance and stochastic wind regimes can erode system inertia, precipitate frequency excursions, spawn harmonic distortion, and complicate reactive-power management. To bolster resilience, contemporary Hybrid Energy Systems (HES) integrate utility-scale Energy Storage Systems (ESS)—including lithium-ion, sodium-sulfur and redox flow batteries; flywheels; supercapacitors; compressed-air and pumped-hydro storage—as the primary spinning-reserve surrogate. These assets enable energy shifting, ramp-rate smoothing, synthetic inertia injection, and black-start capability, thereby curtailing the reliance on diesel gen-sets traditionally deployed for contingency support. Where load demand temporarily outstrips the discharge capacity of the ESS, auxiliary diesel generators may be sequenced in as tertiary reserves; still, their dispatch must be meticulously optimized in view of fuel logistics, marginal operating cost, and CO_2 externalities, with storage always afforded precedence. Notwithstanding sharp cost declines, global capital flows remain insufficient to align with Paris-compatible trajectories. The International Renewable Energy Agency (IRENA) reported that renewable investments reached roughly USD 1.4 trillion in 2022—an 8% uptick year-on-year—yet must quintuple to approximately USD 5 trillion annually to cap the mean temperature rise well below 2°C [4]. Bridging this finance gap will require robust policy instruments—such as carbon-pricing regimes, green bonds, and blended-finance vehicles—and the overhaul of market architectures to appropriately monetize flexibility and environmental attributes. Concurrently, grid codes, forecasting algorithms, and power electronic interfaces must evolve to accommodate low-inertia dynamics, suppress subsynchronous oscillations, and maintain power quality under increasingly volatile meteorological conditions. The technical and economic promise of a renewable-centric grid is unequivocal. Still, its realization hinges on integrated planning, adaptive operation, and continuous optimization across the generation–transmission–distribution nexus.

A Hybrid Renewable Energy System (HRES) represents an integrated topology in which two or more heterogeneous renewable energy resources—such as PV arrays, wind, small hydro units, biomass generators, or geothermal modules—are electrically coupled to deliver a power supply that is inherently more reliable and more energy-efficient than any single-resource Renewable Energy System (RES). Figure 1.1 juxtaposes the operating characteristics of a conventional RES with those

	HRES	RES
Components	Two or more renewable energy sources	One renewable energy source
Efficiency	High	Medium
Reliability	High	Medium
Initial cost	High	Low
Scalability	Easy	Depends on the source
Flexible distribution	High	Medium
Grid dependency	Low	High

Figure 1.1 Differences between HRES and RES

of an HRES, clearly illustrating the complementary nature of hybridized resources: temporal production gaps in one source (e.g., nocturnal PV deficits) are compensated by surplus generation in another (e.g., elevated nocturnal wind output). The overarching objective of an HRES is to capitalize on the strengths of each constituent technology—high solar conversion efficiency during peak irradiance, superior wind capacity factors offshore, dispatchable biomass generation—while mitigating their respective intermittency and seasonal constraints so that load requirements are met continuously, with acceptable power-quality indices and minimal curtailment [5]. Recent breakthroughs in power-electronic converters (SiC/GaN-based multilevel inverters and bidirectional DC–DC stages) and the advent of high-resolution meteorological/load-forecasting algorithms have amplified the techno-economic attractiveness of the hybrid paradigm. Concomitantly, precipitous declines in levelized capital costs for PV modules, wind turbines, and advanced battery storage have lowered the barrier to entry, rendering HRES deployment both technically viable and economically competitive. Converging with the irreversible global shift toward carbon neutrality, Smart Grids (SG) have been rolled out at scale to accommodate bi-directional power flows, real-time monitoring, and autonomous control of distributed hybrid assets. The International Energy Agency (IEA) estimates that SG implementation can curtail transmission-and-distribution losses by approximately 5%–15% worldwide [6]. Empirical data corroborate these projections: In the United States, grid modernization initiatives reduced aggregate losses from 7% to 6% over the 2010–2020 decade [7], whereas in Haryana, India, smart-metering and feeder reconfiguration schemes trimmed losses from 25% to 15% within five years [8]. Collectively, the synergy between HRES and SG infrastructures forms a pivotal cornerstone in the global quest for Net-Zero emissions.

HRES optimizes the use of locally available resources while driving down the levelized cost of electricity. In both grid-connected and stand-alone configurations, they offer the operational agility required to satisfy load profiles that vary according to geography and climate. Modern HRES are usually engineered by co-locating two or more complementary renewable technologies. Widely adopted topologies include PV–wind, PV–hydropower, wind–hydropower, and solar–thermal–biomass schemes, each selected to exploit offsetting generation patterns and thereby improve supply security and overall conversion efficiency Figure 1.2. Among these, the PV-wind

Figure 1.2 Some common types of hybrid renewable energy systems (HRES)

pairing dominates due to its global resource availability: photovoltaic modules harvest irradiation during daytime hours, whereas wind turbines frequently deliver peak output at night or under overcast skies, thereby mitigating meteorological intermittency and supporting near-continuous production. A generic HRES architecture comprises: renewable generators as the primary source of electrical power; an energy-storage subsystem—flywheels, battery energy-storage systems (BESS), pumped-storage hydro, hydrogen fuel cells, etc.—to buffer short- and long-term fluctuations; and a forecasting and supervisory control layer that orchestrates dispatch and maintains power-quality compliance (Figure 1.3) [5,9,10]. Most reported HRES operate in islanded mode, reflecting their intended deployment in remote or insular locations where grid extension is impractical. Nevertheless, each functional block may be tailored—substituting, for example, super-capacitors for flywheels or integrating tidal generation in coastal sites—to align the plant with indigenous resources and maximize techno-economic performance, as illustrated conceptually in Figure 1.4.

In modern HRES, the ESS functions as the primary spinning reserve, instantaneously absorbing supply–demand imbalances and deferring the need for thermal backup. Diesel generators are therefore relegated to an emergency role: they deliver firm, continuous power when all other resources are exhausted, yet their fossil-fuel dependence and high specific emissions are antithetical to carbonization goals. Contemporary supervisory strategies thus target explicit minimization of diesel run-hours while maximizing renewable utilization and maintaining system reliability. Compared with a conventional synchronous grid, HRES supervision is inherently more complex. Legacy grids draw on a limited palette of dispatchable plants—such as coal, oil or gas—whose output can be scheduled in large, predictable blocks. By contrast, a hybrid plant must orchestrate a heterogeneous portfolio of photovoltaic arrays, wind turbines, hydro units, and multi-hour battery packs whose production is driven by stochastic meteorological conditions. As

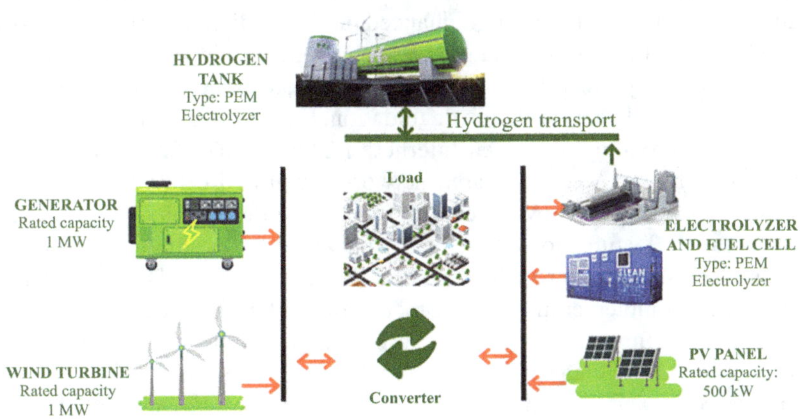

Figure 1.3 Energy Management System (EMS) for HRES Wind, PV, and BESS [5,9,10]

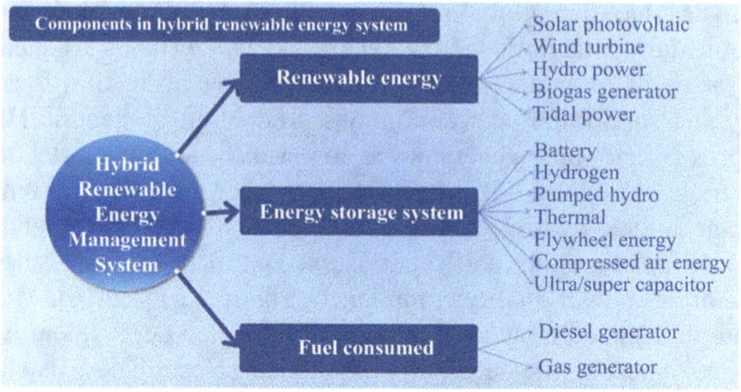

Figure 1.4 Components in HRES

a result, the management architecture is decentralized and adaptive, distributing intelligence to the edge so that individual resources can autonomously negotiate operating points while a central optimizer coordinates global objectives. The monitoring tier ingests high-resolution telemetry—voltages, currents, weather parameters, state-of-charge, vibration and thermal signatures—then applies advanced analytics to benchmark efficiency and flag incipient faults. At the control tier, closed-loop routines regulate inverter setpoints, battery charge–discharge profiles and curtailment commands to maintain frequency, voltage and power-quality indices within statutory limits. Above this lies a multi-objective optimization layer that solves rolling, mixed-integer formulations to determine generation schedules, reserve allocations and market bids. Inputs include short-term load forecasts, probabilistic renewable energy forecasts, day-ahead electricity prices, ancillary service tariffs, and equipment maintenance windows. Digital enablers substantially elevate this hierarchy. Artificial intelligence and machine learning models, recurrent and temporal convolutional neural networks, provide minute-to-hour forecasts for demand and variable generation. Reinforcement-learning-enhanced model predictive control continuously updates dispatch policies as new data arrive, coping gracefully with forecast error and asset-state uncertainty. Deep-learning classifiers mine vibration, infrared and partial-discharge streams to detect insulation degradation, bearing wear or hotspot formation long before catastrophic failure. An Internet of Things (IoT) fabric underpins pervasive sensing. Low-power, wide-area networks such as LoRaWAN and NB-IoT, permit the cost-effective deployment of thousands of nodes. Meanwhile, edge analytics extracts salient features locally, reducing backhaul bandwidth and latency while preserving cybersecurity through Transport Layer Security (TLS)-based mutual authentication. Complementing this, Blockchain and broader Distributed-Ledger Technology (DLT) furnish an immutable, transparent substrate for settlement. Smart contracts automatically clear peer-to-peer energy trades, tokenizing the provision of frequency response and spinning reserve, and guarantee non-repudiation of metering data—thereby enabling self-governing micro-markets within and across interconnected microgrids. The integration of Artificial Intelligence (AI), IoT and DLT yields pronounced benefits: improved forecast accuracy, tighter real-time control, reduced operating costs, enhanced cyber resilience and new revenue streams from ancillary service participation. Consequently, investment in advanced HRES management systems delivers not only robust financial returns but also significant social value—accelerating the penetration of renewable energy, displacing fossil fuel consumption, and contributing measurably to global climate mitigation objectives.

Smart Grid technology constitutes the digital backbone of contemporary Hybrid Renewable Energy Systems, furnishing the intelligence, interoperability and real-time visibility required to integrate variable renewable energy sources seamlessly into legacy power networks. Through a multilayer communications fabric—spanning field bus protocols, IEC 61850 substation automation, and wide-area IP backbones—the SG enables high-speed, bidirectional exchange of telemetry and control signals among generators, storage assets, protection relays, and market platforms. Synchrophasor data, advanced metering infrastructure (AMI) readings, and dense sensor arrays stream continuously to supervisory applications, allowing operators to forecast RES output, track net demand, and orchestrate assets with sub-second granularity. The consequent improvement in situational awareness translates directly into enhanced dispatch efficiency, reduced curtailment, and tangible economic value for stakeholders across the supply chain. AI and Machine Learning (ML) algorithms are now deeply embedded in this cyber-physical architecture. Gradient-boosted tree ensembles, recurrent neural networks and physics-informed digital twins ingest historical weather data, satellite irradiance imagery, turbine Supervisory Control and Data Acquisition (SCADA) records and socio-economic demand drivers to produce high-fidelity forecasts of load and renewable generation on horizons ranging from minutes to days. These predictions feed into model-predictive control schemes that continuously adjust inverter setpoints, schedule charge/discharge cycles for the BESS, and issue ancillary-service bids in wholesale markets. As a result, wind, solar, and small-hydro units operate in a fully autonomous yet coordinated manner, maximizing renewable penetration while minimizing both spilled energy and fossil-fuel backup. Battery storage, empowered by SG intelligence, has emerged as the pivotal resource for balancing the stochastic nature of RES. Lithium-ion, sodium-sulfur, and emerging flow-battery chemistry provide fast-acting frequency regulation, ramp-rate smoothing, and peak-shaving capabilities that previously required expensive spinning reserves from thermal plants. Optimization algorithms within the SG environment evaluate battery degradation costs, round-trip efficiency, and price signals to determine economically optimal dispatch trajectories. This not only enhances system reliability and power-quality indices but also unlocks stacked revenue streams—capacity payments, frequency-response contracts, and energy arbitrage—that improve the bankability of HRES projects. Beyond operational gains, the SG's two-way information infrastructure catalyses new market paradigms. Prosumer households equipped with rooftop PV and behind-the-meter storage participate actively in trans-active energy platforms, trading excess generation peer-to-peer under blockchain-verified smart contracts. Microgrids have the capability to operate independently during emergencies, ensuring an uninterrupted power supply to essential systems. Moreover, they play a vital role in restoring power by providing black-start functionality when reconnected to the main grid. Demand-side management programs leverage real-time price signals and ML-based occupancy detection to shift flexible loads—such as Electric Vehicle (EV) charging; Heating, Ventilation, and Air Conditioning (HVAC); and industrial processes—thus flattening net-load profiles and deferring costly network reinforcements. Taken together, these advances create a virtuous cycle: sophisticated data analytics optimize the utilization of storage, which in turn raises the effective hosting capacity for renewable energy, driving further investment in green generation. Comparative techno-economic analyses consistently demonstrate that well-managed HRES, underpinned by SG functionality, achieve a lower levelized cost of electricity (LCOE) and higher resiliency than systems dependent on a single renewable resource or conventional stand-alone grids. In essence, the Smart Grid is no longer merely an incremental upgrade; it is the enabler of a flexible, decarbonized, and customer-centric electricity ecosystem, accelerating the transition to sustainable energy futures worldwide. ESS have become a cornerstone of power-system resilience and sustainable economic growth. As highlighted in [11], coupling ESS with HRES markedly mitigates the intermittency of solar and wind generation. Within an HRES framework, ESS can be dispatched under multiple operating modes—frequency regulation, voltage support, spinning-reserve substitution, and peak-load shaving—thereby extracting greater value from each stored kilowatt-hour. The economic upside is equally compelling. By absorbing surplus generation during off-peak windows and releasing it when the locational marginal price spikes, an ESS can arbitrage energy to elevate revenue streams or provide firm capacity when grid contingencies arise [12]. Such temporal shifting capability is essential for maintaining supply–demand equilibrium and safeguarding system reliability whenever renewable output falls short of the forecast. Technology selection remains application-specific. Comparative assessments weigh gravimetric and volumetric energy density, calendar life, depth-of-discharge tolerance, round-trip efficiency, and total cost of ownership to identify the optimum battery chemistry. While conventional lithium-ion packs dominate today's deployments, inherent constraints—finite cycle life, safety concerns at high state-of-charge, and effective efficiencies in the 85%–95% range—are catalyzing interest in emerging media such as vanadium redox flow batteries, zinc–bromine hybrids, and green-hydrogen storage, each promising extended durability and larger usable capacity. At the plant level, the Power Conversion Unit (PCU) anchors ESS integration. Its heart, the power-electronics converter (PEC), must translate DC bus energy into grid-compliant AC with minimal total harmonic distortion, strict power factor control and fault-ride-through capability. Precise PEC specification—topology, semiconductor device choice and modulation strategy—therefore dictates overall safety, efficiency and grid code compliance. Beyond bulk storage, ESS deployment accelerates parallel markets: it stabilizes voltage for electric-vehicle fast-charging corridors, underpins microgrid black-start services, and enhances the bank's ability of high-renewable portfolios. Properly managed through advanced battery-management systems and predictive maintenance analytics, ESS is positioned to address long-duration backup requirements, unlock higher renewable penetrations, and deliver the operational stability indispensable to next-generation HRES.

HRES have become a focal point of contemporary energy engineering research, with investigations spanning resource modeling, optimization, storage integration and niche micro-hybrid applications. Recent literature converges on the development of adaptive HRES frameworks that dynamically rebalanced generation shares in response to real-time meteorological inputs, load variations, and market signals. By coupling high-resolution weather data with stochastic or robust optimization techniques, these models can continuously fine-tune the contribution of solar PV, wind, biomass, small hydro, and auxiliary thermal units, thereby increasing renewable penetration while preserving system reliability. Within the optimization domain, significant effort has been channelled into the creation of metaheuristic and AI-assisted solvers—such as genetic algorithms, particle swarm optimization, non-dominated sorting genetic algorithm 2 (NSGA-II) and machine-learning-enhanced forecast engines—that minimize the life-cycle cost of energy, unmet-load probability and greenhouse-gas emissions simultaneously. Commercial platforms such as HOMER Pro as well as bespoke research codes now embed these algorithms, enabling researchers to rapidly iterate through thousands of design permutations and perform multi-objective trade-off analyses that account for fuel price volatility, demand uncertainty and component degradation. Parallel to advances in modeling, energy-storage research focuses on elevating both the performance and techno-economic appeal of BESS and alternative long-duration storage options. Investigations span material discovery (e.g., high-entropy electrolyte formulations, silicon-dominant anodes), system-level enhancements (novel thermal management loops, modular flow-battery stacks) and holistic cost benchmarking against incumbent lithium-ion technology. Comparative studies highlight the importance of operational constraints including allowable depth-of-discharge, round-trip efficiency, and end-of-life recyclability. These factors significantly influence the selection of storage chemistry, which in turn affects the optimal system architecture, capital expenditure, and overall project lifespan. Research into micro-hybrid configurations—sub-100 kW clusters designed for households, agricultural holdings, and small enterprises—has expanded the applicability of HRES beyond utility-scale projects. Hybrid energy systems often include components such as rooftop PV arrays, small wind turbines, battery banks with capacities ranging from 20 to 80 kWh, and backup diesel or Liquefied Petroleum Gas (LPG) generators. These systems are interconnected using bidirectional converters and smart inverters, enabling smooth transitions between islanded (off-grid) and grid-connected operations. Field trials report renewable fractions above 70%, diesel savings exceeding 60% and payback periods of under seven years in favorable policy environments, underscoring the economic viability of decentralized hybrid supply. Case-study evidence further substantiates the practical merit of these concepts. Reference [13] conducted a comprehensive techno-economic and environmental assessment for residential electrification in Bostegan Village, Hormozgan Province, Iran. Using the HOMER optimization suite and the algorithmic flow chart depicted in Figure 1.5 of the original work, the authors identified a hybrid configuration comprising PV, wind, an AC/DC converter, a lead-acid battery bank, and a standby diesel generator. The system achieved LCOE of 1.058 USD/kWh with a renewable fraction of 64%. The Step-wise Weight Assessment Ratio Analysis—Weighted Aggregated Sum Product Assessment (SWARA–WASPAS) decision-making approach effectively prioritized design alternatives, demonstrating adaptability through a comprehensive

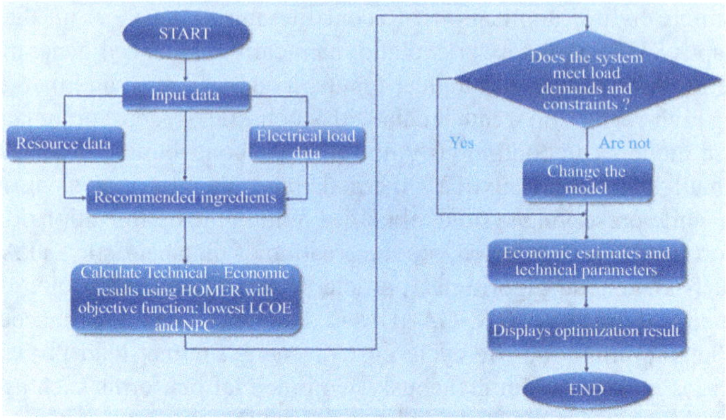

Figure 1.5 The data analysis workflow diagram of HOMER

27-scenario sensitivity analysis. This analysis accounted for dynamic factors such as fluctuating fuel prices and the unpredictable nature of renewable resource availability, ensuring robust and informed decision-making. In the worst-case projection, the LCOE rose to 1.343 USD/kWh, yet the hybrid option remained cost-competitive. The integration of a 50 MW electrolyzer for surplus power utilization demonstrates the capacity to produce roughly 64,363 tons of hydrogen annually. This approach highlights the potential for creating valuable opportunities through power-to-hydrogen conversion, supporting energy storage and sustainable fuel solutions. A separate investigation, detailed in [14], examined solar–biomass hybridization for remote Indian communities where high-voltage grid extension is uneconomic. India's abundant agricultural residues and strong solar energy potential offer a promising foundation for sustainable energy solutions. By integrating parabolic-trough solar fields with biomass boilers within a shared Rankine cycle, the study explores an innovative approach to harnessing these resources efficiently. This combination not only optimizes energy production but also aligns with environmental goals by utilizing renewable and locally available materials. The authors concluded that system attractiveness will rise as steam-based thermal-storage technologies (molten salts, phase-change materials) mature, facilitating 24/7 dispatch. Cogeneration systems, which produce electricity, process heat, and chilled water simultaneously, offer significant advantages for medium- and small-scale installations. These systems achieve energy-use efficiencies that are 15%–25% higher than those of stand-alone biomass plants. This makes cogeneration an effective and sustainable solution for providing electrification, cooling, and heating in rural and peri-urban areas. Battery Energy Storage System sizing is commonly characterized by several key parameters, including rated power and energy capacity, minimum and maximum state of charge (SOC), initial SOC levels, round-trip efficiency, and associated capital costs. These factors are critical in determining the performance, feasibility, and economic viability of BESS solutions across various applications [15–18]. Early-stage evaluations frequently exclude specific battery chemistry details to simplify calculations and reduce uncertainty in parameters, making it a practical method for initial assessments. When advancing to the detailed engineering stage for site-specific applications, it becomes crucial to account for the distinct characteristics of various energy storage technologies. Chemistries such as lithium-ion, vanadium redox flow, and sodium-sulfur (NaS) batteries present unique capital expenditure profiles, thermal management requirements, replacement schedules, and end-of-life disposal considerations. These factors can significantly influence financing strategies, permitting processes, and long-term operational risks. Furthermore, challenges such as limited cycle life under high-depth-of-discharge conditions, ventilation and fire safety concerns, and intricate recycling logistics highlight the importance of incorporating technology-specific modeling into high-precision feasibility analyses. Contemporary research in HRES provides a comprehensive framework that encompasses adaptive resource modeling, advanced optimization techniques, integrated storage solutions, and detailed case study validation. These efforts collectively contribute to the broader goal of achieving widespread integration of renewable energy. However, significant challenges remain, including the need to account for storage degradation within dispatch algorithms, establish standardized and cyber-secure communication protocols for micro-hybrid controllers, and develop regulatory mechanisms that adequately address the externalities associated with fossil fuel-based backup systems. Exploring sector coupling, such as power-to-heat, power-to-mobility, and power-to-chemicals, is essential for advancing resilient, low-carbon energy systems globally. By integrating these sectors, we can unlock greater flexibility and create new revenue streams, ensuring a more sustainable and adaptable energy infrastructure for the future. Flywheel Energy Storage (FES) technology provides significant advantages, including high power density, rapid response times, and the ability to cycle indefinitely without degradation [18]. These features make it particularly effective for applications such as fast frequency regulation and black-start support in power systems. By leveraging inertial energy, flywheels can quickly restore power to de-energized generators and synchronize them with the grid. However, continuous operation introduces challenges such as parasitic load and accelerated self-discharge. Additionally, the need for a specialized vacuum enclosure increases initial investment costs. Rotor burst containment and robust safety interlocks are critical for ensuring operational safety in dynamic systems. To address over-generation scenarios, excess renewable energy output is managed through a dump-load assembly. This approach prevents over-frequency excursions, albeit with some conversion losses. These mechanisms collectively contribute to maintaining dynamic stability during rapid changes in load or fluctuations in renewable energy generation. Techno-economic assessments of Hybrid Renewable Energy Systems HRES emphasize two primary metrics: the Net Present Cost (NPC) and LCOE [13,15,17]. NPC accounts for all discounted costs over the system's lifecycle, including capital, replacement, and operation and maintenance expenses, offering a comprehensive view of long-term affordability. On the other hand, LCOE calculates the average cost per kilowatt-hour of energy produced by dividing the total discounted costs by the total discounted energy output. This metric allows for effective comparisons across different technologies, such as PV-battery versus PV-FES hybrids. By minimizing both NPC and LCOE, decision-makers can achieve configurations that balance competitive electricity pricing with financial sustainability, aiding in optimal sizing, dispatch strategies, and component selection for robust and cost-efficient HRES implementations.

Applying optimization techniques can help HRES enhance performance, save costs, increase reliability, maximize the utilization of renewable energy, and provide flexible responses to changes in renewable energy sources, supporting decision-making for operators. These methods are categorized and grouped into major classes, including artificial intelligence techniques, iterative methods, and software tools. Each major group is further subdivided into smaller branches, as presented in Figure 1.6. In [16], two methods, the Genetic Algorithm (GA) and Particle Swarm Optimization (PSO), are employed to determine the optimal component sizing of HRES, including configurations such as PV–wind–biomass and PV–wind systems. This analysis helps assess the economic feasibility of deploying HRES by minimizing the cost of energy (COE) and ensuring reliability through the assumption of a maximum value for the loss of power supply probability (LPSP). The PSO method showed better results than GA in achieving lower COE and higher performance in both energy management strategies: Cycle Charging Strategy (CCS) and Load Following Strategy (LFS). The Energy Management System (EMS) proposed in [19] utilizes the Differential Evolution (DE) algorithm to optimize the performance of a microgrid integrated with renewable energy sources. The microgrid includes distributed generators, photovoltaic panels, and battery storage. This system is also connected to the main utility grid to exchange power when the purchase and sale costs are favorable. EMS is optimized to reduce the operating costs of the microgrid and maximize profits from power exchanges. The results, analyzed using MATLAB®, indicate that the DE algorithm outperforms in terms of quality and convergence speed. Study [20] introduces a dual optimization method for developing a short-term wind power forecasting tool that combines artificial neural networks with PSO and genetic algorithms. The models utilize PSO to tune the neural network parameters and then apply either

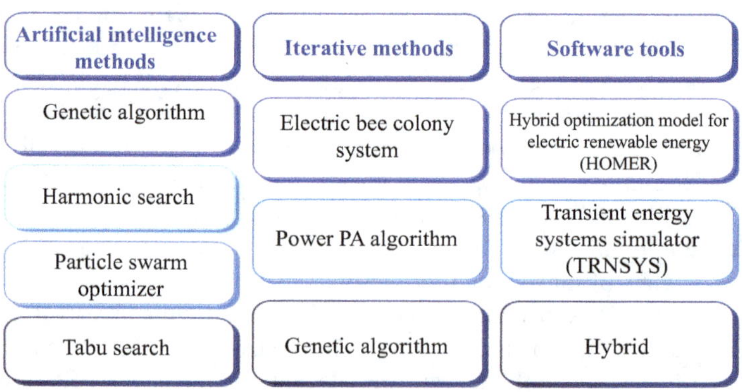

Figure 1.6 Common optimization methods in HRESs

PSO or genetic algorithms to further optimize the PSO parameters, thereby improving forecast accuracy. Testing at the Tuy Phong wind farm in Binh Thuan, Vietnam, showed improved accuracy and broad applicability to other wind farms.

The HOMER platform is widely recognized for its capability to conceptualize and optimize microgrids and Hybrid Renewable Energy Systems [13,15,16]. It enables designers to integrate conventional generators, renewable energy sources, storage technologies, and demand-side management into a cohesive optimization framework. Through multi-objective simulations, HOMER identifies the least-cost system configurations while analyzing lifecycle economics and environmental impacts. Its advanced reporting and visualization tools help stakeholders evaluate trade-offs among capital costs, fuel savings, emissions reduction, and system reliability, thereby simplifying the deployment of complex hybrid energy architectures. For real-time supervisory control, a two-layer strategy has been introduced that combines the Modified Elephant Herding Optimization Algorithm (MEHOA) with Tabu Search (TS) [21]. In the outer loop, MEHOA generates a lookup table mapping instantaneous power imbalances to potential control actions. The inner loop optimizes proportional-integral (PI) controller gains using TS, ensuring rapid stabilization of frequency and voltage deviations. This approach, implemented in MATLAB/Simulink®, outperforms traditional PI controllers as well as Genetic Algorithm and Particle Swarm Optimization methods, achieving faster settling times and minimal overshoot. Voltage regulation and reactive power compensation are addressed using a photovoltaic–wind hybrid array interfaced with a Distribution Static Synchronous Compensator (D-STATCOM) controlled by an Artificial Neural Network (ANN) [22]. Simulations conducted across three operating scenarios show significant improvements, including a reduction in real power losses from 380 MW to 323 MW and an increase in bus voltage from 1.056 pu to 1.083 pu. The ANN-controlled D-STATCOM proves to be an effective solution for maintaining power-quality standards in smart grid environments. The Vietnam Transmission Efficiency Project, as assessed by the World Bank in May 2024, has achieved significant enhancements to the country's high-voltage transmission infrastructure across its four key economic regions [23]. Through upgrades such as reinforcing critical substations, reconductoring heavily loaded transmission corridors, and implementing advanced monitoring systems, the project has boosted transmission capacity by approximately 15%. Additionally, it has reduced average operation and maintenance costs per transmitted megawatt-hour by nearly 20%. These improvements have also enhanced reliability indices, facilitating the integration of variable renewable energy sources while preventing an estimated 95,000 tons of greenhouse gas emissions annually. Vietnam's power sector is undergoing significant digital modernization, supported by advanced technologies such as SCADA upgrades, phasor measurement units, and asset-management platforms. These developments lay the groundwork for enhanced efficiency and reliability across the nation's energy infrastructure. Simultaneously, the Vietnam Renewable Energy Development Project has successfully attracted private investments into large-scale wind, solar, and small hydro projects, boosting renewable energy to nearly 10% of the country's total installed capacity. Vietnam's favorable maritime conditions also hold substantial potential for offshore wind farms and floating solar installations.

Additionally, the dynamic coastlines in central provinces present opportunities for tidal and wave energy technologies, pending advancements in cost-effectiveness and technological solutions.

World Bank assessments released in May 2024 confirm that the Vietnam Transmission Efficiency Project has measurably upgraded the country's high-voltage backbone across the four principal economic regions [23]. By reinforcing critical substations, reconductoring heavily loaded corridors and deploying state-of-the-art monitoring systems, the program has elevated transmission capacity by roughly 15% while trimming average operation-and-maintenance expenditure per transmitted MWh by close to 20%. Reliability indices have improved commensurately, thereby enabling large-scale absorption of variable renewable generation and avoiding an estimated 95,000 tons of greenhouse gas emissions each year. The digital infrastructure deployed under the project—SCADA upgrades, phasor measurement units, and advanced asset-management platforms—also establishes a foundation for the broader digital modernization of Vietnam's power sector. Running in parallel, the Vietnam Renewable Energy Development Project has catalyzed private capital inflows into utility-scale wind, solar, and small hydro assets, increasing the renewable share to almost 10% of the nation's installed capacity. Favorable maritime conditions provide latent potential for offshore wind arrays and floating photovoltaic plants, while energetic coastlines in the central provinces offer promising—but as yet untapped—resource envelopes for tidal and wave technologies once cost and technology barriers are overcome. Vietnam's electricity sector development strategy, outlined in Decision No. 500/QD-TTg (PDP8) promulgated in 2023, establishes a comprehensive framework for achieving carbon neutrality by 2050. This plan spans the period from 2021 to 2030, with a long-term vision extending to 2045. Key interim milestones include enhancing energy security, accelerating the deployment of renewable energy, and reducing reliance on coal. By 2030, wind and solar energy are expected to contribute approximately 31% of the nation's electricity demand, increasing to around 60% by 2050. Natural gas is identified as the primary transitional fuel for the 2020s, gradually being replaced by low-carbon hydrogen starting after 2040. In addition to this, the strategy incorporates measures such as establishing inter-regional renewable-energy service hubs, advancing electrification programs for rural, mountainous, and island areas, and importing electricity from Laos to diversify the energy supply. These integrated policies, combined with Vietnam's resource advantages, have positioned the country as a key player in the global clean-energy value chain. HRES, which combine resources like wind, solar, small hydro, storage, and backup generation, are well-suited to Vietnam's geographical diversity and dispersed settlement patterns. These systems are particularly effective in delivering dependable and climate-resilient power to remote areas, such as highland communes and archipelagic districts, where extending the traditional power grid is either technically challenging or economically unfeasible. By implementing HRES, Vietnam can promote inclusive development while simultaneously advancing its national goals for decarbonization.

Vietnam's transition to a sustainable energy future hinges on the successful integration of HRES into microgrids. By adopting these systems, the nation can reduce the environmental impact of power generation while aligning with global climate goals. This shift not only protects natural ecosystems but also drives economic growth by fostering opportunities within the clean-energy sector. As the modernization of Vietnam's power infrastructure progresses, the demand for skilled professionals in advanced energy technologies will grow, making workforce development a critical priority. However, regulatory challenges remain a significant hurdle. Vietnam currently lacks a unified framework to facilitate seamless interconnection between microgrids and the national grid. Establishing clear technical standards, simplifying permitting processes, and creating bankable tariff structures are essential steps to encourage private investment. Fiscal incentives, such as tax breaks and concessional financing, would further accelerate the adoption of HRES by reducing financial barriers. By implementing these policy measures, Vietnam can ensure that microgrids are integrated efficiently and cost-effectively, thereby paving the way for a reliable and sustainable energy system that supports the nation's broader energy-transition goals.

Bach Long Vi, positioned strategically at the entrance of the Gulf of Tonkin, stands as Vietnam's most remote territory within the region. Its unique location not only reinforces the nation's sovereignty but also acts as a catalyst for economic development. The island boasts significant potential for tourism, maritime services, and defense logistics. However, its progress has historically been constrained by reliance on a diesel-powered electricity system, which is both costly and logistically challenging. Recognizing the need for sustainable growth and reduced carbon emissions, a transformative solution was implemented in early 2022. An islanded HRES was introduced, combining solar panels, wind turbines, and advanced battery energy storage technologies. This innovative system addresses the island's increasing power demands while significantly reducing dependence on conventional fossil fuels, paving the way for a greener and more efficient future. The deployment of the microgrid architecture has led to extensive optimization studies focusing on enhancing technical efficiency and reducing lifecycle costs. Reference [24] introduced a robust design featuring a 500 kW PV array, a 1 MW wind turbine generator (WTG), a 2 MWh 500 kW BESS, and two 1 MW diesel generators (DG) for backup. To further reduce reliance on diesel fuel and utilize surplus renewable energy, the system incorporated a hydrogen pathway. Excess power was converted into hydrogen via an electrolyzer, stored, and later transformed into electricity through fuel cells. Three alternative configurations were analyzed: Plan 1 – PV/WTG + BESS + DG; Plan 2 – PV/WTG + Hydrogen storage/fuel cell + DG; Plan 3 – PV/WTG + BESS + Hydrogen storage/fuel cell + DG. A Mixed-Integer Quadratic Programming (MIQP) scheduler was developed to optimize unit commitment, dispatch, and storage operations while accounting for uncertainties in renewable energy resources during typical summer and winter conditions. Among the configurations, Plan 3 emerged as the most effective, delivering the lowest operating expenditure (OPEX), the highest renewable energy integration, minimized curtailment of excess energy, and reduced greenhouse gas (GHG) emissions. The microgrid currently operational on Bach Long Vi Island exemplifies an advanced HRES system, integrating diverse technologies for optimized performance. This system includes 500 kW of PV capacity, a 1 MW WTG, a 2 MWh BESS with a 500 kW discharge rate, dual 1 MW diesel generators, and a hydrogen subsystem. Key metrics highlight its effectiveness: an NPC of

9,262,712 USD and a LCOE at 0.1118 USD/kWh, and annual CO_2 emissions reduced to 584,343 kg. Additionally, the system achieves an 88% renewable energy penetration rate, minimizes electricity losses to 27.1%, and reduces reliance on diesel generators to just 7.54%. This hybrid-storage configuration demonstrates significant economic and environmental benefits, showcasing reduced dependence on fossil fuels and alignment with Vietnam's decarbonization goals. The success of the Bach Long Vi microgrid provides a scalable model for electrifying remote islands, emphasizing how well-designed HRES systems can enhance energy security, lower costs, and contribute to environmental sustainability. This approach serves as a replicable framework for similar regions aiming to transition to cleaner energy solutions (Figure 1.7). Phu Quy Island, an archipelago of ten islets located 120 km southeast of Phan Thiet, spans an area of 16.5 km² and is home to approximately 27,500 residents across 5,677 households [25]. Known for its exceptional wind conditions, with an annual mean wind speed exceeding 9 m/s, the island presents significant potential for HRES. Currently, the microgrid on the island is powered by a combination of diesel generators (70%) and wind turbines (30%). However, with electricity demand projected to grow at an average annual rate of 21%, there is a pressing need for a more sustainable and cost-effective energy dispatch strategy to meet future requirements. The optimization framework detailed in Reference [26] is a comprehensive approach designed to improve system performance by integrating economic, operational, and stability considerations. It operates through three key stages: (1) Load Aggregation – This step involves clustering hourly demand data using the Kmax–Kmin algorithm to create representative load blocks. By doing so, the complexity of energy demand patterns is reduced, enabling more effective analysis and planning. (2) Wind Speed Classification – High-quality anemometric data undergoes clustering with the same algorithm to categorize wind speed series. For datasets that are incomplete or noisy, approximate groupings are developed to address these limitations and ensure reliable classification. (3) Mixed-Integer Optimization – A deterministic linear programming model is implemented in Lindo, leveraging a scenario matrix that integrates load

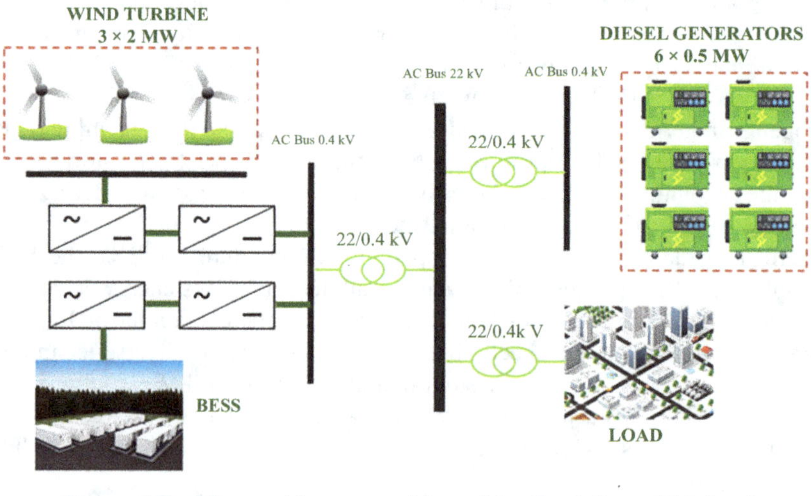

Figure 1.7 Microgrid power grid used for Bach Long Vi Island

blocks with wind speed classifications. The primary goal is to minimize diesel fuel costs while maintaining adherence to generator constraints, reserve margins, and power-balance requirements. Concurrently, the framework strives to maximize wind energy utilization. Even with the challenges of sparse and partially corrupted meteorological and load data, particularly for Phu Quy, the algorithm achieves robust convergence to near-optimal solutions. This leads to enhanced scheduling and overall system performance, even in cases where a mathematically exact optimum may not be attainable. Phu Quy Island has the potential to significantly transform its energy landscape by adopting a wind-led, battery-buffered power system. Compared to the 2018 baseline, wind energy generation could increase by 81.69%, while diesel fuel consumption and associated CO_2 emissions are projected to decrease by 31.23%. If this optimized dispatch plan is applied during 70% of the island's annual operational hours, diesel savings may reach up to 125 billion VND. This represents a practical and impactful step toward redefining the island's power system operations. To bolster grid stability, innovative solutions involving BESS have been proposed. Two complementary control strategies have been evaluated: Ramp-Rate Control, which smooths the net-load profile and mitigates frequency fluctuations; and SOC Recovery, which modulates diesel output to maintain the BESS energy capacity during periods of low wind generation. Dynamic simulations performed in MATLAB/Simulink confirm that both methods effectively reduce power imbalances, minimize diesel cycling, and ensure voltage and frequency remain within statutory limits. Among these, the SOC-recovery approach demonstrates slightly better fuel efficiency under conditions of high wind variability. Phu Quy Island can successfully transition to a hybrid energy system that primarily utilizes wind power, complemented by advanced technologies such as load clustering, mixed-integer dispatch optimization, and adaptive controls for BESS and diesel generators. This approach not only decreases dependence on fossil fuels but also cuts operational expenses, reduces environmental harm, and ensures the island's growing electricity needs are met efficiently.

Ly Son Island, located in Quang Ngai Province along Vietnam's central coastline, holds significant geopolitical importance and is notable for its abundant solar and wind energy potential (Figure 1.8). The proposed HRES for Ly Son Island integrates PV generation with the existing diesel power infrastructure to meet the island's electricity demand of approximately 1.5 MW, which currently serves 3,700 consumers [27]. This system aims to enhance energy supply reliability, reduce diesel fuel consumption, and lower operational costs. The microgrid design was simulated using PowerWorld Simulator, analyzing performance under both peak-load and light-load conditions. Key design considerations included the capacity of the PV arrays, inverter specifications, diesel dispatch strategies, and reserve margins. The total capital costs were calculated, covering civil works, equipment procurement, and land-use compensation. Operational costs were assessed annually, taking into account diesel fuel price trends, routine maintenance, and component replacements. Simulation results demonstrated that the PV-diesel hybrid system could achieve annual operating cost savings of approximately VND 2,118,259,780. Over 20 years, these savings are projected to reach VND 42,365,195,608. This outcome meets the economic performance benchmarks for public-sector energy initiatives in Vietnam. The study confirms that

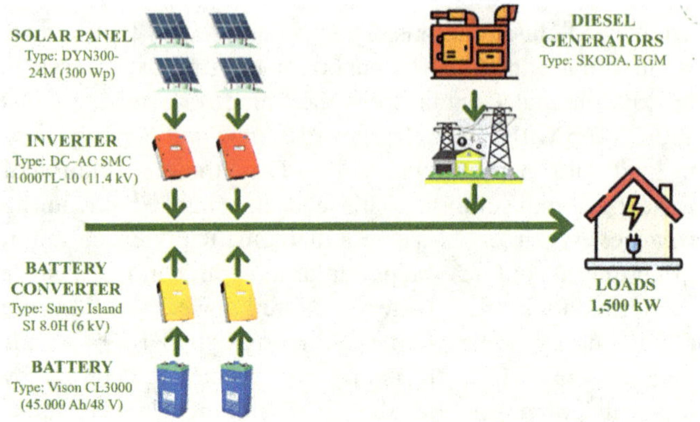

Figure 1.8 Microgrid power grid used for Ly Son Island

transitioning Ly Son Island to a solar-augmented HRES is both technically viable and financially beneficial. This shift supports local energy resilience while aligning with national goals for renewable energy integration.

Con Dao Island, located within Ba Ria-Vung Tau Province, is home to approximately 7,000 residents and a growing influx of tourists. This surge in tourism has led to a significant increase in electricity demand each year. Currently, the island relies almost entirely on two diesel power stations with a combined capacity of 8.61 MW, alongside a small-scale 30 kWp rooftop photovoltaic pilot project. The dependence on diesel power comes with substantial challenges, including high fuel costs, logistical complexities, and a reliance on carbon-intensive energy sources tied to external supply chains. Over the first five months of 2017, the diesel stations generated 6.19 GWh of electricity, with 5.70 GWh delivered to end-users. As tourism continues to expand, energy consumption is expected to rise in parallel, further straining the island's existing infrastructure and highlighting the need for sustainable energy solutions. The Floating Solar and Microgrid Project proposed for Con Son's Quang Trung Reservoir represents an innovative approach to renewable energy integration optimized for local conditions. At its core, the project features a 16 MWp (DC) floating photovoltaic array mounted on buoyant pontoons. This design not only maximizes land-use efficiency but also offers additional benefits such as reducing module operating temperatures, shading the water surface to mitigate evaporation, and potentially improving water quality. The array is anchored using a robust mooring system engineered to withstand typhoon-class wind and wave conditions. To ensure energy reliability and accommodate high levels of renewable penetration, the project incorporates a 40 MWh lithium-ion Battery Energy Storage System paired with a 4 MW/AC bidirectional Power Conversion System (PCS) capable of grid-forming operations. The modular design allows for scalability, enabling future expansions of both the floating solar array and the energy storage system as demand grows. Additionally, plans are in place for a 50 MW offshore wind farm, which is set to complement the system in a subsequent phase, further enhancing the renewable energy capacity. The entire setup is designed to operate efficiently over a 25-year service life, adhering to international IEC and IEEE standards for islanded microgrids. By leveraging successful international case studies, this project aims to provide a sustainable energy solution tailored to the unique resource characteristics of Con Son. The Microgrid-Energy-Management System (M-EMS) proposed for the Con Dao project is designed with a multi-layered operational structure to ensure efficient and resilient energy management. It comprises three primary components: (1) Energy Management System (EMS) – this layer focuses on real-time optimization of energy dispatch, setting state-of-charge targets for batteries, and allocating spinning reserves. It ensures that energy resources are used effectively to meet demand while maintaining system reliability. (2) Grid-Forming System (GFS) – the GFS is responsible for maintaining voltage and frequency stability, especially during islanded operations or under weak-grid conditions. This ensures uninterrupted energy supply and system stability in challenging scenarios. (3) Central Communication Layer (CCL) – utilizing IEC 61850 MMS and DNP3 protocols over a redundant fiber-optic communication backbone, the CCL facilitates secure and reliable data exchange between field devices and control centers. This cyber-secure framework enhances system monitoring and coordination. The control hierarchy integrates advanced features such as model-predictive scheduling, ramp-rate smoothing, and contingency handling routines, all rigorously tested through hardware-in-the-loop simulations to validate performance. A techno-economic analysis conducted using HOMER Pro software highlights the significant benefits of the M-EMS. The LCOE shows a substantial reduction compared to diesel-only systems, while the renewable energy fraction exceeds 80% in the base scenario. Sensitivity analyses on variables such as fuel prices, battery replacement costs, and load growth further confirm the robustness of the project. The proposed architecture and optimization strategies make the M-EMS a compelling solution for sustainable energy management on Con Dao Island (Figure 1.9) [28].

The proposed program for Con Dao outlines a strategic two-phase approach to transition from diesel-powered electricity generation to HRES. Phase 1 involves implementing a grid-tied photovoltaic array integrated with the existing diesel grid to supply power to ten households and two public offices. This phase will serve as a pilot project or "living laboratory," collecting data on power quality, fuel-offset ratios, and user satisfaction over a year. The results will be benchmarked against predetermined criteria to assess viability [28]. If Phase 1 proves successful, Phase 2 will scale the solution across the island over three years. This expansion will equip 270 households, 72 government facilities, and 90 small- and medium-sized enterprises with rooftop or carport photovoltaic systems, achieving a total capacity of approximately 1.17 MWp. The initiative focuses on sectors experiencing the most significant electricity shortages—manufacturing, commerce, and tourism—ensuring maximum socio-economic benefits while reducing peak diesel demand. Projected outcomes include an annual reduction of about 961 tons of CO_2 emissions and a decrease in government fuel-price subsidies by approximately VND 727 billion once the program is fully implemented. The initiative described focuses on a collaborative financial model involving state budgets, concessional lenders, and beneficiary co-investors to fund renewable energy projects. Capital-cost recovery is managed through a blended tariff system designed to balance affordability with investor returns. Technical

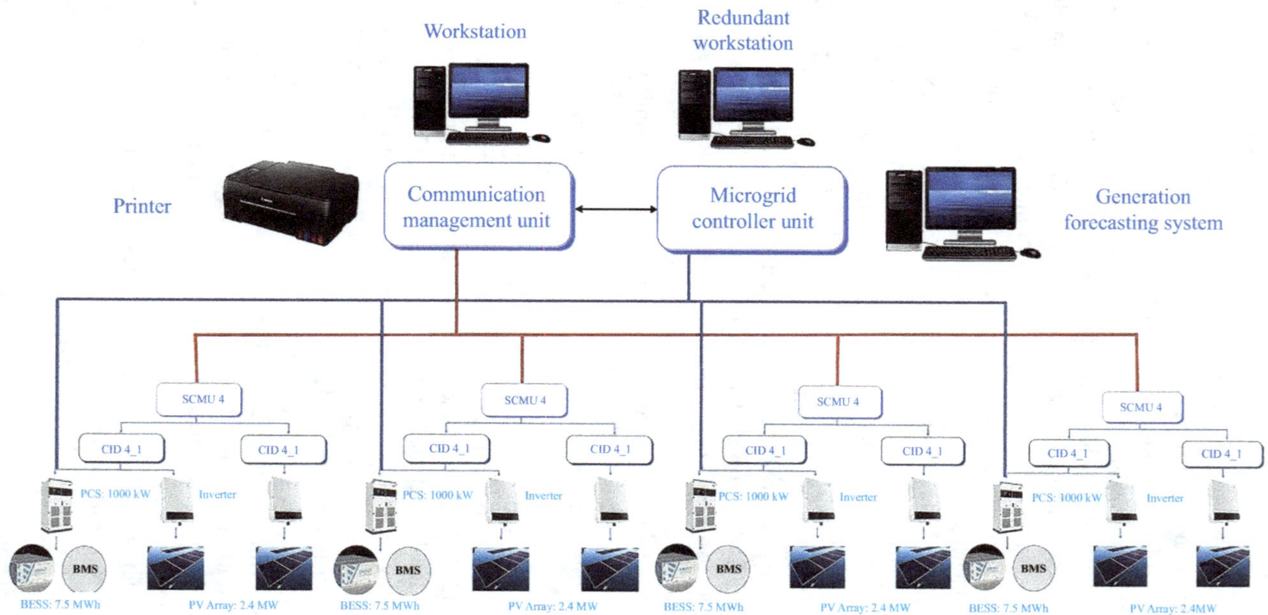

Figure 1.9 M-EMS proposal scheme for Con Dao Project

governance is based on established standards and algorithms to ensure interoperability, cybersecurity, and grid stability as renewable energy sources like wind turbines and photovoltaic systems gain prominence (Figure 1.6). This approach addresses challenges posed by high renewable energy penetration, such as grid instability during severe faults or fluctuations in wind speeds and solar irradiance. By adhering to modern grid codes, renewable energy units are required to remain connected and provide network support similar to that of conventional generators under all conditions. The initiative not only contributes to Vietnam's decarbonization goals but also offers a scalable model for other diesel-dependent islands, enhancing air quality, supply resilience, and regional sustainability efforts.

The book is organized into 13 chapters, as illustrated in Figure 1.10. Chapter 1 provides an overall introduction. Chapter 2, authored by Mohammad Abdul Mannan and colleagues, examines how integrating large-scale wind turbines and PV plants influences system stability. Its primary objective is to develop design procedures and control strategies for modern wind and PV installations, then model and evaluate an HRES under grid-fault conditions and under pronounced variations in wind or solar input.

Renewable power sources are inherently intermittent, their output rising and falling unpredictably with changing weather and climatic conditions. In grid-connected applications, this variability can be smoothed by dispatching complementary generators; however, stand-alone systems must rely on energy-storage devices—an option that significantly increases capital costs. Chapter 3, authored by Van Ga Bui and colleagues, introduces an off-grid Solar–Wind–Biomass Hybrid Renewable Energy System (SWB-HRES) capable of autonomous operation without conventional batteries. The chapter addresses: System architecture and performance assessment of the SWB-HRES, biomass buffering via a refuse-derived-fuel (RDF) production chain, on-site hydrogen generation integrated with the hybrid plant, advanced gasification for high-quality syngas achieved by enriching the air feed with O_2 and injecting H_2 into the gasifier's product stream, fuel-delivery strategy and optimal operating window for a syngas–hydrogen internal-combustion engine, and a supervisory control scheme that maximizes overall system efficiency and reliability. Special emphasis is placed on the RDF gasifier and the syngas–hydrogen

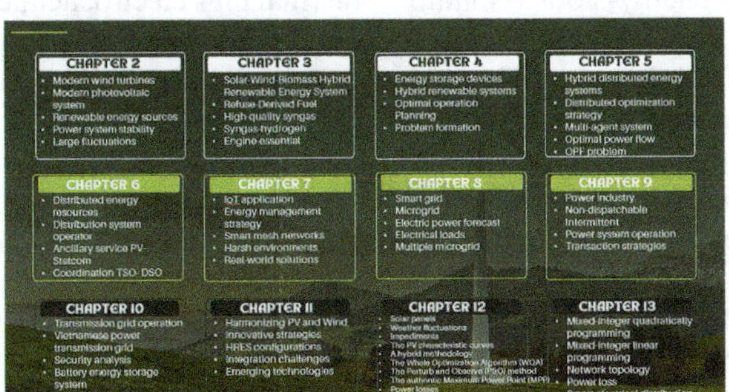

Figure 1.10 Book outline and keywords

engine—two pivotal yet commercially uncommon components—providing an in-depth technical discussion of their design, operation, and integration within the SWB-HRES.

Chapter 4, written by Dinh Duong Le and co-authors, explores cutting-edge strategies for sizing and dispatching utility-scale energy-storage assets in hybrid renewable power systems. The material is organized into six sections: Introduction, problem statement and mathematical formulation, a proposed optimization framework for storage sizing and operational scheduling, and representative case studies. Results and critical analysis, concluding remarks, and future research directions.

Chapter 5, authored by Tung Lam Nguyen and colleagues, develops a distributed-optimization framework for hybrid, distributed energy systems. The optimal power flow (OPF) problem is decomposed into cooperative sub-problems and solved by a network of autonomous agents, all of which are orchestrated within a multi-agent architecture. Practical viability is demonstrated through hardware-in-the-loop (HIL) experimentation, which subjects the algorithm to real-time operating conditions. The chapter is organized as follows: Introduction, casting the OPF into a generic distributed optimization problem, a multi-agent solution strategy for the OPF, hardware-in-the-loop validation and experimental results, and conclusions.

Global decarbonization efforts are accelerating investment in the energy transition, driving down reliance on fossil fuels and curbing carbon emissions. While environmentally beneficial, the rapid growth of distributed energy resources (DERs) in medium- and low-voltage feeders presents significant operational challenges: thermal congestion, voltage/frequency excursions, and limited real-time observability for operators. Closer coordination between Transmission System Operators (TSOs) and Distribution System Operators (DSOs) is therefore recognized as a pivotal lever for maintaining system security. This chapter begins with a comprehensive survey of TSO–DSO coordination schemes, including their typologies, current deployment status, and the practical barriers that hinder large-scale roll-out. The discussion then pivots to advanced control and energy-management algorithms that exploit the dual active- and reactive-power capabilities of PV-STATCOM units to deliver ancillary services—specifically voltage and frequency regulation—thereby reinforcing overall grid stability both day and night. Key innovations include synergistic control of real and reactive power to extend PV-STATCOM support beyond daylight hours and integration of hybrid energy-storage solutions (thermal plus electrochemical) to optimize PV-STATCOM dispatch and maximize the economic value of photovoltaic generation. Chapter 6, authored by Hong Lam Le and colleagues, provides an in-depth study entitled "Coordination of PV-STATCOM and Hybrid Energy-Storage Systems for Reactive-Power Management in Distribution Networks." The chapter addresses TSO–DSO interaction frameworks for ancillary-service provision, PV-STATCOM concepts, capabilities, and application domains, a distributed control architecture enabling coordinated Volt/Var and frequency support, detailed simulation case studies validating the proposed strategy's effectiveness, and critical analysis of results, with emphasis on the performance gains delivered by hybrid storage in augmenting

PV-STATCOM functionality. Together, these contributions outline a scalable pathway for leveraging PV-STATCOMs and hybrid storage to enhance grid resilience in high-DER environments.

Chapter 7, authored by Dinh Thanh Ngo and collaborators, investigates cutting-edge energy-management strategies for Internet-of-Things deployments, focusing on smart mesh networks that draw power from hybrid renewable sources. The chapter offers a structured, in-depth examination of methods for maximizing energy efficiency under the demanding conditions typical of remote or inhospitable sites. Its principal sections are: The energy challenge in IoT; optimization techniques for low-power operation; design considerations for harsh-environment deployments; case study; energy-saving principles demonstrated in practice; implementation lessons and real-world engineering solutions; conclusions; and future research directions. Through this framework, the authors deliver both theoretical foundations and field-validated guidance, equipping readers with practical tools for deploying resilient, energy-optimized IoT systems in even the most adverse settings.

Chapter 8, authored by Sonia Leva and colleagues, builds on extensive forecasting research for both renewable generation and electrical demand, drawing empirical insight from two living laboratories—the MG2Lab microgrid at Politecnico di Milano (Italy) and the Bad River Band community microgrid in northern Wisconsin (USA). The chapter unpacks the challenges and solutions associated with embedding hybrid renewable resources into the evolving landscape of smart grids and microgrids. Key themes include smart grid architectures in which prosumers, e-mobility, DERs, and bidirectional communications cooperate seamlessly; microgrids as keystone elements of the smart grid paradigm, featuring distributed generators (micro-turbines, fuel cells, PV), energy-storage devices (flywheels, supercapacitors, battery banks), and controllable loads such as electric vehicles; control-oriented microgrid design enabling autonomous, optimized power-flow management in both grid-connected and islanded modes; and a conceptual diagram illustrating a smart grid topology composed of multiple interconnected microgrids. The discussion is structured around the following focal areas: cost-function formulation for hybrid RES planning, design-phase integration with smart grid and microgrid architectures, operational integration, including dispatch and flexibility services, control challenges spanning voltage, frequency, and energy-management layers, conclusions, and outlook. By combining forward-looking methodologies with real-world demonstrators, the authors present a roadmap for building sustainable, resilient, and highly efficient energy systems.

Rising electricity demand and the imperative to curb CO_2 emissions are accelerating the deployment of renewable generation. Yet the non-dispatchable, intermittent nature of wind and solar introduces significant operational uncertainty. One promising remedy is to partition the grid into smaller, self-managed regions—microgrids—and to interconnect them so that power imbalances caused by forecast errors or equipment faults can be shared. As these clusters proliferate, they are becoming integral building blocks of the wider power system, reshaping traditional operating practices. To keep such a network both economical and sustainable, transaction strategies that optimise microgrid benefits against operating costs are essential. Chapter 9, by Ngoc Thien Nam Tran and colleagues, presents a comprehensive optimization framework for energy exchange among multiple microgrids across three temporal layers: intraday (time-domain control), day-ahead scheduling, and real-time balancing. The model accommodates wind, solar, and battery storage and defines coordinated mechanisms for both the energy and reserve markets so that real-time deviations from renewable forecasts can be swiftly corrected. The chapter is organized as follows: introduction, system architecture and problem statement, proposed optimization methodology, case study and simulation results, and conclusions. Through this structure, the authors deliver a practical roadmap for the cost-effective and reliable operation of interconnected microgrids in the face of renewable variability.

The rapid expansion of renewable generation is reshaping transmission-grid operations worldwide. Chapter 10, authored by Thanh Viet Dinh and co-workers, offers a series of real-world case studies that quantify these effects on Vietnam's high-voltage network. The investigation covers: Transmission-line congestion scenarios, N-1 contingency performance, system-stability margins, voltage-profile deviations, and short-circuit duty increases. In addition, the authors present a detailed study of secondary-frequency regulation provided by a BESS installed in the Central Highlands transmission corridor, illustrating how storage can mitigate renewable variability and enhance grid resilience. Chapter structure: Impacts of rising renewable penetration on Vietnam's transmission grid, secondary-frequency regulation with BESS. Through data-driven analysis and field evidence, the chapter delivers practical insights for operators facing the twin challenges of accommodating large-scale renewable energy and maintaining system security.

Chapter 11, authored by Gabriela Nicoleta Tanasiev and colleagues, surveys the evolving art of pairing photovoltaic and wind technologies within Hybrid Renewable Energy Systems (HRES). The authors examine the complementarities, technical hurdles, and cutting-edge solutions that make PV–wind hybrids a cornerstone of a cleaner, more resilient energy future. Chapter outline: Introduction, fundamentals of photovoltaic systems, fundamentals of wind energy systems, hybrid renewable energy systems (HRES), synergies and challenges in combining PV and wind, system design and optimization, economic and environmental assessments, future trends and emerging technologies, and conclusions. By blending theoretical foundations with practical design guidance, the chapter provides a comprehensive roadmap for engineers, policymakers, and researchers dedicated to unlocking the full potential of PV–wind hybrid configurations.

In legacy distribution networks, fault currents flow in only one direction; therefore, their protection schemes were engineered for strictly unidirectional power. The surge in DERs, however, is turning these traditionally passive, single-sourced feeders into active, multisourced systems. Many grid codes have already been revised to require greater DER participation during fault events. This transition introduces several protection challenges: The fault-current magnitude can differ dramatically between grid-connected and islanded operation, allowing certain faults to persist undetected; photovoltaic and wind plants are inverter-interfaced and therefore mimic neither the magnitude nor the phase characteristics of synchronous machines. Their fault-current phase angle is governed by control algorithms, which can confuse conventional directional relays. In islanded mode, the modest short-circuit contribution of inverter-based DERs may fall below the pickup of traditional overcurrent protection. Chapter 12, by Kuo Lung Lian and co-authors, tackles a different but equally pressing DER topic: maximizing power extraction from partially shaded PV arrays. The authors propose a hybrid maximum-power-point-tracking scheme that couples the Whale Optimization Algorithm (WOA) with the classic Perturb-and-Observe (P&O) method. By combining WOA's global search capability with P&O's local fine-tuning, the strategy accelerates tracking, curbs power loss, and boosts overall energy yield under complex shading patterns. Simulation results confirm the hybrid algorithm's superiority over conventional MPPT techniques in maintaining PV performance despite partial shading conditions.

Chapter 13, authored by Nang Van Pham and colleagues, develops optimization models for reconfiguring smart distribution networks with embedded hybrid renewable systems to minimize power losses. Two mathematical formulations are proposed: Mixed-Integer Quadratically Constrained Programming (MIQCP) and Mixed-Integer Linear Programming (MILP). Both approaches capture the discrete nature of switch positions while respecting electrical constraints, providing alternative pathways for large-scale, real-time network optimization. Chapter outline: Introduction, formulation based on mixed-integer nonlinear programming, MIQCP formulation, MILP formulation, case studies, results and discussion, and conclusions. Through comparative case studies, the authors highlight the trade-offs between solution quality and computational speed, offering practical guidance for distribution operators seeking loss-minimized topologies in renewable-rich grids.

This introductory chapter sets the stage for a comprehensive, 13-chapter volume on HES, tracing the forces that are reshaping the world's power sector and outlining the technologies, control strategies, and optimization techniques required to meet surging demand sustainably. After framing the global context—population growth, rapid urbanization, stricter climate targets—the authors argue that no single renewable resource can deliver the security, affordability, and resilience today's grids require. Instead, multi-resource hybrids that combine solar, wind, biomass, hydro, diesel backup, and a widening palette of energy-storage options must be orchestrated through smart grid intelligence, advanced forecasting, and AI-driven energy management. The chapter reviews breakthroughs in power electronics, battery chemistries, hydrogen pathways, blockchain-enabled transactive markets, and IoT sensing that collectively make high-renewable penetration feasible. It then previews the book's structure: Chapters 2–4 tackle system stability, battery-free off-grid hybrids, and large-scale storage sizing; Chapters 5–7 explore distributed optimization, PV-STATCOM-based ancillary services, and energy-aware IoT mesh networks; Chapters 8–10 examine forecasting in living-lab microgrids, multi-microgrid energy trading, and the impact of renewable on Vietnam's transmission grid; Chapter 11 dissects PV–wind synergies, Chapter 12 introduces a whale-optimization–enhanced MPPT for shaded PV arrays, and Chapter 13 presents MIQCP and MILP reconfiguration models for loss-minimized distribution feeders. Together, these contributions offer researchers, engineers, and policymakers a state-of-the-art roadmap for planning, operating, and optimizing hybrid renewable systems that can underpin the next energy transition.

References

[1] Ritchie H and Rodés-Guirao L. Peak Global Population and Other Key Findings from the 2024 UN World Population Prospects; 2024. *Our World in Data.*

[2] Dechamps P. The IEA World Energy Outlook 2022 – A Brief Analysis and Implications. *European Energy & Climate Journal.* 2023;11(3):100–103.

[3] Yolcan OO. World Energy Outlook and State of Renewable Energy: 10-Year Evaluation. *Innovation and Green Development.* 2023;2(4):100070.

[4] Krishna KS and Kumar KS. A Review on Hybrid Renewable Energy Systems. *Renewable and Sustainable Energy Reviews.* 2015;52:907–916.

[5] Hossain MS, Madlool NA, Rahim NA, *et al.* Role of Smart Grid in Renewable Energy: An Overview. *Renewable and Sustainable Energy Reviews.* 2016;60:1168–1184.

[6] Bayindir R, Colak I, Fulli G, *et al.* Smart Grid Technologies and Applications. *Renewable and Sustainable Energy Reviews.* 2016;66:499–516.

[7] El-Hawary ME. The Smart Grid—State-of-the-Art and Future Trends. *Electric Power Components and Systems.* 2014;42(3–4):239–250.

[8] Aydın Ö and Demir Z. Smart Grid Integrated with Hybrid Renewable Energy Systems. *Eskişehir Technical University Journal of Science and Technology A: Applied Sciences and Engineering.* 2019;20:120–131.

[9] León Gómez JC, De León Aldaco SE, and Aguayo Alquicira J. A Review of Hybrid Renewable Energy Systems: Architectures, Battery Systems, and Optimization Techniques. *Eng.* 2023;4(2):1446–1467.

[10] Phan TM, Nguyen TT, Duong MQ, *et al.* Optimal Design and Operation of Battery Energy Storage Systems in Renewable Power Plants to Reach Maximum Total Electric Sale Revenues. *Neural Computing and Applications.* 2024;36(20):12061–12082.

[11] Hannan MA, Wali SB, Ker PJ, *et al.* Battery Energy-Storage System: A Review of Technologies, Optimization Objectives, Constraints, Approaches, and Outstanding Issues. *Journal of Energy Storage.* 2021;42:103023.

[12] Almutairi K, Hosseini Dehshiri SS, Hosseini Dehshiri SJ, *et al.* Use of a Hybrid Wind–Solar–Diesel–Battery Energy System to Power Buildings in Remote Areas: A Case Study. *Sustainability.* 2021;13(16):8764.

[13] Mohaghegh MR, Heidari M, Tasnim S, *et al.* Latest Advances on Hybrid Solar–Biomass Power Plants. *Energy Sources, Part A: Recovery, Utilization, and Environmental Effects.* 2021;47(1):4901–4924.

[14] Ishraque MF, Shezan SA, Nur J, *et al.* Optimal Sizing and Performance Investigation of a Solar-Wind-Battery-DG Based Hybrid Microgrid System Applicable to the Remote School of Bangladesh; 2020. Authorea Preprints.

[15] Linn S and Ya AZ. Solar/Wind/Diesel Hybrid Energy System with Battery Storage for Rural Electrification. *International Journal of Scientific Engineering and Technology Research*. 2014;3:2172–2176.

[16] Yusuf SS and Mustafi NN. Design and Simulation of an Optimal Mini-Grid Solar-Diesel Hybrid Power Generation System in a Remote Bangladesh. *International Journal of Smart Grids*. 2018;2(1):27–33.

[17] Combe M, Mahmoudi A, Haque MH, *et al.* Cost-Effective Sizing of an AC Mini-Grid Hybrid Power System for a Remote Area in South Australia. *IET Generation, Transmission & Distribution*. 2019;13(2):277–287.

[18] Lê Tuân DMQ, Nam NB, and Tân NV. Optimization of Generation Scheduling Using Differential Evolution Algorithm. *The University of Danang-Journal of Science and Technology*. 2021;19(4):90–94.

[19] Viet DT, Phuong VV, Duong MQ, *et al.* Models for Short-Term Wind Power Forecasting Based on Improved Artificial Neural Network Using Particle Swarm Optimization and Genetic Algorithms. *Energies*. 2020;13(11):2873.

[20] Kumaravel S and Ponnusamy V. An Efficient Hybrid Technique for Power Flow Management in Smart Grid with Renewable Energy Resources. *Energy Sources, Part A: Recovery, Utilization, and Environmental Effects*. 2024;46(1):12721–12741.

[21] Chandrasekaran K, Selvaraj J, Amaladoss CR, *et al.* Hybrid Renewable Energy-Based Smart Grid System for Reactive Power Management and Voltage Profile Enhancement Using Artificial Neural Network. *Energy Sources, Part A: Recovery, Utilization, and Environmental Effects*. 2021;43(19):2419–2442.

[22] Vietnam Overview. Accessed: 10 Nov 2024. Available from: https://www.worldbank.org/en/country/vietnam/overview.

[23] Zhu Z, Weng Z, and Zheng H. Optimal Operation of a Microgrid with Hydrogen Storage Based on Deep Reinforcement Learning. *Electronics*. 2022;11(2):196.

[24] Quoc MN, Linh ND, Van Anh NH, *et al.* Danh Gia Tiem Nang Tich Hop Pin Nhien Lieu Hydro Vao Lui Dien Microgrid Bach Long Vy. *Journal of Military Science and Technology*. 2023;86:27–37.

[25] Phuong NH, Cuong VV, Au NN, *et al.* A New Proposed Operating Mode of Diesel-Wind Power Generation System for Phu Quy Island. *The University of Danang-Journal of Science and Technology*. 2021;19(5):29–34.

[26] Phuong NHV, Nam NB, Luat HV, *et al.* Strategy for Power Control of Battery Storage System: Phu Quy Island Study Case. *The University of Danang-Journal of Science and Technology*. 2021;19(4):12–17.

[27] Phuong MN. Giai Phap Xay Dung He Thong Pin Mat Troi Ket Hop Nguon Diesel Hien Co Cap Dien Cho Huyen Dao Ly Son – Quang Ngai [Master's Thesis]. *Truong Dai hoc Bach Khoa, Dai hoc Da Nang*; 2013.

[28] Con Dao Given an Opportunity to Accelerate Itself. Accessed: 01 Oct 2024. Available from: https://en.evn.com.vn/d6/news/Con-Dao-given-an-opportunity-to-accelerate-itself-66-163-3750.aspx.

Chapter 2

Stability analysis of hybrid power systems including wind turbine and solar PV systems

Shaikh Tawhidul Islam Anik[1], Shameem Ahmad[1,2], Md. Rifat Hazari[1], Effat Jahan[1], Chowdhury Akram Hossain[1] and Mohammad Abdul Mannan[1]

Microgrids integrating photovoltaic (PV) and wind energy systems provide a sustainable and efficient means of energy generation. Nevertheless, stability of the hybrid microgrid system is of utmost importance under dynamic operating conditions and variable environmental factors. The dynamics of stability present a significant challenge in designing the controlling operation, which needs to handle various energy inputs and system disturbances. In this chapter, therefore, a streamlined yet sturdy controller is designed for ensuring the stability and efficient operation of a hybrid microgrid. Different scenarios have been evaluated, including power changes for the PV and wind systems, fault conditions, and recovery capabilities from the fault. The results have shown that the proposed controller maintained a 400 V steady DC bus voltage, though there were significant variations in power generation. Detailed total harmonic distortion (THD) analysis for both inverter and grid has been presented. For all the cases, THD results were between 1% and 5% and most importantly before fault it was 3.37% and after when it stabilizes within 0.1 s it became 3.83%. This clearly points out the viability of the controller to effectively solve the issues related to hybrid microgrid stability.

2.1 Introduction

A microgrid is a localized aggregate of electricity sources and loads that typically runs connected to a traditional centralized grid but can also disconnect and function independently in "island mode." This flexibility is what makes microgrids vital for energy resilience, particularly in areas that are prone to outages or that have no access to a central grid [1]. Combining several generation units, including solar panels, wind turbines, and batteries, into a microgrid allows for more efficient and reliable operation.

[1]Faculty of Engineering, American International University-Bangladesh, Bangladesh
[2]Department of Electrical and Electronic Engineering, BRAC University, Bangladesh

Microgrids can help in decreasing reliance on big energy suppliers by lowering the energy resources. By diversifying the energy sources, microgrids can reduce dependence on a single power source, thereby improving operational stability and providing a more sustainable energy solution [2].

The dynamics of a microgrid system are very complicated in maintaining stability. This is further complicated when different energy sources are integrated, for example, solar panels, wind turbines, and batteries, along with a normal grid. The output of these renewable sources is highly unpredictable, often causing changes in voltage and frequency that can jeopardize the stability and reliability of electricity supply [3]. Stability in microgrids is essential for not only uninterrupted power supply but also safeguarding the infrastructure and ensuring the proper functioning of each component [4]. Hence, advanced management and control strategies need to be designed and implemented to guard a microgrid from disturbances and keep it running properly in various conditions [5].

2.1.1 Background studies

Some previous studies have examined the stability and control of renewable energy systems but often fall short of considering hybrid systems with both PV and wind power on a comprehensive level under dynamic conditions. Jithin K. *et al.* investigated the stability of a DC microgrid with single and multiple PV units alongside batteries, using multi-loop control mechanisms to assess the impact of DER penetration on system stability. However, the paper focuses on DC microgrids and does not extensively cover the integration and stability analysis involving both wind power and solar PV in a hybrid system [6]. Hanwen Gu *et al.* developed a virtual filter-based stability control for DC microgrids connected to unbalanced grids. Though it addresses stability in terms of virtual impedance and unbalanced grid conditions, the study does not include fault analysis [7]. Gibin Mathew P. *et al.* investigated the effect of non-linear consuming devices on generated pulse width modulation (PWM) signals and voltage distortion and last THD in the microgrid [8]. However, the study did not explore fault analysis. Anusuya *et al.* explored real-time stability enhancement techniques for islanded microgrids to evaluate stability by load disturbance classification and frequency oscillation. The paper concentrates on islanded microgrid stability without explicitly examining the impacts and stability analysis of combined wind and solar PV systems under varying environmental conditions [9]. Shridhar S. Khule *et al.* have reviewed the stability of microgrids and proportional-integral (PI) control in an attempt to cover the interaction between individual control modes for voltage and frequency regulation. However, the paper does not deal with the complexity associated with simultaneous wind and solar PV integration, nor does it address THD investigations and fault analyses comprehensively [10]. M. Vijay K. *et al.* have brought forth an artificial neural network (ANN)-genetic algorithm (GA) optimization strategy for improved power quality and reduced THD in AC microgrids. The study does not delve into various power-sharing operations that are vital for testing the stability of the system. Zongwen Zhang *et al.* examined the stability of an islanded PV-battery DC microgrid under different operating points of the PV curve with small-signal modeling for stability assessment. The work only covers

DC microgrids and does not consider the effects of a wind power source integrated with PV, which is one of the major areas of investigation in hybrid microgrid analysis [12]. Yajuan G. *et al.* developed a harmonics suppression control strategy for grid-tied microgrids using resonant controllers. The method addresses harmonic suppression but does not consider system stability under varying power [13]. Ram Krishan *et al.* have examined the analysis of stability and control strategies of DC microgrids with constant power loads and proposed a unique virtual negative impedance-based control loop for stability enhancement. Although it answers most of the stability challenges posed by the constant power loads in DC microgrids, hybrid renewable power sources are not examined, and total harmonic distortion related to the variable load is not thoroughly analyzed [14]. Jithin K. *et al.* analyze the stability of interconnected DC microgrids when units are disconnected, using a modified system matrix algorithm. The paper does not explore hybrid systems with integrated solar and wind sources or their effects on stability [15]. Padmavathi L. has presented a harmonic compensation method using a harmonic observer and feed-forward control to improve the power supply quality. However, the study focuses only on low-voltage microgrids and THD correction [16]. Mohd. Brado Frasetyo *et al.* focused on using a synchronverter algorithm in an islanded wind farm microgrid to enhance stability, simulating traditional generator controls. However, the study limits its focus to wind farms and does not include an analysis of solar integration or variable environmental conditions [17]. Moussa Saadati Toularoud *et al.* developed a hierarchical control strategy for hybrid microgrids to manage voltage and frequency stability, employing distributed control methods. But it does not discuss fault analysis or real-world scenarios involving dynamic interactions between different renewable energy sources [18]. Rashad M. Kamel *et al.* proposed a new technique for enhancing the transient stability of a standalone microgrid using a superconductor fault current limiter specifically with wind generation systems. It focuses narrowly on wind generation systems using a specific technology for fault management and does not address the broader stability impacts in microgrids that integrate both wind and solar power [19].

The mentioned works do not go deeply into the dynamic control complexities associated with the management of both wind turbines and solar PV systems in hybrid microgrid operations. They do not discuss integrated control strategies for dealing with simultaneous variations of wind speed and solar irradiance comprehensively. In addition, these works do not carry out detailed discussions of fault management or THD at the various points of fluctuating power output that are essential to the stability and efficiency of the system. This work fills that gap by introducing an advanced control strategy that dynamically manages energy input from both wind and solar sources while simultaneously improving the system's resilience to changing environments and unpredictable operational actions. The proposed innovative controller determines renewable input fluctuations in real-time to ensure the microgrid remains functional and stable. Main contributions of this chapter are the following:

1. To introduce a robust controller that seamlessly handles both wind and solar power inputs by optimizing power distribution based on real-time environmental data.

2. To develop and validate a hybrid microgrid model using MATLAB®/Simulink® that simulates the microgrid's behavior under various operational scenarios, demonstrating the effectiveness of the proposed control strategies.
3. To perform a critical analysis on impacts of large-scale renewable energy systems into the grid system.

2.2 Modeling

The microgrid model comprises PV panels and wind turbines that are interfaced through power conditioning units into a 400 V DC bus, as depicted in Figure 2.1. The PV panels utilize a boost DC–DC converter with a Maximum Power Point Tracking (MPPT) controller for maximum power extraction, while the connection with wind turbines is through a rectifier followed by another boost converter. A battery storage subsystem is also included in the setup connected to the DC bus by a bi-directional DC–DC converter, the charge controller of which performs both charging and discharging functions. This guarantees steady energy availability even with unequal generation and demand. The DC power is converted to AC power by an inverter and filtered through an Inductor-Capacitor-Inductor filter for harmonic attenuation, after which it is stepped up through a transformer for the grid or local load use. Circuit breakers protect the entire system and facilitate the integration with the AC load and grid for safe and reliable operations. Important parameters of the system have been given in Table 2.1.

2.2.1 *PV panel*

The PV system is designed to develop 10 kW at peak efficiency. It uses an MPPT algorithm to optimize output under different solar conditions. From the panel voltage, the power is boosted up to the desired 400 V DC for the microgrid via a boost DC–DC converter. The performance of the PV array can be represented by (2.1) [20].

$$I = I_{ph} - I_o\left(e^{\frac{q(V+IR_s)}{nkT}} - 1\right) - \frac{V + IR_s}{R_{sh}} \tag{2.1}$$

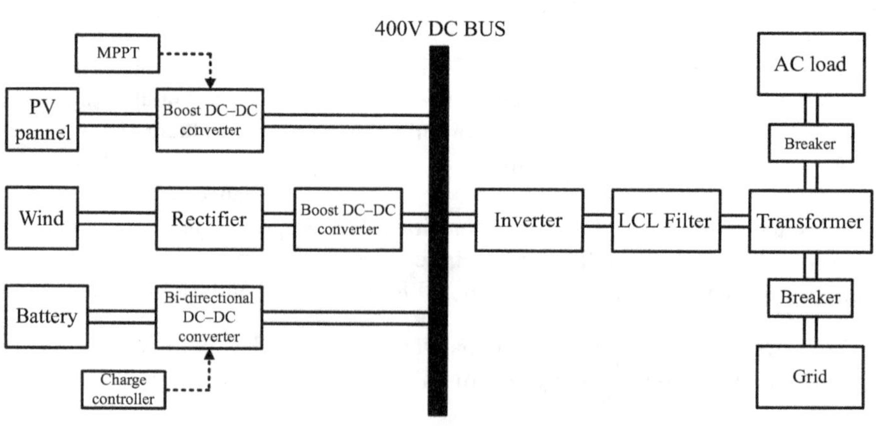

Figure 2.1 Microgrid system block diagram

Table 2.1 *Components and parameters of the microgrid system*

Components	**Ratings**
PV panel	10 kW
Wind turbine	10 kW
Battery	384 V, 28.5 Ah
DC bus voltage	400 V
Inverter	10 kW, 400 V
Transformer	10 kVA, 400–230 V
Grid	230 V, 50 Hz

Where, I is the output current of the PV cell, I_{ph} is the photocurrent generated by solar irradiance, I_o is the diode reverse saturation current, q is the electron charge, V is the output voltage of the PV cell, R_s is the series resistance, R_{sh} is the shunt resistance, n is the diode ideality factor, k is Boltzmann's constant, and T is the absolute temperature in Kelvin.

2.2.2 *Wind turbine*

The 10 kW wind turbine, which is rectifying wind energy to electrical power and boosting it to produce 400 V DC for the microgrid, ensures a consistent electricity power supply, thereby increasing the overall reliability and efficiency of the system. The power equation for wind turbine is given in (2.2).

$$P_{\text{wind}}(t) = \frac{1}{2}\rho A C_P v_w^3(t)\eta_w \tag{2.2}$$

Where, ρ is the air density (kg/m^3), A is the swept area of the wind turbine blades (m^2), C_p is the power coefficient of the wind turbine, $v_w(t)$ is the wind speed at time t (m/s), and η_w is the overall efficiency of the wind turbine system.

2.2.3 *Battery*

The system employs a 384 V, 28.5 Ah battery connected to a DC–DC converter, in charging or discharging modes as demanded. Charge controllers include SOC management and power management, which are crucial to optimizing battery efficiency and battery life and are calculated by (2.3) and (2.4) [21].

$$SOC(t) = SOC(t) + \frac{1}{C}\int I dt \tag{2.3}$$

$$P_B(t) = \eta_{\text{charge}} P_{\text{charge}}(t) - \frac{P_{\text{discharge}}(t)}{\eta_{\text{discharge}}} \tag{2.4}$$

Where, $SOC(t)$ is the state of charge of the battery at time t, C is the battery capacity (Ah), I is the battery current, $P_B(t)$ is the battery power, $P_{\text{charge}}(t)$ and $P_{\text{discharge}}(t)$ are the charging and discharging powers, respectively, and η_{charge} and $\eta_{\text{discharge}}$ are the charging and discharging efficiencies of the battery.

2.2.4 Inverter

The inverter converts the microgrid's bus 400 V DC voltage to grid 400 V AC voltage for synchronization. The inverter is rated for 10 kW and requires high reliability and efficiency for conversion operations. The basic operation of grid-connected inverters can be described with the help of the power equation (2.5).

$$P = VI \cos(\phi) \tag{2.5}$$

Where, P is the active power delivered by the inverter, V is the RMS voltage, I is the RMS current, and ϕ is the phase angle between voltage and current.

2.2.5 AC load

The microgrid system aims to feed AC loads of 6 kW. Loads may range from household or commercial appliances that need a steady and reliable supply of 230 V AC from the transformer output.

2.2.6 Grid

The grid interface of the system is done via a 230 V, 50 Hz connection protected by a circuit breaker, which guarantees safety and reliability. The grid connection allows one to either feed excess energy to the grid or draw energy in case of low production and storage status in the microgrid. The power equation associated with the grid is given in (2.6).

$$P_g(t) = P_L(t) - \sum(P_{PV}(t), P_{wind}(t)) \tag{2.6}$$

Where, $P_g(t)$ is the power exchanged with the grid at time t, $P_L(t)$ is the load demand, $P_{PV}(t)$ is the power generated by the PV system, and $P_{wind}(t)$ is the power generated by the wind turbine.

2.3 Controller

The advanced control architecture shown in Figure 2.2 ensures operating stability and efficiency for grid-connected inverters in microgrids. The PI controllers are implemented for fine control of the dq0 components of voltage and current, achieving remarkable performance beyond conventional controllers. By securing control over these parameters, the PI controllers allow fast and smooth response toward variations in load and generation and give stability against various operating conditions.

2.3.1 Voltage and current control

The controller is operating PI controllers for controlling the voltage and current D and Q components. In the direct segment, it ultimately brings about precise control over active power, while in the quadrature segment, it does so over reactive power. Thereby, by manipulating both these components independently, we can say that yet another very important task is achieved, which is the power delivery and voltage

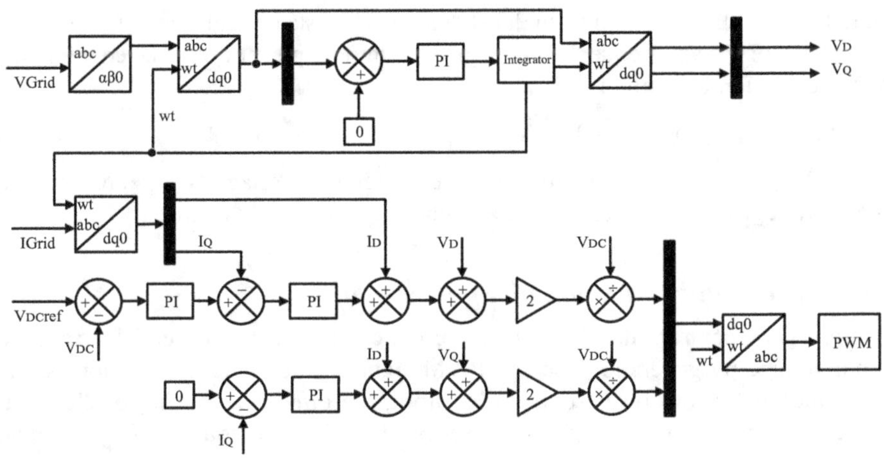

Figure 2.2 DQ-based dual-loop vector control with PI regulators of microgrid system

stabilization across the interface of the entire grid. Voltage dynamics with error integration and direct current control can be found by (2.7) and (2.8), respectively.

$$\frac{dV_{DC}}{dt} = \frac{1}{C}(I_Q - I_{DC} + K_P\,(V_{DCref} - V_{DC}) + K_I \int (V_{DCref} - V_{DC})\,dt) \tag{2.7}$$

$$I_D = (\frac{V_{DC} - V_{DCref}}{R}) + K_P\,(V_{DCref} - V_{DC}) + K_I \int (V_{DCref} - V_{DC})\,dt \tag{2.8}$$

where, V_{DC} = DC bus voltage, C = DC bus capacitance, I_D = Direct axis current component, I_Q = Quadrature axis current component, V_{DCref} = DC bus reference voltage, K_P = Proportional gain, and K_I = Integral gain.

2.3.2 Integration and error correction

An integrator continues to accumulate the error from the voltage control, measuring the difference between actual and desired voltage levels continuously. This error is then sent to the PI controllers to make adjustments to the control actions, thereby minimizing any long-term drift or steady-state error. The mathematical expression for error calculation is given in (2.9).

$$u(t) = K_P e\,(t) + K_I \int_0^t e(\tau)d\tau \tag{2.9}$$

where, $\int_0^t e(\tau)d\tau$ accumulates past errors to correct long-term change.

2.3.3 PWM generation

The PWM generator generates signals for control of the switching of the inverter's power electronic devices based on the outputs of the PI controllers. The PWM signal governs the timing and duration of the switches' on-and-off states, synthesizing the desired AC output waveform from the DC input. This is a critical step in synchronizing with the phase and frequency of the grid for smooth power flow. The equation for PWM generation is given in (2.10).

$$V_{PWM} = V_{dq}\cos(\omega t) + V_{dq}\sin(\omega t) \tag{2.10}$$

Where, V_{PWM} is the pulse-width modulated reference voltage, V_{dq} represents the dq-axis voltage components, ω is the angular frequency, and t is time.

2.3.4 Harmonic reduction and power quality

With precision control, the harmonic distortion can be drastically reduced from the output voltage by generating specific PWM signals. The lower the harmonics, the higher the quality of power supplied, ensuring the effective working of electrical appliances while fulfilling stipulations set by the grid. The equation for harmonic reduction via filtering is given by (2.11).

$$V_{\text{out}} = V_{\text{in}}(1 - e^{-\frac{t}{RC}}) \tag{2.11}$$

Where, V_{out} is the output voltage after filtering, V_{in} is the input voltage, R is the filter resistance, C is the filter capacitance, and t is time.

A well-optimized control of the DC link voltage helps to maximize power conversion efficiency while minimizing the risks of transient overvoltage that may lead to system failure. In this context, PWM techniques further enhance the output from the inverter, which leads to a marked reduction of voltage harmonics. This reduction of harmonics is paramount in the prevention of power quality problems that may basically impact the microgrid as well as the main grid. These characteristics turn the controller into not just a functioning necessity but a value-adding entity for the energy flow management, enhancing grid resilience and meeting strict electrical specifications. The impact that this advanced control scheme may cause is not limited to operating control but is also useful in terms of aiding grid integration toward sustained stability and reliability of power supply systems.

2.3.5 Stability analysis

Figure 2.3(a) and (b) shows the step response and Bode plot of the controller, respectively. The step response signifies system behavior in the event of an abrupt change in inputs. The response starts at 1 and tends to steadily decrease, which represents a stable system free from overshoot and oscillations. The smooth settling behavior indicates that the controller is indeed capable of sustaining stability in dynamic operational conditions without major perturbations.

The Bode diagram depicts the frequency response of the controller, with its magnitude plot displaying a constant low-frequency gain followed by a gradual attenuation at higher frequencies, consequently providing attenuation of high-frequency disturbances. In the phase plot, 0° is nearly attained, suggesting little distortion in the low frequency, followed by a steady decrease that reflects the delay expected at the higher frequencies. The concurrence of these two diagrams implies a well-tuned controller with adequate margins in phase and gain to sustain robust stability.

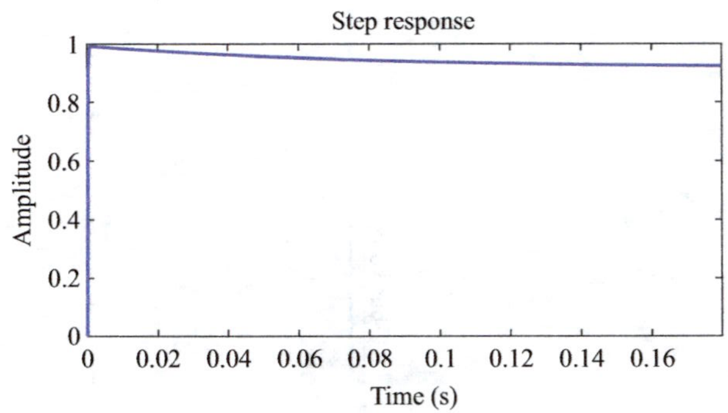

(a)

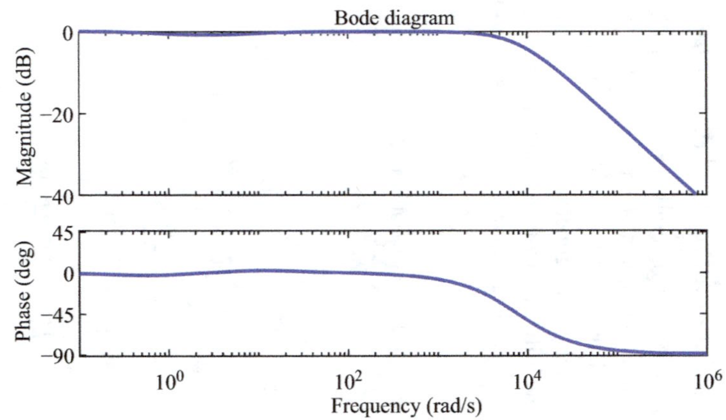

(b)

Figure 2.3 (a) Step response of controller and (b) Bode diagram of controller

2.4 Result analysis

2.4.1 PV, wind power variation at normal operating condition

Figure 2.4(a) and (b) depicts the power generation profiles of PV and wind energy over 4 s. The PV output essentially begins at around 10 kW before it drops sharply to 6 kW at 1 s, then down to 4 kW at 2 s and 2 kW at 3 s, after which it remains constant. On the other hand, wind power generation seems to gradually rise from a lower value, passing from a low initiation value through 2 kW around 1 s, then registering approximately up to 4 kW at the 2 s mark, 4 kW at 3 s, and from 3 to 4 s, it is

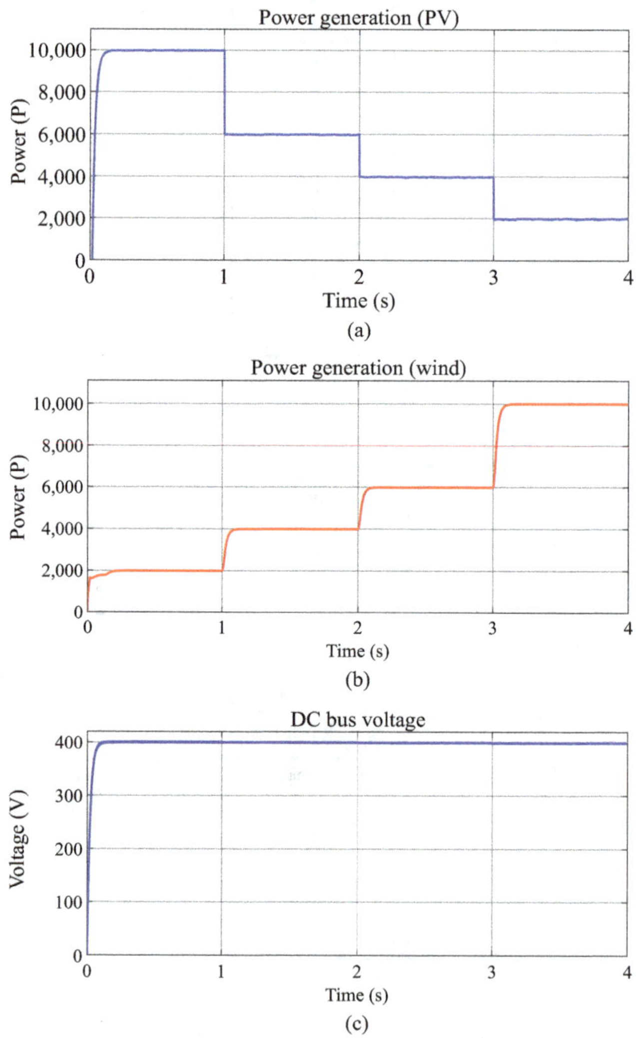

Figure 2.4 (a) Power generated by PV, (b) power generated by wind turbine, (c) DC bus voltage

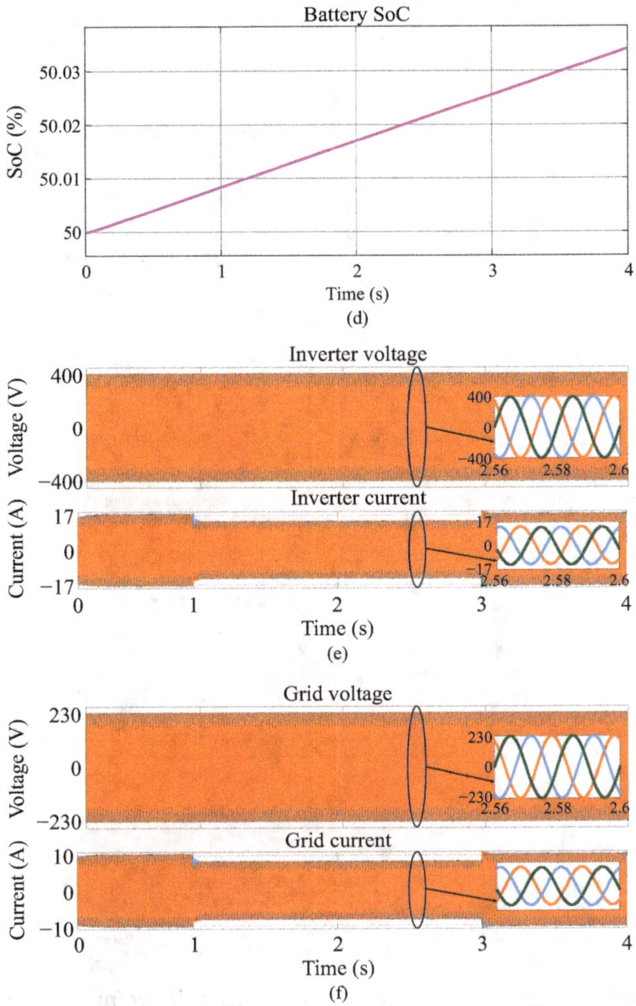

Figure 2.4 (Continued) (d) battery's state of charge, (e) inverter voltage and current, and (f) grid voltage and current

10 kW. Figure 2.4(c) shows the DC bus voltage of the whole operation that is 400 V constant, and Figure 2.4(d) shows the battery state of charge that is charging during the operation period. Figure 2.4(e) shows the inverter's voltage and current. Inverter voltage is 400 V pure sinusoidal and current decreased at the 1 s mark as power generation decreased. Figure 2.4(f) shows the grid voltage that is 230 V pure sinusoidal and current which also decreases at the 1 s mark but again increases at 3 s the mark.

2.4.1.1 THD analysis of inverter and grid

Figure 2.5(a) and (b) shows the THD results of the inverter during the 0–1 s operation period. For voltage, it is 1.42% and for current it is 1.35%. Figure 2.15(c) and (d)

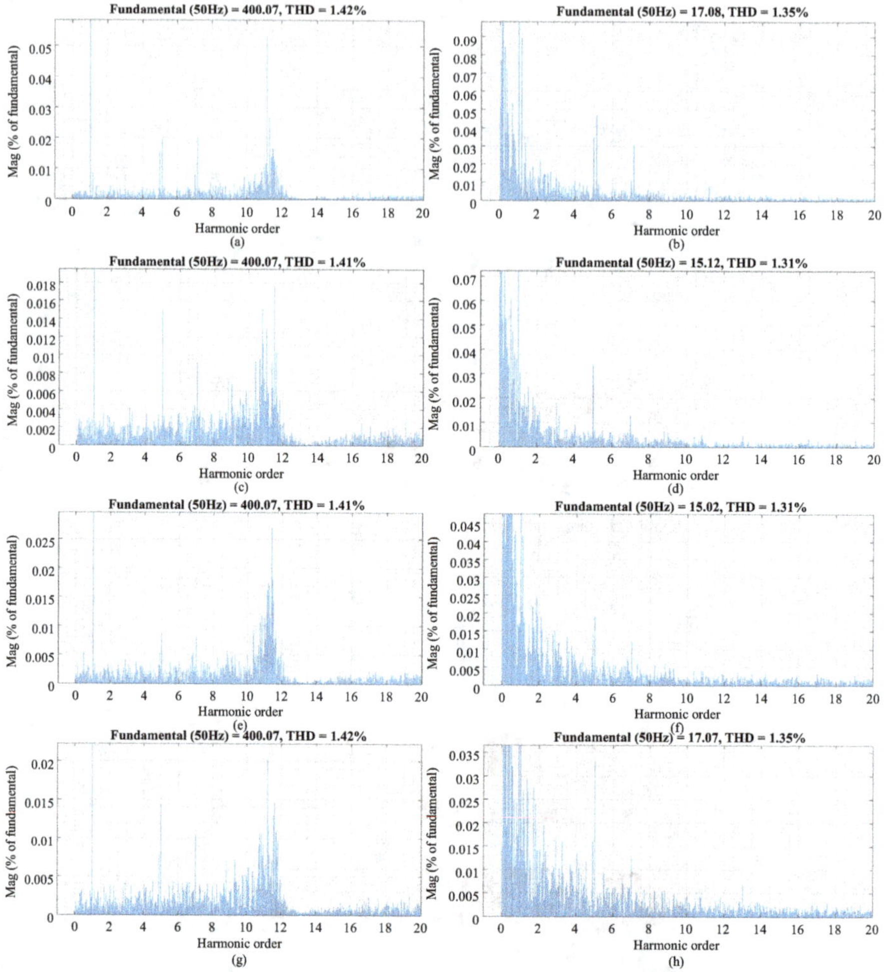

Figure 2.5 (a) THD of inverter voltage (0–1 s), (b) THD of inverter current (0–1 s), (c) THD of inverter voltage (1–2 s), (d) THD of inverter current (1–2 s), (e) THD of inverter voltage (2–3 s), (f) THD of inverter current (2–3 s), (g) THD of inverter voltage (3–4 s), and (h) THD of inverter current (3–4 s)

shows the THD for a 1–2 s period and for voltage and current it is 1.41% and 1.31%, respectively. Figure 2.5(e) and (f) shows the results for the time period 2–3 s and these results are same as for the 1–2 s period. Finally, back at the 3–4 s period, the results become similar as the 0–1 s period of time that is 1.42% for voltage and 1.35% for current, and these are shown in Figure 2.5(g) and (h).

Figure 2.6(a) and (b) shows the THD results of grid during 0–1 s operation period. For voltage it is 1.58% and for current it is 2.66%. Figures 2.6 (c) and (d) shows THD for 1–2 s period and for voltage and current it is 1.58% and 1.68%,

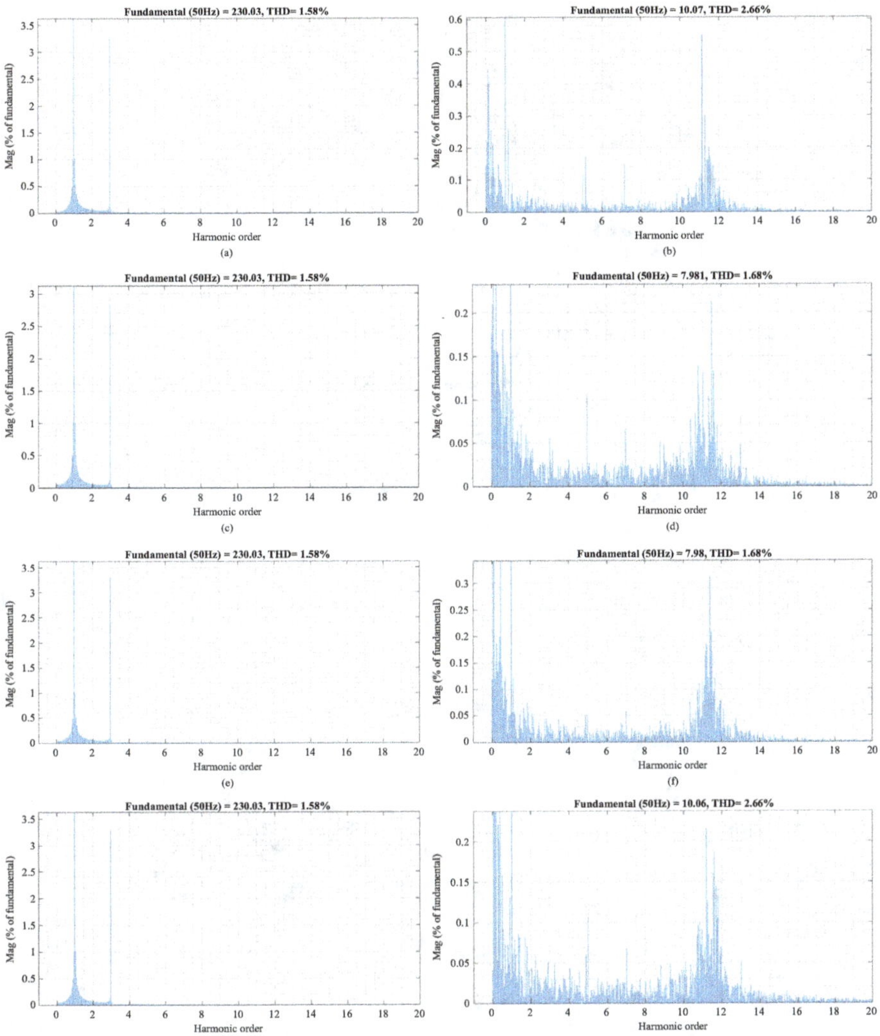

Figure 2.6 (a) THD of grid voltage (0–1 s), (b) THD of grid current (0–1 s), (c) THD of grid voltage (1–2 s), (d) THD of grid current (1–2 s), (e) THD of grid voltage (2–3 s), (f) THD of grid current (2–3 s), (g) THD of grid voltage (3–4 s), and (h) THD of grid current (3–4 s)

respectively. Figure 2.6(e) and (f) shows the results for the time period 2–3 s and these results are same as for the 1–2 s period. Finally, back at 3–4 s period the results become similar as the 0–1 s period of time which is 1.58% for voltage and 2.66% for current and these are shown in Figure 2.6(g) and (h).

2.4.2 PV, wind power variation during fault

A fault was applied to the system at the 1 s mark to test the system's reliability. Figure 2.7(a) shows the grid voltage with the fault applied. At 1 s, voltage becomes

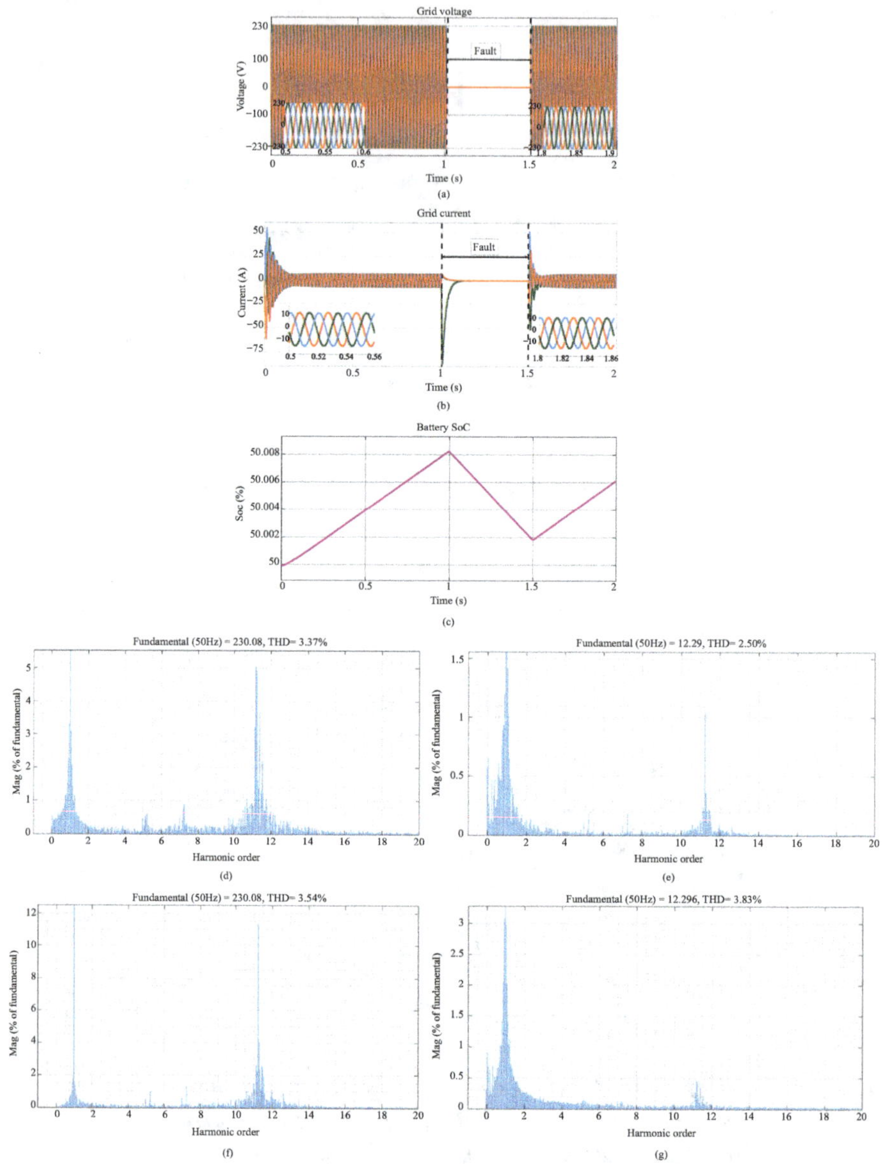

Figure 2.7 (a) Grid voltage with fault applied, (b) grid current with fault applied, (c) battery state of charge, (d) THD of grid voltage before fault, (e) THD of inverter current before fault, (f) THD of inverter voltage after fault, and (g) THD of inverter current after fault

0 V as the circuit breaker disconnects the grid after it detects a fault. Again, when the fault is removed at the 1.5 s mark, the operation becomes normal. Figure 2.7(b) shows the grid current with the fault applied. Because of the fault, there is a sudden

Table 2.2 Comparative study with the proposed system and existing works

References	Control method	THD of voltage (%)	THD of current (%)
Gibin Mathew P. *et al.* [8]	Static load-flow-based harmonic control	33.77	35.22
M. Vijay K. *et al.* [11]	ANN-based control	18.32	18.32
Yajuan G. *et al.* [13]	Feed-in grid current suppression loop	9.68	2.97
Padmavathi L. [16]	Feed-forward control	6.07	1.75
Proposed system	DQ-based dual-loop vector control with PI regulators	1.41	1.31

increase in the grid current at the 1 s mark. Figure 2.7(c) is the battery state of charge, which is discharging when the fault occurred during the 1–1.5 s period. Figure 2.7(d) and (e) shows the THD of grid voltage and current before the fault, respectively. The values are 3.37% for voltage and 2.5% for current. Figure 2.7(f) and (g) shows the THD of grid voltage and current after fault, respectively. The values are 3.54% for voltage and 3.83% for current. Table 2.2 shows the THD results of some previously developed models and also the proposed model. The table compares THD values from four studies, highlighting that Gibin Mathew P. *et al.* reported the highest THD at 33.77%, followed by M. Vijay K. *et al.* with 18.32%, Yajuan G. *et al.* with 9.68%, and Padmavathi L. with 6.07%. The proposed system demonstrates superior performance by maintaining a consistently low THD of 1–4% across all cases, showcasing its effectiveness in improving power quality.

2.5 Summary

This chapter discusses the modeling, control, and stability analysis of a hybrid microgrid integrating PV panels, wind turbines, and battery storage on a 400 V DC bus. Analytical results highlight the system's ability to maintain stable operating conditions with less than 4% THD in both inverter and grid outputs for all operating segments. The system has been tested by varying the power of PV as well as wind turbines (2 kW, 4 kW, 6 kW, and 10 kW). When PV power was 10 kW and wind 2 kW, THD results of inverter voltage current were 1.42% and 1.35%, respectively. For the grid, it was 1.58% and 2.66%. When PV power decreased to 6 kW and wind increased to 4 kW, THD results for inverter's voltage and current were 1.41% and 1.31%. At that period, THD for grid were 1.58% and 1.68%. When power became 12 kW combined, the results became the same as the first operating period. THD slightly increased at the time of fault application; before fault, it was 3.37% for voltage and 2.5% for current, and after fault, it became 3.54% and 3.83%. The fault analysis validates the reliability of the system, where it is shown that it effectively recovers and synchronizes to the grid after disturbance. The battery actively supports at the time of fault, therefore making the system robust. An advanced controller developed for the dynamic power management and stabilization has been validated using simulations and provides, therefore, a robust and efficient framework for hybrid microgrid operation.

In the future, this particular study will be extended to a multi-microgrid system that investigates stability, power sharing, and grid synchronization of multiple interconnected microgrids under different operating conditions. Furthermore, a laboratory prototype will be built for the validation of the proposed control strategy and for assessing actual performance against the simulations.

References

[1] Eklas H., Ron P., and Ramazan B., "Implementation of hybrid energy storage systems to compensate microgrid instability in the presence of constant power loads," in *Proceedings of the IEEE International Telecommunications Energy Conference (INTELEC)*, 2015.

[2] Xuesong Z., Tie G., and Youjie M., "An overview on microgrid technology," in *Proceedings of the IEEE Conference*, Tianjin, China, 2015.

[3] Mostafa F., Sebastian K., Claudio A. C., Kankar B., and Thomas L., "Battery energy storage system models for microgrid stability analysis and dynamic simulation," *IEEE Transactions on Power Systems*, vol. 33, no. 2, pp. 2301–2310, 2018.

[4] Guohong W., Seiya I., and Hang Y., "DC voltage stabilization in direct current/alternating current hybrid microgrid by cooperative control of multiple energy storages," in *Proceedings of the IEEE Conference*, Sendai, Japan, 2018.

[5] Sukumar M. and Aquib J., "Autonomous battery storage energy system control of PV-wind based DC microgrid," in *Proceedings of the IEEE Conference*, Delhi, India, 2018.

[6] Jithin K., Mayadevi N., Hari Kumar R., and Mini V. P., "The effect of multiple PV and battery penetration on stability of DC microgrid with single bus topology," in *Proceedings of the IEEE International Conference on Power Systems*, Thiruvananthapuram, India, 2021.

[7] Hanwen G., Xintong L., Hao Q., Yueming J., Zaibin J., and Yinan X., "A virtual-filter-based stability enhancement control for grid-connected converter in DC microgrids under unbalanced grid conditions," *International Journal of Electrical Power and Energy Systems*, vol. 161, pp. 110224, 2024.

[8] Gibin Mathew P., Twinkle T., Miranda T., Juliya S., and Gopakumar P., "Harmonic analysis of microgrid operation in islanded mode with nonlinear loads," in *Proceedings of the IEEE International Conference on Computer Communication and Informatics*, Coimbatore, India, January 2016.

[9] Anusuya A., Jayashri R., and Eliathamby A., "Real time stability enhancement for islanded microgrids," in *Proceedings of the IEEE International Conference on Power Systems and Green Energy Conference*, Sydney, Australia, 2017.

[10] Shridhar S. K., Pooja D. U., and Sharad W. M., "Microgrid stability issues and PI control technique," in *Proceedings of the IEEE International Conference*

on Information, Communication, Engineering and Technology, Pune, India, 2018.

[11] M. Vijay K. and Priyanka C., "Power quality enhancement in AC microgrids using integrated genetic algorithm in ANN," in *Proceedings of the IEEE International Conference on Electronics, Communication and Aerospace Technology*, Greater Noida, India, January 2021.

[12] Zongwen Z., Fei G., Boshen Z., and Pengfeng L., "Stability analysis of islanded PV-battery DC microgrid considering different operating points of PV curve," in *Proceedings of the IEEE Power System and Green Energy Conference*, Shanghai, China, 2024.

[13] Yajuan G., Wei F., Jinghang Lu, Josep M. G., and Juan C. V., "A novel grid-connected harmonic current suppression control for autonomous current sharing controller-based AC microgrids," in *Proceedings of the IEEE International Conference on Power and Energy Systems*, Aalborg, Denmark, 2018.

[14] Ram K., Rohith Y., and M. Pavan K., "Stability assessment and simplified control strategy for DC microgrids with constant power load," in *Proceedings of the IEEE Power and Energy Society General Meeting*, Washington, USA, 2024.

[15] Jithin K., Mayadevi N., Hari Kumar R., and Mini V. P., "Stability analysis of interconnected DC microgrid during disconnection of microgrid units," in *Proceedings of the IEEE Asia Pacific Conference on Innovation in Technology*, Mysuru, India, 2024.

[16] Padmavathi L., "Harmonic compensation method for low-voltage small-scale islanded microgrids," in *Proceedings of the IEEE International Conference on Power, Control & Embedded Systems*, Prayagraj, India, January 2023.

[17] Mohd Brado F., Fransisco Danang W., and Eka F., "Islanded wind farm microgrid stability control using synchronverter algorithm," in *Proceedings of the IEEE International Power Electronics Conference (IPEC-Himeji - ECCE Asia)*, Himeji, Japan, 2022.

[18] Moussa Saadati T., Mohammad Khoshhal R., Sajad B., and Amir Hossein S., "A hierarchical control approach to improve the voltage and frequency stability for hybrid microgrids-based distributed energy resources," *Energy Reports*, vol. 10, pp. 2693–2709, 2023.

[19] Rashad M. K., "New technique for improving transient stability of standalone microgrid using superconductor fault current limiter," *Journal of Engineering Research*, vol. 12, pp. 160–172, 2024.

[20] Roni S., Rohit R., Bikram B., Chinmoy M., Bilash B., Jamerul I. *et al.*, "Mathematical modeling of solar photovoltaic cell using MATLAB/Simulink and study under different climatic condition," *International Journal of Engineering Research and Applications*, vol. 13, no. 6, pp. 51–57, 2023.

[21] Marwa G., Mouna R., and Lotfi K., "Power dispatch strategy for interconnected microgrids based hybrid renewable energy system," *International Journal of Renewable Energy Research*, vol. 8, no. 2, pp. 839–847, 2018.

Chapter 3

Solar–Wind–Biomass Hybrid Renewable Energy System with hydrogen production

Van Ga Bui[1] *and Thi Minh Tu Bui*[1]

The Solar–Wind–Biomass (SWB) configuration is a suitable option for a Hybrid Renewable Energy System (HRES) with hydrogen production, offering significant techno-economic and environmental advantages as a multi-generation system utilizing renewable energy for sustainable development. However, further research is needed to optimize system component capacities and refine biomass energy conversion technologies, including biomass storage, gasification, and internal combustion engines fueled by a syngas–biogas–hydrogen mixture.

For the SWB grid-connected HRES with hydrogen production to achieve economic parity with a grid-connected system without hydrogen production, the price of green hydrogen must reach \$5/kg. In comparison to the CO_2 emissions from producing 74.235 kWh of grid electricity in Vietnam, the annual CO_2 emission reductions are 39 tons for the SW-HRES without hydrogen production, 50 tons for the SW-HRES with hydrogen production, 207 tons for the SWB system without hydrogen production, and 218 tons for the SWB system with hydrogen production. The cost of electricity (COE) for a standalone HRES with hydrogen production at a 5% annual capacity shortage rate is \$0.097/kWh, whereas for a grid-connected system, it is \$0.039/kWh.

Refuse-derived fuel (RDF) presents an effective solution for biomass storage, transportation, and energy conversion. The optimal excess air ratio (ER) for RDF biomass gasification ranges from 0.3 to 0.4, and enriching the oxidant with 30% and 50% oxygen increases the calorific value of syngas by 60% and 90%, respectively. In a syngas-diesel dual-fuel engine, power derating ranges from 33% to 24% with port injection (PI) and is reduced to 12.6% with direct injection (DI) using a twin-injector system. Power derating in syngas engines can be mitigated by enriching syngas with biogas or hydrogen, although higher hydrogen or biogas content increases NO_x emissions. For spark-ignition engines, the optimal hydrogen and biogas contents in the syngas mixture are 20% and 30%, respectively. In dual-fuel engines using DI, output power can match that of diesel engines when the hydrogen content in the syngas mixture reaches approximately 60%. Additionally, power derating in syngas engines can be controlled by increasing fuel calorific value through oxygen-enriched air supplied

[1]The University of Danang – University of Science and Technology, Vietnam

to the gasifier. For dual-fuel engines using PI, power derating is 22.10%, 9.57%, and 4.73% when the oxygen content in the oxidizing air is 10%, 30%, and 50%, respectively.

3.1 Introduction

The effects of global warming have manifested earlier than scientists had predicted. Meteorological centers worldwide recorded unprecedented atmospheric temperatures in 2024 compared to previous years. Numerous climate change summits have urged nations to immediately reduce greenhouse gas (GHG) emissions to prevent global temperatures from rising beyond 1.5°C above pre-industrial levels. However, achieving this goal is increasingly challenging.

In alignment with the objectives outlined in the Paris Agreement, achieving the global climate targets requires a 45% reduction in CO_2 emissions by 2030 and the attainment of net-zero emissions by the year 2050, relative to 2010 levels. In practice, global CO_2 emissions saw a slight decrease during the COVID-19 pandemic, but have since resumed their previous trajectory. At COP29 in 2024, countries were called upon to significantly reduce GHG emissions to mitigate global warming. Transitioning from fossil energy to renewable energy is the central strategy for achieving the net-zero goal.

Currently, no single renewable energy source can entirely replace the fossil fuels used around the world. Therefore, a combination of different renewable energy sources is essential. The integrated production of electricity and green hydrogen from various renewable sources can be considered a common model for future energy systems. Hydrogen is a sustainable fuel with significant potential for widespread applications in electricity generation, transportation, industrial production, and everyday life. It is projected that hydrogen energy could meet 11% of global energy demand by 2025 and increase to 34% by 2050 [1,2]. Consequently, the integration of hydrogen in a multi-generation renewable energy system is critical for advancing the energy transition.

A major limitation of renewable energy sources lies in their inherent intermittency. The integration of diverse energy resources within an HRES offers a robust approach to mitigating this challenge, thereby improving the consistency and overall efficiency of hydrogen production processes [3–6]. Numerous studies have investigated hydrogen production utilizing hybrid solar-wind power systems [7–10]. These investigations consistently demonstrate that hybrid systems markedly improve hydrogen production efficiency [11–13]. For instance, Akyuz *et al.* employed Hybrid Optimization of Multiple Energy Resources (HOMER) software to analyze hydrogen production efficiency in hybrid renewable systems, reporting efficiencies as high as 60%, which exceed those of standalone solar or wind energy systems [7]. Similarly, Okonkwo *et al.* evaluated the techno-economic characteristics of a hybrid solar-wind renewable energy system integrated with hydrogen production, also using HOMER software. Their findings confirmed that hybrid systems achieved superior hydrogen production efficiencies compared to systems relying solely on individual renewable sources [10].

HRES multi-generation producing electricity and hydrogen is a key emerging trend in recent renewable energy research [14,15]. The incorporation of water electrolyzers into HRES for hydrogen production has gained prominence as an effective strategy, offering numerous advantages in terms of energy efficiency and reductions in environmental emissions [16–18]. Bukar *et al.* [18] highlighted that grid-connected HRES configurations optimize equipment capacity utilization, resulting in lower hydrogen production costs. For off-grid HRES, storing renewable energy as hydrogen has proven advantageous [19,20]. Hydrogen generated from these systems can serve as fuel for the transportation sector or be blended with other renewable fuels to power generators, thereby ensuring stable energy output in HRES [21,22].

One of the key criteria for evaluating the efficiency of hybrid renewable energy systems (HRES) with hydrogen production is the fuel cost. The cost of hydrogen produced from HRES depends on various factors, including renewable energy potential, electricity prices, and equipment costs specific to the system's installation location. For example, hydrogen production costs from HRES that combine solar and wind energy are estimated at approximately \$2/kg in Chile and Argentina [23]. However, studies in other regions report significantly higher costs, such as \$6.2/kg in the United States and \$4.64/kg in Morocco [24–27]. The investment recovery period for HRES integrated with hydrogen production is generally less than four years [28]. Reducing hydrogen production costs within these systems can be achieved by optimizing the selection of equipment components and improving overall system efficiency. Despite these measures, experts project that the average cost of green hydrogen will stabilize around \$3.7/kg [29], which remains higher than the anticipated cost of hydrogen produced from fossil fuels, estimated at \$3/kg by 2050 [28].

From an engineering standpoint, the capacity factor is a key performance metric used to evaluate the operational efficiency of power generation systems. It is quantitatively expressed as the ratio between the actual energy generated over a given time period and the theoretical maximum output, assuming continuous operation at rated capacity during the same interval [30]. Among various renewable energy technologies, biomass-based systems generally demonstrate the highest capacity factors, typically ranging from 35% to 94% [31]. In comparison, wind energy systems exhibit capacity factors between 17% and 40%, whereas photovoltaic (PV) systems tend to have lower capacity factors due to their dependency on solar irradiance [32].

From an environmental standpoint, the life-cycle GHG emissions of power generation equipment vary significantly across renewable energy sources. Biomass power generation systems exhibit the highest emissions, reaching up to 650 g-CO_2-eq/kWh, followed by solar power at a maximum of 300 g-CO_2-eq/kWh, and onshore wind power at 124 g-CO_2-eq/kWh [28]. However, the environmental efficiency of biomass requires a more comprehensive assessment, particularly regarding its integrated role in solid waste management.

When biomass is processed into fuel, such as syngas, and subsequently converted into electricity using internal combustion engines or gas turbines, net GHG emissions can approach zero. Conversely, if biomass is left untreated, it naturally degrades into biogas, primarily composed of methane (CH_4), a GHG with a global warming potential 20 times greater than that of carbon dioxide (CO_2). A promising approach for converting biomass into energy involves processing it into RDF and subsequently converting it into syngas through gasification [33,34].

To meet the Net-Zero strategy goal [35], the International Energy Agency projects that renewable energy sources are expected to supply 90% of global energy consumption by 2050 [36]. Recently, Vietnam announced its Eighth Power Development Plan, targeting over 70% of the national power supply from renewable energy sources by 2050. In the transportation sector, many countries had outlined roadmaps to phase out fossil-fuel-powered vehicles in favor of electric, fuel cell, or renewable-fuel vehicles. The transition to electric vehicles underscores the critical importance of the energy sources used to generate electricity. The reduction of GHG emissions from vehicles will only be meaningful if the electricity is generated from renewable sources. If electricity continues to be produced from fossil fuels, pollution and GHG emission issues merely shift from the vehicle's point of use to the electricity generation site.

Most renewable energy systems are, in fact, grid-connected due to their high dependence on weather and seasonal factors [37,38]. In the absence of fossil energy, HRES should be utilized to reduce dependence on energy storage systems [38,39]. HRES can potentially lower energy costs by up to 30% compared to traditional renewable energy systems [40]. Furthermore, HRES components can be optimally selected to reduce equipment costs [41]. HRES is an optimal choice for remote and off-grid areas or regions with limited grid capacity [42]. Tropical agricultural countries have advantages in solar, wind, and biomass energy. Combining these sources in an HRES, referred to as the Solar–Wind–Biomass Hybrid Renewable Energy System (SWB-HRES), can positively contribute to energy transition strategies, green economic development, circular economies, and the achievement of net-zero goals.

This overview highlights that HRES integrated with hydrogen production is a prominent trend in contemporary renewable energy research and applications. The efficiency of hydrogen production is highly dependent on the renewable energy potential at the deployment site. Most published studies in this domain focus on hybrid solar-wind power systems, while relatively few have explored the integration of biomass power into HRES with hydrogen production. Utilizing biomass for electricity generation offers dual benefits: recovering energy from solid waste and alleviating the burden of solid waste management. This approach also significantly reduces GHG emissions, a pressing issue in many developing countries.

The limited use of biomass in HRES can be attributed to the complexities involved in the storage and conversion of biomass into electricity. Besides, the equipment needed for this conversion is not yet widely available on the market. This chapter focuses on in-depth research of biomass integration into a multi-generation HRES that produces electricity and hydrogen.

3.2 Schema of SWB-HRES

Figure 3.1 illustrates the SWB-HRES with hydrogen production. Biomass is divided into two streams: easily degradable biomass (3), which is used to produce biogas in an anaerobic digester (4), and hardly degradable biomass (1), which is processed into

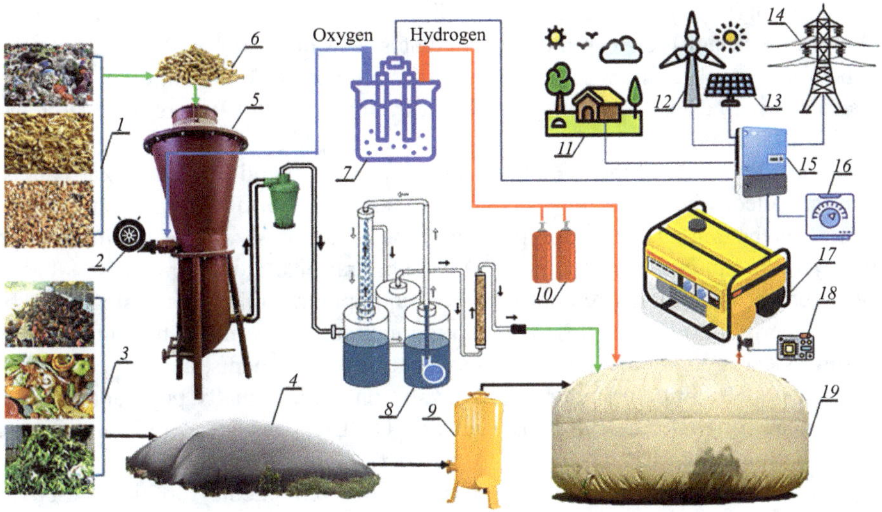

1. Hard degradable material
2. Air blower
3. Easy degradable material
4. Biogas digester
5. Gasifier
6. RDF
7. Electrolyzer
8. Syngas filter
9. Biogas filter
10. Hydrogen canister
11. Electric load
12. Wind turbine
13. PV solar panels
14. Grid
15. Converter
16. Generator
17. ECU
18. Gas mixture tank

Figure 3.1 Schematic of the SWB-HRES

RDF (6) and then gasified into syngas in a gasifier (5). When the output capacity of the solar power system (13) and wind power system (12) exceeds the load demand (11), the surplus energy is directed to an electrolyzer (7) to generate hydrogen and oxygen. The produced hydrogen can be stored in tanks (10) for use as fuel in transportation or blended with biogas and syngas in the gas storage bag (19) to create a fuel mixture suitable for a stationary internal combustion engine coupled with a generator (17).

The oxygen generated during electrolysis is used to enrich air oxidant introduced into the gasifier via a blower (2), enhancing the quality of the syngas. In instances where solar and wind power outputs are insufficient to meet load demands or during supply interruptions, the internal combustion engine (17) coupled with generator generates electricity to ensure a stable and continuous power supply. Thus, the internal combustion engine-generator set acts as an energy storage system within the SWB-HRES.

Due to variation in compositions of biogas, syngas, and hydrogen fuel, along with the continuously varying engine load required to meet electricity demand, the fuel-supply control system (18) for the internal combustion engine in the SWB-HRES should be specifically designed. Electricity generated from various sources is connected to the load and grid (14) via a converter (15). The Energy Management System (16) coordinates the operation of all components of the SWB-HRES.

Since syngas and biogas contain impurities that can be detrimental to fuel-utilizing equipment, they undergo thorough purification before storage in gas bags

(19). The syngas exiting the gasifier is directed to a filtration system (8), which includes a cyclone filter to remove particulates, a water scrubber to eliminate tar, and bentonite-zeolite filters to remove sulfur compounds. Meanwhile, biogas from the anaerobic digester is passed through a filtration column (9) to remove H_2S and other harmful impurities.

In the SWB-HRES, many components—including wind turbines, solar panels, converters, electrolyzers, hydrogen storage tanks, and control systems—are widely available on the market. These components come in various sizes to accommodate the capacity requirements of different HRES configurations. However, biomass-based power generation systems remain relatively unpopular, limiting the integration of biomass energy into HRES [43]. This limitation is primarily due to the bulky nature of biomass energy systems, their complex control requirements, and the challenges associated with traditional biomass storage solutions. Additionally, popular optimization software for HRES calculations, such as HOMER, does not specifically include detailed biomass modules.

Therefore, the research and development of components for biomass-based power generation systems are essential. This effort will not only support the broader adoption of SWB-HRES but also contribute to the enhancement and expansion of HRES optimization software capabilities. In the first part of this chapter, we simulate and analyze a case study of SWB-HRES with hydrogen production on Ly Son Island (Vietnam), where grid power capacity is limited. The latter part of this chapter presents our research findings on biomass-based power generation systems, including the gasification process and internal combustion engines fueled with syngas–biogas–hydrogen.

3.3 Materials and methods

3.3.1 Case study location

The case study was conducted on Ly Son Island, located at 15.382°N, 109.098°E. Ly Son is an island district in Quang Ngai Province, situated 15 nautical miles from the mainland, originating from Sa Ky Port (see Figure 3.2). The island spans an area of approximately 10 km^2 and is home to a population of over 21,000 people. Of these residents, around 60% of households rely on fishing as their primary livelihood, while 30% are engaged in agriculture.

Ly Son is influenced by the tropical monsoon climate typical of the tropical oceanic region, characterized by high temperatures and humidity. The island receives an average of approximately 2,430 hours of sunshine annually, with peak solar radiation intensity occurring between 10:00 AM and 2:00 PM. The average wind speed is around 1.5 m/s, peaking during the northeast monsoon season (October to June) at 5–10 m/s, and occasionally reaching 30–40 m/s, primarily in October. These climatic conditions provide Ly Son with significant renewable energy potential, promoting sustainable development. Figure 3.3(a) and (b) presents the average solar radiation and wind speed data for Ly Son Island, derived from NASA's datasets.

In the district, solid waste primarily consists of household waste and agricultural by-products. Collected waste is transported to the solid waste treatment plant

Figure 3.2 Geographic location of Ly Son Island

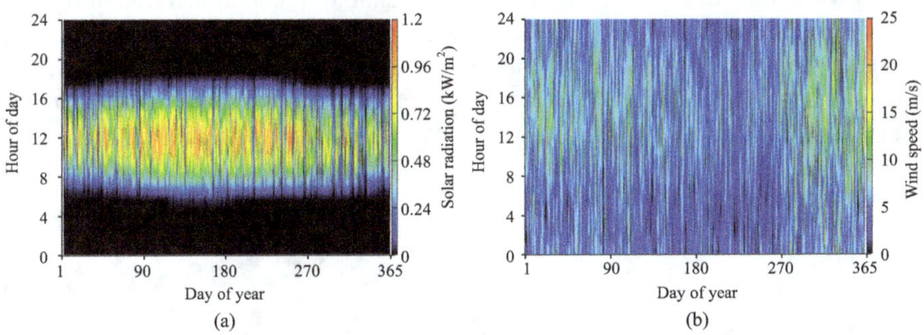

Figure 3.3 Solar radiation (a) and average wind speed (b) on Ly Son District Island

for processing. The solid waste treatment plant on Ly Son Island has a maximum capacity of 25 tons/day. On average, 22–25 tons/day of household waste is collected and processed, with peak volumes reaching 26–30 tons/day.

Ly Son Island is connected to the national grid via a 22 kV submarine cable. However, due to the limited grid capacity, the integration of renewable energy sources is necessary. In this study, we utilized HOMER software to optimize the components of the grid-connected HRES on Ly Son Island. Simulations were conducted for two scenarios: one without hydrogen production and another incorporating hydrogen production.

3.3.2 Simulation method

In this book chapter, the HOMER software was employed to model and simulate the HRES, while ANSYS Fluent was utilized to perform computational fluid dynamics (CFD) simulations of the gasification process and to analyze the performance of an internal combustion engine operating on a syngas–biogas–hydrogen fuel blend.

3.3.2.1 HOMER software

Effective modeling and simulation tools are essential for designing HRES, as they enable researchers and engineers to evaluate technical and economic performance under various conditions. Among various available modeling tools, the HOMER software stands out due to its computational robustness and adaptability to a wide range of applications. Developed by the National Renewable Energy Laboratory (NREL), HOMER enables comprehensive simulation, techno-economic evaluation, and multi-scenario optimization of hybrid energy systems by systematically analyzing diverse configurations and operational conditions. Its advanced simulation capabilities allow researchers and practitioners to assess system reliability, life-cycle costs, and environmental impacts across diverse conditions. The modular framework and user-friendly interface of HOMER make it particularly suitable for simulating complex systems that involve renewable energy integration.

In this study, we utilize HOMER software to simulate and analyze the SWB renewable energy system, both with and without hydrogen production. The software's capacity to conduct detailed sensitivity analyses enables the identification of optimal configurations based on varying technical and economic parameters. By leveraging HOMER's comprehensive simulation features, this work aims to provide valuable insights into the feasibility and performance of HRES, thereby contributing to their broader adoption in energy planning and policy.

PV output power

In HOMER software, the output power of PV array is calculated by the following equation [80]:

$$P_{PV} = Y_{PV} f_{PV} \left(\frac{\bar{G}_T}{\bar{G}_{T,STC}} \right) [1 + \alpha_P (T_c - T_{c,STC})] \tag{3.1}$$

where

- Y_{PV}: Rated capacity of the PV array (power output under standard test conditions (STC)) [kW]
- f_{PV}: PV derating factor [%]
- $\bar{G}_T$: Solar radiation incident on the PV array in the current time step [kW/m^2]
- $\bar{G}_{T,STC}$: Incident radiation at STC [1 kW/m^2]
- α_P: Temperature coefficient of power [%/°C]
- T_c: PV cell temperature in the current time step [°C]
- $T_{c,STC}$: PV cell temperature under STC [25°C]

The photovoltaic (PV) derating factor represents a correction coefficient utilized in the HOMER software to adjust the nominal output of a PV array, thereby accounting for performance reductions under real-world operating conditions. These adjustments are necessary because manufacturers typically specify PV module output under STC, which assumes an irradiance of 1 kW/m^2, a cell temperature of 25°C, and the absence of wind. However, actual operating environments often deviate from STC, particularly with cell temperatures frequently exceeding 25°C under full sunlight. To reflect these deviations, temperature-related losses are incorporated into the model. In cases where temperature effects are not explicitly considered, HOMER assumes a zero temperature coefficient for power output, simplifying the estimation process.

Wind turbine output power

In the HOMER simulation environment, the power output of a wind turbine is determined through a structured three-step methodology. Initially, the software estimates the wind speed at the hub height of the turbine. Subsequently, it calculates the theoretical power output corresponding to this wind speed, assuming standard air density conditions. Finally, HOMER adjusts the computed power output to account for variations in actual air density [44]. The wind speed at the turbine's hub height is commonly derived using the logarithmic wind profile model:

$$U_{hub} = U_{anem} \frac{\ln\left(\frac{z_{hub}}{z_0}\right)}{\ln\left(\frac{z_{anem}}{z_0}\right)} \tag{3.2}$$

where

- U_{hub}: Wind speed at the hub height of the wind turbine [m/s]
- U_{anem}: Wind speed at anemometer height [m/s]
- z_{hub}: Hub height of the wind turbine [m]
- z_{anem}: Anemometer height [m]
- z_0: Surface roughness length [m]
- $\ln(\cdots)$: Natural logarithm

The anemometer height refers to the elevation above ground level at which wind speed measurements are recorded. Since wind velocity generally increases with altitude, it becomes necessary to adjust the measured data when the anemometer height differs from the wind turbine's hub height. In practice, meteorological assessments often use an anemometer height of approximately 10 meters. However, for wind energy applications, measurement instruments are typically positioned at greater heights to better align with the hub height of wind turbines, which generally ranges from 25 to 100 meters. The closer the anemometer is positioned to the intended hub height, the more accurate the estimation of wind resource availability becomes for turbine performance analysis.

The surface roughness length is a critical parameter used to quantify the aerodynamic characteristics of the terrain, influencing wind flow patterns near the ground surface and affecting vertical wind speed profiles. For smooth surface $z_o = 10^{-5}$ m, in the surface with few trees, few buildings, $z_o = 0.25$ m. In the city center with tall buildings, $z_o = 3$ m.

Once HOMER estimates the wind speed at the hub height, it utilizes the turbine's power curve to determine the corresponding power output under standard temperature and pressure (STP) conditions. If the wind speed falls outside the operational range specified in the power curve—specifically below the cut-in speed or above the cut-out threshold—the turbine is assumed to generate no power. Power curves are generally provided by manufacturers based on STP conditions to standardize performance metrics. To better reflect actual atmospheric conditions, HOMER adjusts the estimated power by applying a correction factor based on the ratio of actual to standard air density, as defined in the following equation [44]:

$$P_{WTG} = \left(\frac{\rho}{\rho_0}\right) \cdot P_{WTG,STP} \tag{3.3}$$

where

- P_{WTG}: Wind turbine power output [kW]
- $P_{WTG,STP}$: Wind turbine power output at STP [kW]
- ρ: Actual air density [kg/m^3]
- ρ_0: Air density at STP (1.225 kg/m^3)

Electric power from biomass

The overall system of electricity production from biomass consists of the biomass feedstock, gasifier, internal combustion engine (ICE), and generator. The output power of the system can be generally described by:

$$P_b = P_{biomass} \cdot \eta_{gasi} \cdot \eta_{ICE} \cdot \eta_{gen} \tag{3.4}$$

where

- $P_{biomass} = M_{biomass} \cdot Q_{LHVbiomass}$ (kW)
- $M_{biomass}$: Biomass consumption rate (kg/s)
- $Q_{LHVbiomass}$: Heating value of the biomass (kJ/kg)
- η_{gasi}: Efficiency of the gasification process
- η_{ICE}: Efficiency of the internal combustion engine
- η_{gen}: Efficiency of the generator

The above equation provides a comprehensive framework for calculating the electric power produced from a biomass-gasifier-internal combustion engine-generator system. By considering the efficiencies at each stage of the conversion process, the model can accurately simulate the performance and output of the biomass energy system.

Hydrogen production from electrolysis

$$M_{H_2} = \frac{P \cdot t \cdot \eta_{elec}}{HV_{H_2}} \tag{3.5}$$

where

- M_{H_2}: Amount of hydrogen produced during t hours (kg)
- P: Input electric power to the electrolyzer (kW)
- t: Time of operation (hours)
- η_{elec}: Efficiency of the electrolyzer
- HV_{H_2}: Heating value of hydrogen (approximately 33.33 kWh/kg)

The efficiency of the electrolyzer, indicating how effectively it converts electrical energy into chemical energy stored in hydrogen. The efficiency of electrolyzers typically ranges from 60% to 90% depending on the specific type of electrolyzer and its operating conditions. Alkaline electrolyzers that are commonly used for large-scale hydrogen production and are known for their robustness and long lifespan have an efficiency range from 60% to 80%. Proton exchange membrane (PEM) electrolyzers having a higher efficiency are more responsive to fluctuations in power supply, making them suitable for renewable energy applications, and have an efficiency range from 70% to 90%. Solid oxide electrolyzers operating at high temperatures can achieve higher efficiencies, especially when using waste heat from industrial processes, and have an efficiency range from 70% to 90%.

3.3.2.2 ANSYS Fluent software

ANSYS Fluent is a prominent software platform specializing in CFD, developed by ANSYS Inc. It is extensively employed for modeling complex physical phenomena, including fluid flow behavior, heat transfer processes, and multi-species chemical interactions. Utilizing the Finite Volume Method (FVM) as its numerical foundation, Fluent solves fundamental governing equations of fluid mechanics, such as the Navier–Stokes equations, the continuity equation, the energy conservation equation, and species transport formulations [45]. These equations are numerically discretized over a computational mesh comprising finite control volumes.

ANSYS Fluent uses fundamental equations of fluid dynamics and thermodynamics to solve various physical and engineering problems. The equation system is primarily based on the Navier–Stokes equations and incorporates additional equations for heat transfer, species transport, turbulence, and multiphase flows, depending on the application. Below are the key equations used in Fluent:

Continuity equation

$$\frac{\partial \rho}{\partial t} + \nabla \cdot (\rho \vec{v}) = 0 \tag{3.6}$$

where

- ρ is the fluid density.
- $\vec{v}$ is the velocity vector.
- t is time.

Navier–Stokes equation

$$\frac{\partial (\rho \vec{v})}{\partial t} + \nabla \cdot (\rho \vec{v} \vec{v}) = -\nabla p + \nabla \cdot \boldsymbol{\tau} + \rho \vec{g} + \vec{F} \tag{3.7}$$

where

- p is pressure.
- $\boldsymbol{\tau}$ represents the stress tensor (includes viscous effects).
- $\vec{g}$ is the gravitational acceleration.
- $\vec{F}$ is any additional body force (e.g., electromagnetic forces).

Energy conservation equation

$$\frac{\partial (\rho E)}{\partial t} + \nabla \cdot (\vec{v}(\rho E + p)) = \nabla \cdot (k \nabla T) + \Phi + S_h \tag{3.8}$$

where

- E is the total energy per unit mass.
- k is the thermal conductivity.
- T is the temperature.
- Φ accounts for viscous dissipation.
- S_h represents heat sources.

Species transport equation

$$\frac{\partial(\rho Y_i)}{\partial t} + \nabla \cdot (\rho \vec{v} Y_i) = \nabla \cdot (\vec{J}_i) + R_i + S_i \tag{3.9}$$

where

- Y_i is the mass fraction of species i.
- $\vec{J}_i$ is the diffusion flux.
- R_i is the net production rate due to chemical reactions.
- S_i includes any additional source terms.

Fluent supports a variety of models tailored to specific applications, such as turbulence (e.g., $k - \varepsilon$, $k - \omega$), multiphase flows (e.g., Eulerian-Eulerian), combustion (e.g., finite-rate chemistry and species transport), and heat transfer (e.g., conduction, convection, radiation). It also provides advanced meshing techniques, adaptive refinement, and parallel computing capabilities to handle large-scale simulations [81].

The software is applied across numerous industries, including automotive (e.g., combustion engine optimization), aerospace (e.g., aerodynamic design), energy (e.g., gasification processes, wind turbines), and chemical processing (e.g., reactor design). Its robustness and flexibility make it a critical tool for engineering design, process optimization, and environmental impact analysis. Fluent has been extensively used in scientific research to analyze fuel combustion efficiency, pollutant emissions, and thermal management, providing insights to improve performance and sustainability.

ANSYS Fluent is extensively used for modeling and simulation of gasification processes in gasifiers, helping to optimize design and operation for efficiency and environmental compliance. Gasification, a thermochemical process that converts solid fuels (e.g., coal, biomass) into syngas, involves complex multiphysics phenomena, including fluid dynamics, heat and mass transfer, and chemical reactions. ANSYS Fluent provides robust tools to simulate various stages of gasification: Drying and Pyrolysis (decomposition of solid fuel into volatiles and char); Oxidation (partial combustion of volatiles and char to produce heat for the process); Reduction (endothermic reactions between char and gases to generate syngas).

ANSYS Fluent is widely utilized in the modeling and analysis of combustion processes within internal combustion engines, providing engineers and researchers with insights into the performance and efficiency of engine systems. Its advanced capabilities enable the simulation of complex physical and chemical interactions in combustion chambers, allowing optimization of fuel usage and emissions control. ANSYS Fluent includes a variety of combustion models tailored to different engine types and operating conditions: Non-Premixed Combustion (suitable for diesel engines, where fuel and air mix during combustion); Premixed Combustion (applied in spark-ignition engines with pre-mixed air-fuel mixtures); Partially Premixed Combustion (handles scenarios involving both premixed and non-premixed combustion zones); Finite-Rate Chemistry Models (resolve detailed chemical kinetics to simulate pollutant formation like NO_x and soot) [45].

ANSYS Fluent provides robust tools to simulate turbulence-chemistry interaction via the models such as the Eddy Dissipation Model (EDM) and Eddy Dissipation Concept (EDC) for resolving turbulence-chemistry interaction, which is critical for capturing realistic combustion dynamics in ICEs. Besides, ANSYS Fluent simulates the formation of pollutants, including NO_x, CO, and unburnt hydrocarbons, using detailed reaction mechanisms. These capabilities assist in meeting stringent regulatory requirements for cleaner engines.

3.4 Simulation of HRES

3.4.1 Standalone SWB-HRES without hydrogen production

3.4.1.1 System components

Figure 3.4 illustrates various configurations of the HRES without hydrogen production. The electrical energy sources comprise solar power (S), wind power (W), and syngas-fueled generators (B), all of which are direct current (DC) power sources. The DC power is converted to alternating current (AC) to supply loads via a converter. This approach simplifies power grid synchronization when utilizing separate power sources. Additionally, a battery (Bat) serves as an energy storage device.

In Configuration 1, the system incorporates all components: solar (S), wind (W), syngas-fueled generators (B), and battery (Bat). Configuration 2 excludes the syngas-fueled generators (B), while Configuration 3 omits the battery (Bat). The system provides electrical power to an AC load with an average consumption of 40 kWh/day

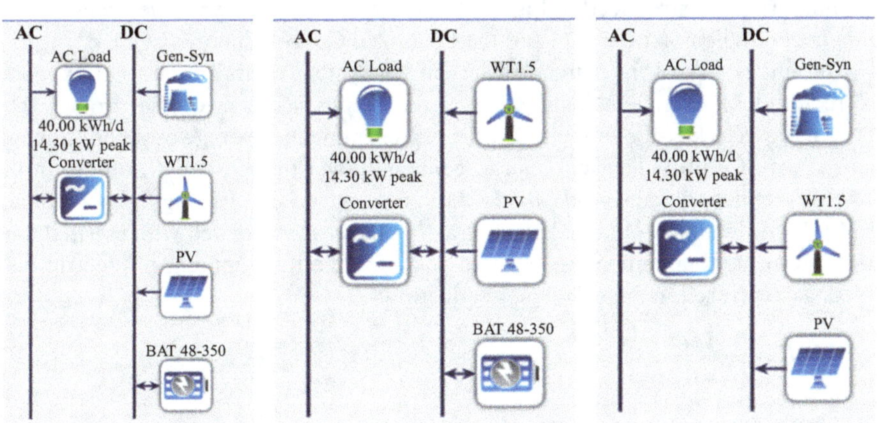

Figure 3.4 Different configurations of HRES without hydrogen production

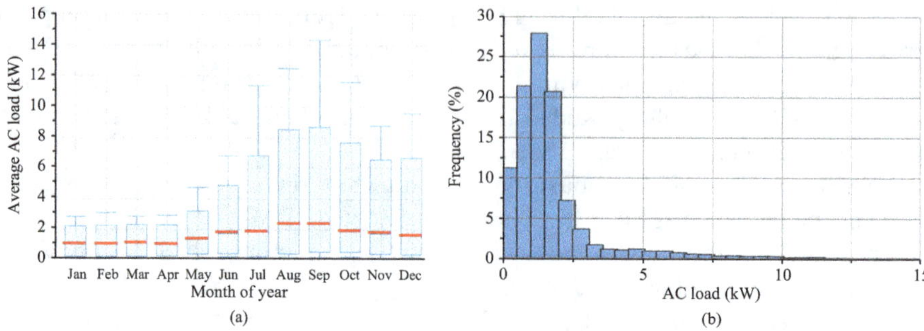

Figure 3.5 Average monthly load capacity (a) and load frequency (b)

and a peak capacity of 14.3 kW. Figure 3.5(a) presents the monthly variation of average power. Power consumption tends to increase during the summer months and at the end of the year. Figure 3.5(b) depicts the load frequency, with the highest load frequency occurring within the power range of 1 kW to 1.5 kW.

3.4.1.2 Simulation results

Effect of the system configuration

When configuring the HRES and providing a comprehensive breakdown of equipment costs, operating expenses, maintenance costs, lifespan, and annual capacity shortage, HOMER will conduct simulations and present results ordered by Net Present Cost (NPC). This includes crucial information regarding equipment capacity, electricity output from each source, excess electricity, and more. Economically, HOMER generates results for NPC, Initial Capital Cost (ICC), Cost of Energy (COE), Return on Investment (ROI), Simple Payback, and other relevant metrics [44].

In this study, the costs associated with each component in the system, along with operating and maintenance costs, are based on current prices in Vietnam, converted to USD. Table 3.1 summarizes the capacities of the equipment in the HRES based on the configurations introduced in Figure 3.4. For economic analysis, we focus on two fundamental indicators: the ICC and the Levelized Cost of Energy (LCOE).

The initial capital investment of an energy system refers to the total expenditure required for complete installation at the commencement of the project. The LCOE serves as a key economic indicator, representing the average cost per kilowatt-hour (kWh) of useful electrical energy generated over the system's lifetime. In the HOMER simulation framework, the LCOE is computed by dividing the net annualized COE production—obtained by subtracting the cost associated with thermal load supply from the total annualized system cost—by the total amount of electric load served, as expressed in the subsequent equation:

$$COE = \frac{C_{ann,tot}}{E_{served}} \tag{3.10}$$

where

- $C_{ann,tot}$: Total annualized cost of the system (\$/year)
- E_{served}: Total electrical load served (kWh/year)

Table 3.1 indicates that with the same annual capacity shortage level of 2%, if the HRES includes all components (Configuration 1), the COE is $0.102/kWh. If the system lacks a generator (Configuration 2), the COE is $0.10/kWh; conversely, if the system lacks batteries (Configuration 3), the COE rises to $0.22/kWh. This demonstrates that the energy storage system is crucial to the COE produced by the HRES.

The difference in energy prices between HRES including batteries and excluding batteries can be elucidated by analyzing the power variation chart of the generation sources throughout the day. When the HRES excludes batteries (Figure 3.6(a)), the generator must operate continuously, adjusting its load mode according to demand. This scenario not only reduces the generator's efficiency but also increases operating hours, ultimately shortening the equipment's replacement cycle. Conversely, when the HRES includes batteries, the generator operates only when battery charging is necessary, specifically when the State of Charge (SOC) falls below a predetermined minimum threshold (Figure 3.6(b)). This operational mode allows the generator to function at its rated load, enhancing the overall efficiency of the engine-generator unit and leading to a reduced COE.

Figure 3.7(a) illustrates that when the battery's SOC remains within operational limits, the battery supplies supplementary power to stabilize the system's output, allowing the generator to remain inactive. However, when the SOC drops below the minimum threshold, the generator activates to restore the SOC to a fully charged state. Figure 3.7(b) indicates that the battery supplies primary power from 4:00 to 6:00

Table 3.1 Equipment capacity and economic efficiency of different HRES configurations

Shortage (%)	PV (kW)	WT (kW)	Gen (kW)	Bat. (kWh)	Conver. (kW)	Init. capital ($)	COE ($/kWh)
2	4	1.5	17	62.8	7.96	24.487	0.102
2	12	1.5	0	126	12	22.081	0.1
2	4	1.5	17	0	7.96	17.389	0.22

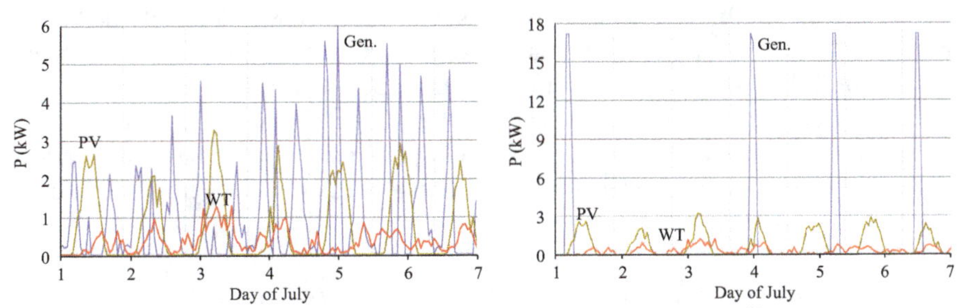

Figure 3.6 Variation in the power output of the sources S, W, and B during the early days of July when the HRES excludes battery (a) and includes battery (b)

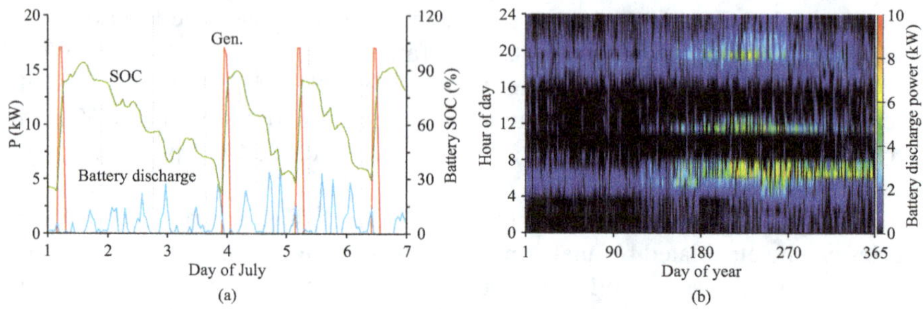

Figure 3.7 *Variations in generator power output, power supplied by the battery to the HRES, and the battery's SOC in early July (a); hourly power output variations from the battery throughout the year (b)*

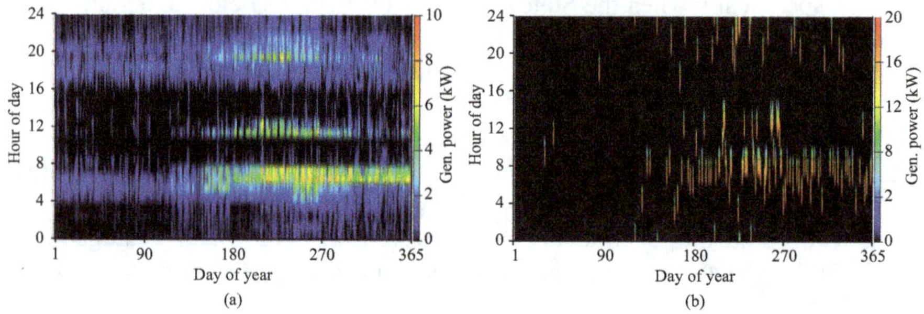

Figure 3.8 *Variation of the generator power output by hour during the day in the case of HRES excluding batteries (a) and including batteries (b)*

and from 17:00 to 21:00, periods when solar power is unavailable and load demand remains high.

Figure 3.8(a) and (b) compares the generator's power output variation over the year for HRES systems including and excluding batteries. In the absence of a battery, the maximum power output is limited, yet the generator operates continuously, resulting in constant power fluctuations. In contrast, when the HRES includes a battery, the generator provides rated power and operates at a lower frequency, particularly during the latter months of the year. During months with lower power demand, the battery can accumulate excess energy generated from wind and solar sources to compensate for shortfalls during peak load times (Figure 3.9(b)). During peak months (from July to October), when load demand is high, the energy accumulated from wind and solar sources becomes insufficient, necessitating the generator to charge the battery. In scenarios where the system lacks a battery, the generator must operate even during months of low power demand (Figure 3.9(a)).

Impact of the quality requirements of the power source

In addition to the influence of the HRES structure on energy costs, these costs are also affected by the operating regime of the power sources and the quality requirements of the energy served. The quality of the energy served is indicated by the annual

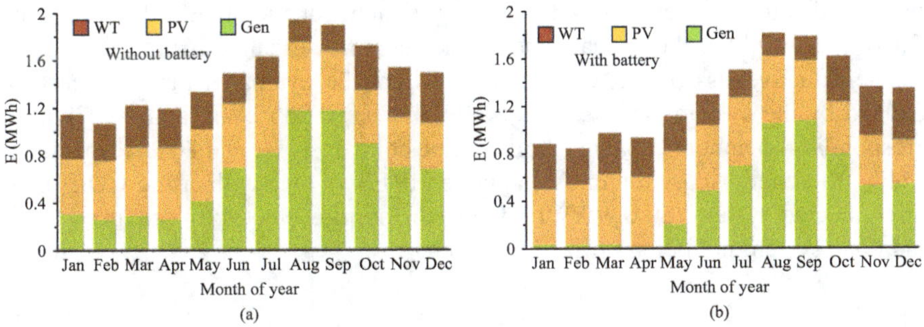

Figure 3.9 Comparison of the electric energy generated from different sources of the HRES excluding batteries (a) and including batteries (b)

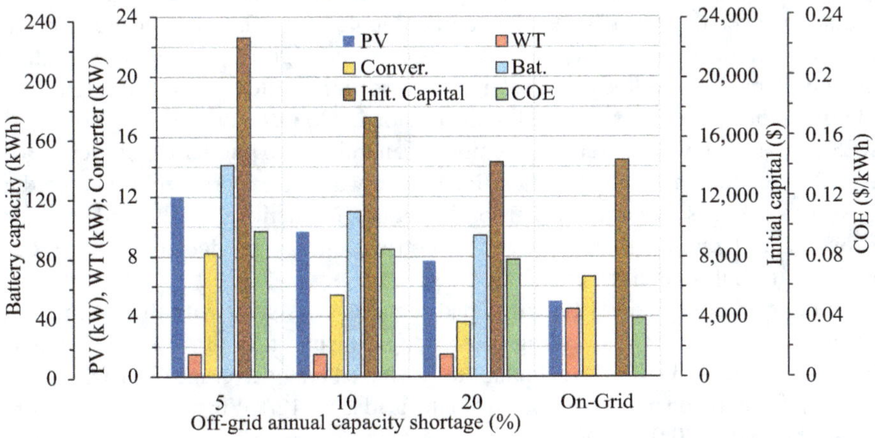

Shortage (%)	PV (kW)	WT (kW)	Bat. (kWh)	Conver. (kW)	Init. Capital ($)	COE ($/kWh)
5	12	12	141	8.25	22,618	0.097
10	9.68	9.68	110	5.43	17,304	0.085
20	7.7	7.7	94.2	3.64	14,292	0.078
On-Grid	5	5	0	6.65	14,443	0.039

Figure 3.10 The impact of the annual capacity shortage on equipment capacity and energy costs of the HRES

capacity shortage; a lower shortage rate signifies that the power source capacity can more effectively meet varying load demands. A lower annual capacity shortage rate necessitates a larger system capacity, which can lead to excess electricity generation during periods of low load, ultimately increasing energy costs.

Figure 3.10 illustrates the variation in the capacity of components within the HRES and the corresponding energy costs for annual capacity shortage rates of 5%, 10%, and 20%. In all three scenarios, the wind turbine capacity remains constant at 1.5 kW. It is observed that the capacities of the other components (PV, battery, converter) decrease as the annual capacity shortage rate increases. This reduction in capacity leads to a decrease in the initial investment for the HRES, dropping from \$22,618 to \$14,292 as the annual capacity shortage rate escalates from 5% to 20%.

The effectiveness of these configurations can be evaluated through the COE produced by the system. The COE is \$0.097/kWh, \$0.085/kWh, and \$0.078/kWh, corresponding to annual capacity shortage rates of 5%, 10%, and 20%, respectively. Notably, the energy cost for the standalone HRES is higher than the COE of the grid-connected system (\$0.039/kWh) due to the surplus electricity produced by the standalone system not being utilized for profit.

3.4.2 Grid-connected SWB-HRES without hydrogen production

3.4.2.1 Configuration of HRES

As depicted in Figure 3.10, the LCOE associated with the grid-connected HRES is notably lower than that of its standalone counterpart. This discrepancy primarily stems from the sizing strategy of the standalone configuration, which necessitates sufficient capacity to satisfy peak demand conditions, often leading to significant underutilization during periods of reduced load. Conversely, the grid-connected HRES is designed with operational flexibility, enabling it to purchase electricity from the grid when demand exceeds generation capacity and to export surplus energy during periods of excess generation. This bidirectional interaction with the grid enhances the overall utilization efficiency of the electrical energy produced by the HRES, thereby contributing to a lower LCOE.

In the grid-connected HRES, the load remains consistent with that of the standalone system, featuring an average electricity consumption of 40 kWh/day and a peak capacity of 14.3 kW. The calculations are based on the average electricity prices in Vietnam, where the purchase price from the grid is \$0.12/kWh, and the selling price back to the grid is \$0.094/kWh.

Figure 3.11 presents the configurations of the grid-connected HRES without hydrogen production. This HRES integrates three sources of electricity generation: solar power, wind power, and syngas generators. Within this system architecture, bidirectional energy exchange with the utility grid is enabled. Specifically, when the generation capacity of the HRES falls short of the load demand, the deficit is compensated by importing electricity from the grid. Conversely, during periods of generation surplus—when the system's output exceeds consumption requirements—the excess electrical energy is exported back to the grid, thereby enhancing overall energy utilization and economic performance. Thus, in principle, there is no surplus electricity, and the reduced capacity of the equipment results in lower energy costs for the grid-connected HRES compared to standalone systems.

3.4.2.2 Simulation result

In the grid-connected HRES, the generator functions as an active energy source rather than merely a backup, as is the case with standalone HRES. In the proposed configuration, all power sources are DC to simplify system synchronization. These power sources supply the load and connect to the grid via a converter, which serves as the bottleneck of the energy system.

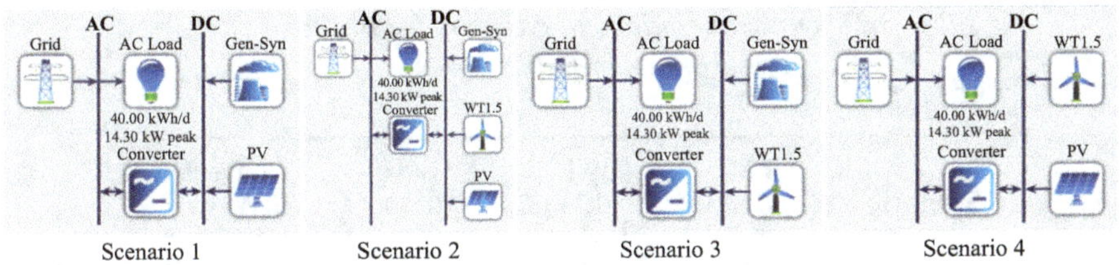

Figure 3.11 Grid-connected HRES configurations without hydrogen production

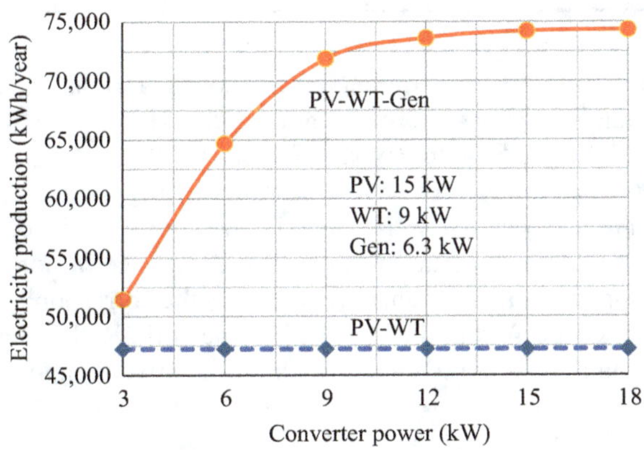

Figure 3.12 Variation in the electricity output generated by the HRES according to the converter's capacity

Figure 3.12 illustrates the annual variations in electricity output generated by the HRES based on the converter's capacity. The electricity output from photovoltaic (PV) and wind turbine (WT) sources varies annually due to climatic and weather conditions, which are beyond adjustment. The electricity produced by the generator, however, is modulated according to load requirements. When the converter capacity is limited, the electricity output from the generator is constrained to ensure that the total capacity of the power sources does not exceed the converter's capacity. As shown in Figure 3.12, with PV (15 kW), WT (9 kW), and generator (6.3 kW) components, the HRES operates at full capacity when the inverter capacity exceeds 15 kW.

Table 3.2 presents the equipment capacity for different HRES configurations designed to generate the same electricity output of 74,235 kWh/year, as provided by HOMER software. In the HRES configurations depicted in Figure 3.11, the generator capacity is selected to meet peak load requirements. Consequently, the system's efficiency is suboptimal if the HRES is not grid-connected or does not utilize excess electricity for other applications, such as heat generation or hydrogen production. Table 3.2 indicates that when the system employs three energy sources (solar, wind, and biomass), the capacities of the components are reduced, resulting in an optimal converter capacity of only 16.7 kW, compared to 22.5 kW for the solar and biomass (SB) case or 25.4 kW for the solar and wind (SW) case.

Table 3.2 Optimal equipment capacity for different configurations of HRES

Scenario	PV (kW)	WT (kW)	Gen (kW)	Converter (kW)
SB	29.5	0	6.3	22.5
SWB	15	9	6.3	26.7
WB	0	18	6.3	21.8
SW	22.5	6	0	25.4

The economic viability of different HRES configurations is assessed in this section. To evaluate the economic efficiency of the HRES, we select three key parameters: ICC, ROI, and Simple Payback Period (SPP).

The ROI serves as a key economic metric, reflecting the annualized financial benefit in relation to the system's initial capital expenditure. In the context of energy system analysis, it is defined as the mean annual difference in nominal cash flows over the project's operational lifespan, normalized by the corresponding difference in upfront investment costs. Within the HOMER framework, ROI is determined using a predefined computational expression, which quantitatively captures the system's cost-effectiveness over time:

$$ROI = \frac{\sum_{i=0}^{R_{proj}} (C_{i,ref} - C_i)}{R_{proj}(C_{cap} - C_{cap,ref})} \tag{3.11}$$

where

- $C_{i,ref}$: Nominal annual cash flow for base (reference) system
- C_i: Nominal annual cash flow for the current system
- R_{proj}: Project lifetime in years
- C_{cap}: Capital cost of the current system
- $C_{cap,ref}$: Capital cost of the base (reference) system

The ROI considers both the bank's interest rate and the inflation rate. In this case study, the estimated bank loan interest rate for renewable energy investments is 6% per year, while the estimated inflation rate is 4% per year. Thus, the ROI can be interpreted as the net interest rate received by the investor.

The SPP represents the duration, typically expressed in years, required for the cumulative net cash flow—resulting from the economic difference between the proposed system and a baseline configuration—to transition from negative to positive. This parameter provides a straightforward estimation of the time needed to recover the additional capital investment associated with the evaluated system relative to the reference case.

In this study, the base case system is identified as the one with the lowest ICC, which, in this instance, is the grid.

Figure 3.13 compares the economic efficiency of four grid-connected HRES configurations designed to produce a total annual electricity output of 74,235 kWh

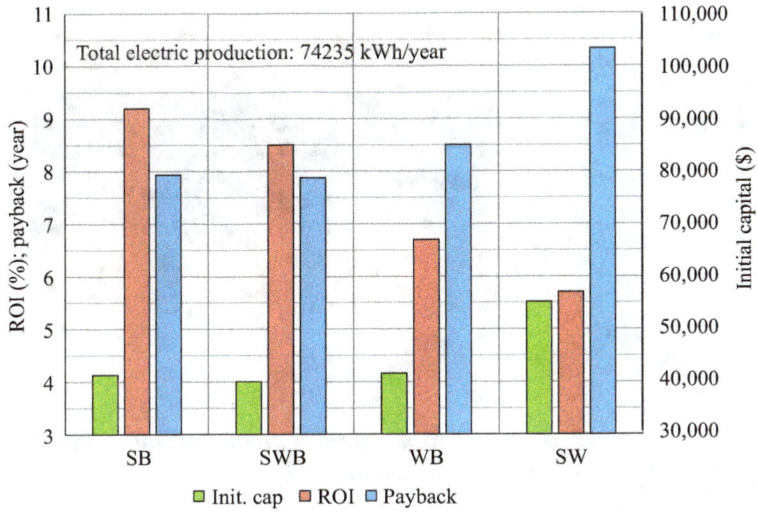

Scenario	Init. Cap ($)	ROI (%)	Payback (year)
SB	41,279	9.2	7.93
SWB	40,046	8.5	7.87
WB	41,593	6.7	8.5
SW	55,105	5.7	10.33

Figure 3.13 Comparison of the economic efficiency of different configurations of HRES (annual capacity shortage 2%)

without hydrogen production, as analyzed by HOMER software. The initial investment for the SB, SWB, and WB systems is approximately the same, around $41,000. The ROI for the SB, SWB, and WB options is 9.2%, 8.5%, and 6.7%, respectively. Meanwhile, the initial investment for the SW system exceeds $55,000, with an ROI of only 5.7%. The payback period for the SB, SWB, and WB options is 7.93, 7.87, and 8.5 years, respectively. In contrast, the payback period for the SW option is as long as 10.33 years. Thus, the HRES system with syngas generators gives better economic efficiency than the HRES system without this component.

3.4.3 Grid-connected HRES with hydrogen production

3.4.3.1 Configuration of HRES

In the grid-connected HRES with hydrogen production, the electricity generated by the system primarily supplies the load, with a portion allocated for hydrogen production, while any excess electricity is sold back to the grid. In this study model, the load consumes only **15 kWh/day**, and the maximum load capacity is **5.36 kW** (as shown in Figure 3.14(a)). The power distribution chart for the load resembles that of Figure 3.15, although with lower power values. The selected hydrogen load is set at a maximum of 6 kg/day. However, since hydrogen production only occurs when there

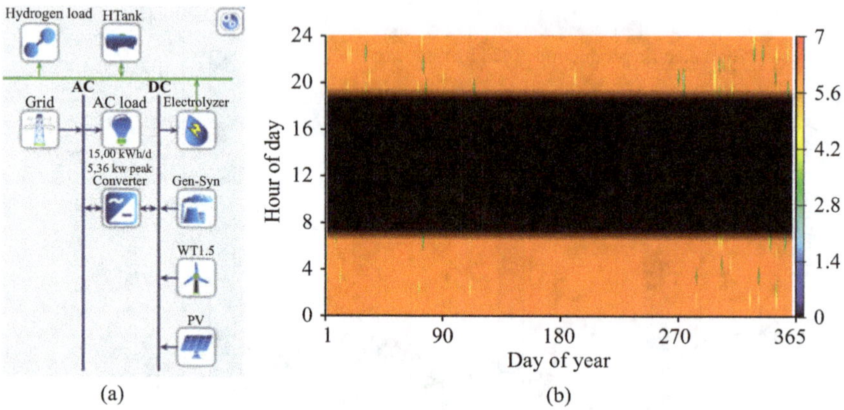

Figure 3.14 Grid-connected HRES with hydrogen production diagram (a) and the operating schedule of the generator (b)

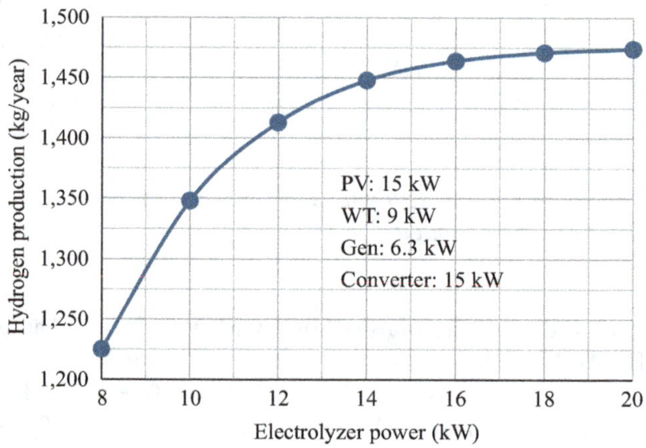

Figure 3.15 Variation of hydrogen production according to the electrolyzer capacity

is surplus electricity, and the generated hydrogen is stored in a tank for external use, it is not essential for the system to meet the hydrogen load requirements at all times. Consequently, the maximum unmet hydrogen load can be adjusted to 100%.

The generator operates exclusively at night, from 7:00 PM the previous day to 6:00 AM the following day (illustrated in Figure 3.14(b)). The total electricity output generated by the HRES is 74,235 kWh/year, averaging 200 kWh/day. Given that the total capacity of the HRES significantly exceeds the load capacity, the system rarely requires purchasing electricity from the grid for supplementation. However, due to the electricity consumption of the electrolyzer, the surplus electricity capacity is nearly zero.

3.4.3.2 Simulation result

Figure 3.15 illustrates the variation in hydrogen production relative to the power of the electrolyzer in the HRES, which is configured with a photovoltaic (15 kW), wind turbine (9 kW), generator (6.3 kW), and converter (15 kW). When the electrolyzer operates at a lower power level, surplus electricity is sold to the grid, resulting in decreased hydrogen production. The simulation results indicate that with this HRES configuration, when the electrolyzer power reaches 20 kW, hydrogen production achieves a maximum value of 1,475 kg/year.

Figure 3.16(a) presents the variations in power output from the generation sources during the early days of June. Unlike the HRES without hydrogen production, the generator operates at full capacity during its active hours, thereby enhancing its performance and simplifying the engine control system. The power generated by the generator remains stable throughout the year. Wind power output decreases during the middle months of the year but increases in the early and late months, while solar power exhibits the opposite trend. This combination of three power generation sources contributes to stabilizing the overall output of the HRES.

With the HRES configuration of PV (15 kW), WT (9 kW), Gen (6.3 kW), Converter (15 kW), and Electrolyzer (20 kW), HOMER gives simulation results of the annual power generation by the HRES components PV (23,533 kWh), WT (231,45 kWh), and Gen (27,478 kWh) (Figure 3.16(b)). As shown in Figure 3.16(b), the total annual power generation sums to 74,235 kWh. This electricity is utilized to meet a yearly load of 5,475 kWh and to produce 1,475 kg of hydrogen. The electricity purchased from the grid is 79 kWh/year, while the electricity sold back to the grid is 66 kWh/year.

Hydrogen production from the HRES is directly proportional to the electricity supplied to the electrolyzer. The simulation calculations estimate hydrogen output based on the efficiency of the electrolysis process. The electrical energy required to electrolyze 1 kg of hydrogen ranges from 50 to 55 kWh. Figure 3.17(a) illustrates the variation in hydrogen production during the early days of June. Figure 3.17(b)

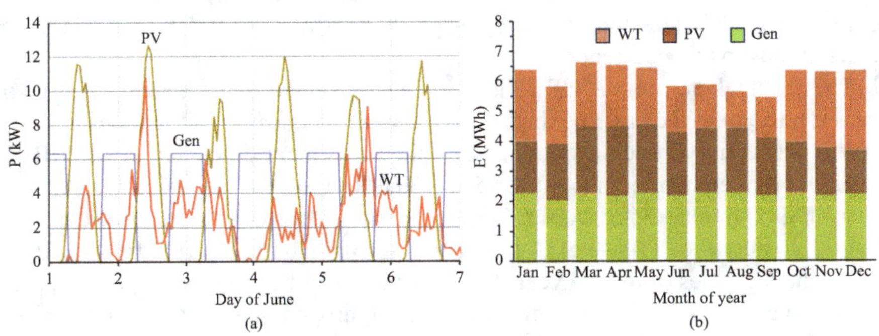

Figure 3.16 Variations in power output from the generation sources in the early days of June (a) and the electricity output generated by various sources in HRES (b)

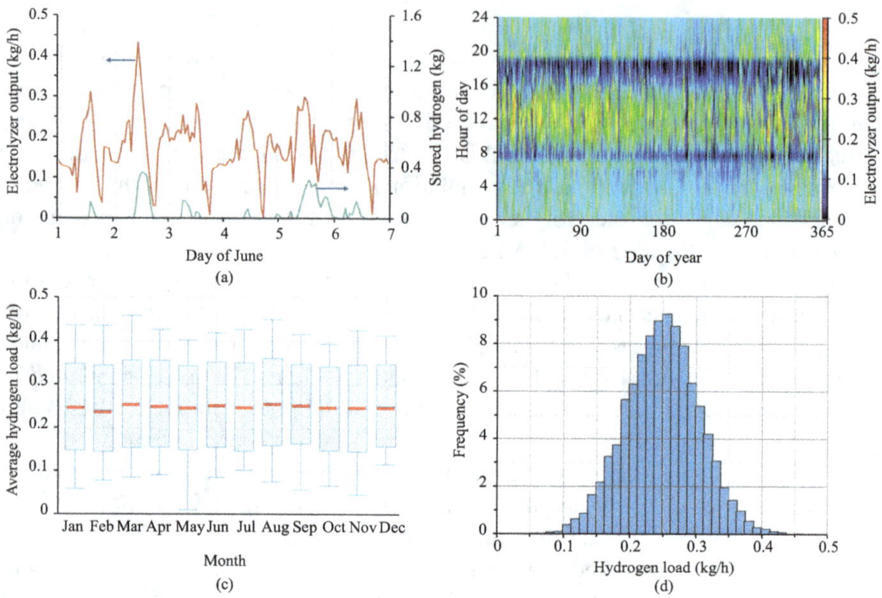

Figure 3.17 Variation of hydrogen production within the first week of June (a), and every hour within a year (b); variation of average hydrogen load monthly (c) and frequency distribution of hydrogen load (d)

presents the hourly variation in hydrogen production throughout the year. Hydrogen production decreases in the early morning and late afternoon when the generator is not operational. During these times, solar energy diminishes, and the electrical supply from the generator has not yet compensated for this decrease. The average monthly variation in hydrogen output aligns with the system's electricity production, as shown in Figure 3.17(c). During months characterized by low electricity output, such as August and September, hydrogen production also experiences a decline. Figure 3.17(d) indicates that a hydrogen production rate of 0.25 kg/h is the most frequently observed occurrence.

The profit taken from the HRES is derived from two primary sources: the sale of excess electricity to the grid and the sale of hydrogen. Figure 3.18 illustrates the relationship between hydrogen production and the investment efficiency of the HRES. The price of green hydrogen is set at $5/kg. The ROI increases almost linearly with hydrogen production, rising from 5.6% for an annual hydrogen output of 915 kg to 6.5% for an output of 1,474 kg. Correspondingly, the payback period for the HRES investment decreases from 10.3 years to 8.97 years as hydrogen production increases.

To achieve the same levels of electricity and hydrogen production, various HRES configurations can be employed. Table 3.3 presents four distinct HRES configurations capable of annually generating 74,235 kWh of electricity, which includes the energy required to produce 1,475 kg of hydrogen. If a generator is not used, the PV capacity is 22.5 kW, the wind turbine capacity is 15 kW, and the electrolyzer capacity is

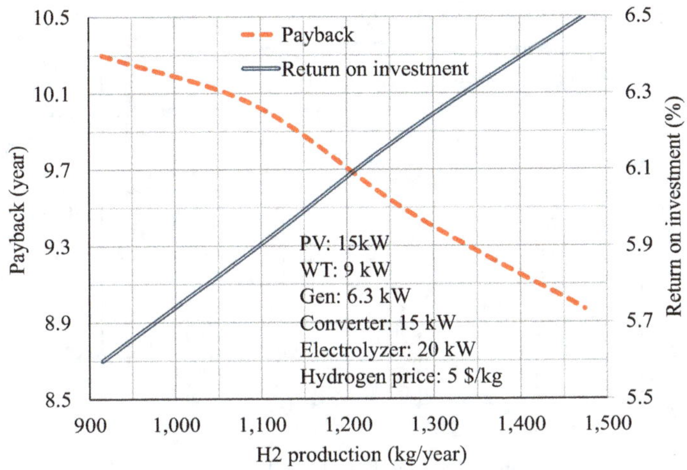

H2 (kg/year)	ROI (%)	y (year)
915	5.6	10.3
1,097	5.9	10.03
1,271	6.2	9.48
1,474	6.5	8.97

Figure 3.18 Variation of HRES investment efficiency according to hydrogen output

Table 3.3 HRES configuration to produce 1,475 kg of hydrogen per year, total power output 74,235 kWh per year

Scenario	PV (kW)	WT (kW)	Gen (kW)	Converter (kW)	Electrolyzer (kW)
SB	29.5	0	6.3	15	23
SMB	15	9	6.3	15	20
WB	0	19.5	6.3	15	20
SW	22.5	15	0	15	30

30 kW. The capacity of the converter is 15 kW for all cases. If a 6.3 kW generator is integrated into the SWB system, the PV capacity becomes 15 kW, the wind turbine capacity is 9 kW, and the electrolyzer capacity remains at 20 kW. If the SB and WB options are used, the PV and wind turbine capacities are 29.5 kW and 19.5 kW, respectively. These configurations highlight the flexibility of the HRES in meeting energy production requirements while optimizing investment efficiency.

The configuration of HRES significantly impacts its economic performance. Figure 3.19 illustrates the economic efficiency of various HRES configurations that yield the same annual electricity and hydrogen output. The SW option exhibits the lowest economic efficiency, requiring an initial investment of $76, 700, yet it offers a modest ROI of only 4.1% per year, resulting in a payback period of 11.86 years. Conversely,

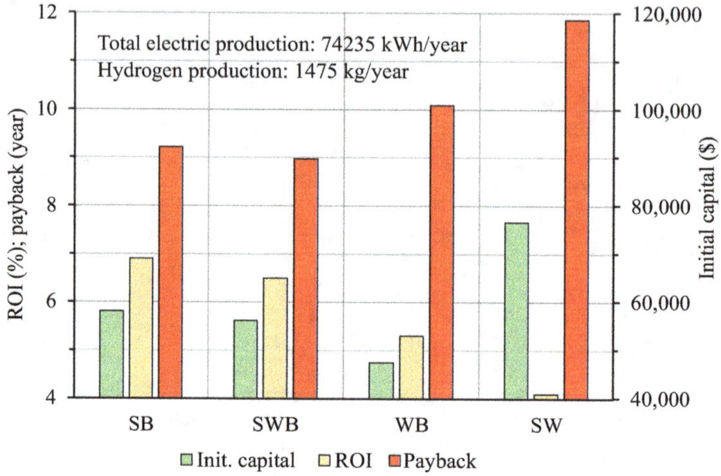

Scenario	Init. capital	ROI	Payback
SB	58,040	6.9	9.21
SWB	56,140	6.5	8.97
WB	47,491	5.3	10.09
SW	76,600	4.1	11.86

Figure 3.19 Comparison of the economic efficiency of various configurations of HRES

the SB option necessitates the lowest initial investment at \$47, 491, but it also provides low returns, with an ROI of 5.3% per year and a payback period of 10.09 years. Thus, both the WB and SW options are not economically favorable.

In comparison to the SB option, the SWB option requires an initial investment approximately \$2,000 lower and has an ROI that is 0.4% lower; however, it benefits from a shorter payback period by 0.24 years. Additionally, the waste management process for wind turbines post-project life cycle is simpler and less costly than that for solar panels. Wind power offers more consistent output than solar energy, ensuring that even if the syngas generator faces maintenance issues, wind power can still provide some capacity during nighttime. Therefore, the SWB option is well-suited for a renewable energy system focused on hydrogen production.

The economic efficiency of the HRES with hydrogen production is compared to a scenario without hydrogen production, where the system generates the same amount of electricity. The initial investment for the HRES system with hydrogen production is \$56,140, which is higher than the \$39,540 required for the system without hydrogen production, due to the inclusion of an electrolyzer and hydrogen storage tank. In the first operational scenario, a portion of the electrical energy generated by the HRES is allocated to serve the local load demand, while the excess is directed toward hydrogen production via electrolysis. In contrast, under the second scenario, the surplus electricity is exported to the utility grid. The overall economic viability of the HRES is significantly influenced by the market price of the green hydrogen produced and the prevailing electricity tariffs, both of which determine the comparative advantage between the two energy utilization strategies.

Figure 3.20 presents the cumulative cash flow variations of the project at green hydrogen prices of \$4.5/kg, \$5/kg, and \$5.5/kg. The intersection of the cumulative cash flow curve for the current system and the reference system (which utilizes grid electricity without hydrogen production) indicates the payback period for the investment. The Net Present Value (NPV) over the system's operational lifetime represents the discounted value of future cash flows and is computed as the difference between the NPC of the reference (base case) configuration and that of the proposed system. A positive NPV indicates that the evaluated system yields greater economic benefit relative to the base case, thereby signifying a financially advantageous investment.

According to the HOMER simulation results depicted in Figure 3.20, when the system does not produce hydrogen, the present worth is \$57,555, with an ROI of 8.5% per year and a payback period of 7.83 years. If the cost of green hydrogen is set at \$4.5/kg, the present worth decreases to \$45,037, with an ROI of 5.2% per year, both lower than in the scenario without hydrogen production, leading to a longer payback period of 10.68 years.

In contrast, if the price of green hydrogen rises to \$5.5/kg, the present worth increases to \$74,057, surpassing the present worth of the HRES without hydrogen

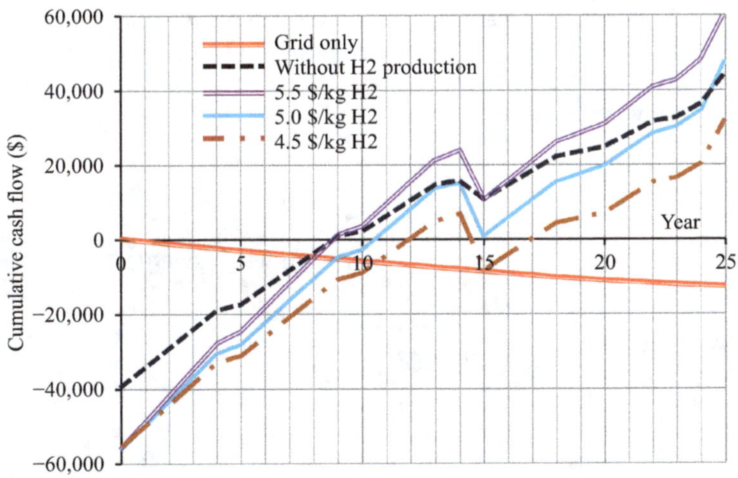

H2 price (\$/kg)	Init. capital (\$)	Present worth (\$)	ROI (%)	y (year)
4.5	56,140	45,037	5.2	10.68
5		59,577	6.5	8.79
5.5		74,057	7.8	7.08
Without H2	39,540	57,555	8.5	7.83

Figure 3.20 The impact of green hydrogen prices on the cumulative cash flow variation of the project

production, despite a lower ROI and a longer payback period. To achieve an equivalent present worth to the case without hydrogen production, the price of green hydrogen must be \$5/kg. According to the International Council on Clean Transportation, green hydrogen costs range from \$3.5/kg to \$5.5/kg. Therefore, the simulated price of green hydrogen appears reasonable within this context.

3.4.4 Environmental effectiveness assessment of HRES

The environmental efficiency of the HRES is evaluated in comparison to grid electricity usage. The reduction of GHG emissions depends on the configuration and outputs of the HRES. When using an engine fueled by syngas from biomass, the GHG emission reduction efficiency must also take into account (CH_4) consumption. Additionally, when the HRES produces hydrogen, the effectiveness of GHG reduction should consider the substitution of fossil fuels with the generated hydrogen. The following section analyzes the effectiveness of GHG emission reduction across different HRES scenarios, as illustrated in Figure 3.21.

- **GHG emissions from grid electricity usage**: The average CO_2 emission level from electricity generation in Vietnam is 521 g/kWh. Therefore, to produce 74,235 kWh/year, the Vietnamese electricity generation system emits approximately 39 tons of CO_2 annually.
- **Case of the SW system without hydrogen production**: In this scenario, when the HRES relies solely on solar and wind power without utilizing syngas generators or producing hydrogen, the CO_2 emissions reduction is equivalent to the emissions from grid electricity generation, resulting in a reduction of 39 tons of CO_2 (the negative sign indicates reduced emissions).

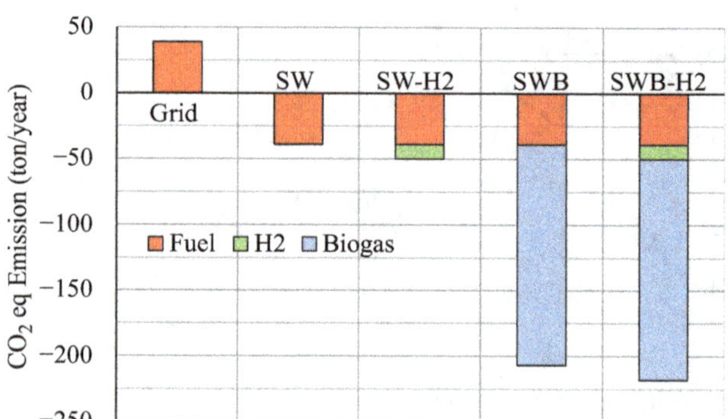

	Grid	SW	SW-H2	SWB	SWB-H2
Biogas	0	0	0	−168	−168
H2	0	0	−11	0	−11
Fuel	39	−39	−39	−39	−39

Figure 3.21 Comparison of CO_2 emissions according to different scenarios of the HRES configuration

- **Case of the SW system producing hydrogen**: In energy terms, 1 kg of hydrogen is equivalent to 3 kg of gasoline. The CO_2 emissions from gasoline combustion are 2 kg/liter, which corresponds to 2.5 kg CO_2/kg of gasoline. Consequently, replacing gasoline with 1 kg of hydrogen results in a CO_2 emissions reduction of 7.5 kg. By producing 1,475 kg of hydrogen annually, the system reduces CO_2 emissions by 11 tons per year. Thus, the total annual reduction in CO_2 emissions provided by the system compared to grid electricity is 50 tons of CO_2.
- **Case of the SWB without hydrogen production**: The syngas generator produces 27,478 kWh of electricity yearly, with an overall engine generator efficiency of approximately 20%. Assuming that the engine operates on biogas, which averages 60% CH_4 and 40% CO_2 by mass, the average calorific value of biogas is 35 MJ/kg. Each year, the engine consumes 14 tons of biogas, which would otherwise be released into the atmosphere. Given that the GHG effect of CH_4 is 20 times that of CO_2, the equivalent CO_2 emissions from the CH_4 released amount to 168 tons. Therefore, the system achieves a GHG emissions reduction equivalent to 207 tons of CO_2 annually.
- **Case of the SWB system with hydrogen production**: In this instance, in addition to reducing emissions by 207 tons of CO_2 annually (as in the previous case), the system also contributes an additional 11 tons of CO_2 reduction by replacing gasoline with 1,475 kg of hydrogen. Consequently, the total reduction in CO_2 emissions each year reaches 218 tons. Moreover, in the HRES with hydrogen production, the power capacities of generation sources can be optimized to meet load demands during peak periods. During off-peak times, excess power is utilized for electrolyzer, resulting in minimal dependence on the grid. In the case study, the SWB-HRES not only supplies electricity to the load but also produces approximately 2 kg of hydrogen daily. This hydrogen output is sufficient to fuel three motorcycles for normal daily operation. Thus, in addition to providing renewable electricity, this HRES system contributes renewable fuel, aiding in the reduction of GHG emissions from transportation.

The simulation results indicate that the SWB-HRES with hydrogen production holds significant economic and environmental potential. To further advance this HRES framework, additional studies on biomass gasification and the control of internal combustion engines fueled by a flexible mixture of syngas, biogas, and hydrogen are essential. These topics will be explored in the following section of the chapter.

3.5 Biomass storage under RDF

3.5.1 RDF production from biomass

Transforming biomass into RDF and subsequently producing syngas through RDF gasification is an effective strategy for solid waste management in rural areas [46]. In its RDF form, biomass becomes significantly easier to transport and store. The concept of RDF was first developed in the early 1970s [47]. Converting waste into energy via RDF offers several practical advantages: promoting the circular economy, reducing the need for landfills, enhancing the efficiency of resource use by transforming waste into usable energy, reducing reliance on fossil fuels [48].

Given the abundance and continuous generation of waste, converting it into energy provides a sustainable alternative to traditional energy sources [49]. The production of RDF involves several mechanical steps, including sorting, shredding, drying, grinding, and extrusion. Recent studies have identified multiple factors influencing RDF quality, which are related to both the production process and the properties of the raw materials. Key process-related factors include:

- **Temperature**: Affects the physical and chemical properties of the material.
- **Compression pressure**: Influences pellet density and quality.
- **Compression time**: Determines the extent of material compaction.
- **Pellet geometry**: Impacts the gasification efficiency.

Research indicates that maintaining a moisture content of 10%–14% yields pellets with higher density and improved quality, as water enhances interparticle contact through Van der Waals forces. Additionally, the biological properties of the biomass significantly affect the calorific value of the pellets. The optimal compression pressure during pellet production is influenced by the condition of the input material, as pellet porosity strongly impacts the biomass gasification process. RDF can be produced in two forms: bulk and pellets (see Figures 3.22(a) and (b)). The pelletized form offers several advantages over bulk RDF, including:

- **Higher material density**: Facilitates more efficient storage and transportation.
- **Improved uniformity**: Enhances the consistency of the fuel properties.
- **Easier transportation and storage**: Reduces logistical challenges.
- **Enhanced gasification**: Promotes more efficient energy conversion during the gasification process.

Through these processes and benefits, RDF represents a viable solution for effective waste management and energy production in rural areas.

3.5.2 Analysis RDF biomass combustion in the atmosphere

The gasification of RDF biomass constitutes a multifaceted thermochemical conversion process, involving intricate interactions between heat transfer, chemical reactions, and phase transformations. To gain a deeper understanding of the gasification mechanism within a gasifier, it is essential to first analyze detailed information

Figure 3.22 Bulk RDF (a) and pellets RDF (b)

about RDF combustion under atmospheric conditions. A simulation of the fuel pellet combustion process was conducted using ANSYS Fluent. The fuel pellet is assumed to have a porous structure containing the components specified in Table 3.1.

The fuel was incorporated into the simulation based on its elemental composition. Using this composition, the air-fuel ratio (r) for a stoichiometric mixture can be calculated as follows:

$$r = \frac{100}{23}\left(\frac{8x_C}{3} + 8x_H - x_O\right) \tag{3.12}$$

where x_C, x_H, and x_O represent the mass fractions of carbon (C), hydrogen (H), and oxygen (O) in the fuel, respectively, as detailed in Table 3.4.

The equivalence ratio (ϕ) for the fuel-air mixture is then determined using the following formula [45]:

$$\phi = \frac{f_r}{1-f} \tag{3.13}$$

where f is mixture fraction. Finally, the Excess Air Ratio (ER) is calculated as:

$$ER = \frac{1}{\phi} \tag{3.14}$$

These calculations provide critical insights into the combustion and gasification behavior of RDF pellets, facilitating a more comprehensive understanding of the gasification process inside the gasifier.

Figure 3.23 illustrates the temperature contours and concentrations of CO, CO_2, CH_4, and H_2 in the RDF-Wood chip flame. It can be observed that the high-temperature region is concentrated in the flame front area, which also corresponds to the high-concentration zone of the combustion products. Under rich mixture conditions, incomplete combustion occurs, leading to the partial conversion of carbon and hydrogen into CO_2 and H_2O. Consequently, the concentrations of combustible gases such as CO, CH_4, and H_2 in the combustion products increase. The concentrations of these species are determined based on the thermodynamic reaction equilibrium. Under stable combustion conditions of the RDF-Wood chip flame, the maximum molar concentrations of CO, CH_4, and H_2 are 18%, 10%, and 6%, respectively, as shown in Figure 3.23.

The variation of temperature and species concentrations along the y-axis of RDF-Rice Husk pellets is presented in Figure 3.24. It can be observed that the maximum

Table 3.4 Fuel Composition in RDF pellet

Fuel pellet	**Fuel composition (% by mass)**			
	C	**H**	**O**	**N**
RDF-Rice husk	0.500	0.060	0.436	0.004
RDF-Solid waste	0.570	0.061	0.350	0.019
RDF-Wood chips	0.533	0.064	0.402	0.001

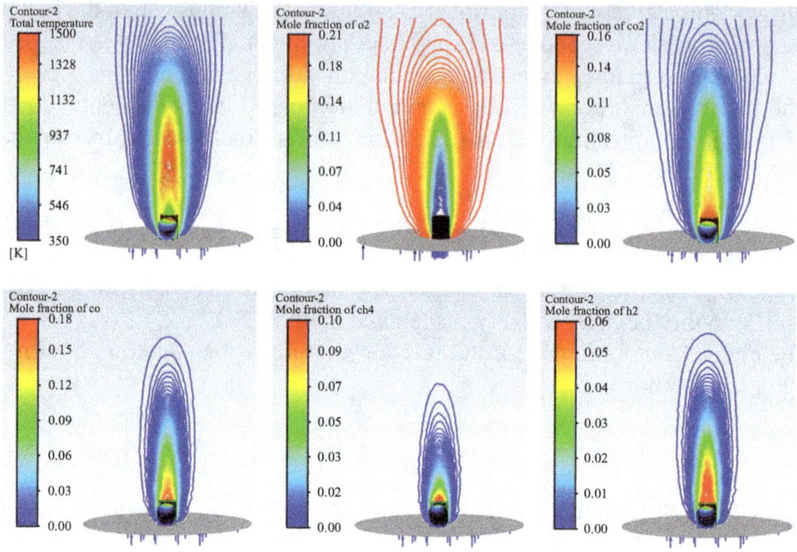

Figure 3.23 Contours of temperature, concentrations of CO, CO_2, CH_4, and H_2 in the RDF-Rice husk Flame on yz direction (ER = 0.3)

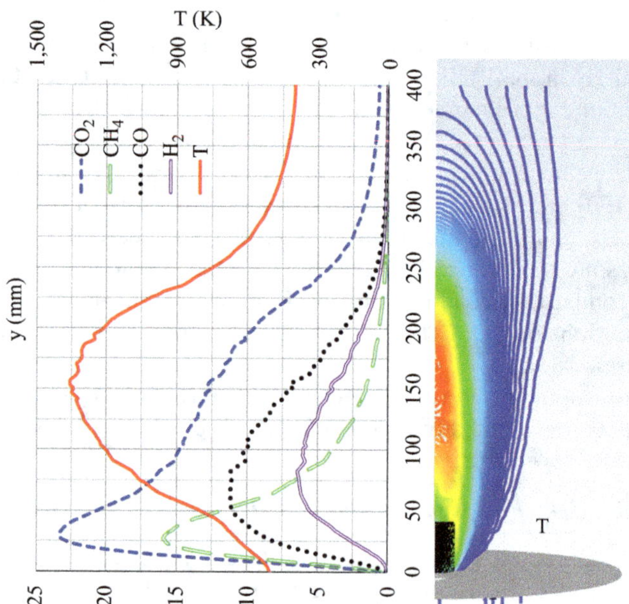

Figure 3.24 Distribution of CO_2, CH_4, CO, and H_2 concentration, and temperature on y-axis of RDF-Rice Husk flame

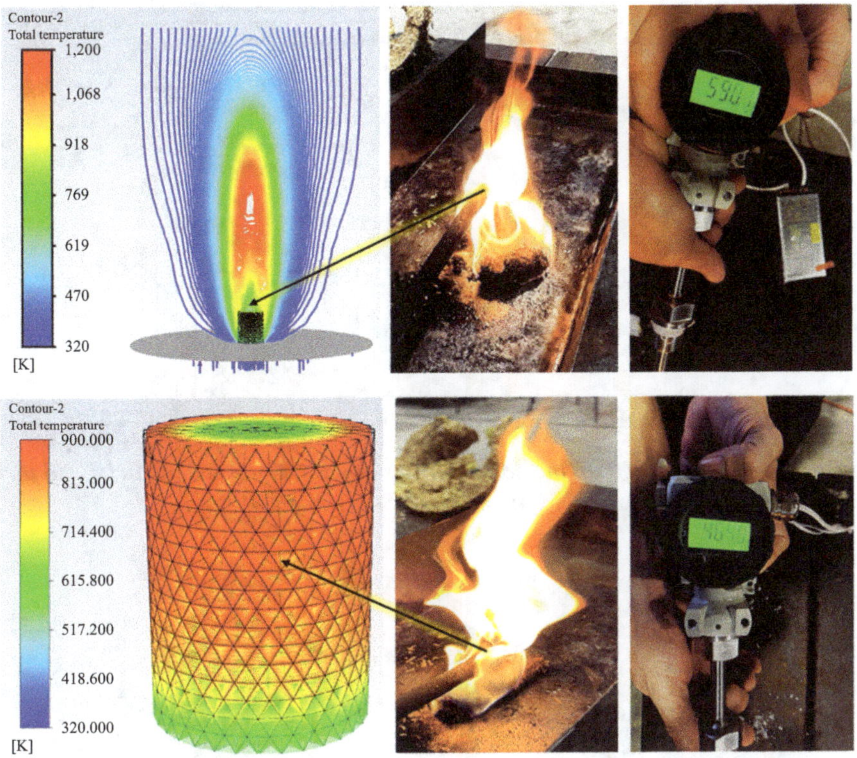

Figure 3.25 Comparison average temperature given by simulation and experiment of RDF-Rice Husk flame

species concentrations are located closer to the RDF pellet compared to the temperature. This can be attributed to the fact that near the RDF pellet, the fuel-air mixture is rich, resulting in a lower combustion temperature but higher concentrations of incomplete combustion products. The gaseous fuel components CO, CH_4, and H_2 tend to reach their maximum concentrations when 25 mm $< y <$ 75 mm, corresponding to a combustion temperature range between 700°C and 1050°C.

Figure 3.25 compares the simulation results with experimental data for the flame temperature and the internal temperature of the RDF pellet. The simulation predicts a flame temperature near the RDF surface of 560°C and a temperature at the pellet center of 470°C. These values closely align with the experimental results, which report a flame temperature of 590°C and a pellet center temperature of 464°C, respectively.

Figure 3.26 specifically illustrates the effects of the ER on the mass composition of species in RDF combustion products and the calorific value of syngas. The species composition is determined by their peak values within the calculation domain. As ER decreases, the content of combustible components increases, leading to a corresponding rise in the calorific value of the syngas.

The calorific value of syngas Q_{syn} is calculated based on the maximum species composition, representing the highest achievable value under the given combustion

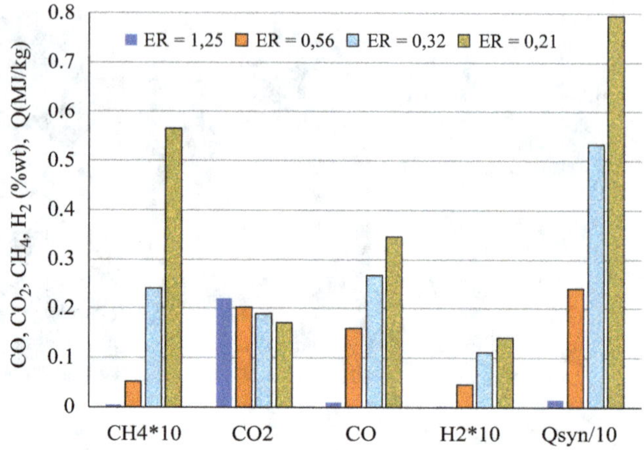

Figure 3.26 Effect of ER on the compositions of CH_4, CO_2, CO, and H_2, and the calorific value of syngas (RDF-Solid Waste)

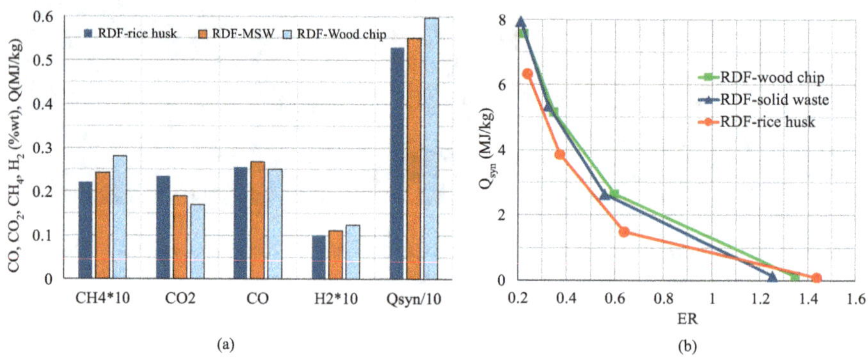

Figure 3.27 (a) Comparison of species concentrations in the flame and the calorific value of syngas in the combustion products of RDF-Rice Husk, RDF-Solid Waste, and RDF-Wood Chips at ER = 0.32. (b) Comparison of the variation in syngas calorific value with ER for the combustion products of RDF-Rice Husk, RDF-Solid Waste, and RDF-Wood Chips

conditions. The calculated Q_{syn} values are 0.15, 2.41, 5.34, and 7.94 MJ/kg, corresponding to ER values of 1.25, 0.56, 0.32, and 0.21, respectively. These values are lower than the syngas calorific values obtained from RDF gasification within a gasifier. This discrepancy may arise because, during open-air combustion, the substances generated at the flame front rapidly diffuse into the surrounding environment, preventing reduction reactions from occurring [50].

Figure 3.27(a) compares the maximum concentrations of species in the combustion products of RDF-Rice Husk, RDF-Solid Waste, and RDF-Wood Chips at an ER of 0.32. Among these three types of RDF, the concentrations of hydrogen

(H_2) and CH_4 are highest in the RDF-Wood Chips flame and lowest in the RDF-Rice Husk flame. As indicated in Table 3.1, the carbon (C) and hydrogen (H) contents in RDF-Rice Husk are lower than those in the other RDF types. Consequently, the composition of combustible substances in the resulting gas is minimal, leading to the lowest syngas calorific value for RDF-Rice Husk among the three types.

Although RDF-Wood Chips have a lower carbon content than RDF-Solid Waste, their hydrogen content is higher. Since the calorific value of carbon monoxide (CO) is significantly lower than that of H_2 and CH_4, the syngas calorific value of RDF-Wood Chips surpasses that of RDF-Solid Waste. Specifically, the syngas calorific values from the combustion gases of RDF-Rice Husk, RDF-Solid Waste, and RDF-Wood Chips are 5.3, 5.5, and 5.9 MJ/kg, respectively.

Figure 3.27(b) illustrates the variation in syngas calorific value from the gasification of RDF-Rice Husk, RDF-Solid Waste, and RDF-Wood Chips. Generally, the calorific value of syngas decreases sharply with increasing ER across all three RDF types. At ER values greater than 1, the calorific value becomes negligible as the RDF fuel is nearly completely combusted. For a given ER value, the syngas calorific value is highest for RDF-Wood Chips, followed by RDF-Solid Waste, with RDF-Rice Husk consistently exhibiting the lowest value among the three types.

3.5.3 Simulation of the gasification process within the gasifier

3.5.3.1 Air oxidant

RDF gasification presents an effective solution to the drawbacks of direct incineration. During gasification, RDF is converted into syngas, which comprises CO_2, CO, H_2, CH_4, H_2O, N_2, and other hydrocarbon gases, along with soot, char, and ash. After appropriate cleaning, syngas can be further utilized to produce liquid hydrocarbon fuels, hydrogen, methanol, ammonia, or as fuel for internal combustion engines [51]. This process offers higher energy recovery efficiency compared to direct incineration, while also producing lower emissions [52]. The oxidizing agent used can be air, steam, oxygen-enriched air, or carbon dioxide [53].

Gasification process

Saidi *et al.* [54] and Chen *et al.* [55] reported that increasing temperature enhances the gasification reaction rate and increases the hydrocarbon content. Down-draft gasifiers [52] can be divided into four distinct zones [56]:

1. **RDF storage zone**: Temperature ranges from 373 K to 423 K.
2. **Pyrolysis zone**: Temperature ranges from 423 K to 973 K.
3. **Combustion zone**: Temperature ranges from 973 K to 1773 K.
4. **Reduction zone**: Temperature ranges from 1073 K to 1373 K.

Syngas is primarily formed in the reduction zone, and its composition can be adjusted by adding steam to this area [56]. Gasification is a complex process that can be simulated using software tools, with ANSYS Fluent being well-documented for studying the gasification process [56–58]. Comparisons between simulation results and experimental data of RDF gasification in gasifiers are presented in various studies [33,59].

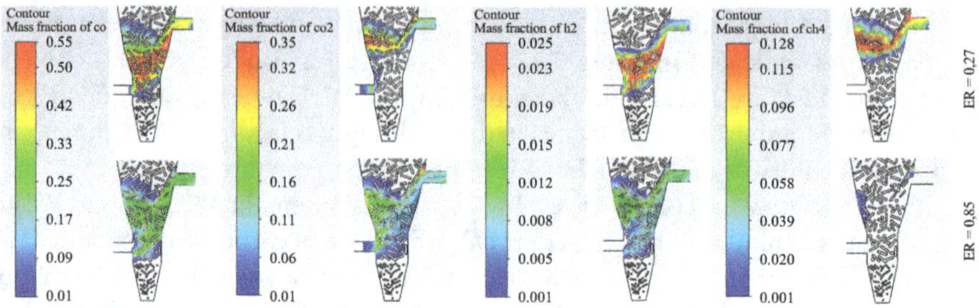

Figure 3.28 Comparison of the contours of CO, CO_2, H_2 and CH_4 concentration in syngas during gasification of coconut shell is with ER = 0.27 and ER = 0.85

Concentration contours

Figure 3.28 presents the spatial concentration contours of key syngas components—CO, CO_2, CH_4, and H_2—resulting from the gasification of coconut shell biomass at excess air ratios (ERs) of 0.27 and 0.85. The simulation reveals that the primary combustion reactions are localized near the air injection zone, encompassing both the oxidation and subsequent reduction regions (highlighted in red in the figure). The highest concentration of carbon monoxide (CO) is observed within the zones of incomplete combustion and active reduction, where CO_2 undergoes conversion to CO. Residual CO_2 accumulates predominantly in the upper section of the reduction region. CH_4 and H_2 are primarily generated within the high-temperature core of the reduction zone, where thermal and chemical conditions are favorable for their formation.

The comparative analysis of the two scenarios with distinct ERs indicates that under fuel-lean conditions (i.e., higher ER), the concentrations of syngas constituents such as CO, CH_4, and H_2 are significantly reduced relative to the fuel-rich case. In high-ER conditions, the oxidation zone expands due to the increased availability of oxygen, promoting more complete combustion of the feedstock. As a result, the gasification zone contracts, limiting the formation of combustible gases. This shift in reaction zones underscores the sensitivity of syngas composition to variations in the air-fuel ratio during biomass gasification processes.

Air–fuel ratio analysis

Under conditions of constant mixture fraction f, an increase in the air-fuel ratio r corresponds to a more fuel-rich mixture. Figures 3.29(a) and (b) illustrate a comparative analysis of the syngas properties—specifically volume concentration, mass concentration, and calorific value—resulting from the gasification of rice husk, wood, and municipal solid waste at a fixed mixture fraction of $f = 0.5$. The respective air-fuel ratios r for rice husk, wood, and domestic solid waste are 1.59, 3.65, and 4.54, indicating significant variations in syngas composition and energy content as a function of feedstock characteristics and mixture richness.

With r ratio of 1.59, rice husk is gasified in a poor mixture, allowing CO_2 and H_2O to convert into CO and H_2, thus reducing CO_2 concentration. Conversely, when

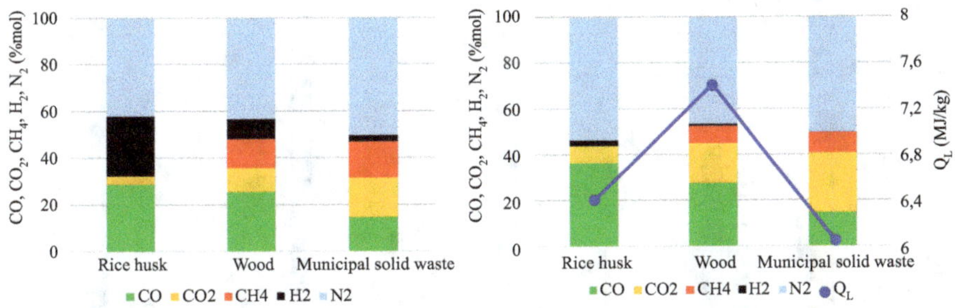

Figure 3.29 Comparison of volume concentration (a), mass concentration and calorific value (b) of syngas obtained from RDF gasification of rice husk, wood, and municipal solid waste (f = 0.5)

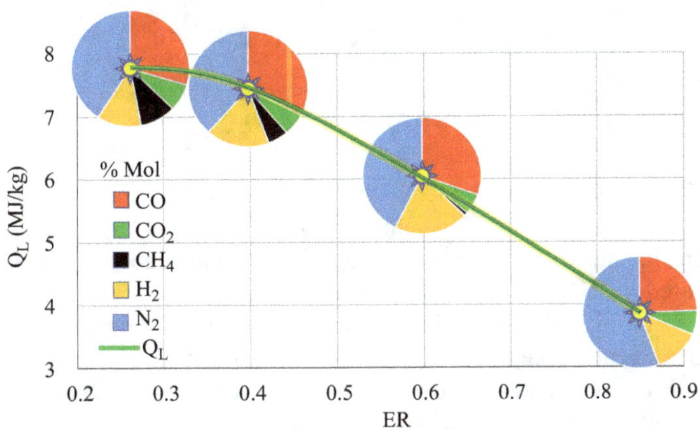

Figure 3.30 Variation of syngas composition and fuel calorific value according to ER during gasification of coconut shell

the mixture is overly rich, as seen with municipal solid waste ($r = 4.54$), the reduction conditions weaken, leading to CO generation due to incomplete combustion and H_2O conversion to CH_4. For wood, with $f = 0.5$, the ER falls within the ideal range of 0.2 to 0.4, resulting in the highest syngas calorific value.

Figure 3.30 presents the relationship between the ER and both the syngas composition and its corresponding calorific value during the gasification of coconut shell biomass. The simulation results indicate that a reduction in ER leads to an increase in CH_4 concentration, whereas the H_2 content decreases. The calorific value of the syngas shows a significant rise as ER decreases from 0.9 to 0.4. However, further reduction in ER beyond this range results in a marked decline in energy content, primarily due to the reduction in H_2 and the increase in CO_2 concentration. These findings underscore the importance of selecting an optimal ER—preferably in the range of 0.3 to 0.4—to maximize the thermal efficiency of the gasification process.

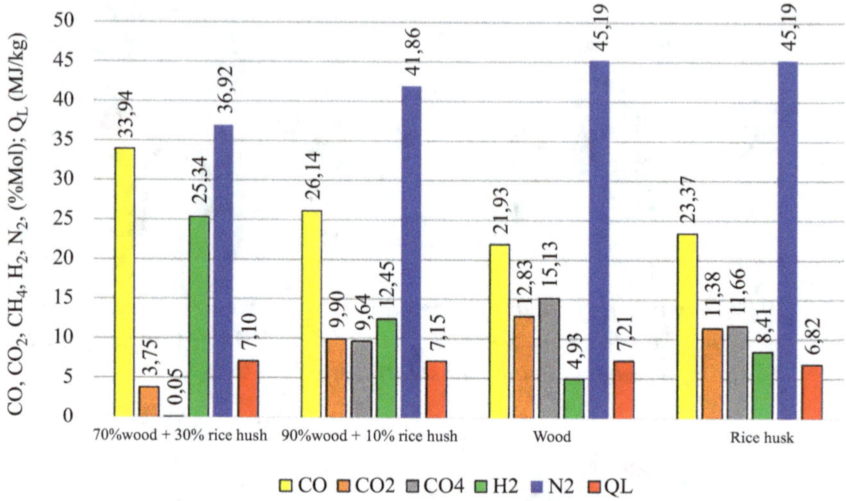

Figure 3.31 Effect of RDF constitution on composition and calorific value of syngas in case of (f = 0.5)

Figure 3.31 illustrates the influence of RDF feedstock composition on the resulting syngas characteristics and its calorific value. The analysis reveals that RDF derived from rice husk yields a syngas with elevated concentrations of CO and H_2 when compared to RDF sourced from wood biomass. In contrast, the CH_4 content is observed to be higher in the syngas produced from wood-based RDF. Furthermore, the lower heating value (LHV) of the syngas obtained through gasification of rice husk RDF is slightly inferior to that produced from wood RDF, with respective values of 6.82 MJ/kg and 7.21 MJ/kg.

When gasifying a mixture consisting of 70% wood RDF and 30% rice husk RDF, the syngas shows an increase in CO and H_2 components, while the CH_4 content remains almost negligible. Reducing the rice husk content to 10% results in a decrease in both CO and H_2, accompanied by an increase in CH_4 concentration. Consequently, the syngas calorific value increases as the proportion of rice husk in the RDF decreases.

The findings suggest that gasifying a blended RDF composed of wood and rice husk results in an overall increase in H_2 concentration within the syngas. This elevated H_2 content contributes to enhanced combustion performance due to its superior flame propagation characteristics. Consequently, utilizing a composite RDF mixture of wood and rice husk is more advantageous for applications requiring high combustion rates, such as in high-speed internal combustion engines, compared to the gasification of these biomass sources individually.

3.5.3.2 Oxygen-enriched air as oxidant

One significant challenge in using syngas as fuel is its low calorific value, which leads to reduced engine power output. The high nitrogen content in syngas significantly contributes to this low calorific value [60]. The LHV of syngas generally falls within the range of 4–6 MJ/kg [61], which corresponds to approximately 10% of the LHV associated with conventional fuels such as natural gas, liquefied petroleum gas (LPG), or gasoline. Nevertheless, due to the significantly lower stoichiometric air requirement for the complete combustion of syngas—estimated at around 10% of that required for traditional hydrocarbon fuels—the resultant reduction in engine output is not linearly correlated with its calorific value. In practice, engine power decreases by approximately 15–20% when operating on syngas compared to diesel engines and by 30–40% compared to gasoline engines [21,50,62].

To enhance engine performance with syngas, fuel enrichment and/or improvements in the engine's volumetric efficiency can be employed. When air is used as the oxidant, the nitrogen content in syngas exceeds 50%. However, employing oxygen-enriched air as the oxidant reduces nitrogen concentration, leading to an increase in combustion temperature. This, in turn, raises the temperature in the reduction zone, enhancing the production of combustible gases in the syngas and improving its calorific value.

Figure 3.32(a) compares the temperature distribution in the combustion and reduction zones when air enriched with 50% oxygen is supplied to the gasifier as the oxidant. The results indicate that oxygen-enriched air increases the maximum combustion temperature and shifts the primary combustion zone toward the opposite side of the air supply nozzle. The maximum combustion temperatures reach 1,600 K with air and 2,200 K with oxygen-enriched air as oxidants.

Figure 3.32(b) illustrates that the maximum concentration of CO in the reduction zone increases to 85% when air enriched with 50% oxygen is used as the oxidant, compared to 60% with air. Similarly, the CO_2 concentration rises from 2.9% with air as the oxidant to 4.2% with oxygen-enriched air. The use of oxygen-enriched air does not significantly affect the CO_2 content in the syngas while maintaining a stable temperature in the gasification zone. However, the N_2 content decreases, which increases the proportion of combustible gases in the syngas.

Figure 3.33 shows that the total mass fraction of combustible gases in the syngas progressively increases with the content of oxygen enrichment in the oxidant. Specifically, the combustible gas fraction is 35%, 41%, 54%, and 66% when air enriched with 0%, 10%, 30%, and 50% oxygen is used, respectively. When the temperature of the reduction zone is not actively controlled, it is influenced by the temperature of the gas mixture in the combustion zone. Enriching the oxidant with oxygen raises the combustion temperature, which in turn increases the temperature in the reduction zone. This rise in temperature shifts the equilibrium of conversion reactions involving CO_2, CO, CH_4, H_2, and H_2O toward thermodynamic equilibrium at higher temperatures [63].

Figure 3.34 illustrates the variation in syngas composition with ER for different contents of oxygen enrichment in the air oxidant. As the oxygen enrichment content increases, the temperature in the gasification zone rises, accelerating the reduction reactions and resulting in higher concentrations of CO, H_2, and CH_4 in the syngas.

Figure 3.35 compares the composition and calorific value of syngas produced from the gasification of RDF-wood chips using air as the oxidant versus air enriched with 30% and 50% oxygen. The results indicate that the calorific value of syngas

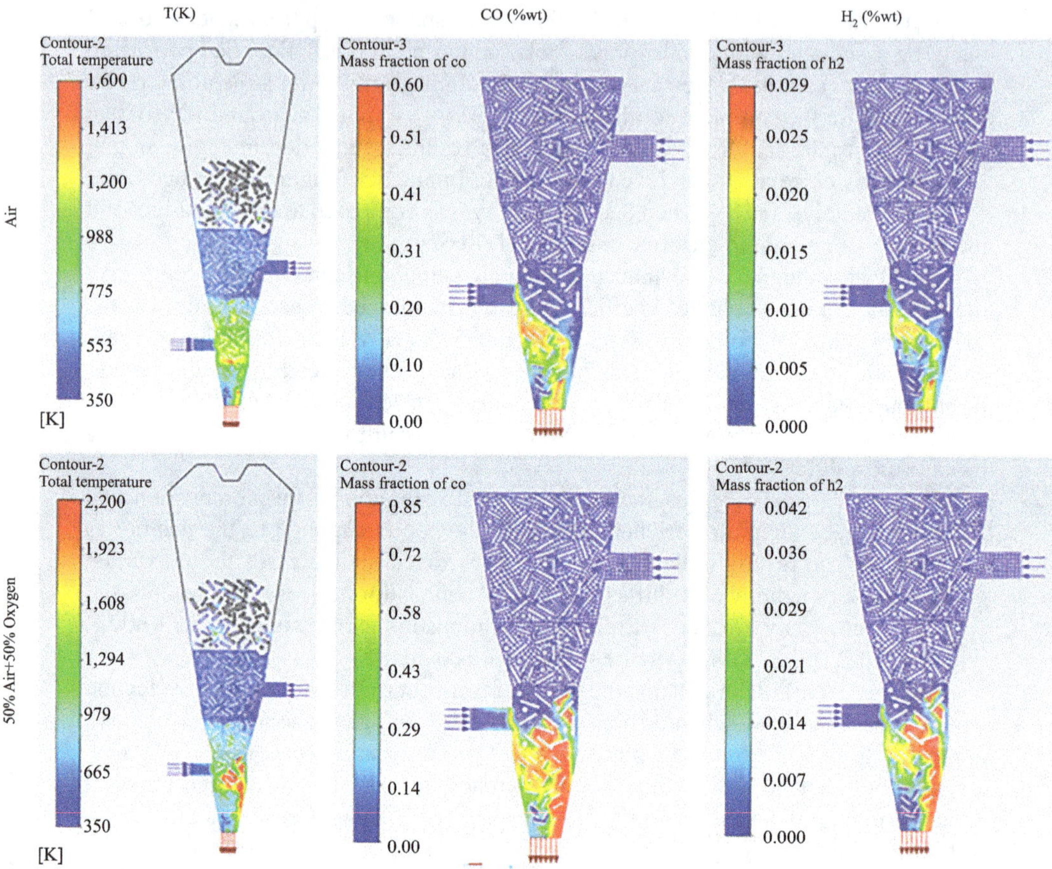

Figure 3.32 Distribution of temperature (a), CO concentration (b), and H_2 concentration (c) in the combustion zone and the reduction zone when using air enriched with 50% oxygen as oxidant (ER = 0.35)

increases by 60% and 90% when air enriched with 30% and 90% oxygen is used, respectively, compared to syngas produced with air as the oxidant. This significant enhancement in calorific value improves the quality of the syngas, thereby enhancing the performance of internal combustion engines.

3.6 Internal combustion engine operating in SWB-HRES

3.6.1 Spark-ignition engine fueled with syngas–biogas–hydrogen mixture

3.6.1.1 Model establishment

The experimental investigation was conducted on a stationary internal combustion engine operating on a biogas–syngas–hydrogen fuel blend, which was retrofitted from

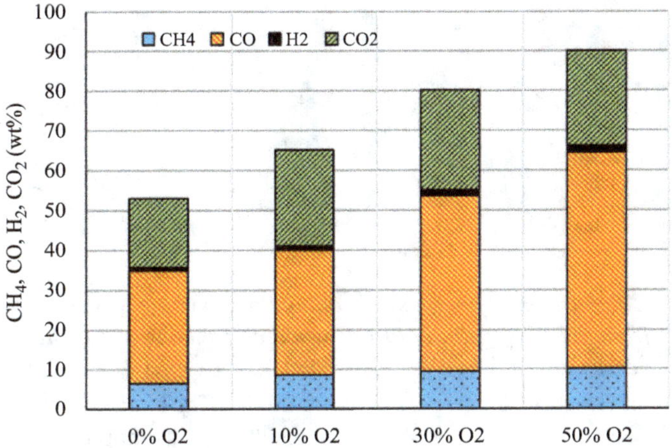

Figure 3.33 Effects of oxygen content enriching air oxidant on syngas compositions (ER = 0.35, temperature at reduction zone 800 K)

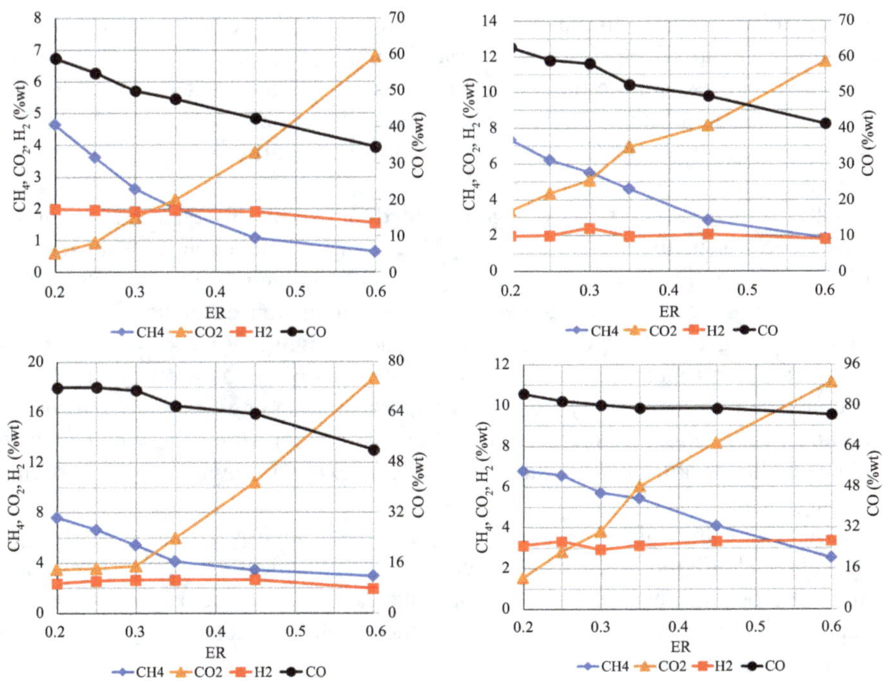

Figure 3.34 Effect of oxygen enrichment content in the air oxidant on the variation of syngas compositions with ER

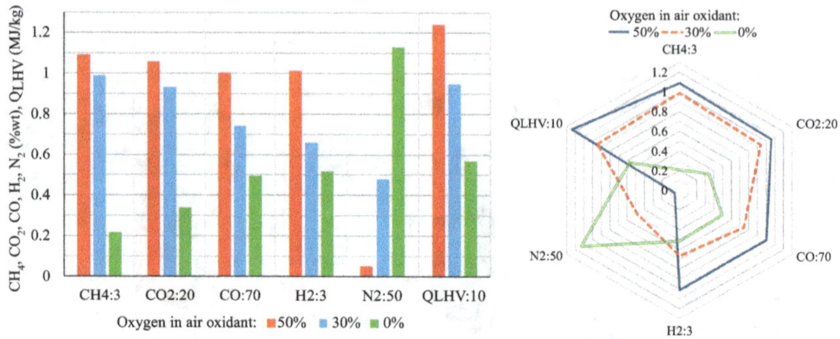

Figure 3.35 Comparison of composition and calorific value of syngas produced from the gasification of RDF-wood chips using air as the oxidant versus air enriched with 30% and 50% oxygen

the conventional Honda GX200 spark-ignition model. The engine features a cylinder bore of 68 mm and a piston stroke of 45 mm, resulting in a compression ratio of 8.5. It employs a traditional carburetor-based fuel supply system and magneto ignition. Under standard gasoline operation, the engine is capable of delivering a maximum output of 4.8 kW at a rotational speed of 3,600 rpm.

Table 3.1 presents the main characteristics of biogas, syngas, and hydrogen. The significant differences in the air-fuel ratio of the constituent fuels in the syngas–biogas–hydrogen mixture pose considerable challenges for the fuel supply system. In the SWB-HRES system, the fuel composition in the syngas–biogas–hydrogen mixture can vary in a large range. Therefore, the fuel supply system must ensure flexibility to maintain an optimal air-fuel equivalence ratio under different engine operating conditions.

Within the syngas–biogas–hydrogen blended fuel, the hydrogen fraction exhibits a high degree of sensitivity to both the air–fuel mixture formation and the subsequent combustion dynamics. This sensitivity significantly influences ignition characteristics, flame propagation behavior, and overall combustion efficiency. Key issues related to hydrogen that require attention include pre-ignition, backfire, rapid pressure rise rates, and increased NO_x concentrations [64,65]. Numerous studies have been conducted on fuel supply systems for spark-ignition engines using hydrogen-containing fuels [66–68]. Adjusting the start of injection timing can help reduce residual H_2 in the intake manifold, mitigating backfire in engines with port fuel injection [69–71].

To examine the influence of the syngas–biogas–hydrogen injection strategy on the air–fuel mixture formation process, four fuel injectors were strategically positioned along the engine's intake manifold, as illustrated in Figure 3.36. The fundamental physicochemical properties of the employed fuels are systematically summarized in Table 3.5.

3.6.1.2 Supplying the syngas–biogas–hydrogen through a single injector

Figure 3.37(a) illustrates the contours of fuel concentration on the symmetric longitudinal section of the cylinder, along with the variation of air flow rate (Q_{air}), fuel flow rate (Q_{fuel}), and equivalence ratio (ϕ) according to the crank angle (φ) when the

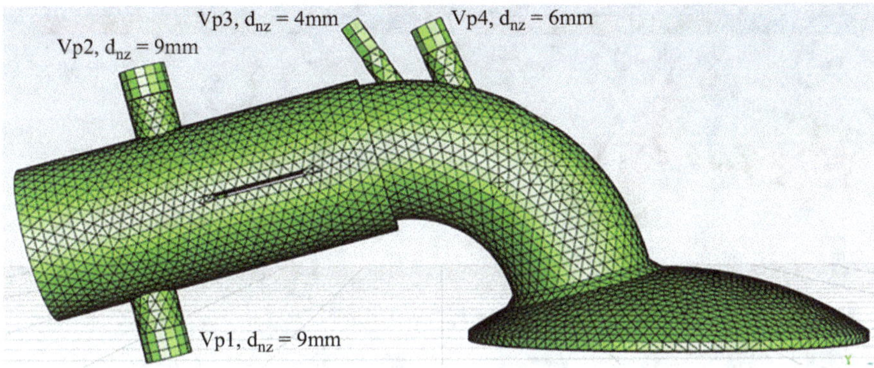

Figure 3.36 Arrangement of syngas–biogas–hydrogen injectors on the engine intake manifold

Table 3.5 Main characteristics of the fuels

	Composition (mol/mol)							
Fuel	**CH_4**	**H_2**	**CO**	**CO_2**	**N_2**	**M (g/mol)**	**m_a/m_f (g/g)**	**V_a/V_f (L/L)**
Biogas	0.7	0	0	0.3	0	24.40	7.98	6.71
Syngas	0.05	0.18	0.20	0.12	0.45	24.64	1.64	1.39
Hydrogen	0	1	0	0	0	2	34.78	2.4

engine is fueled with syngas at a speed of $n = 3{,}000$ rpm. The injection was performed through an injector with a nozzle diameter $d_{nz} = 9$ mm and an injection pressure $p_{inj} = 0.5$ bar. Initiating fuel injection at 10° crank angle (CA) leads to a localized increase in manifold pressure, which momentarily reverses the intake airflow, resulting in a transient negative intake flow. This phenomenon is subsequently offset by an increase in airflow caused by the vacuum effect during the piston's downward motion. When the injection duration is extended to $\varphi_{inj} = 150°$CA, the mixture's equivalence ratio during the compression stroke attains a value of approximately 0.91. However, a considerable fraction of the injected fuel remains in the intake manifold at the conclusion of the intake phase, posing challenges in managing both backfire tendencies and equivalence ratio stability across engine cycles. Under identical injection pressure and nozzle geometry, biogas requires a significantly shorter injection duration of only 38°CA to achieve a stoichiometric equivalence ratio of $\phi = 1$. (see Figure 3.37(b)). The shorter injection duration ensures that all injected fuel is drawn into the cylinder by the end of the intake stroke.

Simulation results reveal that under identical fuel supply conditions (injector diameter $d_{nz} = 4$ mm, injection pressure $p_{inj} = 1$ bar, and injection duration of 130°CA), the equivalence ratio ϕ achieved are 1, 0.2, and 1.7 for biogas, syngas, and hydrogen, respectively (Figure 3.38). Thus, an injector with a small diameter $d_{nz} = 4$ mm is suitable for biogas but results in an excessively rich mixture for

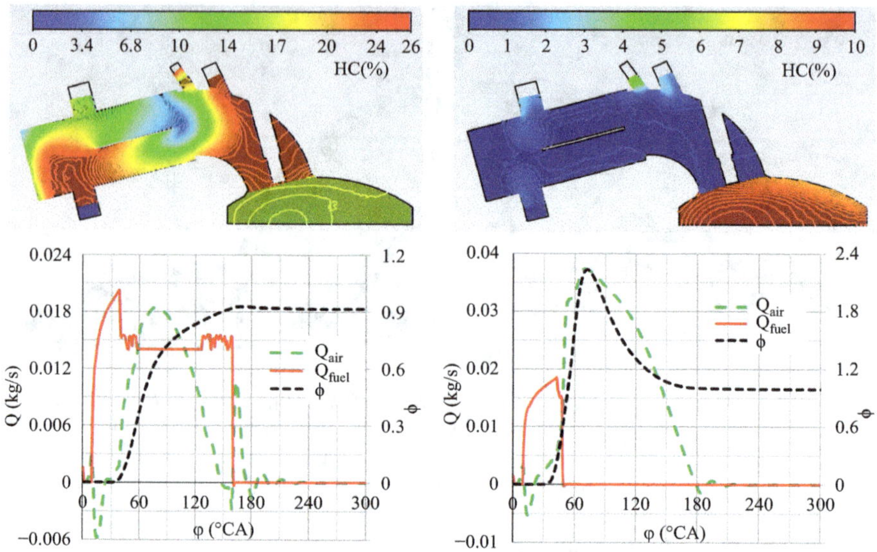

Figure 3.37 Comparison of syngas and biogas injection through the injector of $d_{nz} = 9$ mm ($n = 3{,}000$ rpm, $p_{inj} = 0.5$ bar; (a) Syngas, $\varphi_{inj} = 150°$ CA; (b) Biogas, $\varphi_{inj} = 38°$ CA)

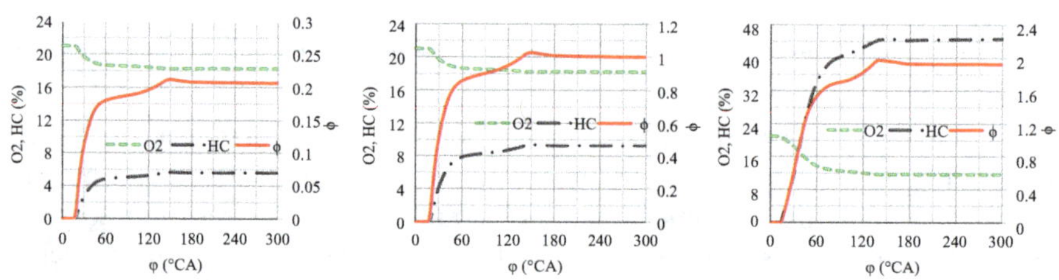

Figure 3.38 Comparison of the mixture formation process for syngas, biogas, and hydrogen through an injector ($d_{nz} = 4$ mm, $p_{inj} = 1$ bar, $\varphi_{inj} = 130°$ CA, $n = 3{,}000$ rpm)

hydrogen and an overly lean mixture for syngas. Conversely, if the injection system is designed for syngas, the injection duration becomes very short when operating on hydrogen, complicating precise control of the injection quantity. If designed for hydrogen, the injection duration for syngas becomes excessively long, making it difficult to achieve a stoichiometric mixture.

To address these challenges, a flexible fuel supply system is essential for the syngas–biogas–hydrogen mixture [72–74]. The system must ensure that the injection duration is neither too short (to allow precise control of the injection quantity) nor too long (to ensure that all injected fuel is fully drawn into the cylinder). A twining injector system could effectively meet these requirements.

3.6.1.3 Supplying the syngas–biogas–hydrogen through a twining injector system

The twining injector system comprises two injectors and a specially designed Engine Control Unit (ECU) that coordinates their operation, ensuring the appropriate fuel quantity is supplied to the engine as operating conditions and fuel composition vary [75–77]. For the spark-ignition engine studied in this work, two injectors are utilized: injector Vp3 with a nozzle diameter of 4 mm and injector Vp4 with a nozzle diameter of 6 mm (see Figure 3.39).

When the fuel mixture consists solely of syngas, the injection angle for injector Vp3 is set at 130°CA, while the injection angle for injector Vp4 is 115°CA. Under these injection conditions, the mixture composition in the combustion chamber at the end of the compression stroke ranges from 0.98 to 1.02. At the end of the intake stroke, the residual hydrocarbon (HC) concentration around the injector in the intake manifold is approximately 6%, which is relatively low compared to the HC concentration in the combustion chamber, around 25%. Therefore, when the engine operates exclusively on syngas, the twining injector system with nozzle diameters of 6 mm and 4 mm proves to be effective. During the compression stroke, the area of high fuel concentration is primarily located at the top of the cylinder, opposite the intake valve. However, by 330°CA, the fuel is nearly uniformly distributed throughout the combustion chamber [78–80].

As the proportion of biogas or hydrogen in the fuel mixture increases, the total injection duration for both injectors decreases. This reduction in injection time occurs in two distinct phases:

1. **Phase One**: The injection duration of injector Vp3 remains constant at its maximum value, while the injection duration of injector Vp4 gradually decreases from its maximum to its minimum value.
2. **Phase Two**: The injection duration of injector Vp4 is fixed at its minimum value, while the injection duration of injector Vp3 decreases to its lowest value, corresponding to a fuel mixture composed solely of biogas or hydrogen.

The transition between these two phases occurs at 70% biogas in the case of a syngas-biogas fuel mixture (Figure 3.40(a)) and at 50% in the case of a syngas-hydrogen fuel mixture (Figure 3.40(b)).

Figure 3.40(a) illustrates the variation of the injection angles for injectors Vp3 and Vp4 as a function of biogas content in the syngas-biogas mixture, aimed at ensuring an equivalence ratio (ϕ) of 1 at the end of the compression stroke. It is evident that the injection angle for injector Vp3 remains stable at 130°CA until the biogas content in the mixture reaches 70%. Beyond this point, the injection angle for Vp3 decreases to 65°CA when the mixture consists of 100% biogas. In contrast, the injection angle for injector Vp4 decreases from 115°CA (corresponding to 100% syngas) to 26°CA (corresponding to 70% biogas), remaining constant at this angle up to 100% biogas.

Figure 3.40(b) similarly presents the variation of the injection angles based on the hydrogen content in the syngas-hydrogen mixture. To maintain an equivalence ratio of 1 at the end of the compression stroke, the injection angle for injector Vp3 remains stable at 130°CA until the hydrogen content in the mixture reaches 50%. After this

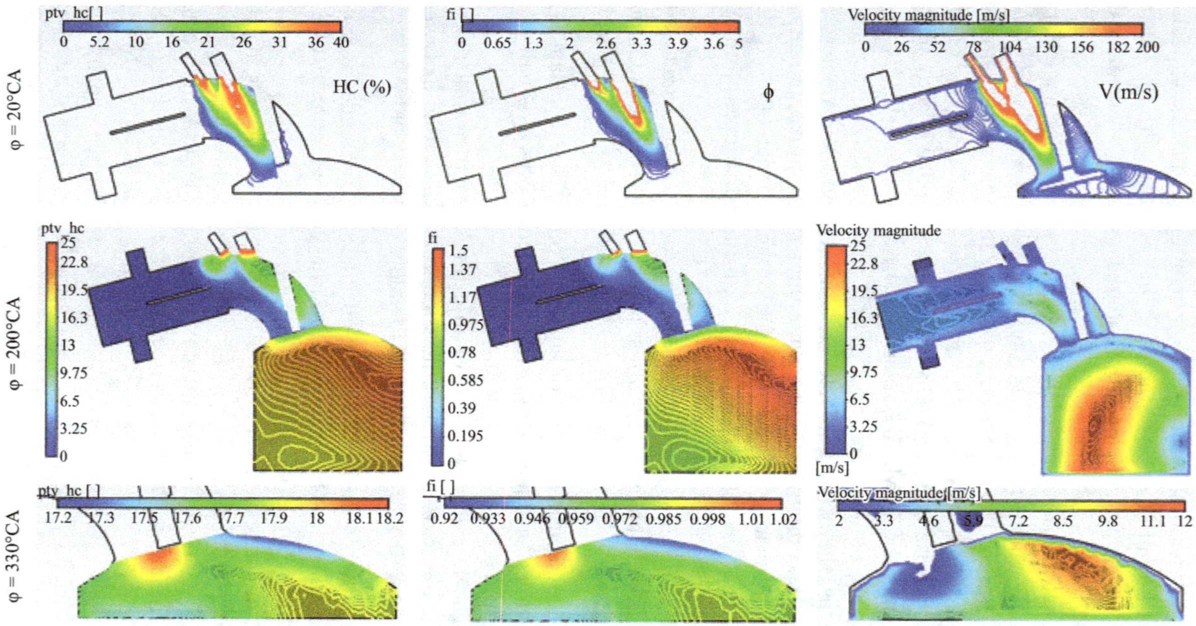

Figure 3.39 *Contour of fuel concentration, equivalence ratio, and velocity magnitude when supplying syngas through injector* V_{p3} $(\varphi_{inj} = 130° CA)$ *and* $V_{p4}(\varphi_{inj} = 115° CA)$, $p_{inj} = 1$ *bar,* $n = 3{,}000$ *rpm*

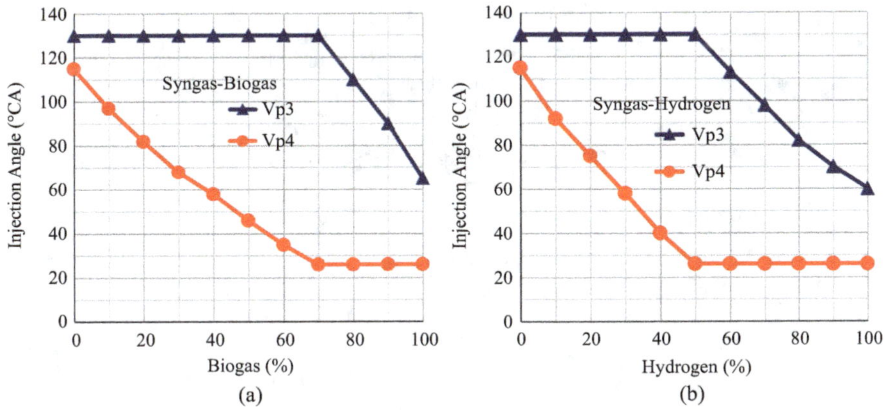

Figure 3.40 Variation of injection angles for injectors Vp3 and Vp4 based on biogas content (a) and hydrogen content (b) in the mixture with syngas to ensure an equivalence ratio of 1 at the end of the compression stroke (n = 3,000 rpm, injection pressure p_{inj} = 1 bar)

point, it decreases to 60°CA when the engine operates entirely on hydrogen. The injection angle for injector Vp4 also decreases from 115°CA (corresponding to 100% syngas) to 26°CA (corresponding to 50% hydrogen), remaining constant at this value up to 100% hydrogen.

This coordinated injection strategy ensures that the opening duration of the injectors remains within the defined limits: a minimum of 26°CA and a maximum of 130°CA. This approach guarantees that the fuel is fully drawn into the combustion chamber while allowing the control system to accurately manage the injected quantity during low-load operating conditions. Under identical injection parameters, the required injection duration follows an ascending order for hydrogen, biogas, and syngas, with hydrogen exhibiting the shortest and syngas the longest duration. As a result, the total injection duration for the syngas–biogas–hydrogen ternary blend remains bounded within the range defined by the syngas–biogas and syngas–hydrogen binary mixtures. Specifically, it does not surpass the duration required for the syngas–biogas blend and is not shorter than that of the syngas–hydrogen configuration.

3.6.1.4 Performance of SI engine fueled with syngas–biogas–hydrogen mixture

A general comparison of combustion characteristics of syngas, biogas and syngas–biogas–hydrogen blends is presented in Figure 3.41. When the equivalence ratio ϕ = 1, CO concentration is less than 0.06%, HC concentration is less than 0.08%. Two important characteristics of the combustion are indicated by the engine cycle work Wi and NO_x concentration. The results showed that Wi reached 183 J/cyc for biogas. Upon a switch to the syngas fueling mode, Wi decreased by more than 30%. The NO_x emission in the case of syngas only approximated 180 ppm, which could be ignored compared to 2,900 ppm in the case of biogas. When 50% of biogas was

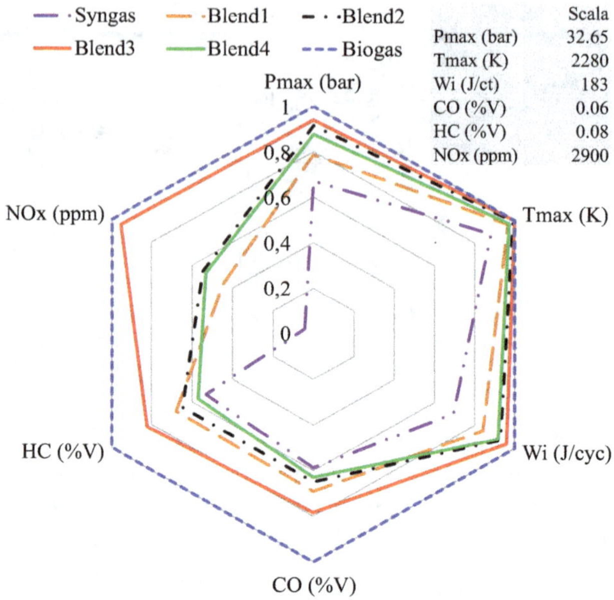

Mixture	Biogas (%)	Syngas (%)	Hydrogen (%)
Blend1	20	60	20
Blend2	40	40	20
Blend3	60	20	20
Blend4	50	50	0

Figure 3.41 Comparison of combustion characteristics of syngas, biogas and syngas–biogas–hydrogen blends (n = 3,000 rpm, ϕ = 1, φ_s =25° CA)

added to syngas (Blend4), NO_x concentration decreased by 41% and the indicated engine cycle work decreased by about 5% compared to the corresponding values of the biogas fueling mode. Therefore, the addition of syngas to biogas leads to a reduction of NO_x much larger than the reduction of Wi. With any fuel composition, Wi increased simultaneously with the increase of NO_x concentration.

Figure 3.42(a) shows the Wi(NO_x) relationship of the engine fueled with a syngas-hydrogen fuel mixture with a different hydrogen content. The results showed that the indicative engine cycle work Wi increased very strongly compared to the increase of NO_x concentration when the hydrogen content was less than 20%. When the hydrogen content exceeds this threshold, Wi increases slowly because the effect of hydrogen in improving combustion is moderate. It has been observed that the formation of nitrogen oxides (NO_x) exhibits an approximately linear correlation with the hydrogen fraction present in the syngas-based fuel mixture. This behavior can be attributed to two primary factors: the elevated LHV of the mixture and the enhancement in combustion characteristics facilitated by hydrogen. Specifically, the inclusion of 50% hydrogen by volume leads to a 22% increase in the Wi, while the NO_x emissions rise by a factor of 20 compared to pure syngas combustion. In contrast, incorporating 20% hydrogen results in a 17% increase in Wi and a sevenfold increase

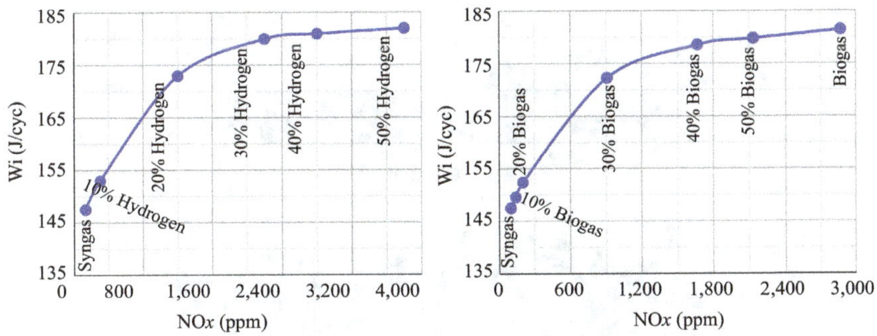

Figure 3.42 Relationship between Wi and NO_x according to hydrogen content (a) and biogas content (b) in the mixture with syngas ($n = 3{,}000$ rpm, $\phi = 1$)

in NO_x concentration. Based on these comparative outcomes, a 20% hydrogen blending ratio can be considered optimal for balancing improved thermal efficiency and acceptable emission levels in syngas-fueled applications.

A similar trend is observed in the case of syngas–biogas fuel blending. As illustrated in Figure 3.42(b), when the biogas proportion in the syngas mixture remains below 30%, the Wi increases at a greater rate than the associated NO_x emissions. However, beyond this 30% threshold, further increases in biogas content result in only marginal improvements in Wi, while NO_x emissions escalate significantly. Based on this trade-off between combustion efficiency and environmental impact, a biogas concentration of approximately 30% in the syngas–biogas blend is identified as an optimal value.

3.6.2 Dual fuel engine fueled with syngas–biogas–hydrogen mixture

3.6.2.1 Dual-fuel mixture formation process

The study was conducted on the dual fuel engine converted from the Vikyno RV165 engine with a cylinder bore of 105 mm, a piston stroke of 97 mm, and a compression ratio of 16.5. The engine operates at 2,000 rpm, 2,400 rpm, and 2,800 rpm. The fuel injection pressure p_{inj} varies in a range of 3 to 9 bar. The properties of fuel components were presented in Table 3.5. The fuel blend is denoted by SxByHz, where x, y, and z are the volume percentage of syngas, biogas, and hydrogen, respectively. The fuel mixture was directly injected into the cylinder via two injectors having a 6-mm nozzle diameter. Figure 3.43 presents the schema of injector disposition [83].

Figure 3.44(a) and (b) reveal the variation of equivalence ratio versus the crank angle using INJ1 and INJ1+INJ2 at the same $\varphi_{start} = 40°$CA and $p_{inj} = 6$ bar. It can be seen that the variation of the equivalence ratio ϕ is strongly depends on fuel compositions. The higher syngas concentration is, the lower ϕ curve is. In the case of single injector INJ1, at 240°CA, the equivalence ratio ϕ are 0.32, 0.58, 0.81, and 1.12 for S100B0H0, S80B10H10, S60B20H20, and S40B30H30, respectively. Near the end of the compression stroke (300°CA), the equivalence ratios of S100B0H0, S80B10H10 are less than stoichiometric value ($\phi = 1$). In the case of INJ1, the

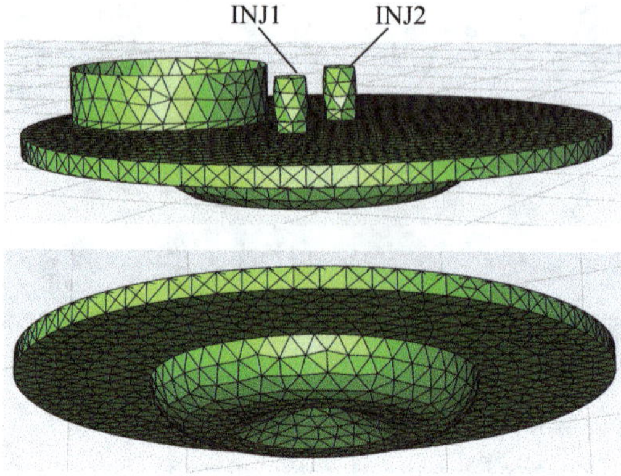

Figure 3.43 The meshing of the calculation space

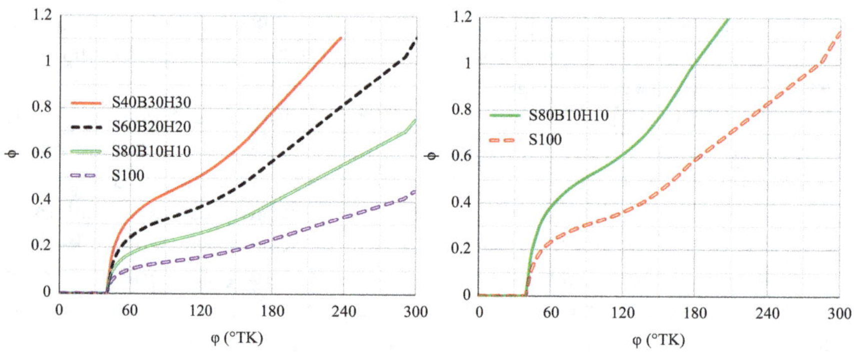

Figure 3.44 The variation of equivalence ratio and in-chamber pressure rise vs. the crank angle when injecting syngas by INJ1 (a) and INJ1+INJ2 (b) at n = 2,400 rpm, p_{inj} = 6 bar, start = 40° CA

equivalence ratio cannot meet the stoichiometric, although the injector opened at the beginning of the intake stroke. In the case of INJ1+INJ2, the crank angle is large enough to adjust the mixture constituent when changing engine speed or load.

When injecting syngas through a single injector, its diameter must be large, and/or the injection pressure must be high. This challenges the fuel compression system and sealing technique for the injector. Furthermore, when increasing engine speed, the available time for fuel supply is shortened. Hence, the injection timing must be advanced to achieve the same equivalence ratio at the same crank angle. The start of injection is a pre-setup parameter related to a given engine speed. Figure 3.45(a) reveals that φ_{start} are 80°CA, 50°CA, and 25°CA when n = 2,000 rpm, 2,400 rpm, and 2,800 rpm, respectively, to achieve equivalence ratio $\phi = 0.75$ at 250°CA. Figure 3.45(b) shows that the higher the engine speed is, the smaller the pressure difference between the PI and the DI is. This result can be attributed to the volumetric efficiency derating as increasing the engine speed. At the same crank angle of 250°CA, the in-cylinder pressure by the DI method are 0.77 bar, 0.85 bar, and 0.89 bar at 2,800 rpm,

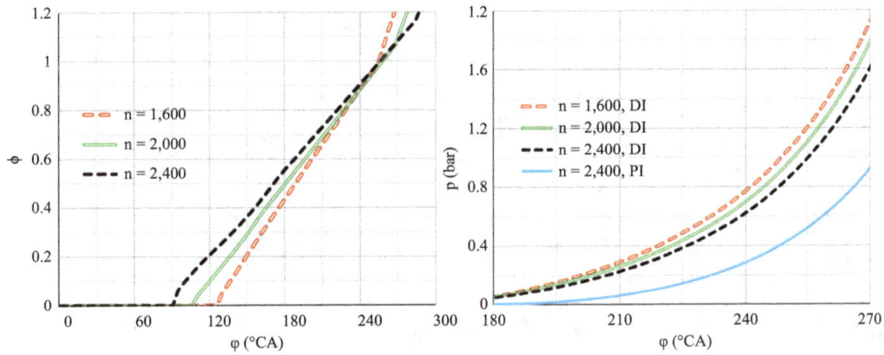

Figure 3.45 The equivalence ratio (a) and in-chamber pressure rise (b) against the crank angle at engine speed of 1,600 rpm, 2,000 rpm, and 2,400 rpm, and at $p_{inj} = 6$ bar (S80B10H10)

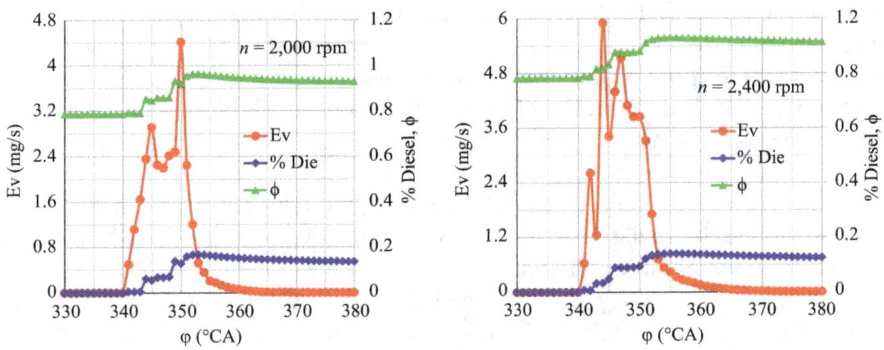

Figure 3.46 Comparing the variation of liquid fuel evaporation rate, diesel concentration, and the equivalence ratio at 2,000 rpm (a) and 2,400 rpm (b) (S80B10H10, INJ1+INJ2)

2,400 rpm, and 2,000 rpm, respectively. For the PI method, it is 0.43 bar at 2,400 rpm. The in-cylinder pressure by DI increases by 100% compared to PI at the same engine speed, suggesting that much more mixtures are fueled into the chamber. Thus, the fuel supply by DI method can improve syngas engine performance.

The engine load is controlled during the operation process to maintain the engine speed at a constant value. As decreasing the engine load, the injection process is finished early, resulting in a decrease in the injected gaseous fuel on the one hand and the derating of the ending compressed pressure on the other hand. Consequently, the engine load by the DI decreases faster than that of the PI. As increasing the engine load, the process is reversed. Therefore, the speed-regulating process of DI fuel supplying method is more convenience than that of PI method.

Figure 3.46 reveals that the diesel is vaporized right after being injected into the hot-compressed air. The evaporation curve rapidly increases in the range of 340°CA to 350°CA. Diesel is wholly vaporized at 360°CA (365°CA) at $n = 2{,}000$ rpm (2,800 rpm). The increase in the engine speed shortens the cycle available time, resulting in a prolonged evaporation process (in °CA). Moreover, under the same

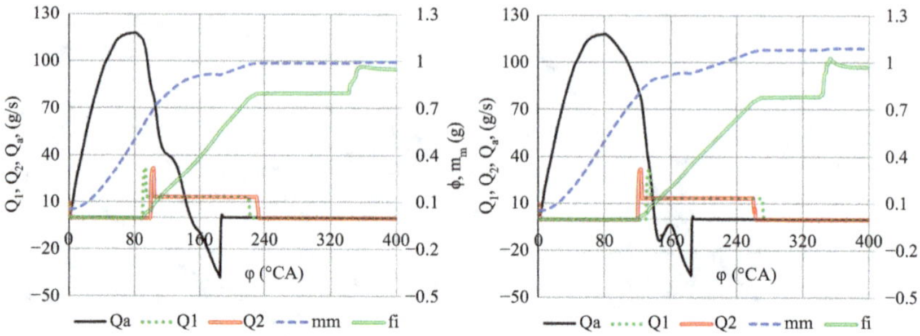

Figure 3.47 The mass flow rate of air and fuel, total mixture mass, and equivalence ratio by INJ2 at $p_{inj} = 6$ bar, $\varphi_{start} = 90°$ CA (a) and $p_{inj} = 6$ bar, $\varphi_{start} = 130°$ CA (b) (S80B10H10)

condition (i.e., $\varphi_{\text{start}} = 10°$ CA, $p_{\text{inj}} = 5$ bar), the contributing equivalence ratio by the injected pilot diesel is 0.15 (0.1) at $n = 2{,}000$ rpm (2,800 rpm), showing that the higher the engine speed is, the smaller the equivalence ratio is. This is because less diesel was fueled. Therefore, the pilot diesel should be controlled according to the rated engine speed to achieve the minimum value that ensures the ignition and the injector lubrication.

Since syngas is directly fueled to the cylinder, the values of φ_{start} and φ_{stop} must be controlled to achieve the desired equivalence ratio. Figure 3.47(a) and (b) presents the mass flow rate of air and fuel, the total mixture mass, and the equivalence ratio at two distinct conditions: $p_{\text{inj}} = 6$ bar and $\varphi_{\text{start}} = 90°$CA (Figure 3.47(a)) versus $p_{\text{inj}} = 6$ bar and $\varphi_{\text{start}} = 130°$CA (Figure 3.47(b)). Start injection of VP2 is phase-shifted 10°CA by VP1. Near the end of the intake stroke, a portion of air is pushed back to the intake manifold, resulting in a negative value of Q_a. The equivalence ratio of the syngas/air mixture is stability at the end of the compression stroke. The total mass m_m of fresh gas increases as retarding injection. It is 1.05 g and 1.22 g corresponding to start injection of VP1 at 90°CA and 130°CA, respectively.

3.6.2.2 Performance of dual fuel engine fueled with syngas–biogas–hydrogen

a. Effects of equivalence ratio on syngas dual fuel engine performance

Figure 3.48 demonstrates the influence of the syngas equivalence ratio (ϕ_{syn}) on the performance and emission characteristics of a dual-fuel engine operating with syngas using the PI method. As ϕ_{syn} increases, NO_x emissions exhibit a slight upward trend, primarily attributed to the rise in combustion temperature. Conversely, CO emissions significantly escalate with higher ϕ_{syn} levels, mainly due to incomplete combustion under fuel-rich conditions. The engine's effective power output corresponding to ϕ_{syn} values of 0.72, 0.75, 0.80, and 0.85 is 8.14 kW, 8.83 kW, 8.87 kW, and 9.23 kW, respectively. When compared to the baseline diesel engine power output of 12.14 kW, the power derating of the syngas-diesel dual-fuel engine utilizing the PI method ranges from approximately 33% to 24% as ϕ_{syn} increases from 0.72 to 0.85.

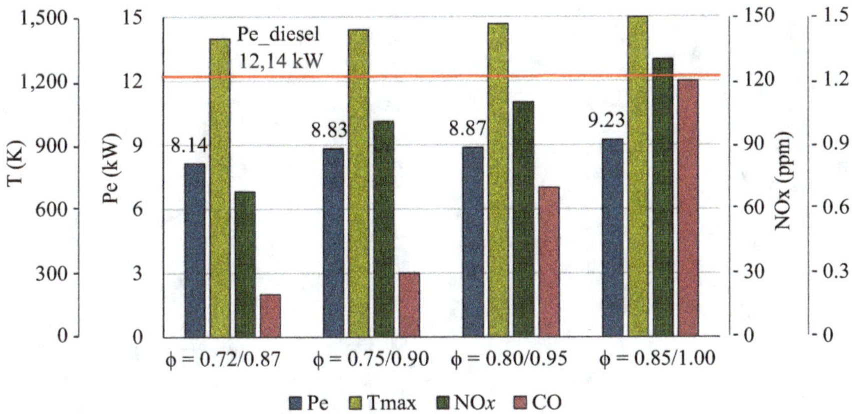

Figure 3.48 Effects of ϕ_{syn} on performance and emission of a dual-fuel engine fueled with syngas by PI method

In practice, the total equivalence ratio ϕ_{tot} for a syngas-diesel dual-fuel engine must remain below the stoichiometric value to ensure successful ignition through diesel pilot injection [73]. Therefore, selecting ϕ_{syn} values in the range of 0.75 to 0.80 and ϕ_{die} values between 0.10 and 0.15 is considered reasonable. Under these conditions, the power derating of the PI syngas-diesel dual-fuel engine is approximately 25% to 30%, which aligns with the 20% to 30% range reported in the literature [81,82]. A comparison of syngas dual fuel engine performance with fuel supply by PI and by DI was presented in References [83–85].

b. Effects of injection regime on dual fuel engine fueled with syngas–biogas–hydrogen

As mentioned earlier, DI significantly increases the fresh fuel-air mixture in the cylinder. The longer the injection duration in the compression stroke is, the greater the amount of mixture is. However, the injection must be finished before the combustion for the fuel-air mixture has enough time to mix well [86]. Figure 3.49(a) presents the effects of start injection on the variation of in-cylinder pressure as dual-fuel engine fueled with S80B10H10. It indicates that the late start of injection increases the in-cylinder pressure, resulting in higher pumping work. However, much more fuel is charged into the cylinder; hence, the indicated engine cycle work increases at the late start of injection. The output power of the engine is 11.71 kW and 12.94 kW, corresponding to start injection angles 90°CA and 130°CA, respectively. When supplying fuel mixture by PI, the output power of the engine is 10.36 kW.

Figure 3.49(b) reveals that φ_{start} and φ_{stop} have a negligible effect on the early stage of HRR on the one hand but a significant influence on the late stage of HRR on the other hand. The later the start of injection is, the longer the injection period in the compression stroke is, leading to the high heat release rate in the late stage. The result could be attributed to the high pressure/temperature before the ignition.

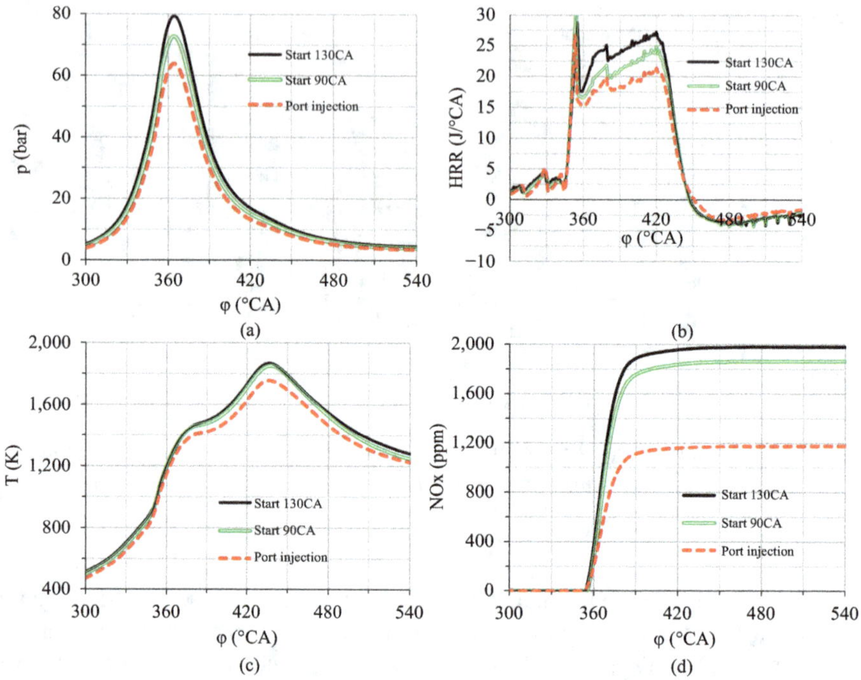

Figure 3.49 Effect of the start of injection on variation of in-cylinder pressure (a), heat release rate (b), temperature (c), and NO_x concentration (d) with crankshaft angle ($n = 2{,}400$ rpm, $p_{inj} = 5$ bar, $\phi_{syn} = 0.75$, and $\phi_{tot} = 0.92$ (S80B10H10))

Consequently, the combustion temperature and NO_x emission increase, as presented in Figure 3.49(c)–(d).

Figure 3.50 presents effects of start and stop injection angles on engine output power and NO_x concentration when the dual fuel engine was fueled with mixture S40B30H30. It can be seen that both engine power and NO_x emissions tend to increase with extended fuel injection duration. Compared to diesel fueling mode, the power of the dual fuel engine increases by 9% when the fuel mixture is injected during the compression stroke, but it decreases by 2.5% and 11% when injection starts at 100°CA and 85°CA, respectively. The NO_x concentration in the exhaust gases is 1,423 ppm, 1,176 ppm, and 874 ppm, corresponding to injection start angles of 185°CA, 100°CA, and 85°CA, respectively.

c. Effects of hydrogen content in the mixture with syngas

Figure 3.51(a) shows that the heat release rate increases with the hydrogen content in the fuel mixture with syngas. The heat release rate in the case of hydrogen fueling mode is nearly double that of the syngas fueling mode. The output power of the dual-fuel engine are 15.35 kW, 11.52 kW, 10.98 kW, and 9.21 kW, corresponding to hydrogen contents in the mixture with syngas of 100%, 50%, 25%, and 0%, respectively (Figure 3.51(b) and (c)).

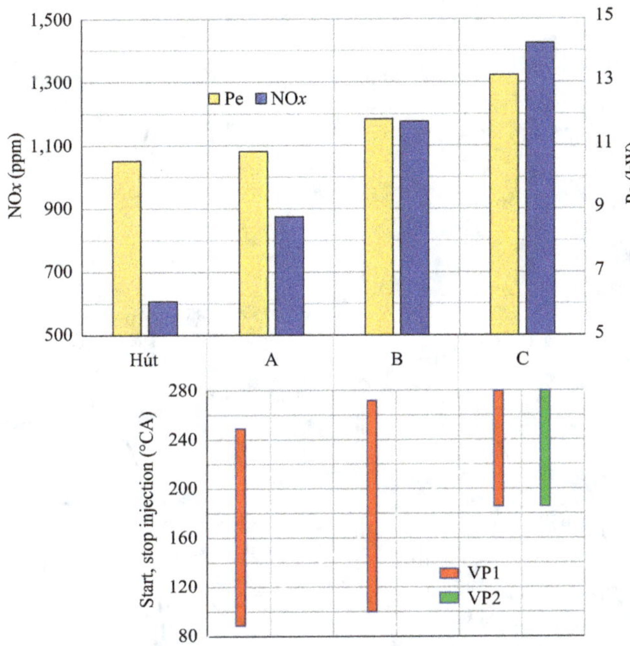

Figure 3.50 Effect of the start of injection on variation of in-cylinder pressure (a), Heat release rate (b), Temperature, (c) and NO_x concentration (d) with crankshaft angle ($n = 2{,}400$ rpm, $p_{inj} = 5$ bar, $\phi_{syn} = 0.75$, and $\phi_{tot} = 0.92$ (S80B10H10))

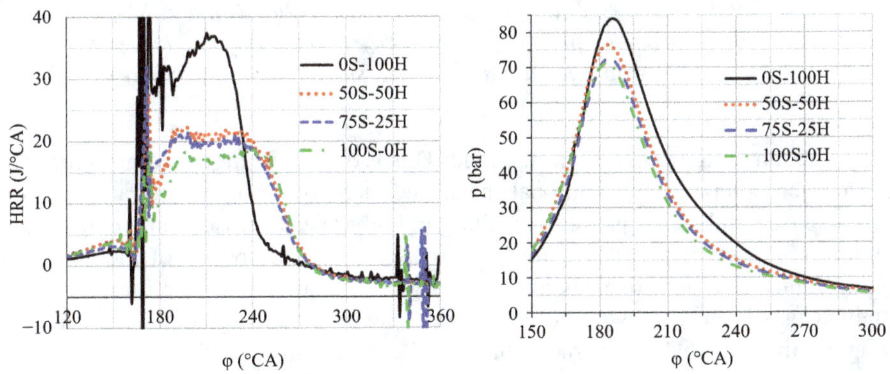

Figure 3.51 Effects of hydrogen content in the mixture with syngas on combustion characteristics ($n = 2{,}400$ rpm, $\phi_{gas} = 0.75$, $\phi_{tot} = 0.92$, $p_{inj} = 5$ bar, Direct injection, start injection at 250° CA, stop injection variable)

Similarly, in the case of fuel supply by PI, the heat release rate of the combustion mixture is significantly improved when syngas is enriched by hydrogen. Figure 3.52(a) shows that the heat release rate in hydrogen fueling mode is more

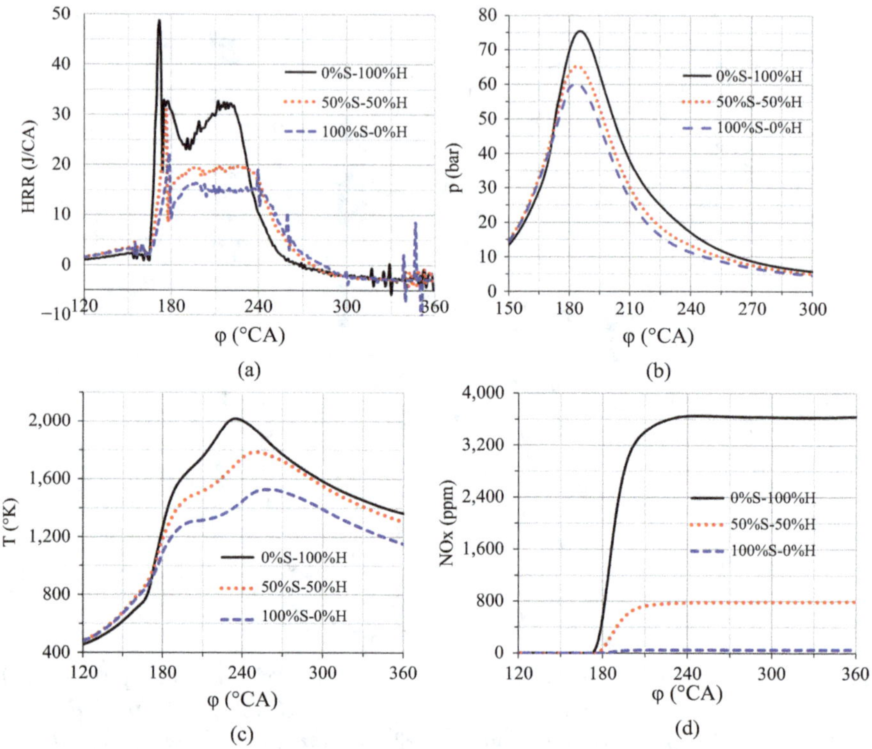

Figure 3.52 Effects of hydrogen content in the mixture with syngas on variation of heat release rate (a), pressure (b), temperature (c), and NO_x emission (d) ($n = 2{,}400$ rpm, $\phi_{gas} = 0.75$, $\phi_{tot} = 0.92$, $p_{inj} = 3$ bar, port injection, stop injection at 250° CA, start injection variable)

than twice that of the syngas fueling mode. This is due to the higher combustion rate of hydrogen compared to traditional fuels. The increased heat release rate causes a more intense combustion process, resulting in higher peak pressure and temperature (Figure 3.52(b) and (c)). When fuel is supplied by PI, the power of the dual-fuel engine running on hydrogen is about 20% greater than that of the diesel engine. When syngas is enriched with 50% hydrogen, the dual fuel engine power decreases by approximately 16% compared to the diesel engine power.

A drawback of using a fuel mixture with a high hydrogen content is the increase in NO_x concentration in the exhaust due to the increase in combustion temperature. Figure 3.52(d) shows that the NO_x concentration reaches up to 3637 ppm in case of hydrogen fueling mode. In contrast, in case of syngas fueling mode, the NO_x concentration is only 53 ppm. When the engine was fueled with syngas enriched with 50% hydrogen, the NO_x concentration reaches 790 ppm.

Figure 3.53 illustrates the variation of dual fuel engine output power with hydrogen content in the mixture with syngas at an engine speed of 2,400 rpm. The fuel mixture was supplied by the PI and by DI methods at an injection pressure of 3 bar. The results show that whether fuel is supplied via PI or DI, the engine power increases

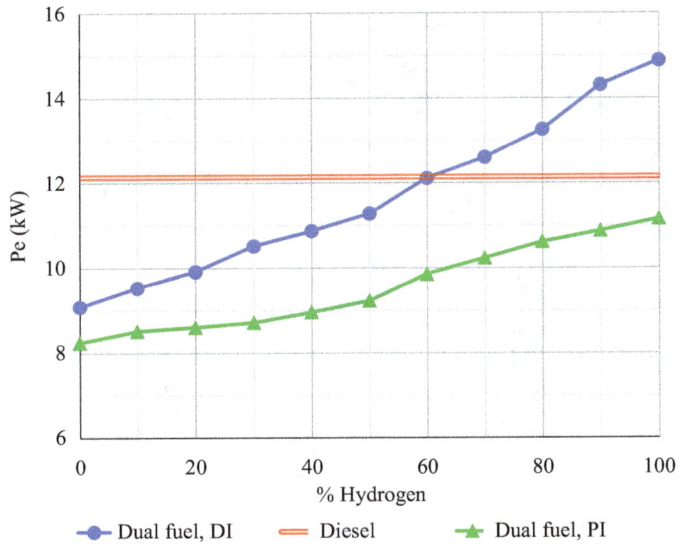

Figure 3.53 Variation of dual fuel engine output power with hydrogen content in the mixture with syngas, fuel supplying by PI method and by DI method with injection pressure of 3 bar (n = 2,400 rpm, $\phi_{gas} = 0.75$, $\phi_{tot} = 0.92$)

with the hydrogen content in the mixture with syngas. Under conditions with a global equivalence ratio limit of 0.92 to ensure that the pilot jet can be ignited, the power of the engine supplied by the PI method is always less than that of the original diesel engine.

When supplying fuel mixture by the DI method, the output power of the dual-fuel engine can match that of the diesel engine, as hydrogen content in the mixture with syngas is around 60%. Exceeding this threshold limit, the power of the dual-fuel engine is greater than that of the original diesel engine. This is an advantage of the dual-fuel engine fueled with syngas enriched by hydrogen supplied through the DI method. Effects of combustion chamber configuration can be found in Reference [87].

d. Effects of oxygen enriched air oxidant supplied to gasifier

The performance of the dual-fuel engine can be further enhanced by improving the quality of the syngas. In the SWB-HRES system, syngas quality can be improved by adding oxygen to the air used for gasification. As the oxygen content in the air, utilized as an oxidant in the gasifier, increases, the calorific value of the fuel rises. This enhancement leads to an increased combustion rate, which subsequently boosts the heat release rate. Figure 3.54 illustrates that the heat release curve exhibits two distinct peaks. The maximum values of these peaks decrease as the oxygen content in the enriched air rises during RDF gasification. Specifically, the heat release rate increases from 15 J/°CA to 23 J/°CA as the oxygen content in the air escalates from 0% to 50%.

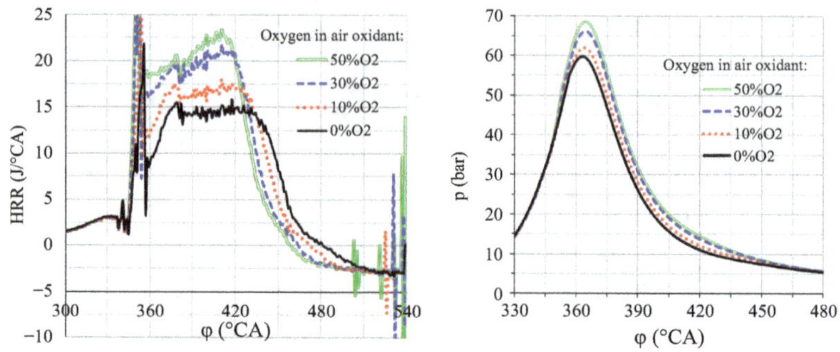

Figure 3.54 Impact of oxygen enrichment in air used as an oxidant in RDF gasification on the heat release rate (a) and in-cylinder pressure (b) of a dual-fuel engine (ϕ_{syn} = 0.8, n = 2,400 rpm)

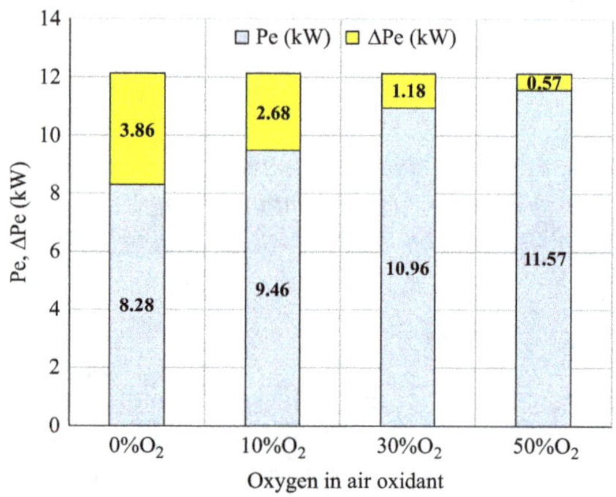

Figure 3.55 Effects of O_2-enriched air content used as oxidant of RDF gasification on syngas dual fuel output power (ϕ_{syn} = 0.8, n = 2,400 rpm, port injection)

The increased heat release rate resulting from higher oxygen content mixed into the air used as the oxidant leads to a rise in peak pressure, subsequently increasing the cycle's indicated work. Figure 3.55 illustrates the engine power outputs are 8.28 kW, 9.46 kW, 10.96 kW, and 11.57 kW corresponding to oxygen content in the oxidizing air 0%, 10%, 30%, and 50%, respectively. In comparison to the diesel engine's power output before modification (12.14 kW), the power output of the dual-fuel syngas-diesel engine decreases, respectively, 31.82%, 22.10%, 9.57%, and 4.73%. This indicates that when gasifying RDF using pure air as the oxidant, the engine's power decreases by approximately 30%, which is consistent with data published in the referenced literature.

Figure 3.56 presents a comprehensive comparison of how the oxygen content in air, used as an oxidant for RDF gasification, influences key combustion parameters in a dual-fuel engine. The figure illustrates the trends in combustion characteristics when the engine utilizes syngas derived from RDF with air enriched by varying levels of oxygen. The aggregated results indicate that parameters such as P_{max}, T_{max}, Wi,

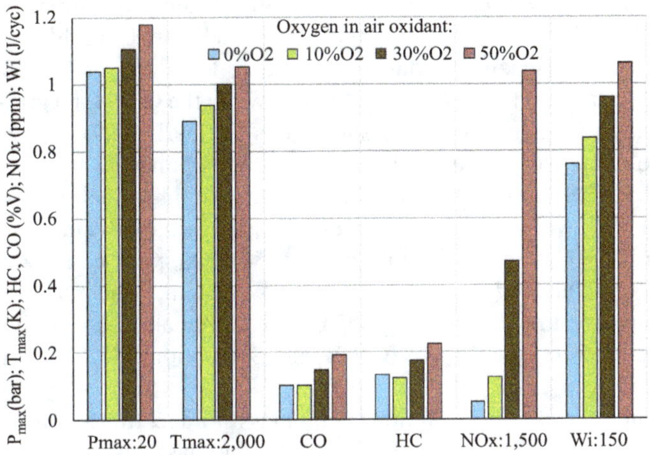

Figure 3.56 Effect of oxygen-enriched air content as an oxidant for RDF on key combustion parameters in a dual-fuel engine ($\phi_{syn} = 0.8$, $n = 2{,}400$ rpm)

and pollutant concentrations of the engine tend to increase linearly with the rise in oxygen content in the enriched air oxidant supplied to the gasifier. This trend can be attributed to the higher heat value of syngas. As a result, the overall combustion quality in the internal combustion engine improves with increasing oxygen content in the air oxidant to the gasifier.

3.7 Conclusion

The findings of this study lead to the following conclusions:

- The SWB configuration is the most suitable option for HRES with hydrogen production. It offers significant techno-economic and environmental advantages as a poly-generation system utilizing renewable energy for sustainable development. However, further research is needed to determine the optimal capacity of each system component and to refine the energy conversion technologies from biomass, including biomass storage, gasification, and internal combustion engines fueled by a syngas–biogas–hydrogen mixture.
- For the SWB grid-connected HRES with hydrogen production to achieve comparable economic effectiveness to a grid-connected system without hydrogen production (in terms of present worth, ROI, and payback period), the price of green hydrogen must reach a threshold of $5/kg.
- In comparison to the CO_2 emissions associated with producing 74,235 kWh of grid electricity in Vietnam, annual CO_2 emission reductions are as follows: 39 tons for the SW HRES without hydrogen production, 50 tons for the SW HRES with hydrogen production, 207 tons for the SWB system without hydrogen production, and 218 tons for the SWB system with hydrogen production.

- The COE for a grid-connected HRES with hydrogen production is lower than that of a standalone HRES. For the same load, the COE for a standalone HRES at annual capacity shortage rates of 5%, 10%, and 20% is $0.097/kWh, $0.085/kWh, and $0.078/kWh, respectively, while the COE for a grid-connected system with a 5% annual capacity shortage is $0.039/kWh.
- The use of a syngas-fueled generator significantly enhances the economic efficiency of HRES. In a grid-connected system with an annual electricity output of 74,235 kWh without hydrogen production, the initial investments for the SB, SWB, and WB systems are approximately $41,000, with corresponding ROIs of 9.2%, 8.5%, and 6.7%. In contrast, the SW system, which lacks a generator, requires an investment exceeding $55,000, yielding an ROI of only 5.7%. The payback periods for the SB, SWB, WB, and SW options are 7.93, 7.87, 8.5, and 10.33 years, respectively.
- RDF provides an effective solution for biomass storage, transportation, and conversion into energy. The optimal ER for RDF biomass gasification ranges from 0.3 to 0.4. When using air as the oxidant, the calorific value of syngas from rice husk RDF and wood chip RDF is 6.82 MJ/kg and 7.21 MJ/kg, respectively. Enriching the oxidant with 30% and 50% oxygen increases the calorific value of syngas by 60% and 90%, respectively.
- Due to significant variations in the air-fuel ratio of the fuel component in syngas–biogas–hydrogen mixture, a twin-injector system is recommended for fuel supply in spark-ignition or dual-fuel engines. The injection duration of the syngas–biogas–hydrogen mixture does not exceed that of the hydrogen-only mixture and is not shorter than that of the syngas-only mixture.
- In a syngas-diesel dual-fuel engine, power derating ranges from 33% to 24% as the equivalence ratio of syngas increases from 0.72 to 0.85. Direct injection of the syngas–biogas–hydrogen mixture via a twin-injector system reduces power loss to 12.6%. Power derating in syngas engines can be mitigated by enriching syngas with biogas or hydrogen; however, NO_x emissions increase with higher hydrogen or biogas content. For SI engines, the optimal hydrogen and biogas contents are 20% and 30%, respectively, in the syngas mixture. In dual-fuel engines using DI, output power can match that of diesel engines when the hydrogen content in the syngas mixture reaches approximately 60%.
- Power derating in syngas engines can also be controlled by increasing the calorific value of the fuel through oxygen-enriched air oxidant supplied to the gasifier. For dual-fuel engines using PI, power derating is 22.10%, 9.57%, and 4.73% when the oxygen content in the oxidizing air is 10%, 30%, and 50%, respectively.

Acknowledgments

This work was supported by the Ministry of Education and Training of Vietnam through the research project "Investigation and Development of an Extended-Range Electric Motorcycle (EREM) Powered by a Hydrogen carrier fuel Engine," Code: B2026-DNA-09.

References

[1] Wisam K Hussam, El Manaa Barhoumi, Abdul-Niby M, and Gregory J Sheard. Techno-economic analysis and optimization of hydrogen production from renewable hybrid energy systems: Shagaya renewable power plant-Kuwait. *International Journal of Hydrogen Energy*. 2024;58:56–68.

[2] Tarhan C and Çil MA. A study on hydrogen, the clean energy of the future: Hydrogen storage methods. *Journal of Energy Storage*. 2021;40:102676.

[3] Nasser M, Megahed TF, Ookawara S, and Hassan H. Performance evaluation of PV panels/wind turbines hybrid system for green hydrogen generation and storage: Energy, exergy, economic, and enviroeconomic. *Energy Conversion and Management*. 2022;267:115870.

[4] Devrim Y and Bilir L. Performance investigation of a wind turbine-solar photovoltaic panels-fuel cell hybrid system installed at İncek region-Ankara, Turkey. *Energy Conversion and Management*. 2016;126:759–766.

[5] Van Ga Bui, Trung Hung Vo, Thi Minh Tu Bui, Le Bich Tram Truong, and Thanh Xuan Nguyen Thi, Characteristics of Biogas-Hydrogen Engines in a Hybrid Renewable Energy System. *International Energy Journal*. 2021;21(4):467–480.

[6] Mohamed Nasser, Tamer F Megahed, Shinichi Okawara, and Hamdy Hassan, Techno-economic assessment of clean hydrogen production and storage using hybrid renewable energy system of PV/Wind under different climatic conditions. *Sustainable Energy Technologies and Assessments*. 2022;52(B):102195.

[7] Akyuz E, Oktay Z, and Dincer I. Performance investigation of hydrogen production from a hybrid wind-PV system. *International Journal of Hydrogen Energy*. 2012;37:16623–16630.

[8] Hasan MM and Genç G. Techno-economic analysis of solar/wind power based hydrogen production. *Fuel*. 2022;324:124564.

[9] Alzahrani A, Ramu SK, Devarajan G, Vairavasundaram I, and Vairavasundaram S. A review on hydrogen-based hybrid microgrid system: Topologies for hydrogen energy storage, integration, and energy management with solar and wind energy. *Energies*. 2022;15:7979.

[10] Okonkwo PC, Mansir IB, Emori W, Radwan AB, *et al.* Utilization of renewable hybrid energy for refueling station in Al-Kharj, Saudi Arabia. *International Journal of Hydrogen Energy*. 2022;47:22273–22284.

[11] Matias Garcia G and Sebastian Oliva H. Technical, economic, and CO2 emissions assessment of green hydrogen production from solar/wind energy: The case of Chile. *Energy*. 2023;278(B):127981.

[12] Khalilnejad A and Riahy GH. A hybrid wind-PV system performance investigation for the purpose of maximum hydrogen production and storage using advanced alkaline electrolyzer. *Energy Conversion and Management*. 2014;80.

[13] Dufo-López R, Bernal-Agustín JL, and Contreras J. Optimization of control strategies for stand-alone renewable energy systems with hydrogen storage. *Renewable Energy*. 2007;32.

[14] Hacımurat Demir. Design and optimization of hybrid renewable energy systems for hydrogen production at Aksaray University campus. *Process Safety and Environmental Protection*. 2024;192:543–556.

[15] Swaminathan P, Ghosh A, Sunantha G, *et al.* A comprehensive review of microbial electrolysis cells integrated with wastewater treatment for hydrogen generation. *Process Safety and Environmental Protection*. 2024;190:458–474.

[16] Gökçek M and Genç MS. Evaluation of electricity generation and energy cost of wind energy conversion systems (WECSs) in Central Turkey. *Applied Energy*. 2009;86(12):2731–2739.

[17] Taipabu MI, Viswanathan K, Wu W, Hattu N, and Atabani AE, A critical review of the hydrogen production from biomass-based feedstocks: challenge, solution, and future prospect. *Process Safety and Environmental Protection*. 2022;164:384–407.

[18] Bukar AL, Chaitusaney S, and Kawabe K. Optimal design of on-site PV-based battery grid-tied green hydrogen production system. *Energy Conversion and Management*. 2024;307:118378.

[19] Kalinci Y, Hepbasli A, and Dincer I. Techno-economic analysis of a stand-alone hybrid renewable energy system with hydrogen production and storage options. *International Journal of Hydrogen Energy*. 2015;40.

[20] Onar OC, Uzunoglu M, and Alam MS. Modeling, control and simulation of an autonomous wind turbine/photovoltaic/fuel cell/ultra-capacitor hybrid power system. *Journal of Power Sources*. 2008;185.

[21] Van Ga Bui, Thi Minh Tu Bui, Van Nam Tran, *et al.* Flexible syngas-biogas-hydrogen fueling spark-ignition engine behaviors with optimized fuel compositions and control parameters. *International Journal of Hydrogen Energy*. 2022.

[22] Van Ga Bui, Thi Minh Tu Bui, Van Giao Nguyen, Van Nam Tran, Le Bich Tram Truong, and Le Hoang Phu Pham, Concept of twining injector system for spark-ignition engine fueled with syngas-biogas-hydrogen operating in solar-biomass hybrid energy system. *International Journal of Hydrogen Energy*. 2022.

[23] Armijo J and Philibert C. Flexible production of green hydrogen and ammonia from variable solar and wind energy: case study of Chile and Argentina. *International Journal of Hydrogen Energy*. 2020;45.

[24] Fopah-Lele A, Kabore-Kere A, Tamba JG, and Yaya-Nadjo I. Solar electricity storage through green hydrogen production: A case study. *International Journal of Energy Research*. 2021;45(9):13007–13021.

[25] Koleva M, Guerra OJ, Eichman J, Hodge BM, and Kurtz J. Optimal design of solar-driven electrolytic hydrogen production systems within electricity markets. *Journal of Power Sources*. 2021;483:229183.

[26] Mokhtara C, Negrou B, Settou N, Bouferrouk A, and Yao Y. Design optimization of grid-connected PV-Hydrogen for energy prosumers considering sector-coupling paradigm: Case study of a university building in Algeria. *International Journal of Hydrogen Energy*. 2021;46(75):37564–37582.

[27] Touili S, Alami Merrouni A, Azouzoute A, El Hassouani Y, and Illah A Amrani. A technical and economical assessment of hydrogen production

potential from solar energy in Morocco. *International Journal of Hydrogen Energy*. 2018;43:22777–22796.

[28] Rawan A Abdelsalam, Moataz Mohamed, Hany EZ Farag, and Ehab F El-Saadany. Green hydrogen production plants: A techno-economic review. *Energy Conversion and Management*. 2024;319:118907.

[29] Agyekum EB. Is Africa ready for green H2 energy takeoff? A multi-criteria analysis approach to the opportunities and barriers of H2 production on the continent. *International Journal of Hydrogen Energy*. 2024;49:219–233.

[30] Park J, Hwan Ryu K, Kim C, *et al.* Green hydrogen to tackle the power curtailment: Meteorological data-based capacity factor and techno-economic analysis. *Applied Energy*. 2023;340:121016.

[31] Naderi S, Banifateme M, Pourali O, Behbahaninia A, MacGill I, and Pignatta G. Accurate capacity factor calculation of waste-to-energy power plants based on availability analysis and design/off-design performance. *Journal of Cleaner Production*. 2020;275:123167.

[32] Bhandari R, Kumar B, and Mayer F. Life cycle greenhouse gas emission from wind farms in reference to turbine sizes and capacity factors. *Journal of Cleaner Production*. 2020;277:123385.

[33] Nguyen Thi TX, Bui TMT, and Bui VG. Simulation and experimental study of refuse-derived fuel gasification in an updraft gasifier. *International Journal of Renewable Energy Development*. 2023;12(3):601–614.

[34] Minh Tung Phung, Van Ga Bui, and Thanh Son Tran. Simulation and Experimental Study on Refuse Derived Fuel Gasification in a Downdraft Gasifier. In: MMMS2022. *Lecture Notes in Mechanical Engineering*; 2024. p. 369–378.

[35] Sahar Ahmadvand and Taraneh Sowlati. The attractiveness of syngas production from forest-based biomass for pulp mills considering carbon pricing and government regulations. *Renewable Energy Focus*. 2023;45:287–306.

[36] Paul Simshauser. The 2022 energy crisis: Fuel poverty and the impact of policy interventions in Australia's National Electricity Market. *Energy Economics*. 2023;121:106660.

[37] Rogelj J, Geden O, Cowie A, and Reisinger A. Net-zero emissions targets are vague: three ways to fix. *Nature*. 2021;591:365–368.

[38] IEA (annual publication). *Clean Energy Innovation*; 2020.

[39] Peker E. *The Wall Street Journal*; 2019.

[40] Sachs JD, Schmidt-Traub G, and Williams J. Pathways to zero emissions. *Nature Geoscience*. 2016;9:799–801.

[41] Fuss S, Canadell JG, Peters GP, *et al.* Betting on negative emissions. *Nature Climate Change*. 2014;4:850–853.

[42] News V. Viet Nam strives to achieve net-zero by 2050 with international support: PM; 2021. Available from: https://vietnamnews.vn/environment/1071075/viet-nam-strives-to-achieve-net-zero-by-2050-with-international-support-pm.html.

[43] Huu Hieu Nguyen, Van Quang Pham, and Van Ga Bui. Optimisation of Solar-Wind-Biomass hybrid renewable energy system in Cu Lao Cham Island. *Journal of Science and Technology-The University of Danang*. 2024;22(11A): 25–31.

[44] Energy H. HOMER2 2.8 Help Manual; n.d. Available from: https://homerenergy.com/pdf/HOMER2_2.8_HelpManual.pdf.
[45] Ansys. Ansys Fluent Theory Guide; n.d. Available from: https://dl.cfdexperts.net/cfd_resources/Ansys_Documentation/Fluent/Ansys_Fluent_Theory_Guide.pdf.
[46] Sireesha Aluri, Akil Syed, Derrick W. Flick, John D. Muzzy, Carsten Sievers, and Pradeep K. Agrawal. Pyrolysis and gasification studies of model refuse derived fuel (RDF) using thermogravimetric analysis. *Fuel Processing Technology*. 2018.
[47] K. Daugherty. An Identification of Potential Binding Agents for Densified Fuel Preparation from Municipal Solid Waste: Phase 1; 1986.
[48] Sever Akdağ A, Atimtay A, and Sanin FD. Comparison of fuel value and combustion characteristics of two different RDF samples. *Waste Management (New York, NY)*. 2015.
[49] Pio DT, Tarelho LAC, Tavares AMA, Matos MAA, and Silva V. Co-gasification of refused derived fuel and biomass in a pilot-scale bubbling fluidized bed reactor. *Energy Conversion and Management*. 2020.
[50] Van Ga Bui, Anh Vu Vo, Tran Ngoc Anh Ho, *et al.* Simulation and Experiment Study of Refuse Derived Fuel (RDF) Combustion in Atmospheric Conditions. *GMSARN*. 2026;2.
[51] Ajay K Dalai, Nishant Batta, Eswaramoorthi I, *et al.* Gasification of refuse derived fuel in a fixed bed reactor for syngas production. *Waste Management*. 2009.
[52] Galvagno S, Casciaro G, Casu S, *et al.* Steam gasification of tyre waste, poplar, and refuse-derived fuel: A comparative analysis. *Waste Management*. 2009.
[53] Halina Pawlak-Kruczek and Malgorzata Frydel Marcin Baranowski. Effect of gasifying agents and calcium oxide on gasification of low-rank coal and wastes. *Transactions of the Institute of Fluid-Flow Machinery*. 2017.
[54] Majid Saidi, Mohammad Gohari, and Ali Talesh Ramezani. Hydrogen production from waste gasification followed by membrane filtration: a review. *Environmental Chemistry Letters*. 2020.
[55] Guanyi Chen, Imtiaz Jamro, Saleem Samo, *et al.* Hydrogen-rich syngas production from municipal solid waste gasification through the application of central composite design: An optimization study. *International Journal of Hydrogen Energy*. 2020.
[56] Prateek Sharma, Subhadip Sen, Pratik N. Sheth, and B. N. Mohapatra. Multizone model of a refused derived fuel gasification: A thermodynamic Semi-empirical approach. *Energy Conversion and Management*. 2022.
[57] Ga BV, Tu BTM, Tung TTH, and Long PD. Simulation of RDF gasification in a downdraft gasifier. In: *Proceedings of the 24th National Conference on Fluid Mechanics*; 2021. p. 18–12.
[58] Ga BV, Dong NV, Tram TLB, and Thanh HN. Comparison of gasification efficiency of solid waste in agriculture production. In: *Proceedings of the 24th National Conference on Fluid Mechanics*; 2021. p. 18–12.
[59] Ga BV, Vu VA, Phung NV, *et al.* Simulation and experimental study of RDF gasification. *Journal of Science and Technology: The University of Danang*. 2023.

[60] Rakopoulos CD and Michos CN. Development and validation of a multi-zone combustion model for performance and nitric oxide formation in syngas fueled spark ignition engine. *Energy Conversion and Management*. 2008;49(10):2924–2938.

[61] Hagos FY, Aziz ARA, and Sulaiman SA. Sulaiman. Trends of Syngas as a Fuel in Internal Combustion Engines. *Advances in Mechanical Engineering*. 2014;6:401587.

[62] Paykani A, Chehrmonavari H, Tsolakis A, Alger T, Northrop WF, and Reitz RD. Synthesis gas as a fuel for internal combustion engines in transportation. *Progress in Energy and Combustion Science*. 2022;90.

[63] Fiore M, Magi V, and Viggiano A. Internal combustion engines powered by syngas: A review. *Applied Energy*. 2020;276:115415–115441.

[64] Harpreet Kaur, Surbhi Gupta, and Arvind Dhingra. Analysis of hybrid solar biomass power plant for generation of electric power. *Materials Today*. 2022;48(5):1134–1140.

[65] SonjaKallio and MonicaSiroux. Hybrid renewable energy systems based on micro-cogeneration. *Energy Reports*. 2022;8(S1):762–769.

[66] Akash Kumar, Shukla K Sudhakar, and Prashant Baredar. Renewable energy resources in South Asian countries: Challenges, policy and recommendations. *Resource-Efficient Technologies*. 2017;3(3):342–346.

[67] Forum WE. Visualizing the world's biggest rice producers; 2022. Available from: https://www.weforum.org/agenda/2022/03/visualizing-the-world-s-biggest-rice-producers/.

[68] Shahbaz M, Al-Ansari T, Aslam M, *et al.* A state of the art review on biomass processing and conversion technologies to produce hydrogen and its recovery via membrane separation. *International Journal of Hydrogen Energy*. 2020;45(30):15166–15195.

[69] Singh S, Singh M, and Kaushik SC. Feasibility study of an islanded microgrid in a rural area consisting of PV, wind, biomass and battery energy storage system. *Energy Conversion and Management*. 2016;128:178–190.

[70] Rakibul Hassan, Barun K Das, and Mahmudul Hasan. Integrated off-grid hybrid renewable energy system optimization based on economic, environmental, and social indicators for sustainable development. *Energy*. 2022;250:123823.

[71] Willian Cézar Nadaleti and Grzegorz Przybyla. Emissions and performance of a spark-ignition gas engine generator operating with hydrogen-rich syngas, methane and biogas blends for application in southern Brazilian rice industries. *Energy*. 2018;154:38–51.

[72] Hagos FY, Aziz ARA, and Sulaiman SA. Syngas (H_2/CO) in a spark-ignition direct-injection engine. Part 1: Combustion, performance and emissions comparison with CNG. *International Journal of Hydrogen Energy*. 2014;39:17884–17895.

[73] Van Ga Bui, Thi Minh Tu Bui, Minh Tien Nguyen, *et al.* Enhancing the performance of syngas-diesel dual-fuel engines by optimizing injection regimes: From comparative analysis to control strategy proposal. *Process Safety and Environmental Protection*. 2024;186:1034–1052.

[74] Zambalov S, Yakovlev I, Ji C, Yang J, and Maznoy A. Effect of dual port-direct injection of syngas in a rotary engine. *Fuel*. 2023;354:129243.

[75] Bates R and Dölle K. Syngas Use in Internal Combustion Engines - A Review. *Advances in Research*. 2017;10:1–8.

[76] Sombatwong P, Thaiyasuit P, and Pianthong K. Effect of Pilot Fuel Quantity on the Performance and Emission of a Dual Producer Gas–Diesel Engine. *Energy Procedia*. 2013;34:218–227.

[77] Mahgoub BKM, Sulaiman SA, and Abdul Karim ZA. Performance study of imitated syngas in a dual-fuel compression ignition diesel engine. *International Journal of Automotive and Mechanical Engineering*. 2015;11:2282–2293.

[78] Malik A and Mohapatra SK. Biomass-based gasifiers for internal combustion (IC) engines: A review. *Sadhana*. 2013;38:461–476.

[79] Banapurmath NR and Tewari PG. Comparative performance studies of a 4-stroke CI engine operated on dual fuel mode with producer gas and Honge oil and its methyl ester (HOME) with and without carburetor. *Renewable Energy*. 2009;34:1009–1015.

[80] Martínez JD, Mahkamov K, Andrade RV, and Silva Lora EE. Syngas production in downdraft biomass gasifiers and its application using internal combustion engines. *Renewable Energy*. 2012;38:1–9.

[81] Uma R, Kandpal TC, and Kishore VVN. Emission characteristics of an electricity generation system in diesel alone and dual fuel modes. *Biomass and Bioenergy*. 2004;27:195–203.

[82] Van Ga Bui, Thanh Xuan Nguyen Thi, and Thi Minh Tu Bui. Simulation of direct injection of a flexible syngas-biogas-hydrogen mixture into the combustion chamber of dual fuel engine. *Journal of Science and Technology-The University of Danang*. 2023;21(4):68–74.

[83] Van Ga Bui, Thi Minh Tu Bui, and Tran Ngoc Anh Ho. Comparison of combustion characteristics of dual fuel engine with syngas supply by port injection and direct injection. *Journal of Science and Technology-The University of Danang*. 2023;21(5):45–51.

[84] Van Ga Bui, Quang Trung Nguyen, Van Dong Nguyen, and Tan Quyen Ho. Direct injection of syngas-biogas-hydrogen mixture for improving Net Zero dual fuel engine. In: *26th National Conference on Fluid Mechanics*; 2024. p. 200–211.

[85] Van Hung Bui, Thi Minh Tu Bui, Minh Tien Le, and Thanh Xuan Nguyen Thi. Performance of Honda GX160 fueled with syngas from biomass gasification. In: *National Conference on Fluid Mechanics*; 2024. p. 338–348.

[86] Van Ga Bui, Thanh Xuan Nguyen Thi, Quang Trung Nguyen, and Phu Nguu Do. Simulation of combustion in dual fuel engine fueled with syngas-biogas-hydrogen mixture. *Journal of Transport and Communication*. 2024;75(2):1213–1226.

[87] Van Ga Bui, Thi Minh Tu Bui, Anh Tuan Hoang, Sandro Nizetic, Thanh Xuan Nguyen Thi, and Anh Vu Vo. Hydrogen-Enriched Biogas Premixed Charge Combustion and Emissions in DI and IDI Diesel Dual Fueled Engines: A Comparative Study. *Journal of Energy Resources Technology*. 2021;143(12):120907.

Chapter 4
Energy storage system operation and planning with hybrid renewable integration

Dinh Duong Le[1], *Nhi Thi Ai Nguyen*[1] *and Alberto Berizzi*[2]

4.1 Introduction

The energy sector is widely recognized as the main contributor to global greenhouse gas emissions and, consequently, a major driver of climate change. In response to the urgent need to limit the long-term rise in average global temperatures to 1.5°C, an ambitious target has been established for this sector to achieve net-zero carbon dioxide (CO_2) emissions by 2050, as outlined in the Net-Zero Emissions Scenario. This framework envisions a fundamental transition in the global energy system, characterized by a substantial shift from fossil-based generation to renewable energy sources (RES). By 2050, it is projected that approximately two-thirds of the total primary energy supply will be derived from renewable resources, including wind, solar, bioenergy, geothermal, and hydropower. Concurrently, electricity is expected to constitute nearly half of total final energy consumption, with around 90% of electricity generation produced from renewable technologies. Among these, wind power and solar photovoltaic (PV) systems are anticipated to collectively contribute up to 70% of total electricity generation capacity [1].

However, integrating a high proportion of RES into existing power systems presents considerable technical and operational challenges, primarily due to the fundamental differences between RES and conventional dispatchable generation. Wind and solar PV generation are inherently variable, as their output fluctuates according to meteorological conditions. This variability spans multiple temporal scales, affecting both short-term (from seconds to hours) and long-term (from months to years) power system operations. In the short term, wind variability is typically associated with transient gusts and lulls, while solar PV generation fluctuates with variations in cloud cover. Although rapid fluctuations over seconds to minutes generally have a negligible effect on grid stability due to their limited magnitude, hourly variations are more significant, particularly in relation to load variations. These fluctuations often

[1]Faculty of Electrical Engineering, The University of Danang – University of Science and Technology, Vietnam
[2]Department of Energy, Politecnico di Milano, Campus Bovisa, Italy

necessitate additional balancing actions and increased system flexibility to manage mismatches between generation and demand.

Over longer time frames, seasonal and inter-annual variability in wind speeds and solar irradiance—shaped by diurnal cycles and broader climatic patterns—affect the adequacy and planning of power systems. While such long-term fluctuations are less critical for daily operational decisions, they play an essential role in long-term system planning [2].

Furthermore, wind and solar generation are subject to inherent uncertainty. Since wind speed and solar radiation are not fully predictable, their generation output at a future period cannot be determined with certainty. The use of forecasts, however, can help mitigate this uncertainty, enabling grid operators to efficiently make decisions regarding generator scheduling, reserve allocation, and preparation for extreme events characterized by unexpectedly high or low renewable generation levels [3].

The degree to which the variability and uncertainty of RES impact power system operations largely depends on their penetration level within the grid. At moderate levels of RES integration, typically less than 15%–20% of total electricity consumption, system operators can generally manage variability through existing flexibility resources. However, when the penetration exceeds 25%, operational complexities increase significantly. Under such circumstances, curtailment of renewable generation may become necessary during periods of low demand to maintain system stability and avoid issues such as frequency deviations, voltage fluctuations, reactive power imbalances, and transmission congestion [4,5].

As outlined in the Net Zero by 2050 roadmap, power supply variability is one of the potential vulnerabilities in the modern energy system. In this case, it is essential to prioritize investments in energy storage technologies, advanced digital solutions, and electricity networks. These solutions are critical to enhancing system flexibility and ensuring the stability and reliability of electricity supply [1]. Indeed, energy storage systems (ESSs) can balance the variability of renewable generation, thereby matching the supply and demand of electricity. Storage systems can be deployed both on the supply side and within transmission networks to improve flexibility in balancing supply and demand across the grids [6]. The potential for ESSs to effectively and efficiently support electric supply, particularly variable renewable generation, has been examined in many studies [7–11]. The cost of ESS technologies has rapidly declined in the last decade and is projected to decline another 50% over the next decade. Several utility-scale projects around the world, such as Hornsdale Power Reserve, a 150 MW (194 MWh) grid-connected ESS co-located with the Hornsdale Wind Farm in South Australia, and the 250 MW (250 MWh) Gateway Energy Storage project in California to help enhance grid reliability and reduce customer energy costs, have demonstrated that ESSs can provide a cost-effective source of flexibility and reliability services for power grids [12,13].

ESSs are capable of providing a wide range of grid support services, which can generally be categorized according to their operational time scales into power services (short-duration) and energy services (long-duration), as outlined in Table 4.1. Among the typical services provided by ESSs are the following functions, extensively documented in the literature [7,14–22]:

Table 4.1 Applications of energy storage [7]

Services	Applications	Charge/discharge duration
Power quality	Transient stability, Frequency regulation, Voltage control	Seconds to minutes
Bridging power	Spinning reserves	Minutes to ~1 hour
Energy management	Energy arbitrage, Load following, Firm capacity, Congestion relief, Upgrade deferral	Hours

Transient stability support: Contributes to maintaining the synchronous operation of the power system during sudden and severe disturbances by rapidly injecting or absorbing power to stabilize system dynamics.

Frequency regulation: Provides rapid response to short-term fluctuations in power balance, thereby supporting system frequency control and preserving operational stability.

Voltage regulation: Supplies or absorbs reactive power to maintain voltage levels within prescribed operational limits across the network.

Spinning reserve: Offers readily available online reserve capacity capable of responding to unexpected changes in generation or demand within a response time typically not exceeding 10 minutes.

Energy arbitrage: Operates by charging during periods of low electricity prices or system demand and discharging during peak periods, thereby enabling load shifting and peak demand reduction.

Load following (balancing): Adjusts power output in real time to track variations in system demand, ensuring continuous alignment between generation and consumption.

Firm capacity contribution: Delivers capacity to meet system peak demand requirements, thereby enhancing resource adequacy and supporting system reliability.

Congestion management: Alleviates transmission constraints by either reducing network flows on overloaded lines or providing alternative power flow routes within the grid to maintain system operational limits.

Infrastructure investment deferral: Reduces stress on transmission and distribution assets, enabling the postponement of costly system upgrades.

Generally, ESSs can be utilized as a solution for generation, transmission, or renewable energy integration issues. When deployed as generation-side assets, grid-scale ESSs are capable of delivering a range of bulk energy services, including firm capacity, energy arbitrage, load following, and various power quality support services such as contingency reserves, area control, frequency regulation, etc. Within transmission networks, strategically selecting locations and capacities of ESS installations can contribute to mitigating network congestion and postponing the need for transmission infrastructure expansion. The potential benefits of deferring such intensive investments are particularly advantageous, given that approximately 50% of the existing generation capacity is utilized only during a relatively small number of peak demand periods.

In power systems with significant renewable energy integration—commonly referred to as hybrid renewable systems—grid-scale ESSs could offer substantial value to mitigate operational challenges posed by the variability and non-dispatchability of renewable generation sources, including [7,14]:

Time-shifting: Energy storage devices can absorb excess energy during periods of high renewable generation, when wind and solar resources are abundant, and subsequently release the stored energy to the grids as needed. This capability helps make renewable generation output both predictable to grid operators and co-temporal to demand. This also helps effectively store the surplus power that may be thrown away to be used when necessary or sold to power markets when the price is high (energy arbitrage). Time shifting functions require ESSs to retain significant amounts of energy over extended duration, typically from hours to days. In this application, the round-trip efficiency of storage technologies is a vital consideration for economic viability, as lower efficiency systems would incur greater energy losses over the storage cycle, thereby diminishing operational and financial benefits.

Ancillary services: These play a vital role in maintaining reliability and operational stability of power systems, particularly in systems with high shares of variable renewable energy. Both power system dispatch and intraday electricity market transactions typically rely on stable generation profiles over short periods of time. However, the inherent short-term variability of renewable generation introduces operational challenges, affecting power quality. These fluctuations require output adjustments from conventional generating units, which results in increased ramping costs. In extreme cases, ramping capabilities of conventional thermal units may be insufficient to track the rapid fluctuations of variable renewable sources, thereby limiting their penetration levels in power systems. Grid-scale ESSs also provide value to renewable integration in short-term variability management. Through short-duration charging and discharging cycles, ESSs can provide essential ancillary services, including frequency regulation and load following, as renewable generation and system load fluctuate. Moreover, ESSs can also deliver fast-response capabilities, enabling them to quickly inject or absorb power to stabilize grids in the event of severe disturbances or emergency conditions. These functions not only enhance system resilience but also support higher levels of renewable integration by mitigating the operational constraints associated with their variability.

Transmission utilization efficiency: Transmission systems are increasingly challenged by the large-scale integration of RES, as existing transmission infrastructure was not originally designed to accommodate such high amounts of energy. Therefore, effective solutions are needed to alleviate network constraints while minimizing adverse impacts on both system reliability and capital and operational costs associated with system expansion and reinforcement. Grid infrastructure upgrades to mitigate congestion will require significant time and financial investment. In contrast, the deployment of grid-scale ESSs presents a flexible and expedient alternative. Storage systems can absorb surplus renewable generation during periods of network congestion and subsequently release the stored energy when transmission capacity becomes available, thereby enhancing transmission utilization efficiency and supporting higher levels of renewable energy integration without immediate large-scale grid expansion.

A fundamental consideration in the deployment of ESSs is identifying appropriate stakeholders with the incentive to invest in such assets. At first glance, renewable energy project owners might appear to be the primary investors, given the operational advantages ESSs can provide. However, in practice, grid operators might have stronger motivation to integrate storage systems, particularly in regulatory environments where system integration responsibilities rest with grid operators rather than generation owners. In some regulatory frameworks, for example, renewable developers are paid for their generation regardless of whether curtailment occurs due to transmission limitations or not. In this case, grid operators are incentivized to deploy ESSs to store and subsequently utilize the curtailed energy that they are paying for. Moreover, transmission and distribution utilities can derive additional value streams from ESS applications, including load shifting, infrastructure upgrade deferral, and the provision of ancillary services. These operational and economic benefits can offset the capital and operational costs associated with ESS deployment, whereas renewable developers may not directly benefit from these applications under current market structures. Given these considerations, ESS investment is generally more economically justified for utilities and grid operators than for individual renewable power producers. An alternative approach could involve establishing collaborative agreements between utilities and renewable developers during the early stages of project development. Such agreements could enable both parties to share operational and financial benefits associated with ESS integration [23].

In general, ESSs can be a promising solution to the challenges associated with high shares of renewable energy integration, as they enhance system flexibility and contribute to both economic efficiency and operational security in modern power systems. ESSs will play a pivotal role in facilitating the transition toward low-carbon and zero-emission energy systems with high renewable penetration. The concept of stationary energy storage is not new. Its application can be traced back to the early 20th century, when power stations were routinely shut down during nighttime hours, and lead-acid batteries were used to supply residual loads on direct current distribution networks. Recognizing the value of storage for enhancing operational flexibility, utility companies began investing in centralized storage facilities. The first large-scale installation, a pumped hydroelectric storage (PHS) plant, was commissioned in 1929. By 2005, more than 200 PHS facilities were deployed globally, providing over 100 GW of installed generation capacity. However, factors such as electricity market deregulation and environmental concerns have resulted in a decline in new PHS investments. At the same time, interest in alternative ESS technologies has evolved, driven by regulatory policy, power quality issues, environmental considerations, etc. Additionally, technological advancement and declining costs in emerging storage technologies have made the deployment of modern ESSs increasingly viable and economically attractive for modern power systems [24].

In this chapter, methodologies for the optimal planning and operational management of power systems incorporating ESSs are discussed. The chapter begins by providing an overview of energy storage applications within power systems, with particular emphasis on their roles in hybrid renewable energy systems. Next, an optimization formulation is presented for optimal operation of ESS in hybrid renewable systems, followed by approaches for optimal planning of ESSs in the systems. Finally, case studies and simulation results are discussed.

4.2 Mathematical formulation for ESS optimal operation

To investigate the operation of ESSs, this chapter formulates an AC Optimal Power Flow (OPF) model, wherein ESSs are utilized to perform time-shifting renewable energy generation. The OPF framework aims to determine the optimal operating point of a power system by minimizing or maximizing a selected objective function, subject to a set of power flow equations and operational constraints. These constraints typically include generator output limits, transmission capacity and stability requirements, voltage magnitude limits, and restrictions associated with mechanical switching devices.

There are many types of OPF formulation, different in their treatment of network constraints, objective functions, and solution techniques. Models that incorporate the full nonlinear AC power flow equations are referred to as AC OPF formulations. In contrast, simplified versions, known as DC OPF, assume constant voltage magnitudes and close to zero voltage angles, allowing for linearized power flow approximations. AC OPF problems can be further classified into full AC OPF, which concurrently optimizes both real and reactive power flows, and decoupled AC OPF, which solves real and reactive power optimization problems separately in an iterative manner to achieve a feasible operating point [25].

4.2.1 *The AC Optimal Power Flow model*

In general, an OPF problem can be formulated as follows [26]:

$$\begin{aligned} \text{Min} \quad & f(x,u) \\ \text{s.t.} \quad & g(x,u) = 0 \\ & h(x,u) \leq 0 \end{aligned} \tag{4.1}$$

The objective function $f(x, u)$ represents the optimization goal that might be minimization of generation cost, minimization of active power losses, minimization of reactive power losses, or maximization of social benefit, depending on specific applications. In this formulation, $g(x, u)$ and $h(x, u)$ represent the system's equality and inequality constraints, respectively. The vector x comprises state variables, including bus voltage magnitudes and angles, branch currents, generator reactive power outputs, and other system-dependent variables. Meanwhile, the vector u contains control variables such as real and reactive power generation, generator voltage setpoints, transformer tap positions, and other adjustable system parameters.

The mathematical nature of the OPF problem is determined by the specific forms of the objective function *f*, and the constraint sets *g* and *h*. Depending on their structure, the OPF problem may take the form of a linear, mixed-integer linear, nonlinear (potentially non-convex), or mixed-integer nonlinear programming (NLP) problem.

4.2.1.1 Objective function

In this AC OPF formulation, the objective is to minimize the total generation cost across all generating units. The cost function for each unit is typically represented as a quadratic function of its real power output, denoted as $C_i(P_{Gi})$. Accordingly, the overall *objective function* can be expressed as follows:

$$\text{Min} \sum_{t=1}^{T} \sum_{i=1}^{ng} \left[c_{2i} \left(P_{Gi}^{t} \right)^2 + c_{1i} P_{Gi}^{t} + c_{0i} \right] = Min \sum_{t=1}^{T} C\left(P_{Gi}^{t} \right) \tag{4.2}$$

where T represents the optimization time horizon, ng is the total number of generating units in the system; c_{0i}, c_{1i}, and c_{2i} are cost coefficients describing the quadratic cost curve of generating unit at bus i; and P_{Gi}^{t} denotes the real power output at bus i during time period t, which serves as a control variable within the OPF framework.

4.2.1.2 Equality constraints

The equality constraints in the AC OPF formulation primarily consist of the active and reactive power balance equations at each bus i for every time period t. The corresponding mathematical expressions are

$$P_i^t = P_{Gi}^t - P_{Li}^t = V_i^t \sum_{k=1}^{nb} V_k^t \left[G_{ik} \cos\left(\theta_i^t - \theta_k^t\right) + B_{ik} \sin\left(\theta_i^t - \theta_k^t\right) \right] + G_{ii} \left(V_i^t\right)^2 \tag{4.3}$$

$$Q_i^t = Q_{Gi}^t - Q_{Li}^t = V_i^t \sum_{k=1}^{nb} V_k^t \left[G_{ik} \sin\left(\theta_i^t - \theta_k^t\right) - B_{ik} \cos\left(\theta_i^t - \theta_k^t\right) \right] - B_{ii} \left(V_i^t\right)^2 \tag{4.4}$$

where

- nb is the total number of buses in the system.
- P_{Gi}^t and Q_{Gi}^t denote the real and reactive power generation at bus i during time period t.
- P_{Li}^t and Q_{Li}^t represent the real and reactive power demand at bus i in period t.
- V_i^t and V_k^t are the voltage magnitudes at buses i and k, respectively, at time t.
- θ_i^t and θ_k^t are the voltage angles at buses i and k, respectively, at time t.
- G_{ik} and B_{ik} correspond to the real and imaginary parts of the ik^{th} element of the bus admittance matrix.

4.2.1.3 Inequality constraints

Inequality constraints define the allowable operating limits of both control and state variables within the system. These include lower and upper bounds on parameters such as real and reactive power generation, voltage magnitudes at buses, line current magnitudes, and generator ramp-up and ramp-down limits.

Real and reactive power generation at each bus i during the time period t are constrained within specified operational limits, as expressed in (4.5) and (4.6):

$$P_{Gi}^{Min} \leq P_{Gi}^{t} \leq P_{Gi}^{Max} \tag{4.5}$$

$$Q_{Gi}^{Min} \leq Q_{Gi}^{t} \leq Q_{Gi}^{Max} \tag{4.6}$$

where P_{Gi}^{Min} and P_{Gi}^{Max} represent the lower and upper limits on real power generation at bus i, and Q_{Gi}^{Min} and Q_{Gi}^{Max} denote the corresponding limits for reactive power generation.

Voltage magnitudes at each bus must also be maintained within acceptable operating ranges, typically defined by strict regulatory and technical standards to prevent voltage instability. This is given by

$$V_{i}^{Min} \leq V_{i}^{t} \leq V_{i}^{Max} \tag{4.7}$$

where V_{i}^{Min} and V_{i}^{Max} are the minimum and maximum allowable voltage magnitudes at bus i.

Additionally, the *thermal limits of transmission lines* are enforced by constraining the square of the current magnitudes on each branch, which simplifies mathematical handling while effectively representing line limits. These constraints are expressed as

$$\left(I_{ij}^{t}\right)^{2} \leq \left(I_{ij}^{\max}\right)^{2} \tag{4.8}$$

$$\left(I_{ji}^{t}\right)^{2} \leq \left(I_{ji}^{\max}\right)^{2} \tag{4.9}$$

where I_{ij}^{t} and I_{ji}^{t} are the current magnitudes flowing from bus i to bus j and from bus j to bus i during period t, respectively, and $I_{ij}^{\max} = I_{ji}^{\max}$ denotes the maximum allowable current for the line.

Operational flexibility is further addressed by imposing *ramp rate limits* on generating units, which restrict the allowable changes in real power output between consecutive time intervals. These are expressed by:

$$P_{Gi}^{t} - P_{Gi}^{t-1} \leq R_{Ui} \tag{4.10}$$

$$P_{Gi}^{t-1} - P_{Gi}^{t} \leq R_{Di} \tag{4.11}$$

where R_{Ui} and R_{Di} are the ramp-up and ramp-down limits of the generating unit located at bus i.

This AC OPF model constitutes an NLP problem due to the inherent nonlinearity and non-convexity of both the objective function and system constraints. While the computational complexity of such problems can be significant, a key advantage of NLP-based OPF formulations lies in their ability to accurately represent the physical and operational characteristics of power systems.

4.2.2 Mathematical model of ESSs

An ESS is characterized by its rated energy capacity, rated charging and discharging power, and associated efficiencies. In this optimization model, the ESS operates to time-shift surplus renewable generation by charging during periods of excess energy—either beyond load demand or transmission capacity—and discharging when additional power is required.

The charging and discharging operations are modeled with two separate control variables: charging power P_{ch} and discharging power P_d, each associated with their respective efficiencies, η_{ch} and η_d. The *energy balance* of the storage system at each time interval is governed by the following equation:

$$B_i^t = B_i^{t-1} + \left(\eta_{chi} P_{chi}^t - P_{di}^t / \eta_{di}\right) \Delta t \tag{4.12}$$

where B_i^t and B_i^{t-1} are the ESS energy levels at bus i in periods t and $t-1$, respectively; P_{chi}^t and P_{di}^t denote real charging and discharging power; η_{chi} and η_{di} re the charging and discharging efficiencies; and Δt is the duration of each time interval.

To ensure operational consistency over the optimization horizon, the final energy level at the end of the scheduling period must return to its initial state, enforced by the energy continuity constraint:

$$B_i^{t=T} = B_i^0 \tag{4.13}$$

where T is the total number of time intervals considered, B_i^0 is the initial energy level of the ESS at bus i, $B_i^{t=T}$ is the final value at the end of the period.

Charging and discharging power limits are imposed as follows:

$$P_{chi}^{Min} \leq P_{chi}^t \leq P_{chi}^{Max} \tag{4.14}$$

$$P_{di}^{Min} \leq P_{di}^t \leq P_{di}^{Max} \tag{4.15}$$

Energy level limits for the ESS are enforced through:

$$B_i^{Min} \leq B_i^t \leq B_i^{Max} \tag{4.16}$$

These constraints ensure that the charging and discharging powers remain within the ESS's rated power capabilities, while the stored energy at each time step respects its available energy limits.

Ramp rate limits for both charging and discharging operations are also incorporated to reflect the dynamic capabilities of the storage system:

$$\left(P_{di}^t - P_{di}^{t-1}\right) - \left(P_{chi}^t - P_{chi}^{t-1}\right) \leq R_{U_{Si}} \tag{4.17}$$

$$\left(P_{di}^{t-1} - P_{di}^t\right) - \left(P_{chi}^{t-1} - P_{chi}^t\right) \leq R_{D_{Si}} \tag{4.18}$$

where $R_{U_{Si}}$ and $R_{D_{Si}}$ denote the ramp-up and ramp-down limits for the ESS at bus i.

Since ESSs are interfaced with the grid via power electronic converters, they can exchange both active and reactive power. The reactive power capability of the ESS does not affect its stored energy and is independently constrained by:

$$Q_{chi}^{Min} \leq Q_{chi}^t \leq Q_{chi}^{Max} \tag{4.19}$$

$$Q_{di}^{Min} \leq Q_{di}^t \leq Q_{di}^{Max} \tag{4.20}$$

where Q_{chi}^t and Q_{di}^t are the reactive power outputs during charging and discharging, respectively, and Q_{chi}^{Min}, Q_{chi}^{Max}, Q_{di}^{Min}, Q_{di}^{Max} denote their corresponding lower and upper limits.

4.2.3 Optimal operation model with storage integration

In conventional OPF formulations without energy storage, the optimization problem is static, with no temporal correlation between consecutive time intervals. Each period (e.g., hourly) is optimized independently based on the system conditions within that interval. However, the integration of ESSs introduces temporal interdependence into the OPF framework. Storage units can be strategically charged during periods of high renewable generation or low electricity prices and discharged during periods of low generation or high market prices. Moreover, the charging and discharging decisions in each period are constrained by the storage system's energy state in the preceding period. As a result, the problem becomes a multi-period optimization, where decisions are coupled across time.

The objective of this optimal operation model is to minimize the total generation cost of all dispatchable generating units and ESSs throughout the scheduling horizon. The objective function is expressed as

$$\text{Min} \sum_{t=1}^{T} \left\{ \sum_{i=1}^{ng} \left[c_{2i}(P_{Gi}^{t})^{2} + c_{1i}P_{Gi}^{t} + c_{0i} \right] + \sum_{j=1}^{ns} \left(c_{dj}P_{dj}^{t} - c_{chj}P_{chj}^{t} \right) \right\} \tag{4.21}$$

where T is the total number of time periods; ng and ns represent the total number of generating units and ESS units, respectively; c_{0i}, c_{1i}, c_{2i} are cost coefficients for generating unit i; and c_{dj} c_{chj} are discharging and charging cost coefficients for storage device j.

To prevent simultaneous charging and discharging of ESSs, a complementary condition is typically enforced. This condition can be managed by leveraging the Locational Marginal Price (LMP) λ_p. Under normal market conditions, if λ_{pi}^{t} is strictly positive, charging and discharging power values P_{chi}^{t} and P_{di}^{t} cannot be simultaneously non-zero. However, in cases where negative LMPs occur—resulting from negative bid offers or congested transmission constraints—an ESS may be incentivized to charge and discharge concurrently. To address this, appropriate fictitious penalty costs for charging and discharging are incorporated to discourage simultaneous operations [27].

The optimization model must satisfy both equality and inequality constraints in each time interval. The equality constraints include:

- *Active and reactive power balance* equations:

$$\begin{aligned} P_{Gi}^{t} - P_{Li}^{t} + P_{di}^{t} - P_{chi}^{t} &= V_i^t \sum_{k=1}^{nb} V_k^t [G_{ik} \cos(\theta_i^t - \theta_k^t) \\ &\quad + B_{ik} \sin(\theta_i^t - \theta_k^t)] + G_{ii}(V_i^t)^2 \end{aligned} \tag{4.22}$$

$$\begin{aligned} Q_{Gi}^{t} - Q_{Li}^{t} + Q_{di}^{t} - Q_{chi}^{t} &= V_i^t \sum_{k=1}^{nb} V_k^t [G_{ik} \sin(\theta_i^t - \theta_k^t) \\ &\quad - B_{ik} \cos(\theta_i^t - \theta_k^t)] - B_{ii}(V_i^t)^2 \end{aligned} \tag{4.23}$$

- *Energy balance and continuity constraints of the ESSs* as described previously in (4.12) and (4.13).

The *inequality constraints* include:

- *Network operational limits:* generation power, voltage magnitude, branch current, and ramp rate limits (4.5–4.11).
- *ESS operational constraints:* charging/discharging power, state of charge, ramp rate, and reactive power limits (4.14–4.20).

This multi-period AC OPF with ESS integration, thus, enables coordinated scheduling of generation and storage resources to minimize system costs while accommodating renewable generation variability and maintaining network reliability.

4.3 Approach for ESS planning in power systems with high renewable penetration

As the deployment of ESSs increases along with the growing integration of RES, determining optimal sizes and sites of these systems becomes essential for maintaining both operational security and economic efficiency. The objective is to allocate the minimum required storage capacity at strategically selected locations, where operational and economic benefits of ESS deployment can be maximized. However, the inherent variability and unpredictability of renewable resources introduce significant challenges for system planning, as conventional deterministic methods are inadequate for explicitly addressing the stochastic behavior of such energy sources. To address these limitations, this section presents a probabilistic planning framework for ESS integration based on a Genetic Algorithm (GA)-driven Probabilistic Optimal Power Flow (POPF) model. This hybrid approach combines the optimization capabilities of evolutionary algorithms with the probabilistic analysis of system uncertainties to simultaneously determine optimal locations and capacities of ESS installations within power systems with high renewable penetration.

Uncertainty is a fundamental consideration in power system planning, influencing decisions related to equipment failures, load demand, and renewable generation. These uncertainties can broadly be classified into two categories:

- *Configuration uncertainties*, which refer to the failure or unavailability of system components such as generators, transmission lines, transformers, or other equipment failures, typically modeled as contingencies.
- *Input uncertainties*, which arise from incomplete or imprecise knowledge of future system parameters, including load demands and renewable generation outputs.

To represent these uncertainties, probability distributions and scenario-based techniques are commonly employed. The choice of uncertainty representation is influenced by the objectives of the analysis, the extent of inherent system uncertainty, and the availability of information related to that uncertainty [28,29].

In the proposed ESS planning approach, input uncertainties are addressed, particularly those associated with load demand and variable renewable generation. This consideration is increasingly important as renewable resources, especially wind power, comprise a growing share of the generation mix. A major challenge in optimization under uncertainty is handling the large uncertainty space, which can make optimization models computationally intensive and difficult to solve. Therefore, appropriate techniques for modeling stochastic variables and efficient optimization algorithms are essential for achieving tractable and reliable planning solutions in power systems with a significant share of renewable generation.

4.3.1 The optimization framework

The optimal placement and sizing of ESSs in this study are performed through a two-stage optimization framework. In the initial stage, candidate locations for ESS installations and the expected values of their optimal capacities are identified using GA in conjunction with the OPF model described in Section 4.2.3. The primary objective in this phase is to minimize the combined ESS investment and total expected generation costs while simultaneously maximizing the expected aggregate output from wind generation and storage facilities. Following this, a probabilistic evaluation on the obtained ESS placement and sizing decisions is conducted in the second stage, providing additional insights for final planning decisions.

In this planning formulation, the objective is to minimize both total generation costs and ESS investment expenses, expressed as:

$$\text{Min}\left\{Z_{ESS}+\sum_{t=1}^{T}\left[\sum_{i=1}^{ng}\left[c_{2i}(P^t_{Gi})^2+c_{1i}P^t_{Gi}+c_{0i}\right]+\sum_{j=1}^{ns}\left(c_{dj}P^t_{dj}-c_{chj}P^t_{chj}\right)\right]\right\} \tag{4.24}$$

where Z_{ESS} represents the daily capital cost of ESSs, determined from the total investment cost Z_o as:

$$Z_{ESS}=I_{rec}Z_o \tag{4.25}$$

and

$$Z_o=\sum_{i=1}^{ns}\left(Z_B B_i^{\max}+Z_R R_i^{\max}\right) \tag{4.26}$$

In these expressions, Z_B [\$/MWh] and Z_R [\$/MW] denote the unit capital costs related to energy and power capacities, respectively, while I_{rec} is the daily capital recovery factor, calculated as [30]:

$$I_{rec}=\frac{r(1+r)^N}{(1+r)^N-1}\frac{1}{N_{day}} \tag{4.27}$$

where N is the ESS lifetime in years, r is the annual interest rate, and N_{day} is the number of days in a year.

The operational limits on ESS charging/discharging power and energy level are updated accordingly:

$$R_j^{Min}\le P^t_{chj}\le R_j^{Max} \tag{4.28}$$

$$R_j^{Min}\le P^t_{dj}\le R_j^{Max} \tag{4.29}$$

$$B_j^{Min}\le B_j^t\le B_j^{Max} \tag{4.30}$$

Additionally, total installed ESS capacity is constrained by upper limits, representing investment budget restrictions:

$$\sum_{j=1}^{ns} R_j^{\max} \leq R_s \tag{4.31}$$

$$\sum_{j=1}^{ns} B_j^{\max} \leq B_s \tag{4.32}$$

The complete planning framework is illustrated in the flowchart of Figure 4.1. In Step 1, optimal ESS locations and expected capacities are determined using the integrated GA and OPF model. GA identifies suitable sites and the number of ESS units to install, while the OPF determines their capacities. GAs are particularly effective in this context due to their ability to manage the non-convex, nonlinear, and discrete nature of ESS placement problems [31–35], where traditional gradient-based methods typically encounter significant challenges.

GA functions based on the principles of natural selection and evolutionary genetics. The algorithm begins with a population of randomly generated candidate

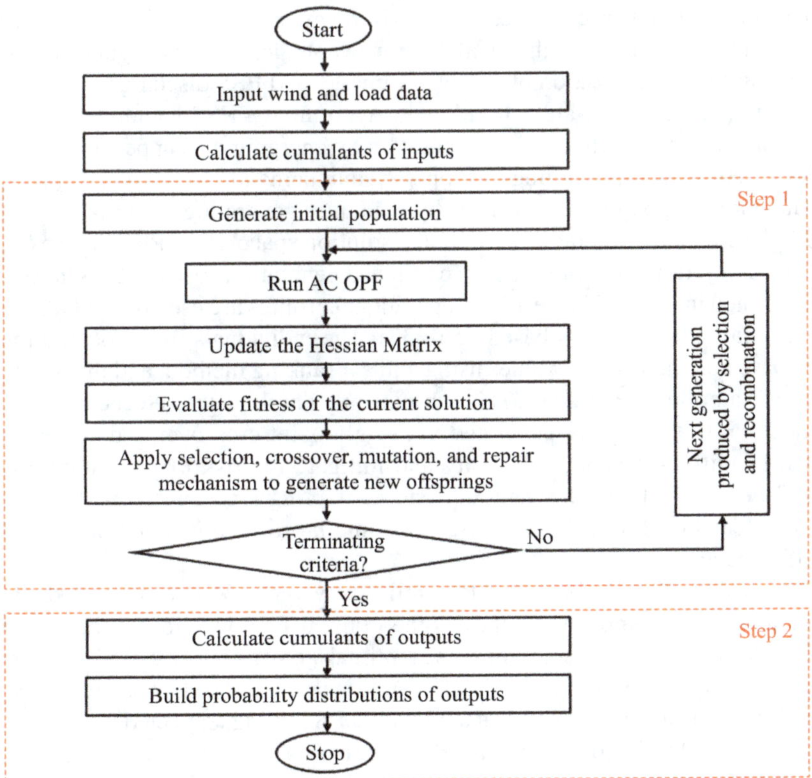

Figure 4.1 Flowchart for ESS planning approach

solutions, each referred to as a *chromosome*. In this context, each chromosome represents a specific configuration of candidate buses for ESS installation, encoded as a binary string. The quality of each chromosome is evaluated through a *fitness function*, which quantifies its effectiveness in contributing to an optimal system solution. Through iterative application of genetic operators such as selection, crossover, and mutation, the algorithm evolves successive generations of candidate solutions. Over multiple iterations, this evolutionary process progressively improves the population's overall fitness [36].

In this particular application, GA is utilized to determine the optimal placement of ESSs with the objective of maximizing the combined output from wind generation and ESSs. The fitness function applied to evaluate each candidate solution is formulated accordingly as:

$$\text{Fitness} = -\frac{\sum_{t=1}^{T}\left[\sum_{i=1}^{nb} P_{Li}^{t} - \left(\sum_{i=1}^{nRG} P_{j}^{t} + \sum_{k=1}^{ns} P_{dk}^{t}\right)\right]}{\sum_{t=1}^{T}\sum_{j=1}^{nRG} P_{RGj}^{t}} \tag{4.33}$$

where P_{Li}^{t} represents the load demand at bus i in period t; P_{j}^{t} is the utilized renewable generation at bus j; P_{dk}^{t} is the discharging power of ESS at bus k; and P_{RGj}^{t} is the available renewable generation at bus j. Here, nb, ns, and nRG denote the number of system buses, ESSs to be installed, and renewable power plants, respectively.

This fitness function evaluates the ratio of renewable and ESS energy utilized to meet system load relative to the available renewable generation. Higher fitness values indicate greater utilization of renewable energy and ESS discharge to serve load, resulting in reduced renewable curtailment. Accordingly, ESS locations determined through GA ensure effective storage operation by charging during periods of surplus wind generation and discharging to supply system demand.

The optimization process initiates by randomly generating an initial population of ESS placement configurations within the solution space. An OPF is then executed for each configuration to minimize the combined generation cost and ESS investment cost as defined in (4.24). The resulting operational profiles are used by the GA to evaluate each candidate's fitness based on (4.33). The individuals in the population are ranked based on their fitness values using a linear ranking method, and the Stochastic Universal Sampling technique [37] is subsequently employed to select candidates for reproduction. Offspring are generated via single-point crossover and subsequently mutated to introduce genetic diversity and mitigate the risk of premature convergence. To maintain feasibility, any infeasible offspring are replaced with randomly generated feasible chromosomes, ensuring compliance with installation constraints for ESS capacity.

This evolutionary cycle is repeated until a convergence criterion is satisfied. The best-performing chromosome identifies the optimal ESS locations, while the OPF simultaneously yields the corresponding optimal power capacity $R^{\max}$ and energy capacity $B^{\max}$, along with their optimal operational schedules. Consequently, Step 1 of the planning framework produces the optimal ESS siting, expected capacities, and operational profiles for given wind and load scenarios.

Subsequently, a probabilistic OPF is conducted to evaluate the operational risks associated with renewable and load uncertainties, particularly the probability of being unable to store surplus renewable energy. The probabilistic formulation and assessment procedure are presented in Section 4.3.2.

4.3.2 Probabilistic OPF

The probabilistic optimization framework employed in this planning approach is founded on the cumulant method. This cumulant-based POPF technique characterizes the statistical behavior of random variables when linearly combined around the operating point obtained from a deterministic OPF solution. The subsequent discussion outlines the formulation of output random variables as linear combinations of input random variables, as well as the procedure for calculating cumulants and deriving the probability distributions of the desired output variables, as detailed in [38–40].

The analysis begins by solving a deterministic OPF problem, wherein the relationship between the vectors of input and output variables is generally expressed as:

$$z = h(x) \tag{4.34}$$

At the optimal operating point, the Karush–Kuhn–Tucker (KKT) conditions must be satisfied. By incorporating the problem variables into the first-order KKT conditions, the following system of nonlinear equations is obtained:

$$F(z, x) = 0 \tag{4.35}$$

Here, $F(.)$ represents the set of equations defining the first-order optimality conditions. To establish the relationship between output and input variables, the full differential of (4.35) is taken, yielding:

$$\boldsymbol{H}_{\mathscr{L}}\Delta z + \Delta x = 0 \tag{4.36}$$

where Δz and Δx denote the incremental changes in the vectors of output and input variables, respectively, and $H_{\mathscr{L}}$ is the Hessian matrix of the Lagrangian function with respect to the output variables z.

Equation (4.36) can be rearranged to express the change in output variables as a function of the changes in input variables:

$$\Delta z = -\boldsymbol{H}_{\mathscr{L}}^{-1}\Delta x \tag{4.37}$$

When input variables of an OPF problem exhibit uncertainty, the OPF is reformulated as a POPF problem. In this case, each element $x_l(l = 1,\ 2,\ \ldots,\ N)$ of the input vector x in (4.34) is treated as a realization of a corresponding random variable $\widetilde{X}_l$. Based on the relationship in (4.37), each element z_l of the output vector z is similarly interpreted as a realization of a random variable $\widetilde{Z}_l$. Accordingly, the relationship between the vectors of output and input random variables can be written as:

$$\begin{bmatrix} \widetilde{Z}_1 \\ \widetilde{Z}_2 \\ \cdots \\ \widetilde{Z}_N \end{bmatrix} = \boldsymbol{M} \begin{bmatrix} \widetilde{X}_1 \\ \widetilde{X}_2 \\ \cdots \\ \widetilde{X}_N \end{bmatrix} \tag{4.38}$$

where $\boldsymbol{M} = -\boldsymbol{H}_{\mathscr{L}}^{-1}$ is the sensitivity matrix obtained from the inverse of the Hessian of the Lagrangian.

The matrix $\boldsymbol{M}$ can be expanded as follows:

$$M = \begin{bmatrix} a_{1,1} & a_{1,2} & \cdots & a_{1,N} \\ a_{2,1} & a_{2,2} & \cdots & a_{2,N} \\ \vdots & \vdots & \ddots & \vdots \\ a_{N,1} & a_{N,2} & \cdots & a_{N,N} \end{bmatrix} \tag{4.39}$$

Subsequently, each output random variable $\widetilde{Z}_l$ can be expressed as a linear combination of the input random variables $\widetilde{X}_1,\ \widetilde{X}_2,\ \ldots,\ \widetilde{X}_N$:

$$\begin{aligned} \widetilde{Z}_1 &= a_{1,1}\widetilde{X}_1 + a_{1,2}\widetilde{X}_2 + \ldots + a_{1,N}\widetilde{X}_N \\ \widetilde{Z}_2 &= a_{2,1}\widetilde{X}_1 + a_{2,2}\widetilde{X}_2 + \ldots + a_{2,N}\widetilde{X}_N \\ &\vdots \\ \widetilde{Z}_N &= a_{N,1}\widetilde{X}_1 + a_{N,2}\widetilde{X}_2 + \ldots + a_{N,N}\widetilde{X}_N \end{aligned} \tag{4.40}$$

Assuming the input random variables $\widetilde{X}_l$ are independent, the rth-order cumulant of each output random variable $\widetilde{Z}_l$ can be computed as:

$$k_{\widetilde{Z}_l^r} = a_{l,1}^r k_{\widetilde{X}_1^r} + a_{l,2}^r k_{\widetilde{X}_2^r} + \ldots + a_{l,N}^r k_{\widetilde{X}_N^r} \tag{4.41}$$

To construct the probability distribution of output random variables, it is assumed that each $\widetilde{Z}_i$ can be decomposed into continuous and discrete components:

$$\widetilde{Z}_i = \widetilde{Z}_{i_c} + \widetilde{Z}_{i_d} \tag{4.42}$$

Here, $\widetilde{Z}_{i_c}$ represents the continuous part formed by continuous distributions of the input random variables, while $\widetilde{Z}_{i_d}$ represents the discrete part derived from discrete distributions.

The continuous part is approximated using a Gram–Charlier series expansion [41], and the combined probability density function (PDF) and cumulative distribution function (CDF) of $\widetilde{Z}_i$ are expressed as

$$f_{\widetilde{Z}_i}(x) = \sum_{i=1}^{\nu} p_i \left[\phi\left(x_{Ni}\right) + \phi\left(x_{Ni}\right) \sum_{i=1}^{\infty} c_l H_l\left(x_{Ni}\right) \right] \tag{4.43}$$

$$F_{\widetilde{Z}_i}(x) = \sum_{i=1}^{\nu} p_i \left[\Phi\left(x_{Ni}\right) + \phi\left(x_{Ni}\right) \sum_{i=1}^{\infty} c_l H_{l-1}\left(x_{Ni}\right) \right] \tag{4.44}$$

In these expressions, ν represents the number of impulses considered; $x_{Ni} = (x - x_i - m_{\widetilde{Z}_{ic}})/\sigma_{\widetilde{Z}_{ic}}$; x_i and p_i are the abscissas and corresponding probabilities of the impulses for the discrete component $\widetilde{Z}_{id}$; and $m_{\widetilde{Z}_{ic}}$ and $\sigma_{\widetilde{Z}_{ic}}$ are the mean and standard deviation of $\widetilde{Z}_{ic}$. The coefficients c_l are computed from the cumulants of $\widetilde{Z}_{ic}$, and $H_l(.)$ denotes the lth-order Hermite polynomial, with computational details available in Reference [42]. Finally, $\phi(x_{Ni})$ and $\Phi(x_{Ni})$ represent the PDF and CDF of standard normal distribution, respectively, as given below:

$$\phi\left(x_{Ni}\right) = \frac{1}{2\pi}e^{\frac{-x^2}{2}} \tag{4.45}$$

$$\Phi\left(x_{Ni}\right) = \int_{-\infty}^{x_{Ni}} \frac{1}{2\pi}e^{\frac{-t^2}{2}}\,dt \tag{4.46}$$

4.4 Case studies

To demonstrate the effectiveness of the proposed ESS operation and planning approaches, modified IEEE test cases with integrated large-scale wind plants are employed, representing power systems with substantial renewable energy integration.

4.4.1 Performance assessment of the ESS optimal operation model

To evaluate the performance of ESSs in power grids with significant renewable energy integration, a case study is carried out using the OPF model described in Section 4.2.3. The test system is based on a modified IEEE 14-bus network (Figure 4.2), where a 400 MW wind farm is connected to bus 2. At this location, an ESS is installed alongside the wind plant to enable energy time-shifting. The system's daily electricity demand is met by conventional generating units and wind generation. Technical parameters of the storage device are provided in Table 4.2, where $B^{\max}$ represents the nominal energy capacity, and $R^{\max}$ indicates the rated power capacity, with

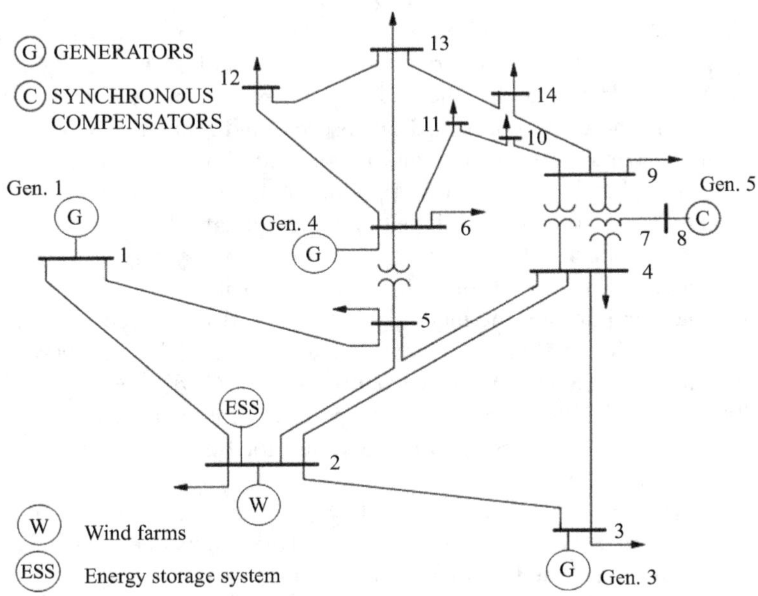

Figure 4.2 IEEE 14–bus test system [31]

Table 4.2 Technical parameters of the ESS

B_0 [MWh]	R^{max} [MW]	B^{max} [MWh]	η_{ch}	η_d
0	100	200	0.85	0.85

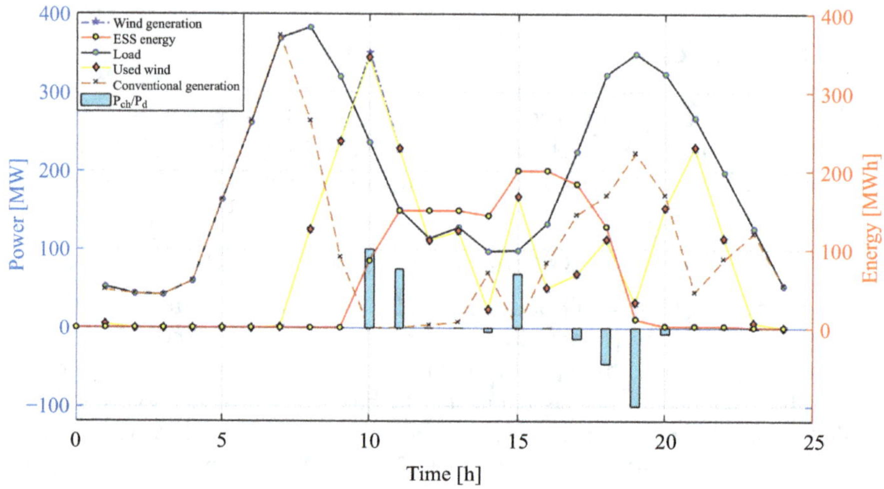

Figure 4.3 Daily optimal operational schedule of the ESS

$P_{ch}^{max} = P_d^{max} = R^{max}$. Using the proposed model, an optimal operational schedule for the storage system over a 24-hour period is generated, as illustrated in Figure 4.3.

As can be seen in Figure 4.3, the ESS is efficiently employed by the model in time-shifting wind energy to supply loads. It absorbs wind generation at periods when there is available wind while the demand is low (e.g., hours 10, 11, and 15). The stored energy will then be released to supply the high demand during peak periods (e.g., hours 17–20). Also noted in this optimal operational schedule is that the ESS operates within its capacities, both power and energy capacity, with greater amounts of power supplied during periods of higher system demand.

For comparison purpose, the model is further tested with three cases: No ESS installed, an ESS with smaller capacity installed ($P_{ch}^{max} = P_d^{max} = 50$ MW, $B^{max} =$ 100 MWh), and an ESS with larger capacity installed ($P_{ch}^{max} = P_d^{max} = 150$ MW, $B^{max} = 300$ MWh). Daily operation of the system without ESS can be seen in Figure 4.4. In Table 4.3, total generation cost and total amount of curtailed wind in different cases are also shown.

Figure 4.4 shows that without the use of ESS, a large amount of wind energy is curtailed (the difference between the blue curve and the yellow one), which is a waste of RES and a loss of revenue for wind farm operators.

As seen in Table 4.3, total system generation cost is significantly reduced with the installation of storage devices. This means its combined operation with the wind farm

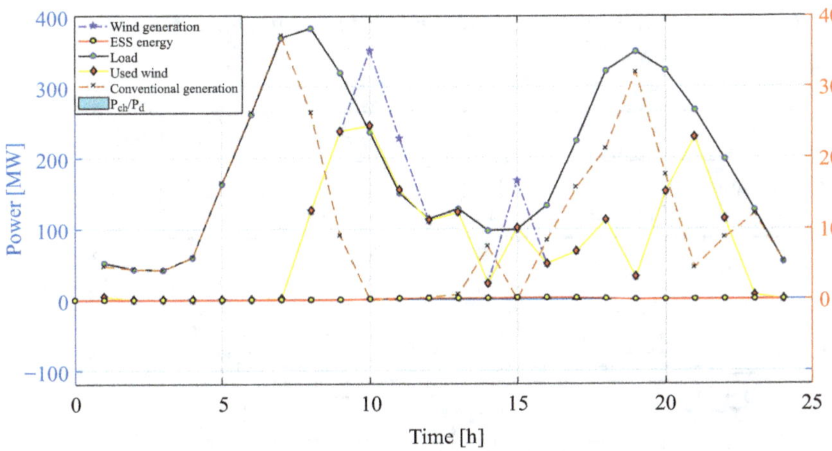

Figure 4.4 Daily optimal operation of the system without ESS

Table 4.3 Generation cost and wind curtailment with different ESS capacities

	No ESS	B^{max}= 100 [MWh] R^{max}= 50 [MW]	B^{max}= 200 [MWh] R^{max}= 100 [MW]	B^{max}= 300 [MWh] R^{max}= 150 [MW]
Generation cost [$]	1.9564e + 05	1.8752e + 05	1.8232e + 05	1.8187e + 05
Curtailed wind [MWh]	246.0474	99.0469	6.7112	0

effectively helps lower system's total generation cost. Larger ESS employed results in lower generation cost. Similarly, installing a larger ESS noticeably decreases the amount of wind energy that has to be curtailed. In this case, a storage system with a capacity of 300 MWh and 150 MW can help eliminate all wind curtailment.

A test with different storage efficiencies is also carried out to study the impact of efficiency on the time-shifting application of storage devices. The resulting system generation costs and corresponding wind curtailment levels are summarized in Table 4.4. Basically, the efficiency of the ESS has little impact on the amount of curtailed wind energy but significantly influences the system's generation cost. Therefore, efficiency remains a critical factor when selecting storage devices for time-shifting applications.

4.4.2 Test on the developed planning approach for ESSs

4.4.2.1 The modified IEEE test system

The proposed model is applicable to power systems incorporating various types of RES. However, due to the greater variability and uncertainty associated with wind generation compared to solar, the method is, without loss of generality, demonstrated

Table 4.4 Generation cost and wind curtailment with different ESS efficiencies

	$\eta = 0.85$	$\eta = 0.75$	$\eta = 0.95$
Generation cost [$]	1.8232e+05	1.8519e+05	1.7898e+05
Curtailed wind [MWh]	6.7112	6.7104	6.7112

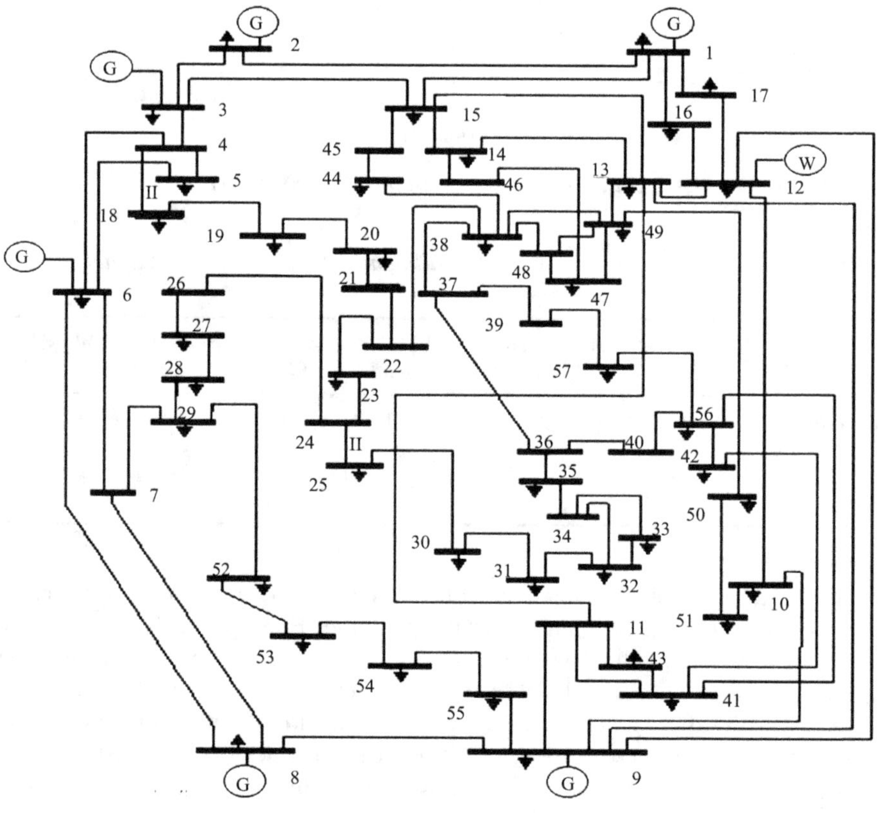

Figure 4.5 IEEE 57-bus test system [31]

on a system with a high level of wind integration. The combined GA-POPF planning framework is tested using a modified IEEE 57-bus network, as illustrated in Figure 4.5.

In this system, electricity demand is met by a combination of conventional generators and wind power. Six conventional generating units are located at buses 1, 2, 3, 6, 8, and 9. A large wind farm with an installed capacity of 450 MW is connected to bus 12, alongside several smaller wind farms situated at nearby buses, bringing the total wind generation capacity to 750 MW. This results in a wind penetration level

Table 4.5 Parameters related to the CAES

Parameter	Description	Value
Z_B [$/kWh]	Per-unit cost of rated energy	5
Z_R [$/kW]	Per-unit cost of rated power	700
N [years]	Lifetime of the system	30
η	Charing/discharging efficiency	0.85
r [%]	Annual interest rate	10

of 46.3%. To enhance the economic value of wind generation and minimize curtailment, ESSs are deployed to shift excess wind power from periods of low demand (and low electricity prices) to periods of high demand (and higher electricity prices). The storage capacities are determined through the OPF solution, with the objective of minimizing total system generation cost while maximizing the use of wind energy. The ESSs considered are based on Compressed Air Energy Storage (CAES) technology, with equal charging and discharging efficiencies, denoted by η. Detailed parameters for the CAES are listed in Table 4.5.

4.4.2.2 Input random variables

In this planning approach, the model treats wind generation and load demand as variable inputs. A long-term probability distribution for wind power is derived from an extended dataset of hourly wind generation over three years. To reflect seasonal variability, the wind data is divided into four seasonal groups, as illustrated in Figure 4.6 [43]. For each season, daily wind profiles are categorized into distinct groups using the K-means clustering algorithm. These clusters are then used to construct a discrete probability distribution for each hour of the day. Specifically, the daily wind profiles assigned to each cluster form individual impulses of discrete distribution, and the probability associated with each impulse is calculated based on the proportion of daily profiles falling within that cluster relative to the total number of daily profiles. By following this procedure for each season, a set of hourly discrete distributions for wind power is obtained. As an example, Figure 4.7 shows the discrete probability distribution of wind generation at bus 12 for hour 16 during Spring, represented by 20 discrete impulses corresponding to 20 clusters.

Figure 4.8 illustrates the daily load profiles across four different seasons [43]. In most existing studies, load is typically represented by a Gaussian distribution for simplicity. However, in this study, to demonstrate the capability of the proposed probabilistic OPF to handle various types of input uncertainties, both Beta and Gaussian distributions are employed for load modeling. Loads at buses 15, 38, 44, 50, and 56 are modeled using Beta distributions, with parameters determined from their expected values and standard deviations, set respectively at 10%, 12%, 9%, 8%, and 11% of those expected values [44]. Meanwhile, the remaining load buses follow a normal distribution, each with a standard deviation of 10% of its expected value. The expected values of total system load and wind power generation for all seasons are shown in Figure 4.9.

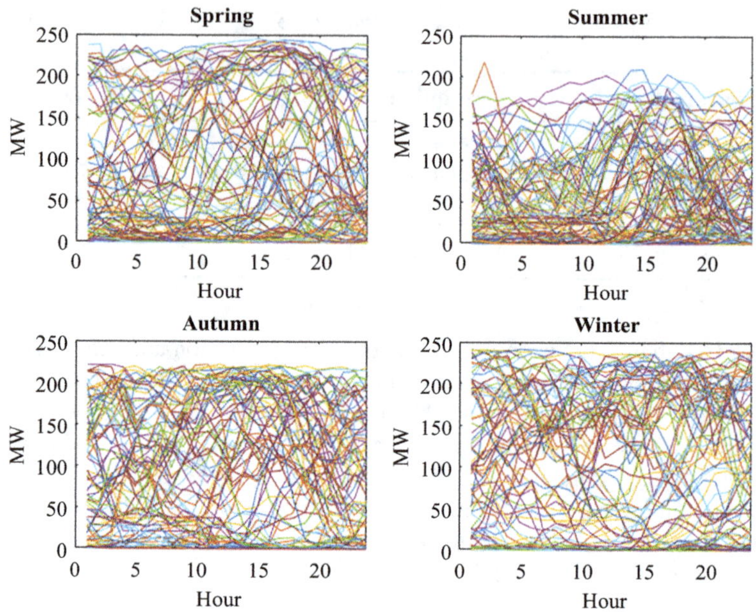

Figure 4.6 Daily wind profiles in a year

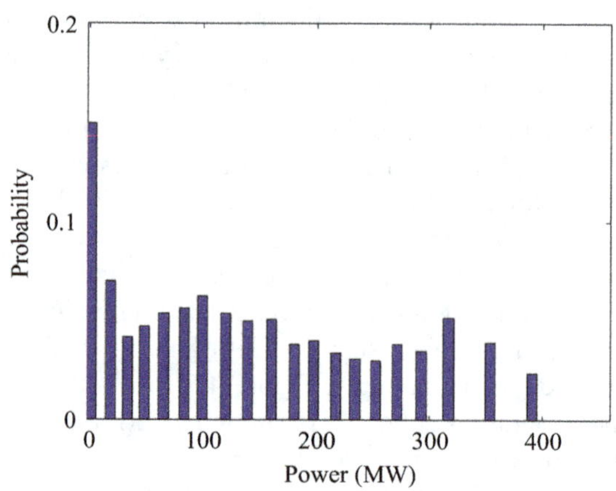

Figure 4.7 Distribution of wind power output at hour 16 in Spring

4.4.2.3 Tests and results

Tests based on the GA-Probabilistic OPF model, presented in Section 4.3.1, are conducted to determine the most cost-effective locations and capacities for ESS installation while minimizing both investment and system generation costs.

The proposed approach is initially implemented using the expected values of the wind and load datasets. The outcomes from the combined GA and deterministic OPF in Step 1 are presented in Table 4.6.

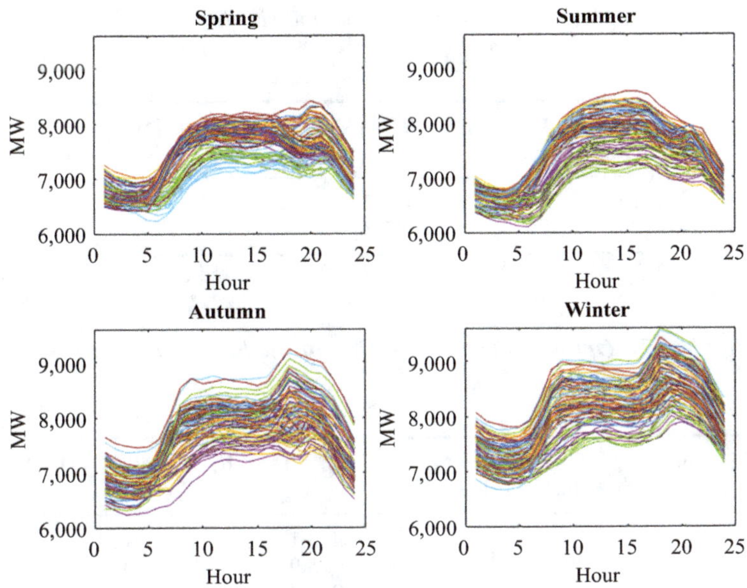

Figure 4.8 Daily load profiles in a year

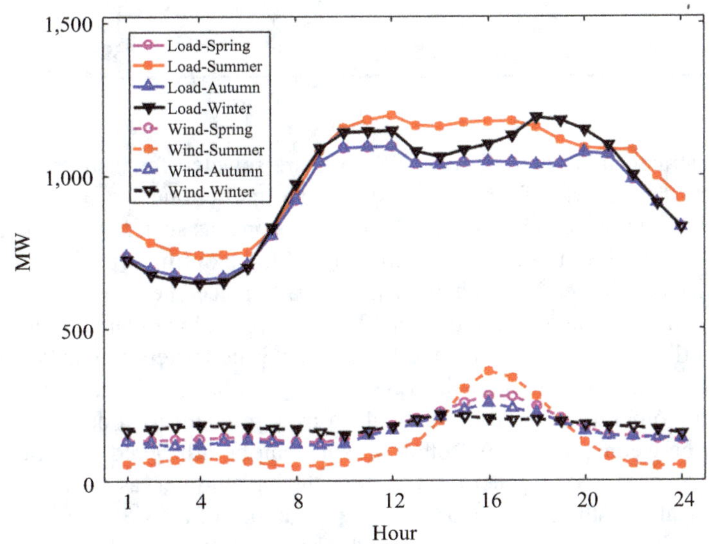

Figure 4.9 Expected wind power output and load by season

To account for seasonal variations in wind generation and load demand, the optimization is also performed separately for each season. The simulation results are provided in Table 4.7, providing the optimal locations of ESSs along with their respective power and energy capacities for each season.

Table 4.6 Optimal location and size for the ESSs

Optimal location (Bus number)	B^{max} [MWh]	R^{max} [MW]
12	495.2	95.6
20	172.2	31.0

Table 4.7 Optimal location and size for the ESSs in tests with four-season data

Season	Optimal location (bus number)	B^{max} [MWh]	R^{max} [MW]
Spring	12	483.1	105.4
	29	155.9	32.5
Summer	12	296.4	67.9
	38	49.5	12.0
Autumn	12	468.9	98.2
	29	157.2	31.1
Winter	12	758.2	144.2
	29	211.9	39.1

As illustrated in Table 4.6, two ESS units are selected for installation, a larger one positioned at the wind farm bus (bus 12) and a smaller one at the load center (bus 29). Similar decisions can be identified for all seasons, as summarized in Table 4.7. Generally, the configuration in Table 4.6 is found to be optimal for three out of four seasons when considered individually, with one ESS placed near the wind generation site and the other located at a major load bus to support local demand and enhance renewable utilization. Specifically, the largest ESS is consistently installed at bus 12 across all seasons, while the smaller ESS is placed at bus 38 during summer and at bus 29 in the other seasons, both recognized as significant load centers. This shift in location between summer and other seasons can be attributed to noticeable differences in load and wind generation profiles during summer, as depicted in Figure 4.9. Additionally, both the power and energy capacities of the ESSs reach their maximum values in winter, corresponding to higher wind availability during off-peak hours. Conversely, during summer, reduced wind generation during off-peak periods leads to lower charging and discharging capacities for the storage systems.

Figure 4.10 presents the optimal operational schedule for wind generation and the ESS at bus 12 during the spring season (corresponding to the Spring scenario in Table 4.7). In this schedule, the ESS is effectively utilized to shift wind power by storing surplus energy during off-peak hours (from hour 1 to 7) and discharging it during peak periods: from hour 9 to 12 and from hour 20 to 22. Notably, all available wind

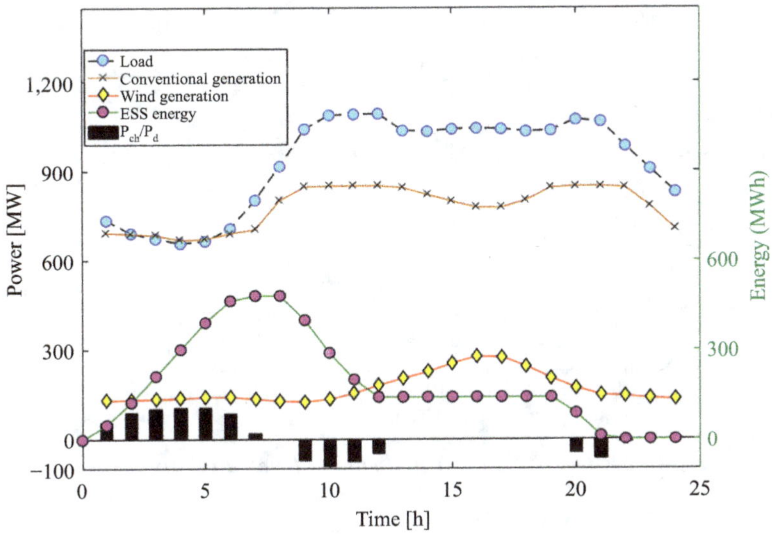

Figure 4.10 The ESS optimal dispatch schedule during Spring

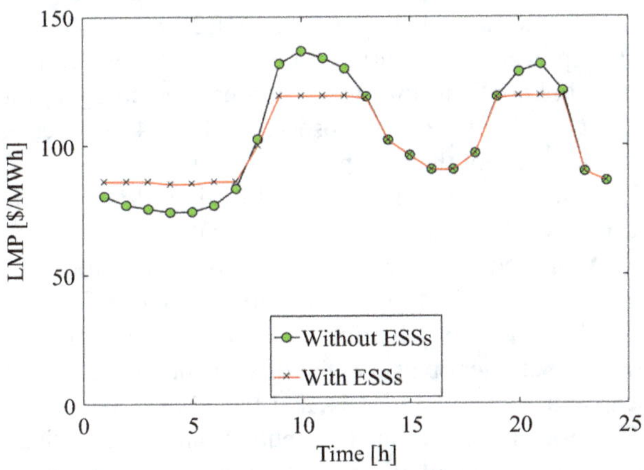

Figure 4.11 Hourly locational marginal prices at bus 12

generation is dispatched without curtailment, which results in a substantial decrease in the need for conventional generation during these high-demand hours.

This operational behavior of the ESS aligns closely with the system's hourly LMP fluctuations, as depicted in Figure 4.11. The ESS charges during lower price periods and discharges when prices rise. Additionally, within the peak periods, the discharge rate corresponds to the price at each hour. For instance, more energy is released at hour 10 when the price is higher, and less at hour 12 when the price declines. The same pattern occurs in the evening peak, with greater discharge at hour 21 than at hour 20, reflecting higher market prices.

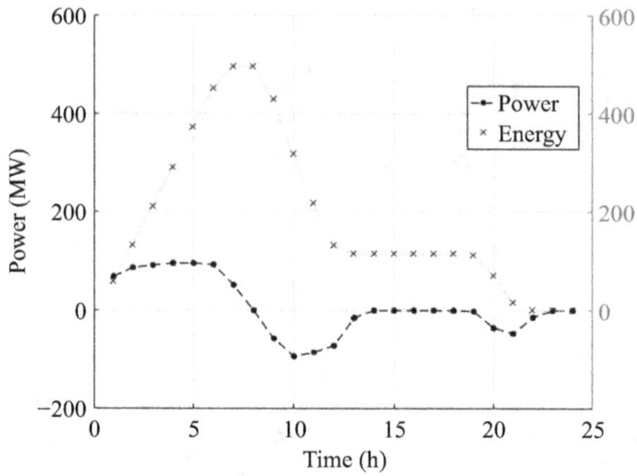

Figure 4.12 The ESS power and energy levels in daily operation

The procedure then proceeds to Step 2, where a probabilistic assessment is applied to the ESS capacities determined in Table 4.6, as detailed in Section 4.3.1.

Figure 4.12 presents the expected hourly power and energy values of the ESS indicated in Table 4.6. As shown in this figure, the ESS reaches its maximum charging and discharging power, $R^{\max}$, during hours 4 and 5, and attains its maximum stored energy, $B^{\max}$, at hour 7, following six consecutive charging intervals. Due to the highest variability in wind and load observed at hour 4, this period is chosen for the probabilistic assessment of the ESS power capacity.

Figure 4.13 illustrates results of the probabilistic OPF, presenting the CDF of the ESS power capacity. The results indicate that, under certain operating scenarios, the capacity value proposed in Table 4.6 may not be sufficient. For instance, if the decision-maker opts to install a 143 MW capacity, it would effectively cover almost the entire range of the probability distribution. However, selecting a lower capacity increases the risk of insufficient power capability to cope with system uncertainties. As an example, capacities of 122, 116, 104, and 102 MW would correspond to coverage of 99%, 95%, 85%, and 75% of the distribution range, respectively. This implies that accepting a 5% risk would allow a reduction in ESS power capacity by approximately 18.9%. Likewise, permitting 15% and 25% levels of risk would result in capacity reductions of around 27.3% and 28.7%, respectively.

The energy capacity of the ESS is evaluated probabilistically, as illustrated in Figure 4.14. From the figure, it can be observed that selecting capacities of 708, 627, 592, and 572 MWh would cover approximately 99%, 95%, 85%, and 75% of the entire probability distribution, respectively. This implies that accepting risk levels of 5%, 15%, and 25% would correspond to reductions in required energy capacity by around 20.4%, 24.9%, and 27.4%, respectively. For example, if a decision maker aims to accommodate 85% of possible wind and load fluctuations, the ESS at bus 12 should be sized at 104 MW for power capacity and 592 MWh for energy capacity.

The simulation results indicate that in order to fully accommodate all possible variations in system uncertainties, the ESS power and energy capacities would need

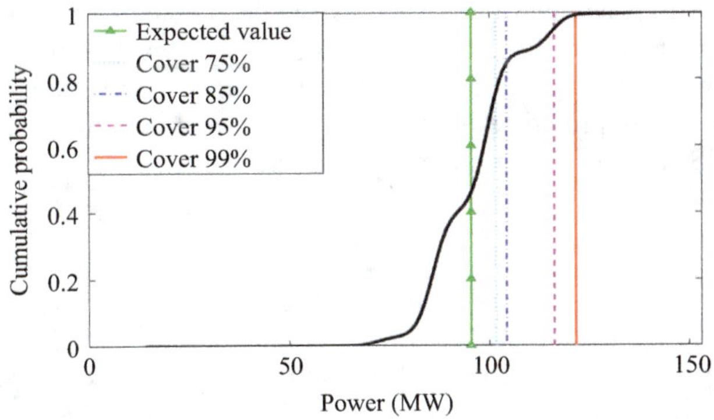

Figure 4.13 CDF representation of the ESS power capacity

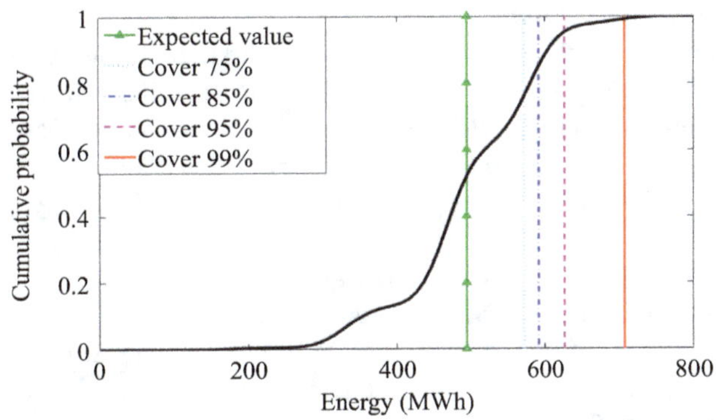

Figure 4.14 CDF representation of the ESS energy capacity

to be significantly larger than their expected values. However, when a certain degree of operational risk is considered acceptable, the required capacities can be substantially reduced. The proposed model is capable of quantifying this trade-off, providing decision makers with flexibility to balance system reliability against investment cost by setting acceptable risk thresholds.

4.5 Conclusions

This chapter describes an optimization framework for the optimal operation of ESSs in power grids with high levels of renewable energy integration. Within this framework, ESSs are utilized to time-shift renewable generation, thereby enhancing the economic value of renewable resources and mitigating curtailment. The model determines optimal operational schedules for ESSs. Test results demonstrate that co-locating storage systems with renewable power plants can effectively improve renewable generation.

Additionally, a planning methodology for ESS placement and sizing is introduced. This method integrates a GA with a POPF in a two-step process to determine the optimal locations and capacities of ESSs while accounting for uncertainties in load demand and renewable generation. The combined GA–POPF approach generates both the optimal siting decisions and the probability distributions of ESS power and energy capacities, providing a valuable decision-making reference for selecting appropriate ESS sizes under varying risk considerations.

References

[1] IEA. *Net Zero by 2050*. 2021.

[2] The European Wind Energy Association (EWEA). *Large Scale Integration of Wind Energy in the European Power Supply: Analysis, Issues and Recommendations*. 2005.

[3] Bird L., Milligan M., and Lew D. *Integrating Variable Renewable Energy: Challenges and Solutions*. 2013.

[4] Mears D., Gotschall H., Key T., and Kamath H. Technology Insights (Firm), EPRI PEAC Corporation, Electric Power Research Institute, United State, Department of Energy, *EPRI-DOE Handbook Supplement of Energy Storage for Grid Connected Wind Generation Applications*. 2004.

[5] DG ENER Working paper. *The Future Roles and Challenges of Energy Storage*.

[6] Hurst K., Springer M., Wikoff H., *et al. Industrial energy storage review*. National Renewable Energy Laboratory, 2024.

[7] Denholm P., Ela E., Kirby B., and Milligan M. The *Role of Energy Storage with Renewable Electricity Generation*. National Renewable Energy Laboratory, 2010.

[8] Zame K. K., Brehm C. A., Nitica A. T., Richard C.L., and Schweitzer III G. D. 'Smart Grid and Energy Storage: Policy Recommendations'. *Renewable and Sustainable Energy Reviews,* 2018, vol. 82, pp. 1646–1654.

[9] Denholm P., Jorgenson J., Hummon M., and Palchak D. *The Impact of Wind and Solar on the Value of Energy Storage*. National Renewable Energy Laboratory, 2013.

[10] Chernyakhovskiy I., Joshi M., Palchak D., and Rose A. *Energy Storage in South Asia: Understanding the Role of Grid Connected Energy Storage in South Asia's Power Sector Transformation*. National Renewable Energy Laboratory, 2021.

[11] Zhao C., Andersen P., Træholt C., and Hashemi S. 'Grid-connected battery energy storage system: a review on application and integration'. *Renewable and Sustainable Energy Reviews*, 2023, vol. 182, **113400**, pp. 1–19.

[12] LS Power. LS power energizes largest battery storage project in the world, The 250 MW Gateway Project in California. *LS Power* (blog), 2020.

[13] AEMO. *Initial Operation of the Hornsdale Power Reserve Battery Energy Storage System, South Australia*. 2018.

[14] Castillo A. and Gayme D. F. 'Grid-scale energy storage applications in renewable energy integration: A survey'. *Energy Conversion and Management,* 2014, vol. 87, pp. 885–894.

[15] Walawalkar R., Apt J., and Mancini R. 'Economics of electric energy storage for energy arbitrage and regulation in New York'. *Energy Policy,* 2007, vol. 35, pp. 2558–2568.

[16] Kintner-Meyer M., Balducci P., Colella W., Elizondo M., Jin C., Nguyen T., *et al. National assessment of Energy Storage for Grid Balancing and Arbitrage: Phase 1*, WECC. Technical Report, U.S. Department of Energy, 2012.

[17] Gayme D. F. and Topcu U. 'Optimal power flow with large-scale storage integration'. *IEEE Trans Power System,* 2013, vol. 28, pp. 709–717.

[18] Su H. I. and El Gamal A. 'Modeling and analysis of the role of fast-response energy storage in the smart grid'. *Proceedings of the 49th Annual Allerton Conference on Communication, Control and Computing*, 2011, pp. 719–726.

[19] Su H. I. and El Gamal A. 'Modeling and analysis of the role of energy storage for renewable integration: power balancing'. *IEEE Trans Power System,* 2013, vol. 28, no. 4, pp. 4109–4117.

[20] Gyuk I., Eckroad S., Mears L., Gotschall H., and Kamath H. *EPRI-DOE Handbook of Energy Storage for Transmission and Distribution Applications*, Technical Report, Electric Power Research Institute (EPRI) and the U.S. Department of Energy, 2003.

[21] Yau T., Walker L., Graham H., Gupta A., and Raithel R. 'Effects of battery storage devices on power system dispatch'. *IEEE Trans Power Apparat System,* 1981, vol. PAS-100, no. 1, pp. 375–383.

[22] Kintner-Meyer M., Jin C., Balducci P., Elizondo M., Guo X., Nguyen T., *et al*. 'Energy storage for variable renewable energy resource integration. A regional assessment for the Northwest Power Pool (NWPP)'. *Proceedings of IEEE/PES Power Systems Conference and Exposition (PSCE)*. 2011, pp. 1–7.

[23] Electric Power Research Institute (EPRI), *Wind Power Integration: Energy Storage for firming and Shaping*, 2005.

[24] Zobaa A. F. *Energy Storage – Technologies and Applications*, InTech, 2013.

[25] Cain M. B., O'Neill R. P., and Castillo A. 'History of optimal power flow and formulations'. *Optimal Power flow Paper*, 2012.

[26] Frank S., Steponavice I., and Rebennack S. 'Optimal power flow: A bibliographic survey I – formulations and deterministic methods'. *Energy Systems,* 2012, vol. 3, no. 3, pp. 221–258.

[27] Castillo A. and Gayme D.F. 'Profit maximizing storage allocation in power grids'. *Proceedings of the 52nd IEEE Conference on Decision and Control,* 2013, pp. 429–435.

[28] Murillo-Sánchez C. E., Zimmerman R. D., Anderson C. L., and Thomas R. J. 'Secure planning and operations of systems with stochastic sources, energy storage, and active demand'. *IEEE Transactions on Smart Grids,* 2013, vol. 4, no. 4, pp. 2220–2229.

[29] Milligan M., Donohoo P., and Omalley M. 'Stochastic methods for planning and operating power systems with large amounts of wind and solar power'.

Proceedings of the 11th Annual International Workshop on Large-Scale Integration of Wind Power into Power Systems as well as on Transmission Networks for Offshore Wind Power Plants, 2012.

[30] Park C. S. *Fundamentals of Engineering Economics*. Pearson Education. New Jersey Inc.; 2004.

[31] Evangelopoulos V. A. and Georgilakis P. S. 'Optimal distributed generation placement under uncertainties based on point estimate method embedded genetic algorithm'. *IET Generation, Transmission and Distribution,* 2014, vol. 8, no. 3, pp. 389–400.

[32] Celli G., Ghiani E., Mocci S., and Pilo F. 'A multiobjective evolutionary algorithm for the sizing and siting of distributed generation'. *IEEE Transaction on Power Systems,* 2005, vol. 20, no. 2, pp. 750–757.

[33] Yang N. and Wen F. 'A chance constrained programming approach to transmission system expansion planning'. *Electric Power Systems Research,* 2005, vol. 75, 2–3, pp. 171–177.

[34] Bakirtzis A. G., Biskas P. N., Zoumas C. E., and Petridis V. 'Optimal power flow by enhanced genetic algorithm'. *IEEE Transaction on Power Systems,* 2002, vol. 17, 2, pp. 229 –236.

[35] Ghofrani M., Arabali A., Etezadi-Amoli M., and Fadali M. 'Energy storage application for performance enhancement of wind integration'. *IEEE Transactions on Power System,* 2013, vol. 28, no. 4, pp. 4803–4811.

[36] Numnonda T. and Annakkage U. D. 'Optimal power dispatch in multinode electricity market using genetic algorithm'. *Electric Power Systems Research,* 1999, vol. 49, no. 3, pp. 211–220.

[37] Zalzala A. M. S. *Genetic Algorithms in Engineering Systems*. Technology & Engineering, 1997.

[38] Schellenberg A., Rosehart W., and Aguado J. 'Cumulant-based probabilistic optimal power flow (P-OPF) with Gaussian and Gamma distributions'. *IEEE Transactions on Power Systems,* 2005, vol. 20, no. 2, pp. 773–781.

[39] Cramer H. *Mathematical Methods of Statistics*. New Jersey: Princeton University Press, 1945.

[40] McCullagh P. *Tensor Methods in Statistics*. London: Chapman & Hall, 1987.

[41] Zhang P. and Lee S. T. 'Probabilistic load flow computation using the method of combined cumulants and Gram-Charlier expansion'. *IEEE Transactions on Power Systems,* 2004, vol. 19, 1, pp. 676–682.

[42] Li G. and Zhang X. P. 'Comparison between two probabilistic load flow methods for reliability assessment'. *Proceedings of IEEE Power Energy Society General Meeting,* 2009, pp. 1–7.

[43] Alberta Electric System Operator (AESO). 10 Min Historical Data for Total Wind Power and Alberta Internal Load. Available from http://www.aeso.ca/gridoperations/20544.html [Accessed 30 December 2024]

[44] Fabbri A., Roman T. G. S., Abbad J. R., and Quezada V. H. M. 'Assessment of the cost associated with wind generation prediction errors in a liberalized electricity market'. *IEEE Transactions on Power System,* 2005, vol. 20, no. 3, pp. 1440–1446.

Chapter 5

A distributed optimization strategy and Hardware-in-the-Loop validation for distributed energy systems

Tung Lam Nguyen[1], *Quoc Tuan Tran*[2] *and Hoang The Tran*[3]

The requirement for coordination of distributed energy resources (DERs) in the future power system provides an incentive to move from the current high level of centralized control to a more distributed control paradigm. In this chapter, an agent-based distributed optimal power flow (OPF) is proposed to optimize the operation of the power system. The OPF problem is built in the general consensus optimization formulation in matrix-based formulation. The agent is then designed to realize the operation of the multi-agent system (MAS) in the cyber-physical system. Agents have the ability to collect local measurement data, communicate with neighbor agents, and implement the Alternating Direction Method of Multipliers (ADMM). Each agent has access to limited information but can give a decision to solve the global problem. The performance is evaluated on the IEEE 9 bus by using a cyber Hardware-in-the-Loop (HIL) platform with a cluster of hardware agents, a real-time simulator OPAL-RT Technologies (OPAL-RT) and a real communication network.

5.1 Introduction

The OPF problem is a fundamental challenge in power system operation, aiming to determine the optimal power generation at each generator while ensuring system reliability and efficiency [1,2]. The goal of OPF is to achieve an optimal operating condition while maintaining power balance and adhering to the physical laws governing the power network. Traditionally, OPF has been addressed using centralized approaches, where a single control entity gathers all necessary system parameters—such as line impedances, network topology, generator cost functions, and load demands—and performs centralized computations to solve the optimization problem.

[1]New York Power Authority, NY, USA
[2]CEA LITEN, INES, Grenoble Alpes University, Le Bourget-du-Lac, France
[3]Future system at Powerco, PowerCo, New Zealand

With the evolution of modern power systems, there is a significant shift towards hybrid energy systems that integrate conventional power plants with DERs, such as photovoltaic (PV) panels, wind turbines, and battery storage systems. The increasing penetration of these distributed energy generators introduces new challenges in system operation, including fluctuations in power generation, bidirectional power flows, and the need for real-time control. These challenges necessitate a transition from centralized OPF methodologies to scalable, decentralized approaches capable of handling the complexity of hybrid power networks.

In this work, we propose a distributed OPF framework tailored for hybrid power systems with DERs. Instead of relying on a single central entity, our approach decomposes the OPF problem into smaller subproblems, each handled by autonomous agents with reduced computational complexity. MAS, which have been extensively utilized in computer science and control engineering, offer an effective paradigm for decentralized optimization. Unlike conventional distributed control architectures, MAS incorporate local intelligence within each agent, enabling them to gather local measurements, process information, and communicate with neighboring agents [3–5]. This decentralized approach is particularly well-suited for hybrid power systems, where distributed energy sources and flexible loads must coordinate efficiently to maintain grid stability and optimize system performance.

The distributed approach to OPF in hybrid systems offers several advantages [6,7]:

1. **Limited data exchange:** Agents only share necessary information with their immediate neighbors, reducing communication overhead and enhancing system robustness.
2. **Reduced computational effort:** Each agent solves a smaller-scale subproblem, leading to lower computational complexity compared to centralized methods, where an increase in grid size results in exponential growth in problem complexity.
3. **Enhanced privacy:** The approach allows entities such as distributed generators (DGs) and consumers to keep sensitive operational data private while still participating in system optimization.
4. **Improved system resilience:** The system is less vulnerable to single-point failures associated with centralized control, making it more adaptable to grid topology and communication infrastructure changes.
5. **Better integration of renewable energy sources:** The decentralized framework allows for seamless integration of renewable energy sources, considering their variability and uncertainty in real-time optimization.

Several studies have explored distributed optimization techniques for OPF in AC hybrid power grids. Comprehensive reviews of these methods can be found in [6–9]. Distributed OPF methods are generally classified into two categories [6]:

1. **Augmented Lagrangian-based decomposition:** This includes techniques such as Dual Decomposition [10,11], ADMM, and Augmented Lagrangian Alternating Direction Inexact Newton (ALADIN) [12].

2. **Decentralized solutions using Karush–Kuhn–Tucker (KKT) conditions:** Examples include Optimality Condition Decomposition (OCD) [13,14] and Consensus + Innovation (C+I) algorithms [15,16].

ADMM has emerged as a widely adopted approach for distributed OPF due to its balance between simplicity and efficiency [17]. Several works have applied ADMM to OPF problems, including region-based decompositions with limited inter-region information exchange [18], adaptive methods to enhance convergence [19], and extensions that handle non-convex OPF formulations via sequential convex approximations [20]. More recently, ALADIN has been introduced as an enhancement of ADMM, achieving faster convergence but requiring more computational effort per iteration and incorporating a centralized update step [21,22]. However, many of these studies focus primarily on theoretical aspects and numerical simulations, without considering real-time implementation challenges such as communication delays and hardware constraints. While some work has explored online ADMM for adapting to grid changes [23], critical aspects—such as interaction between agents, real-time computational delays, and practical deployment—remain underexplored. Bridging this gap is essential for transitioning MAS-based OPF solutions from theory to real-world applications, especially in hybrid power networks.

5.1.1 Contributions of this work

This study presents a practical implementation of distributed OPF using MAS and ADMM, considering real-world operational constraints in hybrid power systems. The key contributions are:

- **Mathematical formulation and software implementation:** A detailed matrix-based formulation of the OPF consensus problem is provided, facilitating efficient computation and numerical implementation. Jacobian matrices are explicitly derived to support optimization solvers.
- **Asynchronous ADMM execution in agents:** The iterative execution of ADMM within each agent is analyzed in the time domain, incorporating real-world factors such as communication delays.
- **Realistic agent design:** Each agent is designed to function autonomously in a practical environment, performing tasks such as (i) exchanging data through a real communication network, (ii) interfacing with local measurement devices to adapt to grid variations, and (iii) solving the OPF problem via distributed computation.
- **Hybrid system optimization:** The proposed approach enables real-time coordination of hybrid energy sources, including renewable generators, battery storage, and conventional power plants.
- **HIL validation:** The MAS framework is deployed on a cyber-HIL platform using Raspberry Pi (RPI) clusters and a real communication network. Real-time interactions between agents and a simulated hybrid power grid (modeled using OPAL-RT) are demonstrated, and the results are compared with centralized OPF solutions to evaluate performance.

The remainder of this chapter is structured as follows:

- **Section 5.2** introduces the general consensus optimization problem and describes the ADMM algorithm for distributed optimization.
- **Section 5.3** presents the OPF problem and its decomposition into a distributed framework using ADMM, with a focus on hybrid power networks.
- **Section 5.4** details the agent-based architecture for distributed OPF, including communication protocols and computational procedures.
- **Section 5.5** evaluates the MAS implementation in a cyber-HIL environment, using a hybrid 9-bus test system to analyze transient behavior and convergence properties.
- **Section 5.6** concludes the chapter with key findings and future research directions.

This work aims to advance the practical implementation of MAS-based distributed OPF in hybrid power systems, bridging the gap between theoretical optimization algorithms and real-world deployment.

5.2 ADMMs for general distributed optimization

5.2.1 Consensus-based distributed optimization

In large-scale optimization problems, particularly those encountered in power systems and multi-agent networks, decision-making is often distributed among multiple subsystems. Consider a scenario where a system consists of K interconnected subsystems, each responsible for optimizing its own objective function while maintaining coordination with its neighboring subsystems. The general form of such a distributed optimization problem can be expressed as

$$\min_{\mathbf{x}} \sum_{k=1}^{K} f_k(\mathbf{x}_k) \tag{5.1}$$

where

- $f_k(\mathbf{x}_k)$ represents the objective function specific to subsystem k.
- $\mathbf{x}_k \in \mathbb{R}^{N_k}$ denotes the decision variables associated with subsystem k. These variables are interdependent, as they may represent shared resources or coupled constraints.

To ensure consistency among subsystems, a set of global variables, denoted by $\mathbf{z} \in \mathbb{R}^N$, is introduced. Each subsystem maintains a local copy of relevant components of $\mathbf{z}$, leading to the following consistency constraint:

$$\mathbf{x}_k = \mathbf{E}_k \mathbf{z}, \quad k = 1, 2, \ldots, K \tag{5.2}$$

where $\mathbf{E}_k$ is a mapping matrix that selects the relevant components of $\mathbf{z}$ for each subsystem. The matrix $\mathbf{E}_k$ is structured such that:

$$(\mathbf{E}_k)_{n,m} = \begin{cases} 1, & \text{if the } n\text{-th component of } \mathbf{x}_k \text{ corresponds to net } m \\ 0, & \text{otherwise} \end{cases} \tag{5.3}$$

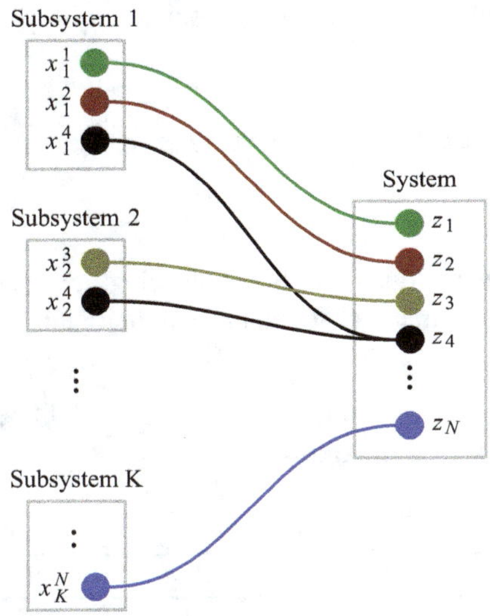

Figure 5.1 Schematic representation of variable dependencies in a distributed optimization problem. Each subsystem maintains a local copy of global variables, ensuring coordination through consensus constraints

Figure 5.1 illustrates the relationship between local and global variables in a distributed system, where different colors represent variables associated with the same global net.

5.2.2 *Alternating Direction Method of Multipliers*

The ADMM is a powerful iterative approach designed to solve distributed optimization problems efficiently. It combines the advantages of dual decomposition and augmented Lagrangian methods, making it well-suited for decentralized optimization [17].

The optimization problem can be reformulated by incorporating local feasibility constraints and enforcing the consensus condition:

$$\begin{aligned} \min_{\mathbf{x},\mathbf{z}} \quad & \sum_{k=1}^{K} f_k(\mathbf{x}_k) \\ \text{subject to} \quad & \mathbf{x}_k \in \mathscr{C}_k, \quad k = 1, \ldots, K \\ & \mathbf{x}_k = \mathbf{E}_k \mathbf{z}, \quad k = 1, \ldots, K \end{aligned} \tag{5.4}$$

where $\mathscr{C}_k$ represents local constraints specific to each subsystem.

To facilitate decentralized computation, an auxiliary variable $\tilde{\mathbf{z}}_k$ is introduced:

$$\tilde{\mathbf{z}}_k = \mathbf{E}_k \mathbf{z} \tag{5.5}$$

allowing the consistency constraint to be rewritten as

$$\mathbf{x}_k - \tilde{\mathbf{z}}_k = 0 \tag{5.6}$$

The augmented Lagrangian function associated with this constraint is formulated as

$$L_\rho(\mathbf{x}, \mathbf{z}, \boldsymbol{\lambda}) = \sum_{k=1}^{K} \left[f_k(\mathbf{x}_k) + \boldsymbol{\lambda}_k^T (\mathbf{x}_k - \tilde{\mathbf{z}}_k) + \frac{\rho}{2} \|\mathbf{x}_k - \tilde{\mathbf{z}}_k\|_2^2 \right] \tag{5.7}$$

where

- $\boldsymbol{\lambda}_k \in \mathbb{R}^{N_k}$ represents the dual variables enforcing the equality constraint.
- $\rho > 0$ is a penalty parameter that enhances numerical stability and convergence speed.

ADMM proceeds by iteratively updating local variables, global variables, and dual variables, as summarized in Algorithm 1.

Algorithm 1: Alternating Direction Method of Multipliers (ADMM)

Initialize $\mathbf{z}^0$ and $\boldsymbol{\lambda}^0$.
repeat
Convergence criteria met
Step 1: Local Variable Update Each subsystem updates its local variables:

$$\mathbf{x}_k^{n+1} = \arg\min_{\mathbf{x}_k} \left[f_k(\mathbf{x}_k) + \boldsymbol{\lambda}_k^{nT} \mathbf{x}_k + \tfrac{\rho}{2} \|\mathbf{x}_k - \tilde{\mathbf{z}}_k^n\|_2^2 \right] \tag{5.8}$$

Step 2: Global Variable Update Global variables are updated to ensure consensus:

$$\mathbf{z}^{n+1} = \arg\min_{\mathbf{z}} \textstyle\sum_{k=1}^{K} \left[-\boldsymbol{\lambda}_k^{nT} \tilde{\mathbf{z}}_k + \tfrac{\rho}{2} \|\mathbf{x}_k^{n+1} - \tilde{\mathbf{z}}_k\|_2^2 \right] \tag{5.9}$$

Step 3: Dual Variable Update The Lagrange multipliers are adjusted:

$$\boldsymbol{\lambda}_k^{n+1} = \boldsymbol{\lambda}_k^n + \rho(\mathbf{x}_k^{n+1} - \tilde{\mathbf{z}}_k^{n+1}) \tag{5.10}$$

Step 4: Iteration Counter Update $n = n + 1$.

5.2.3 *Convergence and practical considerations*

ADMM's iterative structure allows each subsystem to update its local decision variables and dual variables independently, making it well-suited for large-scale distributed optimization. Only the global variable update step requires communication between neighboring subsystems. In practice, consensus is achieved by exchanging local estimates and computing an average, ensuring that

$$\mathbf{z}_n^{n+1} = \frac{1}{K} \sum_{k=1}^{K} \left(\mathbf{x}_k^{n+1} + \frac{1}{\rho} \boldsymbol{\lambda}_k^n \right) \tag{5.11}$$

This decentralized mechanism enables scalable optimization, making ADMM a robust tool for applications in power systems, sensor networks, and multi-agent coordination.

5.3 OPF problem formulation

5.3.1 General formulation of OPF

In power system operations, the OPF problem aims to determine the optimal generation and voltage levels at each bus while satisfying network constraints. Consider an electrical network with N buses and L transmission lines, where the set of buses is denoted as $\mathcal{N} = \{1, 2, \ldots, N\}$ and the set of lines is represented as $\mathcal{L} = \mathcal{N} \times \mathcal{N}$.

To express the power flow equations, we introduce the following notations:

- The subset of generator buses is denoted by $\mathcal{G} \subseteq \mathcal{N}$.
- The complex power generated at bus $k \in \mathcal{G}$ is $s_k^G = p_k^G + jq_k^G \in \mathbb{C}$.
- The complex load power at bus $k \in \mathcal{N}$ is $s_k^L = p_k^L + jq_k^L \in \mathbb{C}$.

The net injected power at each bus k is given by

$$s_k = p_k + jq_k = \begin{cases} s_k^G - s_k^L, & \text{if } k \in \mathcal{G} \\ -s_k^L, & \text{otherwise} \end{cases} \tag{5.12}$$

The vector notation for the system-wide power injections is:

$$\mathbf{s} = \mathbf{p} + j\mathbf{q} \tag{5.13}$$

5.3.1.1 Voltage and current representations

Two key electrical parameters in the network are bus voltages and current injections:

- The voltage at bus k is expressed as $v_k = v_k^{re} + jv_k^{im}$, and the system-wide voltage vector is:

$$\mathbf{v} = \mathbf{v}^{re} + j\mathbf{v}^{im} \tag{5.14}$$

- The current injected at bus k is denoted as $i_k = i_k^{re} + ji_k^{im}$, and the system-wide current injection vector is:

$$\mathbf{i} = \mathbf{i}^{re} + j\mathbf{i}^{im} \tag{5.15}$$

The electrical coupling between buses is described by the network admittance matrix:

$$\mathbf{Y} = \mathbf{G} + j\mathbf{B} \in \mathbb{C}^{N \times N} \tag{5.16}$$

where

- g_{mn} represents the real part (conductance) of the admittance between buses m and n.
- b_{mn} represents the imaginary part (susceptance) of the admittance between buses m and n.

The current–voltage relationship is given by

$$\mathbf{i} = \mathbf{Y}\mathbf{v} = (\mathbf{G}\mathbf{v}^{re} - \mathbf{B}\mathbf{v}^{im}) + j(\mathbf{B}\mathbf{v}^{re} + \mathbf{G}\mathbf{v}^{im}) \tag{5.17}$$

The complex power injection at each bus is related to voltages and current injections by

$$\mathbf{s} = \mathbf{v} \otimes \mathbf{i}^* \tag{5.18}$$

which expands into its real and imaginary components as

$$\mathbf{p} = \mathbf{v}^{re} \otimes (\mathbf{G}\mathbf{v}^{re} - \mathbf{B}\mathbf{v}^{im}) + \mathbf{v}^{im} \otimes (\mathbf{B}\mathbf{v}^{re} + \mathbf{G}\mathbf{v}^{im}) \tag{5.19}$$

$$\mathbf{q} = \mathbf{v}^{im} \otimes (\mathbf{G}\mathbf{v}^{re} - \mathbf{B}\mathbf{v}^{im}) - \mathbf{v}^{re} \otimes (\mathbf{B}\mathbf{v}^{re} + \mathbf{G}\mathbf{v}^{im}) \tag{5.20}$$

To compactly represent these power flow equations, we define:

$$\hat{\mathbf{v}} = \begin{bmatrix} \mathbf{v}^{re} \\ \mathbf{v}^{im} \end{bmatrix}, \quad \mathbf{Z}^p = \begin{bmatrix} \mathbf{G} & -\mathbf{B} \\ \mathbf{B} & \mathbf{G} \end{bmatrix}, \quad \mathbf{Z}^q = \begin{bmatrix} -\mathbf{B} & -\mathbf{G} \\ \mathbf{G} & -\mathbf{B} \end{bmatrix} \tag{5.21}$$

Thus, the total active and reactive power injections in the network are given by

$$\sum_{k=1}^{N} p_k = \hat{\mathbf{v}}^T \mathbf{Z}^p \hat{\mathbf{v}} \tag{5.22}$$

$$\sum_{k=1}^{N} q_k = \hat{\mathbf{v}}^T \mathbf{Z}^q \hat{\mathbf{v}} \tag{5.23}$$

Since power loss in the network corresponds to the difference between total generation and total demand, we obtain:

$$\hat{\mathbf{v}}^T \mathbf{Z}^p \hat{\mathbf{v}} = \sum p^G - \sum p^L = \sum P_{\text{loss}} \tag{5.24}$$

5.3.1.2 Bus-level power balance constraints

For each bus k, the active and reactive power balances are expressed as

$$p_k = \hat{\mathbf{v}}_k^T \mathbf{Z}_k^p \hat{\mathbf{v}}_k \tag{5.25}$$

$$q_k = \hat{\mathbf{v}}_k^T \mathbf{Z}_k^q \hat{\mathbf{v}}_k \tag{5.26}$$

The matrices $\mathbf{Z}_k^p$ and $\mathbf{Z}_k^q$ are structured to contain only the parameters relevant to bus k.

5.3.2 Formulation of the OPF optimization problem

The objective of OPF is to minimize power losses while maintaining system constraints. The problem is formulated as

$$\begin{aligned} \min_{\hat{\mathbf{v}}} \quad & \hat{\mathbf{v}}^T \mathbf{Z}^p \hat{\mathbf{v}} \\ \text{subject to} \quad & P_k^{\min} \le \hat{\mathbf{v}}_k^T \mathbf{Z}_k^p \hat{\mathbf{v}}_k + p_k^L \le P_k^{\max}, \quad k \in \mathcal{G} \\ & Q_k^{\min} \le \hat{\mathbf{v}}_k^T \mathbf{Z}_k^q \hat{\mathbf{v}}_k + q_k^L \le Q_k^{\max}, \quad k \in \mathcal{G} \\ & (v_k^{\min})^2 \le (v_k^{re})^2 + (v_k^{im})^2 \le (v_k^{\max})^2, \quad k \in \mathcal{N} \end{aligned} \tag{5.27}$$

This formulation ensures that generator outputs remain within their operational limits, voltage magnitudes are maintained within acceptable bounds, and power flow constraints are satisfied.

In the following sections, we explore how this OPF problem can be decomposed into a distributed framework using the ADMM, enabling efficient computation across multiple agents in a decentralized power system.

5.4 Multi-agent system for distributed OPF in a cyber-physical framework

The deployment of an MAS facilitates the realization of a distributed OPF solution within a cyber-physical framework. Each agent within the MAS is responsible for managing its local subsystem while coordinating with neighboring agents to achieve a system-wide OPF. To ensure practical implementation, the agents are developed as Python-based software entities with a structured architecture, as depicted in Figure 5.2.

Each agent is designed to perform three fundamental functions:

1. **Interface with the physical system:** The agent gathers real-time measurements from the power system and communicates with local controllers to transmit power set points for the corresponding generator.

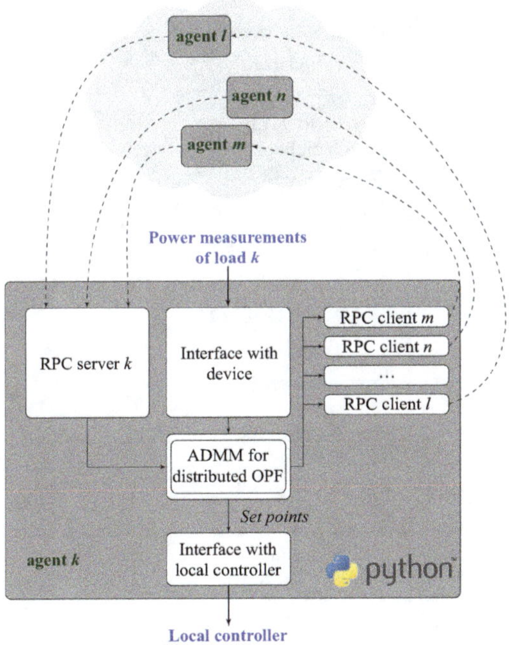

Figure 5.2 Structural composition of an agent implementing the ADMM method for distributed OPF

2. **Communication with neighboring agents:** A remote procedure call (RPC) server-client mechanism enables seamless exchange of information between agents, allowing them to share voltage and power flow estimates.
3. **Execution of the ADMM algorithm:** The agent independently computes the updates for its local OPF subproblem while incorporating shared information from its neighbors, ensuring convergence toward an optimal solution.

Each agent functions as both a server and a client—acting as a server to receive requests from neighboring agents while simultaneously functioning as a client to request information from adjacent agents.

5.4.1 Iterative execution of distributed OPF in time domain

The MAS operates in an iterative manner, continuously updating control decisions based on real-time system conditions. Figure 5.3 illustrates the iterative execution of an agent over time. Each iteration cycle, referred to as a *loop*, begins when an agent receives its local measurements at iteration *0* and concludes once the ADMM

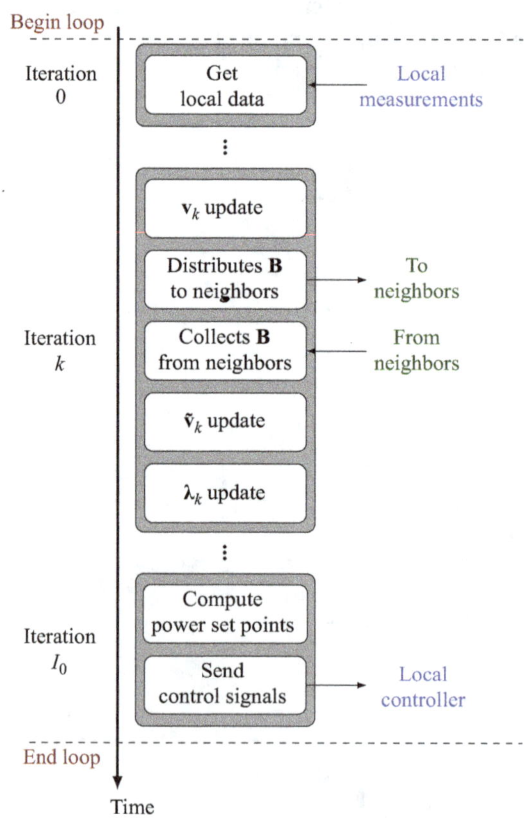

Figure 5.3 Time-domain execution of the ADMM iterative process within an agent

process reaches consensus, allowing the agent to transmit the optimal control signals at iteration I_0.

Within each loop, the agent performs the following sequential steps:

- **State update:** The agent acquires local system measurements, including voltage magnitudes, power flows, and generator states.
- **Computation:** The agent processes the ADMM optimization algorithm, updating its local decision variables based on shared information from neighboring agents.
- **Consensus achievement:** Through iterative updates, the agents progressively refine their solutions until convergence is attained.
- **Control signal transmission:** Once the optimization reaches the prescribed iteration limit I_0, the computed control set points are dispatched to the corresponding generator or control device.

Although agents operate in parallel, their execution is synchronized at each iteration step to ensure that data exchanges occur consistently across all agents. This synchronization is essential for maintaining computational stability and ensuring convergence in a decentralized optimization framework.

The MAS-based implementation provides a scalable and robust mechanism for solving the OPF problem in a decentralized manner. By leveraging real-time data acquisition, distributed computation, and coordinated decision-making, the system effectively manages the dynamic operation of modern power networks while adhering to network constraints and optimizing system performance.

5.5 Cyber-physical HIL experiment

To validate the proposed distributed OPF approach described in Section 5.3 and the MAS framework introduced in Section 5.4, a Cyber-Physical HIL experiment is conducted using the IEEE 9-bus test system [24]. The network topology of the test system is illustrated in Figure 5.4. The system consists of two generators at buses $\{2, 3\}$ supplying power to three load buses $\{5, 7, 9\}$, while bus 1 serves as the slack bus connected to a bulk power system. The objective of the experiment is to assess the ability of the MAS to optimize the generation set points in a distributed manner while dynamically responding to load variations.

To evaluate system performance under dynamic conditions, load variations are introduced over a 1440-second time horizon. The active and reactive power demands are subjected to step changes at 480 seconds (load reduction) and 960 seconds (load increase), as illustrated in Figure 5.5. The MAS is expected to adapt to these disturbances in real time and compute optimal generation set points to minimize power losses while maintaining system stability.

5.5.1 Experimental setup for cyber-HIL simulation

The experimental test bed is constructed as shown in Figure 5.6. The simulation environment consists of both a physical computing infrastructure and a real-time digital

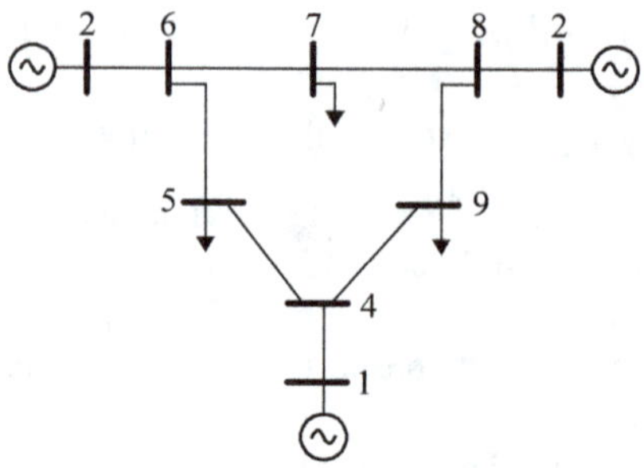

Figure 5.4 IEEE 9-bus test system used for Cyber-HIL validation

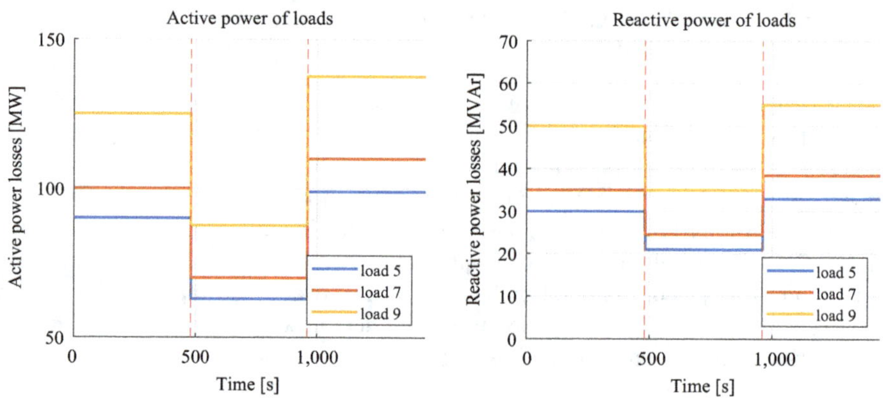

Figure 5.5 Time-domain variations of active and reactive power loads

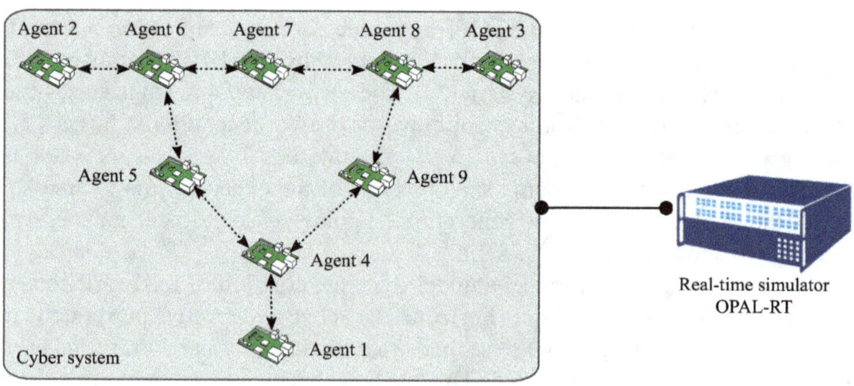

Figure 5.6 Cyber-HIL test bed for distributed OPF implementation

simulator. The power grid model is implemented in MATLAB®/Simulink® and executed in real time using an OPAL-RT simulator, representing the physical layer of the experiment. The cyber layer consists of a networked cluster of RPI devices that implement the MAS framework.

The key components of the test bed include the following:

- **Physical grid simulation:** The power grid model is executed in OPAL-RT, where bus 1 is modeled as a voltage source, and buses 2 and 3 are modeled as controllable current sources.
- **Multi-agent system:** A network of nine Raspberry Pi (RPI 3, Model B, 1GB RAM, 1.2 GHz processor) devices represents the MAS. Each RPI runs an independent agent responsible for solving a local OPF subproblem.
- **Communication network:** Agents communicate over a real network switch using the gRPC protocol, ensuring message exchange between neighbors follows the electrical topology.
- **Cyber-physical interface:** Data exchange between OPAL-RT and the MAS occurs through a UDP (User Datagram Protocol) interface, allowing agents to retrieve load power measurements and send generator set points in real time.

The communication latency between agents is measured and analyzed in Figure 5.7. The observed communication delays are relatively small, with a median latency below 0.05 seconds, ensuring real-time operation of the distributed OPF algorithm.

The computational effort required for solving local OPF subproblems varies across agents, as illustrated in Figure 5.8. The primary computational burden arises from executing the ADMM algorithm to solve the local optimization problem.

5.5.2 Execution of the ADMM-based distributed OPF algorithm

The operation of the MAS can be divided into three distinct phases:

- **Phase 1: Data acquisition.** Each agent retrieves local measurements from OPAL-RT, including real-time load power data at its respective bus.
- **Phase 2: Distributed computation.** Using local data and information received from neighboring agents, each agent iteratively executes the ADMM optimization process. A penalty parameter of $\rho = 25$ is selected to ensure good convergence performance, and each ADMM loop is executed for 1000 iterations.

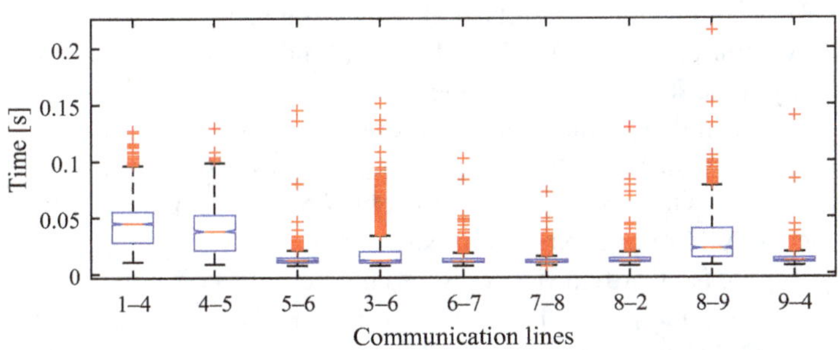

Figure 5.7 Communication latency between agents in the MAS

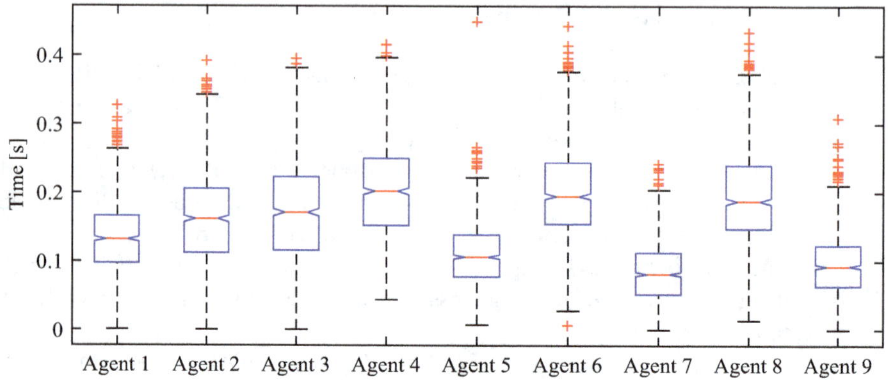

Figure 5.8 Computation time for solving local OPF subproblems in each agent

- **Phase 3: Control update.** At the completion of 1000 iterations, agents 2 and 3 transmit updated generator set points to their respective control systems. The ADMM process is then restarted for continuous optimization.

5.5.3 Experimental results and analysis

Figure 5.9 presents the real-time evolution of active and reactive power outputs at the slack bus (bus 1) and generator buses (buses 2 and 3). Initially, generators operate at a predefined set point of 50 MW and 30 MVAr. The key events in the experiment are marked by vertical lines:

- **Red dashed lines:** Indicate moments when load changes occur (480 s and 960 s).
- **Black dashed lines:** Represent instances when generators receive optimal set points from the MAS (251 s, 756 s, and 1261 s).

The figure compares two data sources:

- Solid blue lines represent power measurements recorded from OPAL-RT.
- Solid gray lines depict computed values from the agents' internal logs.

Despite operating independently and asynchronously, the agents successfully reach ADMM consensus simultaneously, enabling synchronized updates to the generator set points. The MAS effectively responds to real-time system changes, continuously adjusting generator outputs to minimize losses.

Figure 5.10 illustrates the total active power losses throughout the experiment. The losses decrease whenever generators receive updated set points, confirming that the MAS successfully optimizes system performance.

Finally, Figure 5.11 provides insight into the iterative convergence of active and reactive power computations within each ADMM loop. The results indicate that the decentralized ADMM-based approach closely matches the centralized OPF solution, thereby validating the accuracy of the proposed method.

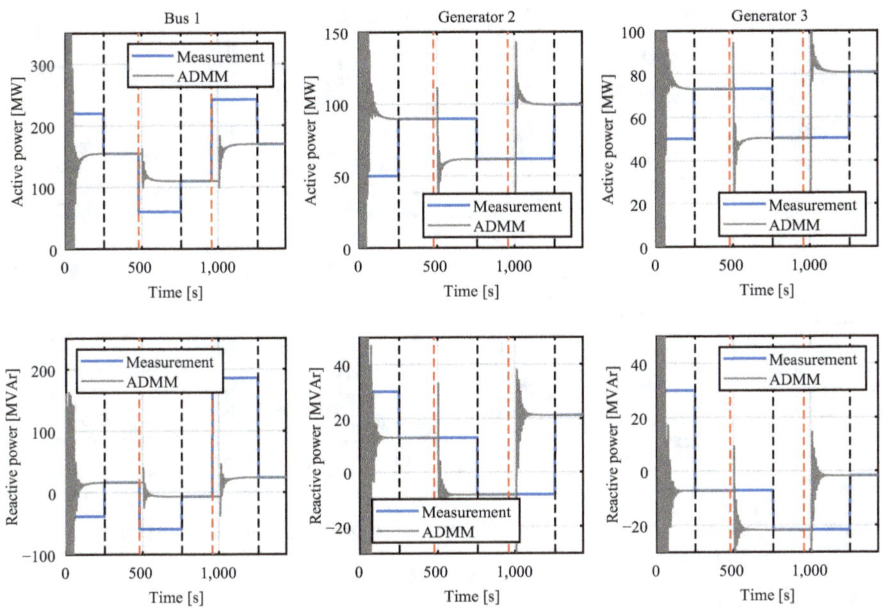

Figure 5.9 Active and reactive power outputs at bus 1 (slack bus) and generator buses (buses 2 and 3)

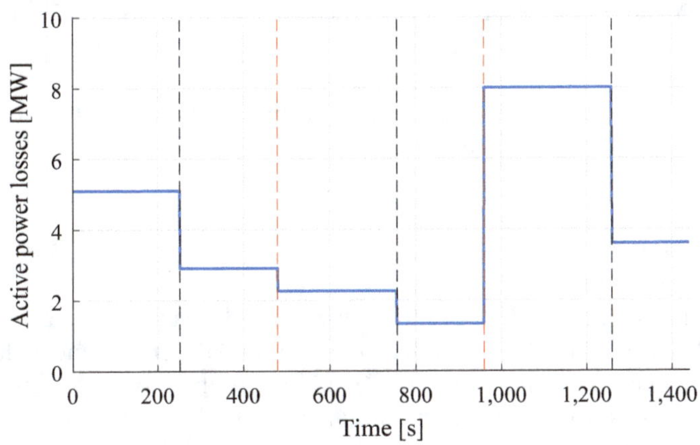

Figure 5.10 Time-domain evolution of total active power losses

5.5.4 Summary of experimental findings

The Cyber-HIL experiment demonstrates the effectiveness of the MAS in optimizing power system operation in a distributed manner. Key takeaways include:

- The MAS dynamically adjusts generator set points in response to changing loads.
- The distributed OPF implementation achieves results comparable to centralized optimization.

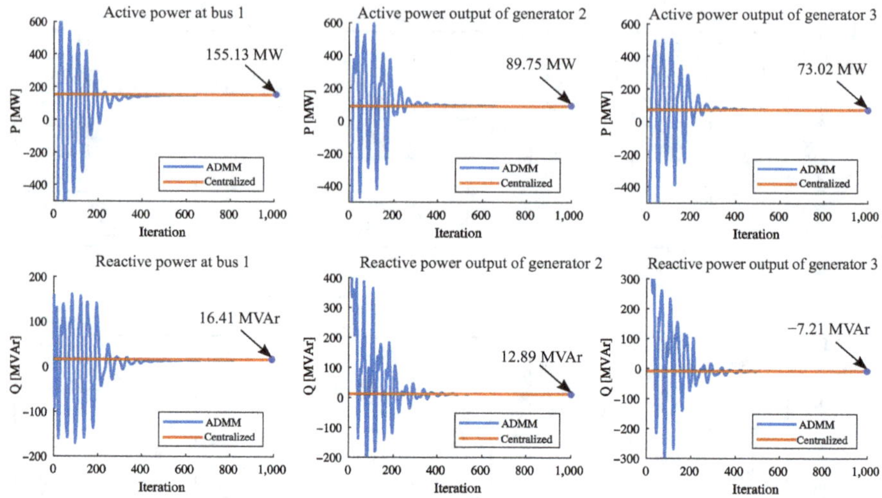

Figure 5.11 Convergence of active and reactive power computations during ADMM iterations

- Communication latencies and computational times are sufficiently low to support real-time operation.
- The ADMM process ensures convergence, allowing agents to collectively determine an optimal network state.

These findings validate the proposed approach as a viable solution for real-time, decentralized power system optimization.

5.6 Conclusion

This chapter presents a distributed OPF strategy for hybrid power systems using MAS and ADMM. The methodology is validated in a cyber-HIL environment, demonstrating its effectiveness in coordinating DERs and improving power system efficiency. Future work will explore advanced communication protocols and the integration of machine learning techniques to enhance real-time adaptability.

References

[1] Frank S, Steponavice I, and Rebennack S. Optimal power flow: A bibliographic Survey I formulations and deterministic methods. *Energy Systems*. 2012.

[2] Wood A, Wollenberg B, and Sheble G. *Power Generation, Operation and Control*. John Wiley & Sons, Inc., Hoboken, NJ; 2014.

[3] Khan MW and Wang J. The research on multi-agent system for microgrid control and optimization. *Renewable and Sustainable Energy Reviews*. 2017;80:1399–1411.

[4] Wang Y, Nguyen TL, Xu Y, *et al.* Distributed control of heterogeneous energy storage systems in islanded microgrids: Finite-time approach and cyber-physical implementation. *International Journal of Electrical Power & Energy Systems*. 2020;119:105898.

[5] Wang Y, Nguyen TL, Xu Y, *et al.* Cyber-Physical Design and Implementation of Distributed Event-Triggered Secondary Control in Islanded Microgrids. *IEEE Transactions on Industry Applications*. 2019;55(6):5631–5642.

[6] Molzahn DK, Dörfler F, Sandberg H, *et al.* A Survey of Distributed Optimization and Control Algorithms for Electric Power Systems. *IEEE Transactions on Smart Grid*. 2017.

[7] Faulwasser T, Engelmann A, Mühlpfordt T, *et al.* Optimal power flow: An introduction to predictive, distributed and stochastic control challenges. *At-Automatisierungstechnik*. 2018.

[8] Capitanescu F. Critical review of recent advances and further developments needed in AC optimal power flow. *Electric Power Systems Research*. 2016;136:57–68.

[9] Wang Y, Wang S, and Wu L. Distributed optimization approaches for emerging power systems operation: A review. *Electric Power Systems Research*. 2017;144:127–135.

[10] Gatsis N and Giannakis GB. Decomposition algorithms for market clearing with large-scale demand response. *IEEE Transactions on Smart Grid*. 2013.

[11] Wang Z, Chen B, Wang J, *et al.* Decentralized Energy Management System for Networked Microgrids in Grid-Connected and Islanded Modes. *IEEE Transactions on Smart Grid*. 2016.

[12] Engelmann A, Jiang Y, Muhlpfordt T, *et al.* Towards Distributed OPF using ALADIN. *IEEE Transactions on Power Systems*. 2018.

[13] Hug-Glanzmann G and Andersson G. Decentralized optimal power flow control for overlapping areas in power systems. *IEEE Transactions on Power Systems*. 2009.

[14] Nogales FJ, Prieto FJ, and Conejo AJ. A Decomposition Methodology Applied to the Multi-Area Optimal Power Flow Problem. *Annals of Operations Research*. 2003.

[15] Kar S, Hug G, Mohammadi J, *et al.* Distributed state estimation and energy management in smart grids: A Consensus+ Innovations Approach. *IEEE Journal on Selected Topics in Signal Processing*. 2014.

[16] Mohammadi J, Kar S, and Hug G. *Distributed Approach for DC Optimal Power Flow Calculations*. 2014;p. 1–11.

[17] Boyd S, Parikh N, Chu E, Peleato B, and Eckstein J. Distributed Optimization and Statistical Learning via the Alternating Direction Method of Multipliers. *Foundations and Trends in Machine Learning*. 2011.

[18] Erseghe T. Distributed optimal power flow using ADMM. *IEEE Transactions on Power Systems*. 2014.

[19] Mhanna S, Verbic G, and Chapman AC. Adaptive ADMM for distributed AC optimal power flow. *IEEE Transactions on Power Systems*. 2019.

[20] Magnusson S, Weeraddana PC, and Fischione C. A Distributed Approach for the Optimal Power-Flow Problem Based on ADMM and Sequential Convex Approximations. *IEEE Transactions on Control of Network Systems*. 2015.

[21] Meyer-Huebner N, Suriyah M, and Leibfried T. Distributed Optimal Power Flow in Hybrid AC–DC Grids. *IEEE Transactions on Power Systems*. 2019;34(4):2937–2946.

[22] Sun G, Li G, Xia S, *et al.* ALADIN-Based Coordinated Operation of Power Distribution and Traffic Networks with Electric Vehicles. *IEEE Transactions on Industry Applications*. 2020;p. 1–1.

[23] Zhang Y, Hong M, Dall'Anese E, *et al.* Distributed controllers seeking AC optimal power flow solutions using ADMM. *IEEE Transactions on Smart Grid*. 2018.

[24] Anderson PM and Fouad AA. *Power System Control and Stability*. Wiley-IEEE Press; 2003.

Chapter 6

Coordination of PV-STATCOM and hybrid energy storage system for reactive power management in distribution system

Le Hong Lam[1], *Tran Thai Trung*[2], *Huu Hieu Nguyen*[1] *and Nguyen Hong Phuong*[2]

The penetration of Renewable Energy Source (RES) has been increasing in recent years; thus, the conventional power plants are replaced. This creates a shortage of crucial Ancillary Services (ASs) that are important for frequency and voltage stability of the system. Today, Distributed Energy Resources (DER) can offer the services in AS through a converter interfaced or Battery Energy Storage System (BESS). However, the ASs usually are defined by Transmission System Operations (TSOs), AS of Distribution System Operation (DSO) is only emerging. Therefore, this chapter discusses the possibility of enabling AS of DSO since the flexibility sources are expanding in the distribution network. The coordination between TSO-DSO is the main challenge due to the way of operating markets and sharing data. Then, the method of using PV-STATCOM and BESS is proposed to enhance the stability of the system, i.e., frequency and voltage. The IEEE 9-bus test system is used to test the robustness of the proposed methodology.

Acronyms

DER Distributed Energy Resources
PV Photovoltaic
BESS Battery Energy Storage System
EV Electric Vehicle
DSO Distribution system
AS Ancillary Service
TSO Transmission System Operation
DSO Distribution System Operation
RES Renewable Energy Source

[1]Faculty of Electrical Engineering, The University of Danang – University of Science and Technology, Vietnam
[2]Department of Electrical Engineering, Eindhoven University of Technology, The Netherlands

VPP Virtual Power Plant
DRF Demand Response Flexibility
DSF Demand Side Flexibility

6.1 Introduction about ancillary services

Balancing electricity supply and demand is the main challenge for grid operators. This is explained by three issues: (i) difficulty in storing energy, (ii) fluctuation in output of RES, and (iii) demand increases and decreases rapidly. Therefore, the grid operator must respond quickly to these fluctuations to maintain the balance of the power system. The AS is an important part of the balancing toolkit, representing the generating capacity that's used to guarantee overall system security.

AS are operated by TSO as services to ensure the reliability of the operation system due to the high penetration of RESs in the power system. At TSO level, AS is clarified into frequency control, voltage control, system management, and system restoration. Currently, these ASs mainly provide synchronous generators driven by conventional fuel power plants. TSOs defines the ASs based on how quickly the response is needed. The cost of producing electricity is decreasing due to the low cost of wind and solar power, but the need to balance it with short response times is increasing. Therefore, in some countries, the speed of integrating RES is hampered by a lack of AS. This challenge is solved depending on each country's specific market. In Vietnam, the National Electricity System and Market Operation Company Limited (NSMO) is responsible for ASs such as approving providers and ensuring reserve capacity. Today, these ASs are provided by hydropower plants and thermal plants. In India, together with the increase of RES capacity, the need for AS is growing. Due to the limitation of regulation, only old and inefficient thermal power plants are used to meet the reserve capacity requirement. In Europe, several AS are available in the market, but all of them rely on conventional and hydropower plants. As more RES are integrated into the power grid, new services need to be introduced to ensure the continuous supply of energy and mitigate power system limitations.

Traditionally the flexibility of the power systems has mainly been assured by large centralized generating resources, e.g., thermal and hydro power plants. However, the changes in power systems occurring nowadays and in the near future, i.e., the high penetration of renewable energy sources and storage capacity, both being DER, increase the overall demand for flexibility but also reduce the supply of flexibility from the traditional sources. Moreover, in many cases, these distributed energy sources will be available at the distribution grid level, due to their reduced size, where they can be exploited for their available flexibility.

In detail, the providers of flexibility at the distribution level comprise the following categories:

- Demand response and electrification of other sectors (e.g., heat),
- Distributed generation (from renewable energy resources, but not exclusively),
- Distributed storage,
- Electric vehicles.

These resources are present in a very large amount at the distribution grid level and, therefore, their individual and independent participation in the AS schemes is very hard to imagine due to the complexity they would introduce. For this reason, the participation of these resources in the AS schemes will be made by aggregating these resources into equivalent flexibility resources. In the following sections, the role of the flexibility assets at the distribution level is described, considering them as aggregated resources.

6.2 Ancillary services that can be provided by distributed assets

In the past, the low-voltage distribution system is designed to distribute electric energy from the centralized power plants to end-users. Today, the rapid growth of DERs such as solar Photovoltaic (PV), diesel generators, BESS, or the new load like Electric Vehicle (EV) requires a change of manner of managing the distribution system. BESS has a large potential for the provision of grid ASs. Nevertheless, providing this grid service must use corrective energy measures and control algorithms and continuously keep their state of charge within limits in order to comply with European regulations. Actually, frequency regulation is one of the most investigated services, both for on-grid and for off-grid scenarios, in order to guarantee quality and security of supply; BESSs are quite investigated for such a task due to their capacity to provide a fast response both for positive regulation provision (BESSs discharge mode) and for negative provision (BESSs charge mode). Some drawbacks also have to be addressed; in particular, due to their limited storage capacity, BESSs could fail during a long period of frequency deviations or periods of biased regulation demand. In particular, with respect to the Grid Codes in place in Europe, the primary frequency regulation contribution for each resource has to be guaranteed for, at least, 30 minutes; in order to cope (or, at least, to limit) the problem, BESSs are typically operated within state of charge (SOC) limits. In literature, the most common approach to this problem is based on corrective BESSs charging or discharging devoted to maintaining the SOC within the desired limits, but the purchase of the energy for such corrective measures increases operation costs. DER can lead to two ways of power flow, which require DSO to be more aware of the distribution network state. The replacement of conventional generators with DER brings great benefits, e.g., voltage profile support, loss reduction, and lower operating cost.

The increase of DER is expected to replace the conventional power plant in the future, which is stated in many reports [1]. However, the lack of conventional power plants will lead to the reduction of ASs, which are responsible for the stability and safe operation of the power system. Therefore, three options are proposed to mitigate this problem: (i) use conventional power plants as AS providers, (ii) BESS operates as a third party, and (iii) aggregate DER connected in the distribution system. Here, the last option represents the decentralized approach in which TSO see DER as Virtual Power Plant (VPP).

The typical ASs in distribution network include the following [2]: (i) distribution network operation, (ii) voltage control, (iii) supply restoration after an outage, (iv) provision of energy to cover network losses, (v) reactive power support, and

(vi) power quality maintenance. Together with the developing of smart grid and the increase of penetration of DER, AS in the distribution network can arise: (i) advanced metering, (ii) demand response, (iii) virtual power plant, and (iv) current congestion mitigation. According to Reference [3], the AS that could be provided from the flexible resources present at the distribution level are:

- Congestion mitigation during the planning phase;
- Primary power reserve;
- Secondary and tertiary power reserve;
- Balancing reserves;
- Reactive power reserve for voltage regulation;
- Real power reserve for voltage regulation;
- Demand response and load rejection;
- Participation in the recovery of the electricity system;
- Availability for the use of intercropping;
- Island operation of a part of the network.

These ASs can be offered for both operation modes, i.e., normal operation and emergency, to address local problems (distribution grid level) as well as global problems (transmission grid level). Here, some services are technical obligations such as primary power reserve, participation in the recovery of the electricity system, availability for use of the intercropping, and island operation of a part of the network. Besides, the remaining services can be offered in the ASs, which the TSOs or DSOs can use to operate the network in safe conditions by solving global or local issues:

- Local ASs: voltage support and congestion management;
- Global ASs: congestion management, primary reserve, secondary reserve, tertiary reserve, balancing resources, and voltage control.

This chapter focuses only on investigating technical requirements to enable ASs for DERs connected to distribution networks.

- **Congestion management:** At this moment, DSOs are required to increase the use of flexible resources while maintaining high levels of reliability and security of supply. This task can only be achieved by improved network capacity planning and congestion management. Conventionally, two methodologies for solving congestion management are re-dispatch and Demand Response Flexibility (DRF). Because this document only considers DER, the DRF methodology is investigated deeply. There are two types of payment of DRF [4]: (i) the Price-based DRF and (ii) the Incentive-based DRF. In the first type, customers can respond to changes in the daily market prices by reducing their consumption, thus reducing their electricity bills. Meanwhile, the second type is supported by the electric companies and grid operators, which incentivize customers by providing compensation for the reduction of the load when needed to solve network congestion. Today, one of the recent Incentive-based DRF is the Demand Side Flexibility (DSF), in which customers have an opportunity to bid for their reduction in consumption according to their preferred time. Therefore, DSOs can provide several services, e.g., peak shifting, demand adjustment, and relieving grid constraints. Moreover, small customers could also benefit from this scheme and reduce their electric bills. Obviously, small customers might not have enough knowledge and experience to participate in this market; thus, a new market player introduced is aggregators that have a target to maximize the flexibility potential for the grid users by aggregating the customer's load.

- **Voltage support:** Today, TSOs and DSO are responsible for controlling the reactive power mismatch in their own networks. Particularly, in Italy, TSO pays for generators for reactive power; meanwhile, there are maybe some penalty charges for DSO or customers for creating additional reactive power mismatch [5]. At the transmission level, an increase in reactive power will cause a voltage to rise, while a decrease will cause a voltage drop along the transmission line. At the distribution level, the variation in demand at the load side is another parameter for the reactive power effect, in which an increase in demand causes a decrease in voltage at the transmission level, while a decrease in demand causes an increase in voltage at the transmission level. At this moment, the devices used to provide voltage/reactive power support are: (i) synchronous generators (the most frequent) and (ii) power electronics-based devices like Synchronous Voltage Condensers and Static Synchronous Compensators or the full AC/AC converters of the RES. Today, the increase in flexible resources provides an opportunity for system operators to use these resources for supporting voltage, such as wind generators or solar panels. These resources can provide reactive power control through converters and inverters that normally require over-sizing the converter or inverter by 10% of the generator rating. It means that the system can provide 46% reactive power at 100% active power or 110% reactive power at zero active power. According to current Italian regulations, intermittent resources cannot participate in supporting voltage, but other countries such as Belgium, France, and the UK allow them to participate in this service. Therefore, the Italian regulator will allow them to participate as well from 2018, in which TSO can ask for reactive power control from DSO. A possible scheme is under discussion where aggregators can communicate directly with the DER, and aggregators can participate too directly in the ancillary market. It should be noted that the price for supporting the reactive power of each type of technique is different, e.g., the price of generators is higher than the price of STACOM and the price of STACOM is higher than the price of aggregators. Therefore, AS of reactive power support plays an important role and must be properly provided. In six ASs, reactive power support is an effective way to ensure the safe operation of the power system [6]. Today, the need for reactive power demand increases in power systems, and the adverse effect is due to its lack; thus, reactive power provision and compensation are of great importance. Reactive power can improve the voltage profile and reduce losses in the network.
- **Frequency reserve:** The main aim of this service, which is to correct sudden imbalances between total production and total demand, includes three types: primary reserve, secondary reserve, and tertiary reserve. From an economic point of view, the frequency reserves are usually yearly contracted or result from daily

auctions, with a pay-as-bid remuneration. In some countries, only the capacity is explicitly remunerated, and automatic activation of balancing energy is paid at the normal electricity spot price. The primary reserve is automatically activated at the European level; the minimum quantity and granularity shall be 1 MW, the validity period shall be 15 minutes, and the full activation time shall be 5 minutes starting from 18 December 2025 [7]. The secondary reserve, which is used for planning 2.5 minutes for the preparation period and 10-minute for the ramping period. Additionally, the minimum quantity of bid is 1 MW, and the full activation time expected is 12.5 minutes [8]. The tertiary reserve is used by TSO to build up suitable margins for the enabled units concerning the minimum/maximum power. DERs connected to distribution networks can be classified into two groups: (i) only go downward and (ii) can go either upward or downward. The first group contains wind generators and solar panels, while the second group contains a combined heat plant and storage system. According to Reference [9], the minimum bid size of the tertiary reserve shall also be 1 MW. Today, this service only allows the generator that has a minimum capacity of 10 MW for Italy, and 5 MW for Nordic countries [10]; thus, in the case that DERs connected to the distribution network can participate in this service, the aggregator is necessary to aggregate the flexible resources managed by this aggregator.

6.3 TSO/DSO coordination market

The most important commodity in the electricity market is active power which is traded in the spot market. TSO has provided suitable conditions for transferring energy in the transmission network. At DSO level, AS with the same characteristic of TSO is used to manage the local system challenges. Nowadays, the unit directly connected to the distribution network can provide balancing services to both TSO and DSO through their coordination scheme. Each coordination scheme is presented by a set of roles, taken up by TSO and DSO, market players. There are five coordination schemes presented so far [11]: (i) Centralized AS market model, (ii) local AS market model, (iii) shared balancing responsibility model, (iv) Common TSO-DSO market model, and (v) Integrated flexibility market model.

Today, the SmartNet project [12] can be considered a well-known project about this work, in which five different coordination schemes [13] have been proposed and organized as follows:

1. **The centralized AS market model in Figure 6.1:** The AS market is operated by TSOs who are also the only buyers. The role of DSOs is very limited and they are only involved in the prequalification process in the case it would be implemented. The system prequalification process can access DERs to offer flexibility in the market and takes place before the clearing of the market. The main aim of this process is to try to ensure that the activation of DERs in the distribution grid by TSOs does not have any negative effect on the distribution system's reliability.
   - Advantage: The operation cost for market operation is low since TSOs are the unique managers of this service and the role of DSOs is limited; thus,

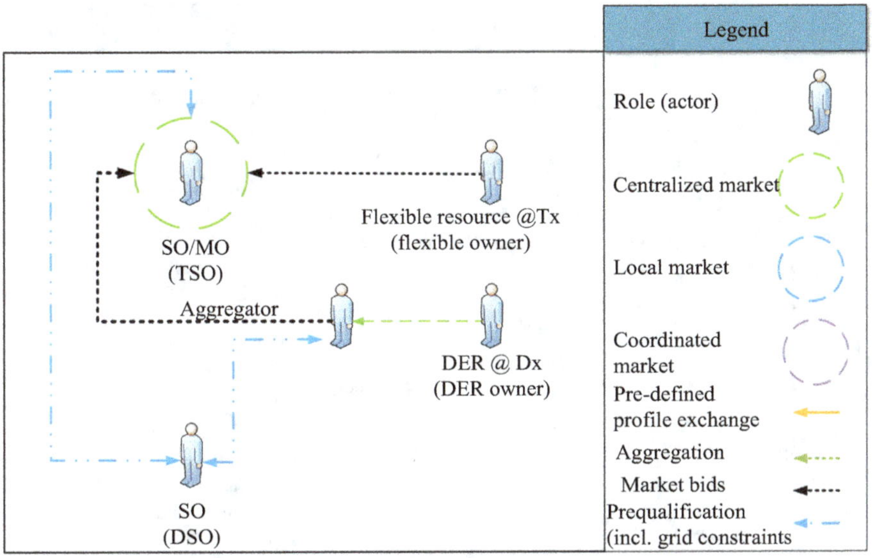

Figure 6.1 The centralized AS market model [13]

little communication between TSOs and DSOs is required. Moreover, this cooperation scheme is relatively simple since it is not too different from the state-of-the-art.

- Disadvantage: The distribution grid constraints may not be respected since there is little communication between TSOs and DSOs in this cooperation scheme. Additionally, DSOs will lose the potential opportunity that could bring to solve local grid challenges.

2. **The local AS market model in Figure 6.2:** There is a separate local market operated by DSOs and used to solve local congestion. Here, DSOs have priority to use DERs from the distribution grids and also are responsible for two tasks: (i) solve local congestion and (ii) aggregate remaining unused bids to pass to the AS operated by TSOs. It should be noted that the communication between DSOs and TSOs in this scheme is necessary because the activation of DERs by DSOs might generate system imbalances; thus, DSOs must notify TSOs of these activations in order to help TSOs in restoring the system's balance.
   - Advantage: The specific local market is beneficial for a DSO because the flexible resources are tailor-made to the needs of a specific DSO.
   - Disadvantage: The costs for procuring flexibilities or setting up the communication are increased because multiple small distribution grids would end up with separate local markets. This issue contains a risk of having multiple illiquid markets operating at higher costs. Furthermore, the more local the market is, the higher the cost becomes for setting it up.
3. **The shared balancing responsible model in Figure 6.3:** There are clear benefits from sharing DER capabilities between TSOs and DSOs, where the DERs

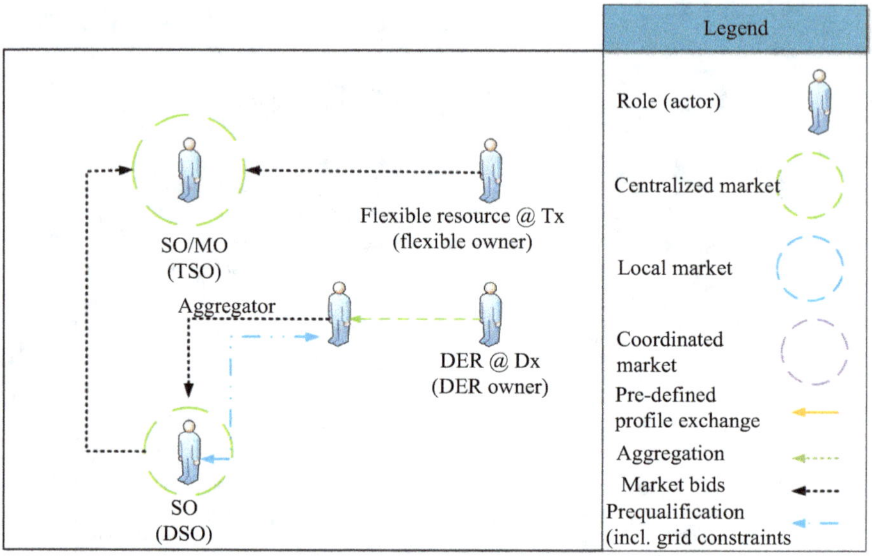

Figure 6.2 The local AS market model [13]

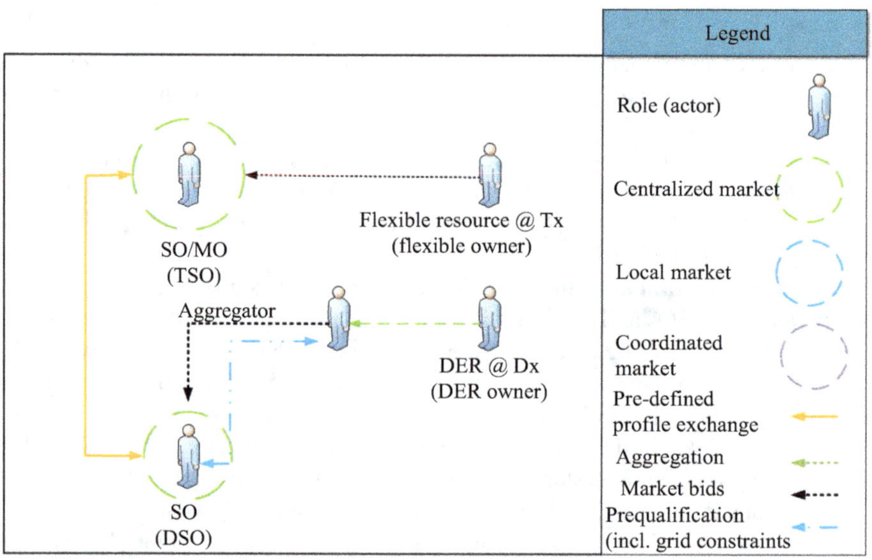

Figure 6.3 The shared balancing responsibility model [13]

connected at the transmission grid level are operated by the TSOs and the DERs connected at the distribution grid level are operated by the DSOs. Notably, the main difference between this coordination scheme and the others is that resources from the distribution grids cannot be offered to the transmission grid; thus, DSOs also exclusively use the flexible resources connected to distribution grids. Here, DSOs will receive the new responsibility of balancing the distribution grid through a schedule defined upfront between a certain TSO and a certain DSO because a TSO cannot access the flexible resources of a DSO. This schedule is determined based on the result of the energy markets or alternatively according to historical data at each TSO–DSO interconnector point.

- Advantage: The liquidity of resulting AS local markets is higher since DSOs take over part of the balancing in distribution grids according to a predefined schedule.
- Disadvantage: Similarly to the local AS market model, the separate local market might risk low liquidity, leading to higher costs of procuring flexibility for balancing purposes. Because the DER can only be used for balancing the distribution network, the TSOs would have fewer flexibility options available while the DSOs would lose the opportunity to sell flexibility resources to the TSOs.

4. **The common TSO-DSO market model in Figure 6.4:** TSOs and DSOs have the same priority to access the flexible resources connected to transmission and distribution grids. The aim of this coordination scheme is to maximize social welfare and decrease the total cost of the whole system. There are two variant models proposed:

   Variant 1 (the extension of the centralized AS market model): The market operator operated by both TSOs and DSOs will collect all bids from transmission and distribution grids and solve only one time; thus, TSOs and DSOs have the same priority to access the flexible resources.

- Advantage: TSOs can share the operational cost with DSOs.

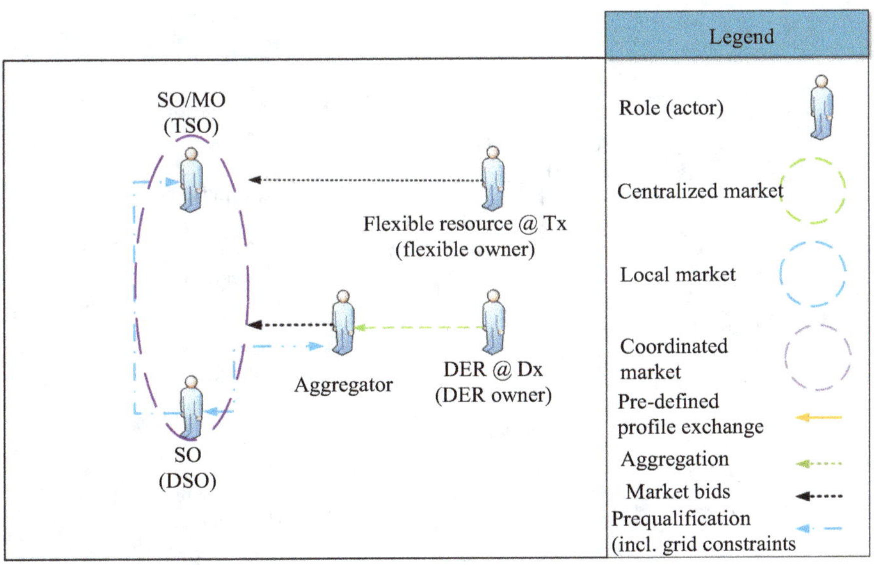

Figure 6.4 The common TSO-DSO AS market model [13]

- Disadvantage: TSOs have to share their data to DSOs. The mathematical model will be difficult to solve when many bids are offered.

Variant 2 (the extension of the local AS market model): The market is organized in a decentralized way containing two different steps. The first one, which solves the local market, is run by DSOs; since this would be only an intermediate step and the remaining bids will be available for the second step, the market participants are not yet informed about the outcomes. After the first step, the TSOs will run the second step of the market which has as its goal the solution of the transmission system issues (reserve, balancing, etc.).

- Advantage: TSOs can share less information with DSOs.
- Disadvantage: The cost of operating a small local market might be relatively high for the DSO.

5. **The integrated flexibility market model in Figure 6.5:** The model introduces a new market for flexible resources connected to transmission and distribution grids; thus, it implies that an independent market operator is introduced to guarantee the neutrality of this coordination scheme. The flexible resources will be allocated to the player with the highest willingness to pay. Moreover, TSOs and DSOs can resell previous contracts that were not used on the market at the same price they contracted the flexibility resources.
   - Advantage: It can provide high liquidity due to the presence of additional buyers of flexible resources. Moreover, TSOs and DSOs can resell the not-used contracts on the market. Therefore, it is beneficial from a global social welfare point of view.
   - Disadvantage: Because system operators might each activate flexible resources that negatively influence each other's positions, leading to

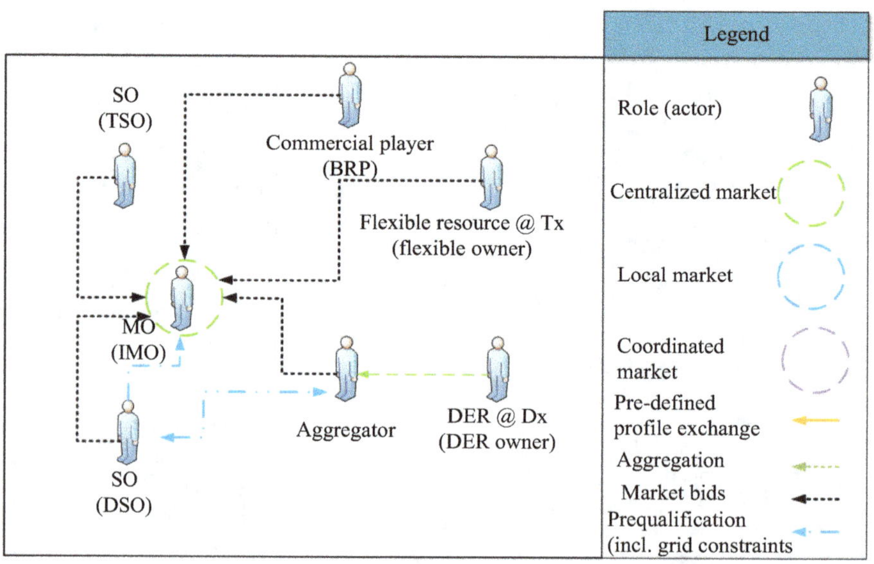

Figure 6.5 The integrated flexible market model [13]

unnecessary costs. It is more complex for TSOs to determine the amount of AS to procure; thus, TSOs need to buy additional capacity upfront to guarantee sufficient resources to control the system balance. The opening of AS for non-regulated parties (new market operators) can hinder the development and liquidity of the coordination scheme.

In the case that the AS market model will be adopted as similarly as the **Centralized AS market model**, TSO is the only buyer for the flexible resources connected to distribution grids; thus, system prequalification organized by DSOs is required to ensure that the activation of resources from the distribution grid by TSOs does not cause additional constraints in the distribution grid. This process should take place before the market clearing. In the case that the AS market model will be adopted as similarly as the **Local AS market model**, DSOs must ensure that only bids respecting local distribution grid constraints are accepted or transferred to TSOs.

In the case that the AS market model will be adopted as similarly as the **shared AS market model**, the flexible resources only can be offered by DSOs. Therefore, the impact of the flexible resources connected to distribution networks is presented by an equivalent provider for each HV/MV interface. In this case, DSO will be responsible for aggregating the flexible resources in their distribution network and representing them as a P-Q plane, including also costs and network constraints because TSO only looks at these flexible resources as a generator at the interface of HV/MV and DSOs are responsible for the security of their networks.

In order to aggregate the flexible resources connected to distribution networks, DSO must have full information in real time about total power installed per source, total load and gross load compensated by DERs, and real-time data for P and Q for all sources. When real-time information is not fully available, updated forecasts and reference production will be used to extrapolate and estimate missing data.

In the case that **a common TSO-DSO AS market model** will be adopted, the first variant activates as the centralized AS market model will be adopted and the second variant follows as the shared AS market model.

In the case that the **Integrated Flexible market model** will be adopted, a new entity introduced has a role in the data management and settlement of the market.

6.4 Data exchange/communication issues

The coordination scheme of TSO-DSO in AS presents new challenges as well as opportunities for communication in terms of communication cost, quality availability, response time, and security [13]. The data management of the new coordination scheme of TSO-DSO must guarantee data privacy, data, and communication security. Moreover, this process must increase transparency, fairness, non-discriminatory access, simplicity, and market competition. Here, the operation of communications is divided into four functional stages: prequalification, procurement, activation, and settlement. The types of data classified by time frame include the following: (i) planning data, which are forecasts, schedule data, and contractual; (ii) measured data, which are real-time and ex-post measurements. In general, the data can be collected and exchanged at the interface of the TSO and DSO's grid. Data such as DSOs realized or expected load, generation flows with detail on the generation technology, information on grid users, and DSOs geographically differentiated information on planned loads (directly from Balancing Response Parties or via TSOs) are needed to validate the available grid capacity before the market can be cleared and also to evaluate the necessity of requesting flexibility for congestion management from distribution grids. Moreover, TSOs and DSOs need information from market parties.

In fact, further detailed data could also be required depending on the role to be performed and which task. The required data [14] are shown in the following:

1. Congestion management: Check whether certain flexibility bids, when activated, could create congestion in their grid.
2. Balancing: Assess the impact of activation of balancing services and other flexibilities on TSO or DSO grids at different time frames (e.g., within the prequalification and day-ahead process, balancing market gate closure time, etc.).
3. Flexibility: Manage the grid properly, considering that the market works as freely as possible.
4. Real-time control and supervision: Maintaining the security of supply for customers and society as a whole; thus, the exchanged data requires a specific level of detail, a specific time frame for delivery, etc., linked to technical needs.
5. Network planning: Supporting the long-term network development process could include simplified electrical grid models, including foreseen and planned grid expansion projects as well as annual demand/generation forecasts per physical TSO-DSO interface. Supporting operational planning could include the year-ahead availability plan, outages, business continuity/emergency plans, and information related to upfront activities for operational security analysis.

The time frame of receiving data is very important because the need for real-time and near-real-time data is required. Additionally, in the case of the AS being opened for flexible resources, the increase in data exchange is the consequence of it; thus, the IT system will be more vulnerable to external threats. Today, many new technologies can be selected for communication, such as the next-generation wireless network (5G) and Internet-of-Things (IoT) to enable a flexible way of collecting information, also from the edges of the network. The Network Function Virtualization (NFV), Software Defined Networking (SDN), and network slicing can be used to offer flexible ASs to stakeholders without dedicated communications networks. Meanwhile, DataHub and Blockchain can be selected to provide easy access to data storage, to enable cross-border data exchange, and to improve security and privacy in communications.

The life cycle of sharing data between TSOs and DSOs in Figure 6.6, including six steps [14]:

1. Measurement: Value that can be read from a measuring instrument;
2. Declaration: Declaration by the concerned party (result of a decision, forecast …);
3. Collection: Actual reading of measuring instruments by physical inspection or remotely;

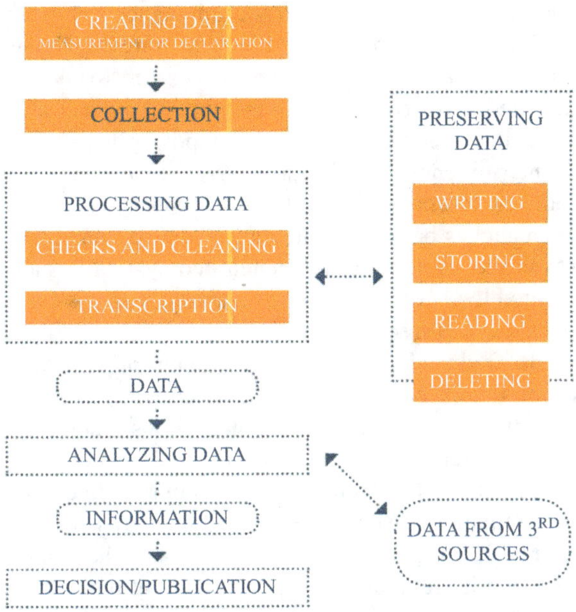

Figure 6.6 Data life cycle [15]

4. Checks and cleaning: No raw data is considered "correct" per se; therefore, it is checked and cleaned/ corrected when necessary;
5. Transcription: Change in format to provide, for instance, standardized data;
6. Analysis data: Could include data aggregation to the appropriate granularity and data interpretation.

6.5 Voltage and frequency regulation as an AS using PV-STATCOM and BESS

The transition from conventional synchronous generators to inverter-based renewable energy sources (RES), such as photovoltaic (PV) and wind generations, introduces significant challenges to the frequency stability of power systems. Synchronous generators inherently contribute to system inertia through their rotating masses, which resist rapid changes in rotor speed, thereby dampening frequency deviations during disturbances. This characteristic, known as "inertial response," provides critical stabilization during sudden events such as loss of generation or abrupt load changes. However, inverter-based resources lack physical inertia, as they are interfaced with the grid via power electronics inverters rather than mechanical coupling. In low-inertia systems, disturbances cause more rapid and pronounced frequency deviations because the grid's ability to resist these changes is significantly diminished. Mathematically, this can be understood through the swing equation:

$$H.\frac{df}{dt} = \Delta P \tag{6.1}$$

where H is the inertia constant and ΔP is the imbalance between generation and demand. A lower inertia constant H amplifies the rate of change of frequency (RoCOF) (*df*), reducing the available time for grid operators to take corrective actions. The issue is exacerbated in distribution systems with high penetration of RES. Such systems often operate in decentralized modes, with power flows that are bidirectional and variable. The intermittency of RES, driven by factors like solar irradiance and wind speed, introduces additional fluctuations in power generation. This variability can result in mismatches between supply and demand, manifesting as frequency oscillations. For example, in a highly PV-penetrated system, a sudden reduction in solar irradiance due to cloud cover can cause an abrupt loss of generation, creating a power deficit and subsequent frequency drop.

Furthermore, the reduced availability of synchronous generation in RES-dominated grids limits the provision of primary frequency response (PFR). Traditional PFR relies on the governor action of synchronous machines to adjust power output proportionally to frequency deviations. The reduced damping and inertia in low-inertia systems lead to higher oscillatory modes in system dynamics. Eigenvalue analysis of power system models shows that critical modes associated with frequency response shift closer to the imaginary axis on the complex plane in low-inertia conditions. This indicates a reduced ability to suppress oscillations, increasing the risk of system instability. In the absence of these machines, the responsibility for frequency regulation shifts to inverter-based resources, which must emulate inertial and governor responses through advanced control strategies. However, the emulation of these characteristics via software introduces latency and challenges in response accuracy.

Addressing these issues requires innovative solutions that can provide synthetic inertia and fast-response capabilities to restore balance and maintain grid reliability. Advanced distributed energy resource technologies, particularly PV STATCOM and battery energy storage systems (BESS), have emerged as critical assets in addressing frequency regulation challenges in low-inertia power systems. PV systems equipped with Static Synchronous Compensators (STATCOMs) utilize inverter technology to provide dynamic reactive power support. Unlike traditional STATCOMs, PV STATCOMs leverage the existing infrastructure of PV inverters, reducing costs and enhancing operational efficiency. During the daytime, PV inverters handle active power generation and reactive power compensation simultaneously. At night, when PV panels are inactive, the inverter can operate solely as a STATCOM, offering reactive power support for voltage stabilization. BESS complements PV STATCOMs by addressing both active and reactive power requirements. BESS with advanced inverter designs can switch between grid-forming and grid-following modes. In grid-forming mode, BESS provides voltage and frequency references, mimicking synchronous machines. In grid-following mode, it adjusts power flow based on grid conditions. This dual functionality enables BESS to provide synthetic inertia, damping oscillations, and smoothing renewable energy fluctuations. In most of the literature, active power (P) predominantly catered to frequency regulation while reactive power (Q) was reserved for voltage regulation. However, a contemporary approach proposes the simultaneous use of both these powers, through a coupled P-Q method, to jointly regulate voltage and frequency, termed as FVPQ. This approach provides the intricacies and benefits of the PV-STATCOM concept coupled with BESS and its potential to revolutionize voltage and frequency regulation, all while ensuring cost-efficiency.

6.5.1 PV-STATCOM—BESS with coupled active and reactive power control

6.5.1.1 Operation modes of PV-STATCOM

PV-STATCOM systems can function under three primary operational configurations: PV mode, partial PV-STATCOM mode, and full STATCOM mode (see Figure 6.7). These modes are selected based on real-time weather conditions and the dynamic requirements of the electrical grid. During daylight hours with sufficient solar irradiance, the system operates in PV mode, prioritizing the injection of maximum active power into the grid to enhance renewable energy utilization. This mode ensures optimal energy harvesting from the PV array.

At night or during periods with negligible solar input, the PV array does not produce active power. In this scenario, the system transitions to full STATCOM mode, wherein the inverter ceases active power delivery but actively contributes to grid stability by providing dynamic reactive power support. This is achieved through the absorption or injection of reactive power (Q), ensuring the voltage profile and system frequency remain within acceptable ranges.

Partial PV-STATCOM mode represents a hybrid configuration that becomes essential when moderate solar power is available but the grid requires reactive support due to transient disturbances, such as voltage sags or load fluctuations. In this mode, the inverter simultaneously handles both active and reactive power components. A

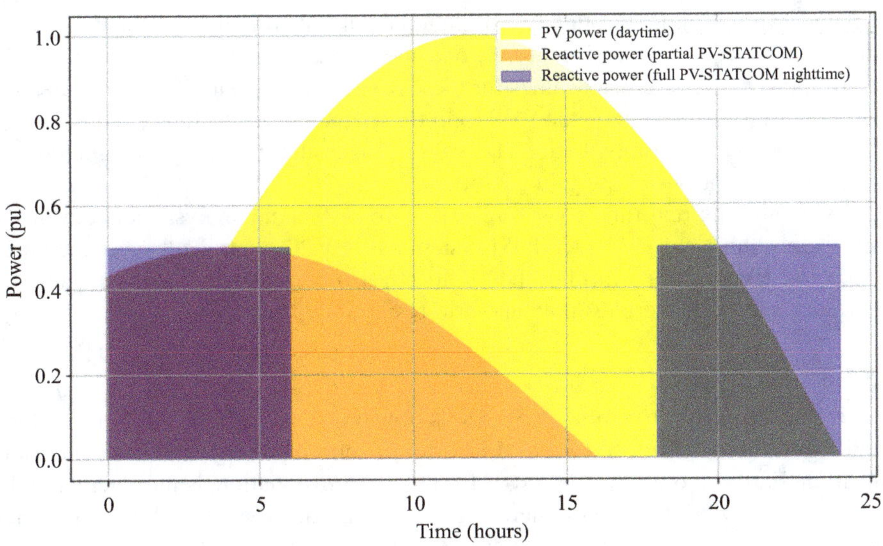

Figure 6.7 PV-STATCOM operation modes

portion of the inverter's capacity is reserved for reactive power compensation, while the remainder continues to inject active power. This balance ensures grid support services are maintained without a significant drop in energy yield.

Such flexible operational strategies allow PV-STATCOM systems to serve dual purposes: renewable energy integration and ancillary grid services. This is particularly valuable in modern power systems with increasing penetration of intermittent renewables.

6.5.1.2 PV-STATCOM control system

The control system of PV-STATCOM, especially under the Frequency-Voltage-Phase-Reactive (FVPQ) control approach, offers greater adaptability compared to conventional methods. Although the PV mode's control structure remains consistent, substantial improvements are observed in the partial and full STATCOM operating modes.

The FVPQ scheme enables reactive power and frequency regulation even during the daytime. Traditional PV systems typically curtail power production entirely during STATCOM operation, but the FVPQ method reallocates curtailed energy for grid stabilization purposes. Both partial and full STATCOM modes within this control architecture are capable of providing active support to frequency control mechanisms by adjusting inverter behavior.

In full STATCOM mode driven by the FVPQ logic, the system activates even under sunlight conditions but with reduced active generation. The curtailed power is instead repurposed for reactive power management and frequency support. This not only enhances grid reliability but also contributes to voltage regulation during disturbances.

The architecture of the FVPQ-controlled PV-STATCOM is typically centered around a Proportional-Integral (PI) current control loop (see Figure 6.8). This loop regulates inverter output through two orthogonal current components: the d-axis current (i_d), which governs active power, and the q-axis current (i_q), which controls reactive power. The reference values for these currents are dynamically generated by outer control loops—namely, the FVP loop for frequency and voltage, and the FVQ loop for reactive power modulation. These references allow for precise regulation of inverter output under varying grid conditions.

By combining real-time solar forecasting, load demand analysis, and advanced control algorithms, the PV-STATCOM system with FVPQ control enhances the functionality of PV inverters, turning them into multi-functional grid assets capable of contributing to both energy supply and grid resilience.

6.5.1.3 BESS control system

The overall control structure of BESS is illustrated in, where Andronov–Hopf Oscillator (AHO) is utilized for GFM inverter to achieve voltage and frequency stabilization. Furthermore, a VIC, derived from the swing equation, is incorporated into the AHO framework to provide inertia support from the BESS to the system. AHO consists of two main parts: (i) a linear resonant LC tank to set the system nominal frequency ($\omega_n = 1/\sqrt{LC}$), and (ii) nonlinear state-dependent voltage and current sources to sustain the oscillation by dynamically modulating their output based on

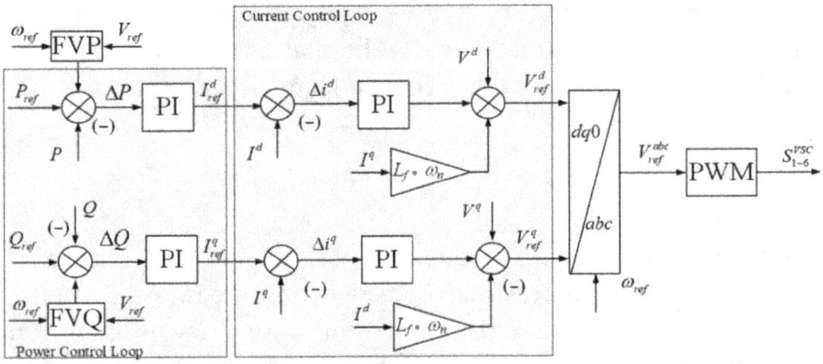

Figure 6.8 PV-STATCOM control system

inductor current or capacitor voltage. These sources ensure oscillation amplitude remains stable, counteracting energy losses and adapting to changes in grid conditions, such as load variations or disturbances. Their nonlinear dynamics enable robust synchronization and decentralized operation, crucial for grid-forming converters in complex power systems. AHO is coupled with the physical electrical power system through voltage (K_v) and current (K_i) scaling factors. This control logic uses measured filter inductor current i_{abc} as an input and its output v_c is used to generate the Pulse Width Modulation (PMW) signals required to drive the converter, with the convergence speed to the steady state of ξ. The dynamics of AHO is governed by the following equations:

$$\begin{aligned}|\dot{V_C}| &= \frac{2\xi}{K_v^2}|V_C|(|V_n|^2 - |V_C|^2) \\ &\quad - \frac{K_vK_i}{3C|V_C|}\cdot(\sin(\psi)(Q - Q_{ref}) + \cos(\psi)(P - P_{ref}))\end{aligned} \tag{6.2}$$

$$\dot{\theta_C} = \omega_n - \frac{K_vK_i}{3C|V_C|^2}(\sin(\psi)(P - P_{ref}) + \cos(\psi)(Q - Q_{ref})) \tag{6.3}$$

According to Reference [16], the generalized equilibrium point of the AHO voltage and frequency is obtained in the steady state when there is no oscillation in the voltage output (i.e., $|\dot{V_C}| = 0$) and the frequency reaches stable value (i.e., $\dot{\theta_C} = \omega_C$), as follows:

$$|V_C| = \frac{|V_n|}{\sqrt{2}}\sqrt{(1 + \sqrt{\begin{aligned}&1 - \frac{2K_iK_v^3}{3C\xi|V_n|^4}(\sin(\psi)(Q - Q_{ref}) \\ &+ \cos(\psi)(P - P_{ref}))\end{aligned}}} \tag{6.4}$$

$$\omega_C = \omega_n - \frac{K_vK_i}{3CV_C^2}(\sin(\psi)(P - P_{ref}) + \cos(\psi)(Q - Q_{ref})) \tag{6.5}$$

The parameter ψ is defined based on the network characteristics. In this work, $\psi =$ is chosen as $\pi/4$ so that both the active and reactive power output of BESSs will have an impact on regulating the grid frequency. It should be noted that, when AHO is operated in GFM mode with VIC, the power set points P_{ref} and Q_{ref} are set to 0. In this work, two internal parameters of AHO, namely nominal voltage V_n and nominal frequency ω_n, are adaptively adjusted by additional terms ΔV_n and $\Delta\omega_n$ to fulfill desired control objectives from a higher control layer, as follows:

$$V_n = \Delta V_n + V_{n0} \tag{6.6}$$

$$\omega_n = \Delta\omega_n + \omega_{n0} \tag{6.7}$$

This work adopts the complex frequency concept for VIC to overcome the limitations of conventional frequency, which considers only steady-state oscillation rates and neglects dynamic behavior like damping or instability. By utilizing complex frequency, which includes both oscillatory (imaginary) and damping (real) components, the proposed approach provides a deeper understanding of system dynamics during transients. The detailed explanation of the complex frequency concept can be found in Reference [17]. Unlike traditional VIC methods that rely on either active or reactive power from BESSs, this work incorporates both active and reactive power, enhancing the accuracy and adaptability of VIC for BESS systems. The detailed formulation for this improved VIC approach integrated into AHO of $BESS_i$ is presented in the following:

$$\Delta P_{VI} = (1-\gamma)J\omega_{CF}\frac{d\Delta\omega_{CF}}{dt} + (1-\gamma)D(\omega_{CF}-\omega_n) \tag{6.8}$$

$$\Delta Q_{VI} = \gamma J\omega_{CF}\frac{d\Delta\omega_{CF}}{dt} + \gamma D(\omega_{CF}-\omega_n) \tag{6.9}$$

In this context, *J* and *D* represent the moment of inertia and the damping factor, respectively, based on the active and reactive power of the *BESS*. These parameters can be modeled as proportional-derivative (PD) controllers, with the gains defined as $K_D = J\omega_{CF}$, $K_P = D$. The parameter γ determines the proportional contributions of active and reactive power required from the BESS to support both frequency regulation and system inertia enhancement. This approach improves practical applicability by leveraging the reactive power capacity in addition to the reserve active power, making the proposed method more efficient and versatile in real-world scenarios.

Finally, the value of ΔV_n and $\Delta\omega_n$ are determined by the following adaptive control law:

$$\frac{d}{dt}\Delta V_n = K_{\Delta V}\left(\sin(\vartheta)(\Delta Q_{VI}) + \cos(\vartheta)(\Delta P_{VI})\right) \tag{6.10}$$

$$\frac{d}{dt}\Delta\omega_n = K_{\Delta\omega}\left(\sin(\vartheta)(\Delta P_{VI}) - \cos(\vartheta)(\Delta Q_{VI})\right) \tag{6.11}$$

The gains $K_{\Delta V}$ and $K_{\Delta\omega}$ define the convergent speed of the adaptive control law. In this work, they are chosen based on trial and error with the trade-off between transient efficiency and system stability.

6.5.1.4 Simulation results and discussion

In this subsection, the effectiveness of the proposed control is validated using the modified IEEE 9-bus [18]. The SG at bus 3 is modeled as a Diesel generator with engine governor and field voltage excitation. The SGs at buses 1 and 2 are replaced with two CBRs (one BESS and one PV system) having the same rated powers (200 MVA), and the inertia constant of SG at bus 3 is reduced from 3.7 to 0.7 to emulate a low-inertia system, as shown in Figure 6.9. To simplify the simulation, CBRs are assumed to interface with the rest of the grid via an LCL filter, a DC-AC converter, and a step-up 13.8/230 kV, 210 MVA transformer, and are controlled by the proposed control method. The detailed specifications of the IEEE 9-bus system can be found in Reference [19]. The control parameters are briefly described as follows:

- CBR capacity: 200 MVA; DC voltage: 30 kV. LCL filter parameters are designed based on Reference [20]: 0.32 mH : 0.192 mH : 167.14 uF.
- The parameters of AHO are chosen following the design process presented in Reference [21].
- VI parameter: $J = 0.9$, $D = 4.5$, $\gamma = 0.4$, $K = -0.8$.
- Adaptive control law: $K_{\Delta V} = 1.51$, $K_{\Delta\omega} = 500$.

It should be noted that the main focus of this work is to develop a VIC method to utilize the CF concept for better frequency regulation and inertia support performance, compared to conventional VIC methods. Therefore, the VIC parameters and adaptive control law are chosen based on a trial-and-error approach. The optimal design for these parameters is out of the scope of this chapter. Three simulation case studies have been conducted in MATLAB®/Simulink® to validate the performance of the proposed control method under different operating conditions: (i) impact of significant load changes and (ii) presence of contingency and loss of generation.

6.5.2 Case study 1: Load changes

In this case study, the goal is to show the improved performance of frequency regulation and inertia response of the proposed VI-AHO over the following: (i) AHO

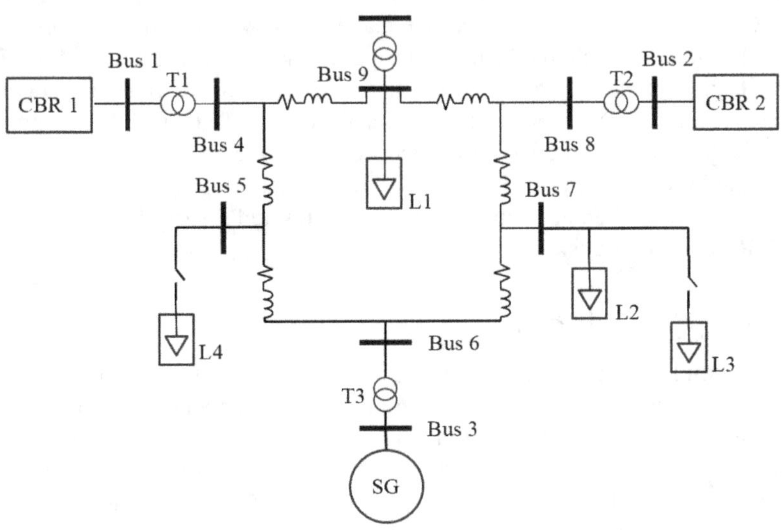

Figure 6.9 Modified IEEE-9 bus test system

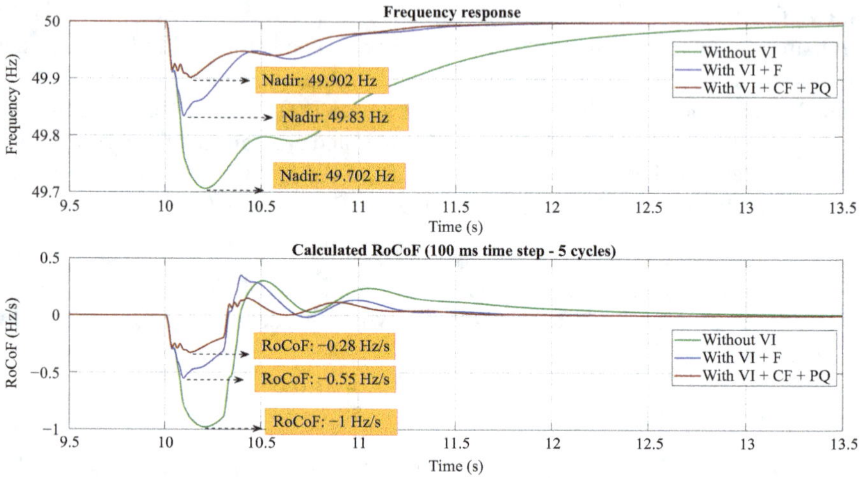

Figure 6.10 Deviation of grid frequency, to compare the proposed VI-AHO versus existing control approaches in case 1

without VIC and (ii) VIC-AHO with frequency measurement by the PLL approach (with VI + F). Initially, the total load power is $190MW + j95MVar$. Then at $t = 10s$, two loads at buses 5 and 7 are connected to the grid. CBR 1 (PV system) and SG are assumed to generate a constant 100 MW power to supply the loads, whereas CBR 2 (BESS) serves as GFM to control the system voltage and frequency and supply the rest of the loads. As can be seen in Figure 6.10, compared to AHO without VI (RoCoF = -1.0 Hz/s), and AHO with VI + F (RoCoF = -0.55 Hz/s), the proposed VIC (with VI + CF + PQ) shows a great improvement in frequency dynamics (RoCoF = -0.28 Hz/s and lowest Nadir of 49.902 Hz). According to the report from the European Network of Transmission System Operators for Electricity [22], all transients with RoCoF higher than 1 Hz/s are not manageable by system operators, posing potential instability problems. With the proposed method, it is possible for CBR to correctly trigger imbalance regulation in time and counter the transient, in contrary to the case without the proposed method. In addition, from Figure 6.11, the proposed method results in smaller active power needed for frequency and inertia support, by using available reactive power. It shows the economical application of the proposed method, especially when applying to CBRs with limited reserved active power like PV and BESS generations.

6.5.3 Case study 2: Contingency

In this case study, the effectiveness of the proposed method is demonstrated under a contingency condition in which a three-phase fault happened at bus 1 ($t = 5$ s). The voltage at bus 1 and power outputs of two CBRs can be seen in Figure 6.12. This severe problem causes overloading on the CBR 1 and forces it to be disconnected from the grid at $t = 5.2$ s. However, CBR 2 with GFM-AHO and the proposed control method automatically adjusts its power output to support the load/generation

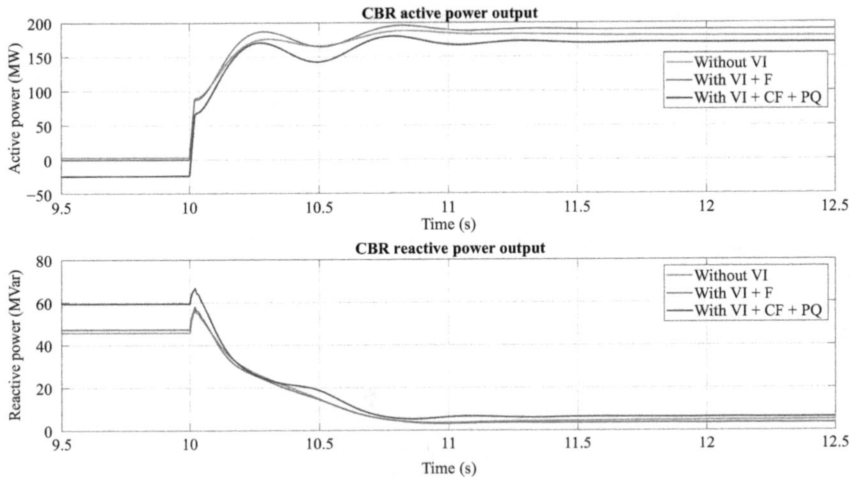

Figure 6.11 Power outputs of GFM-CBR 2 at bus 2

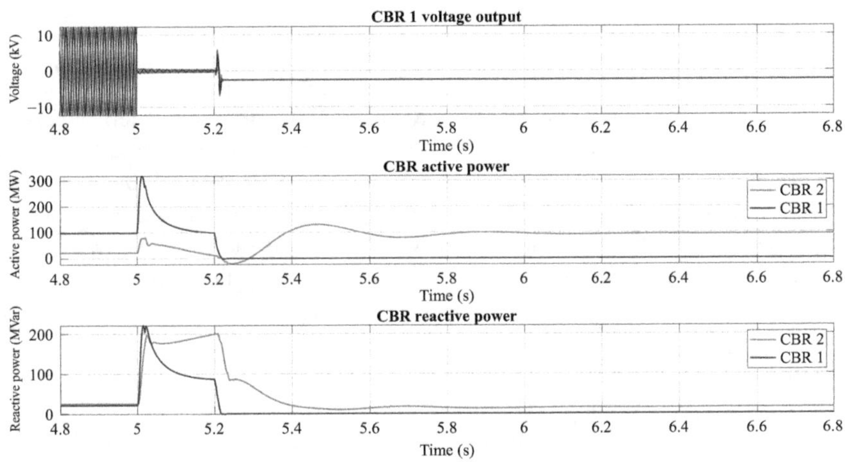

Figure 6.12 Voltage and power outputs of CBRs before and after the fault.

balance while maintaining the frequency stability with a small RoCoF of −0.52 Hz/s, as seen in Figure 6.13 (compared to −1.38 Hz/s in case of without the proposed method). These results prove that the proposed control method has a superior control effect even under severe fault conditions. This method significantly improves the grid frequency and overall system stability.

6.6 Conclusion

This chapter presents the AS for DERs in the distribution network. The need to open the possibility for DERs to participate in AS is necessary since the penetration of RES

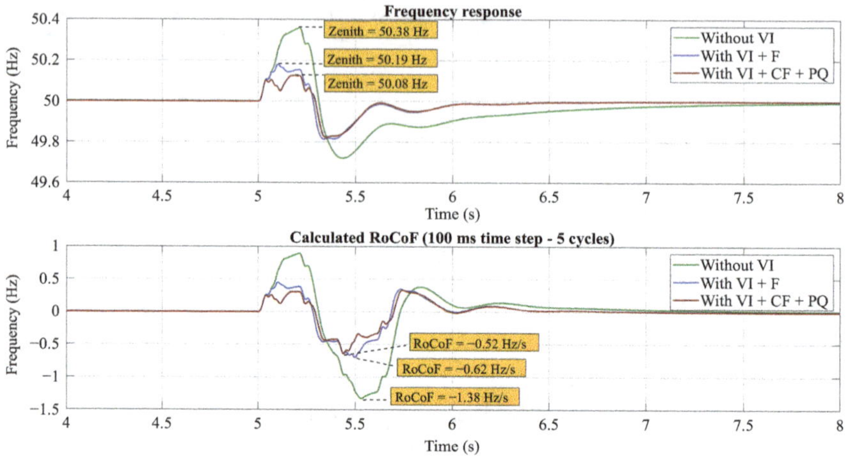

Figure 6.13 Deviation of grid frequency and RoCoF, to compare the proposed VI-AHO versus existing control approaches in case 2

is high in the power system. Normally, the AS is operated by TSO, but RES is located in the distribution network. Therefore, there are five coordination schemes between TSO and DSO to link them since DER is used to support the system. This chapter focuses on the voltage support and frequency reserve of DER in AS by cooperation between PV-STATCOM and BESS. Last but not least, the standard IEEE 9-bus network is used to test the methodology in two cases, and the result shows that this method significantly enhances the grid frequency and overall system stability.

References

[1] Agency TIE. World Energy Outlook 2024; 2024. Available from: https://www.iea.org/reports/world-energy-outlook-2024.

[2] Gubina A. Ancillary services in the distribution network: Where are the opportunities?; 2022. Available from: https://www.project-increase.eu/.

[3] Delfanti M, Galliani A, and Olivieri V. The new role of DSOs: Ancillary Services from RES towards a local dispatch. In: *CIRED Workshop*, Rome; 2014.

[4] Esmat A and Usaola J. DSO congestion management using demand side flexibility. In: *CIRED Workshop* 2016. IET; 2016. p. 1–4.

[5] Ali J, Massucco S, and Petretto G. Reactive power provision to TSO/DSO by aggregators and conventional generators. In: *2017 IEEE International Conference on Smart Grid Communications (SmartGridComm)*. IEEE; 2017. p. 486–491.

[6] Commission UFER, Commission UFER, *et al.* Promoting wholesale competition through open access non-discriminatory transmission services by

public utilities; recovery of stranded costs by public utilities and transmitting utilities. *Order*. 1996;888:24.

[7] ENTSO-E. All TSOs' proposal for the implementation framework for the exchange of balancing energy from frequency restoration reserves with automatic activation in accordance with Article 21 of Commission Regulation (EU) 2017/2195 establishing a guideline on electricity balancing; 2022. Available from: https://consultations.entsoe.eu/markets/afrr_implementation_framework/supporting_documents/20180426_aFRRIF_Implementation_framework.pdf.

[8] ENTSO-E. MARI – First Consultation; 2022. Available from: https://consultations.entsoe.eu/markets/mari-first-consultation-call-for-input/supporting_documents/20171121_%20MARI%20First%20Consultation_Final.pdf.

[9] ENTSO-E. TSOs' proposal for the establishment of common and harmonised rules and processes for the exchange and procurement of Balancing Capacity for Frequency Containment Reserves (FCR) in accordance with Article 33 of Commission Regulation (EU) 2017/2195 establishing a guideline on electricity balancing; 2022. Available from: https://docstore.entsoe.eu/Documents/Network%20codes%20documents/NC%20EB/FCR_Proposal-Article_33_1%20EBGL_20181018_FV.PDF.

[10] TSpA, Terna Grid code, Chapter 4: Dispatching regulation; 2022. Available from: https://download.terna.it/terna/0000/0123/63.PDF.

[11] H2020. Leveraging flexibilities for the European power system; 2022. Available from: https://www.osmose-h2020.eu/.

[12] Migliavacca G, Rossi M, Six D, *et al.* SmartNet: H2020 project analysing TSO–DSO interaction to enable ancillary services provision from distribution networks. *CIRED-Open Access Proceedings Journal*. 2017.

[13] Gerard H, Puente EIR, and Six D. Coordination between transmission and distribution system operators in the electricity sector: A conceptual framework. *Utilities Policy*. 2018;50:40–48.

[14] Lambert E, Morais H, Reis F, *et al.* Practices and architectures for TSO-DSO data exchange: European landscape. In: *2018 IEEE PES Innovative Smart Grid Technologies Conference Europe (ISGT-Europe)*. IEEE; 2018. p. 1–6.

[15] de Jong G, Franz O, Hermans P, *et al*. Tso-dso data management report. TSO-DSO Project Team. Tech. Rep; 2016.

[16] Heins T, Tran T, Raisz D, *et al.* Power control of Andronov-Hopf oscillator based distributed generation in grid-connected microgrids. In: *Advances in Engineering Research and Application: Proceedings of the International Conference on Engineering Research and Applications, ICERA 2020*. Springer; 2021. p. 675–687.

[17] Milano F. Complex frequency. *IEEE Transactions on Power Systems*. 2021;37(2):1230–1240.

[18] Milano F, Alhanjari B, and Tzounas G. Enhancing frequency control through rate of change of voltage feedback. *IEEE Transactions on Power Systems*. 2023.

[19] Tran T, Nguyen K, Bingel F, *et al.* Complex Frequency-Based Virtual Inertia Control for Frequency Regulation in Low-Inertia Power System. Available at SSRN 5059219.

[20] Liserre M, Blaabjerg F, and Hansen S. Design and control of an LCL-filter-based three-phase active rectifier. *IEEE Transactions on industry applications*. 2005;41(5):1281–1291.

[21] Lu M, Dutta S, Purba V, *et al.* A grid-compatible virtual oscillator controller: Analysis and design. In: *2019 IEEE Energy Conversion Congress and Exposition (ECCE)*. IEEE; 2019. p. 2643–2649.

[22] ENTSO-E. Inertia and Rate of Change of Frequency (RoCoF). *ENTSO-E Brussels*, Belgium; 2020.

Chapter 7

Efficient energy management for IoT applications in smart mesh networks for hybrid renewable generation

Thanh Dinh Ngo[1], Van Khoa Pham[2], Fabien Ferrero[3] and Nhan Le Thanh[4]

This chapter introduces an innovative Internet of Things (IoT) system designed for efficient energy management in smart mesh networks, leveraging hybrid renewable energy sources for environmental monitoring. The proposed system, an energy-efficient solution for WSNs using LoRa Mesh technology integrated with the Ad-hoc On-Demand Distance Vector (AODV) routing algorithm, addresses the challenges of energy consumption and connectivity in remote areas with unstable power supplies and harsh conditions. Using solar power supplemented by battery backup, the system ensures reliable data transmission while minimizing energy consumption during severe weather, with devices capable of functioning for up to a month solely on battery backup. Field experiments in Vietnam highlight its effectiveness, showcasing improved energy efficiency and extended operational range, and demonstrate that the proposed system achieves a packet delivery success rate (DSR) of 95%–99%, reduces packet loss to 1%–5%. The solution not only enhances reliability in remote areas without 3G/4G coverage but also supports sustainable IoT applications in hybrid renewable energy systems.

7.1 Introduction

The integration of IoT technologies into hybrid renewable energy systems, such as hydroelectric power plants combined with solar, battery backup, and grid power or wind energy, has revolutionized real-time monitoring and management. Wireless sensor networks (WSNs) play a pivotal role in these systems by enabling the collection and transmission of environmental data, such as rainfall, temperature, and humidity, which are critical for operational efficiency and disaster prevention (e.g., flood

[1]Faculty of Electrical Engineering, University of Danang, University of Science and Technology, Vietnam
[2]Faculty of International Education, HCMC University of Technology and Education, Vietnam
[3]Laboratory of Electronics, Antenna and Telecommunication (LEAT), Universite Cote D'Azur, France
[4]Danang International Institute of Technology, University of Danang, Vietnam

control). However, deploying WSNs in remote and harsh areas, particularly in regions with challenging terrains and no 3G/4G coverage, poses significant challenges in terms of reliability and energy efficiency.

In traditional WSNs, data transmission from the sensor to the gateway faces significant limitations due to factors such as long distances, complex terrain, and harsh environmental conditions. Common transmission technologies like Wi-Fi, Bluetooth, or Zigbee operate effectively only within short ranges—from several tens to a few hundred meters. This poses considerable challenges when deploying the system in remote, mountainous, or hard-to-access areas where the distance between sensor stations and the gateway may exceed the direct communication range, resulting in connection loss or reduced data accuracy.

Moreover, signal degradation or disruption is likely when transmitting through obstructed environments such as dense forests, heavily built-up urban areas, or hilly terrain. Since there is no intermediate relaying mechanism, data is at risk of being completely lost if the sensor lies beyond the direct communication range of the gateway.

Another drawback of traditional systems is their total reliance on direct communication between sensors and the gateway. This necessitates placing gateways near the sensors, which complicates deployment—especially in areas with limited infrastructure. If a gateway malfunctions or fails, all dependent sensor nodes are rendered inoperative, interrupting the data collection process.

To maintain signal transmission across extended distances, sensors in conventional systems need to boost their transmission power, resulting in increased energy usage. This reduces the battery lifespan of the devices and poses challenges for sustained maintenance and operation, especially for sensors reliant on constrained energy sources like batteries or solar panels.

Current technologies such as LoRaWAN often struggle to deliver the necessary performance due to energy limitations and signal attenuation in rugged terrains. Traditional WSNs often suffer from high packet loss rates (PLRs) (up to 20%) and limited coverage in remote areas, leading to unreliable data transmission [1]. Additionally, the energy constraints of battery-powered sensor nodes necessitate innovative solutions to extend operational lifetimes while maintaining performance. In the context of hybrid renewable generation, where hydroelectric plants often operate in remote locations, the need for a robust and energy-efficient WSN becomes even more critical to ensure continuous monitoring and timely decision-making.

This chapter proposes a solution to enhance the reliability and energy efficiency of WSNs for IoT applications in hybrid renewable energy systems. By leveraging LoRa Mesh technology and the AODV routing algorithm, the proposed system addresses the challenges of data transmission in remote areas and harsh conditions. The system's hybrid power approach, combining solar power and battery backup, ensures continuous operation even during solar power outages caused by adverse weather conditions. A four-layer architecture is designed, incorporating a sleep/wake mechanism to optimize energy consumption. The system is evaluated through real-world experiments, focusing on rainfall monitoring, with results demonstrating significant improvements in reliability and energy efficiency.

7.2 Related work

WSNs have been widely researched for IoT applications in various domains, including environmental monitoring, agriculture, and renewable energy systems. Several studies have focused on improving reliability and energy efficiency in WSNs, especially in challenging environments such as remote areas with limited connectivity. Christian Ebi *et al.* [1] proposed a synchronous LoRa Mesh network for monitoring underground infrastructure, such as water levels in sewer systems. Their system achieved packet DSR of 90%–95% in challenging environments with a PLR of 5%–10%. However, their work primarily focused on reliability in confined spaces and did not address energy optimization for battery-powered nodes in remote IoT applications, a critical requirement for hybrid renewable energy systems.

Van Dai Pham *et al.* [2] introduced a method for organizing mesh topology using LoRa technology, demonstrating its potential to extend coverage in WSNs. Their study reported a packet DSR of 90% over a 2 km range but lacked detailed analysis of energy consumption, which is essential for long-term deployments in remote areas. Similarly, Joan Miquel Sole *et al.* [3] implemented a LoRa Mesh library, achieving a DSR of 92% in urban environments. While their work highlighted the scalability of LoRa Mesh, it did not consider the energy constraints of nodes in remote regions without 3G/4G coverage, nor did it explore routing algorithms to enhance reliability.

Routing algorithms for WSNs have also been a focus of research. Douglas de Farias Medeiros *et al.* [4] analyzed the performance of AODV, DSR, and DVR algorithms in ad-hoc networks, showing that AODV outperforms others in terms of packet delivery (95% success rate) and adaptability to network changes. However, their study was conducted in simulated environments and did not address real-world challenges such as terrain obstacles or energy efficiency in LoRa-based networks.

In the context of hybrid renewable energy systems, WSNs have been applied to monitor power parameters [5]. Van-Thuyen Ngo *et al.* [5] developed a WSN for monitoring the energy supply from the hybrid small-scale wind turbine generators and rooftop solar power systems; their system relied on Zigbee, which has a limited range compared to LoRa. Daniel Lundell *et al.* [6] proposed a routing algorithm for LoRa Mesh networks, achieving a DSR of 93% over a 3 km range. Their study included energy analysis, reporting an energy consumption of 150 mJ per transmission at spreading factor SF = 7, but did not implement a sleep/wake mechanism to further optimize energy usage.

Existing solutions often fail to address reliability, coverage, and energy efficiency simultaneously, particularly in remote areas without 3G/4G coverage. While LoRa Mesh and AODV have been studied separately, their integration for energy-efficient IoT applications in hybrid renewable generation remains underexplored. This chapter addresses this gap by proposing a LoRa Mesh network with AODV routing, optimized through a sleep/wake mechanism, to ensure reliable and energy-efficient data transmission in remote hydroelectric plants.

7.3 Problem scope and motivation

7.3.1 Problem scope

In hybrid renewable energy systems, such as hydroelectric plants integrated with solar or battery backup, WSNs are deployed to monitor environmental parameters like rainfall, which is crucial for water resource management and flood prediction. However, these systems often operate in remote areas with the following challenges:

- **Limited coverage**: Traditional WSNs using technologies like Zigbee or WiFi have a short range (100–500 m), making them unsuitable for remote areas without 3G/4G coverage.
- **High packet loss**: In star topology networks, PLRs can reach 20% over distances exceeding 1.5 km, leading to unreliable data transmission [1].
- **Energy constraints**: Sensor nodes in remote areas rely on battery power, and frequent data transmission can deplete batteries quickly, reducing operational lifetimes to mere hours.

7.3.2 Motivation

The motivation for this research stems from the need to develop a reliable and energy-efficient WSN for IoT applications in hybrid renewable energy systems. LoRa technology, known for its extended range and low-power consumption, is highly appropriate for such applications. By integrating LoRa Mesh topology with the AODV routing algorithm, the proposed system aims to:

- Extend coverage in remote areas through multi-hop communication.
- Improve reliability by reducing packet loss and ensuring data delivery.
- Optimize energy consumption through a sleep/wake mechanism, enabling sustainable operation in hybrid renewable energy systems.

7.4 System architecture

7.4.1 Overview

The proposed system is designed to monitor rainfall at hydroelectric plants, supporting hybrid renewable generation by providing real-time data for water management. The architecture of the WSN for rainfall monitoring is structured into four primary layers: **Sensor layer**, **Gateway layer**, **Cloud layer**, and **Application layer**, as illustrated in Figure 7.1. Each layer has a vital function in the acquisition, transmission, processing, and utilization of data to support real-time monitoring and analytics. This layered model ensures the effective collection of environmental data and provides valuable insights for various operational and research purposes.

- **Sensing layer**: This is the foundational layer of the WSN architecture, responsible for collecting rainfall-related measurements. Sensor nodes are deployed across various locations and are designed to operate with minimal power consumption to ensure prolonged functionality, particularly during periods without

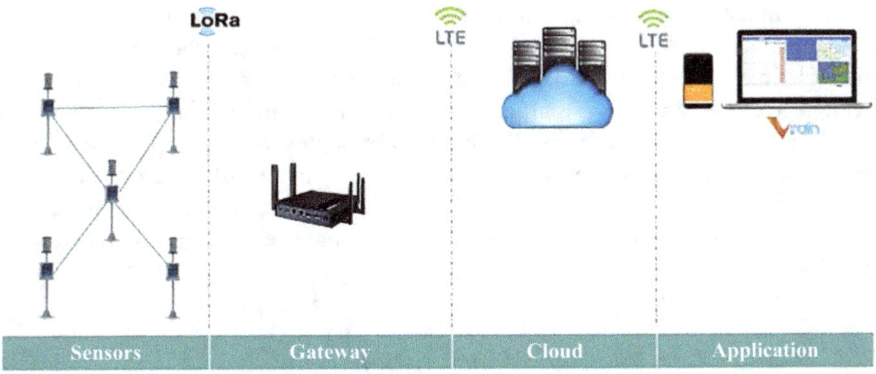

Figure 7.1 *System architecture*

solar or grid-based energy supply. Once data is collected, it is transmitted via LoRa to the gateway layer for further handling.

- **Gateway layer**: Acting as an intermediary between the sensor nodes and the higher-level network infrastructure, the gateway layer is tasked with aggregating data from multiple sensors, performing basic preprocessing, and temporarily storing the information. The processed data is subsequently sent to the cloud layer using the HTTP protocol. By carrying out preliminary processing at this level, the system reduces communication overhead and bandwidth usage. In addition, this layer ensures stable and reliable communication between edge devices and the centralized data processing system.
- **Cloud layer**: The cloud layer manages the task of storing, processing, and analyzing the data received from the rainfall monitoring stations. It provides secure data storage and enables remote, flexible access. Big Data analytics algorithms are deployed within this layer to detect trends, perform predictive modeling, and generate early warnings. The cloud system is also capable of integrating and harmonizing data from multiple sources to improve accuracy and consistency across datasets. The use of cloud computing not only streamlines data management but also enhances computational efficiency, making it suitable for both scientific research and operational needs.
- **Application layer**: At the top of the architecture, the application layer serves end-user needs by delivering processed data in actionable formats. A key feature of this layer is **real-time monitoring**, where users can visualize measurements through intuitive interfaces such as charts and dashboards. Moreover, the system supports early warning functionalities by issuing alerts when parameters exceed predefined safety thresholds—e.g., flood warnings triggered by excessive rainfall levels. The insights provided by this layer support timely and informed decision-making in disaster management, hydropower plant operation, and reservoir regulation. The integration of advanced data analytics tools within this layer further enhances the system's effectiveness in delivering accurate and actionable information.

The implementation of this four-layer architecture not only enables efficient environmental data acquisition and processing but also establishes a robust foundation for intelligent monitoring and decision-support systems. The incorporation of **LoRa Mesh technology** significantly extends the communication range of the network, allowing additional rainfall stations to be deployed in areas previously inaccessible due to weak or non-existent 3G/4G signals. As a result, hydropower plants can achieve more comprehensive spatial distribution of rainfall measurement stations, thereby improving environmental data acquisition, management, and analysis across the entire system.

7.4.2 LoRa Mesh and AODV integration

The system employs LoRa Mesh topology to enable multi-hop communication, extending coverage in remote areas. The AODV routing algorithm is integrated to dynamically establish routes between nodes and the gateway. The route discovery process involves:

- Route Request (RREQ): A source node broadcasts an RREQ packet to find a path to the gateway.
- Route Reply (RREP): The gateway responds with an RREP packet, establishing the route through intermediate nodes.
- Data transmission: Data packets are forwarded along the established route, with a maximum hop count of 5 to limit latency. Each packet is kept compact ($\leq$30 bytes) to optimize LoRa's payload constraints, as detailed in Table 7.1.

The AODV routing algorithm operates based on an *on-demand* principle, meaning that routes are established only when data packets need to be transmitted. AODV combines two essential processes: **route discovery** and **route maintenance**, which together enable dynamic and efficient path management in wireless mesh networks. The routing process can be divided into three primary phases:

- **Route discovery:** When a source node intends to transmit data to a destination node, it starts the route discovery by sending out an RREQ packet to its neighboring nodes. Upon receiving the RREQ, these nodes either reply with an RREP packet if they possess a valid route to the destination or forward the RREQ to their own neighbors. Once the RREQ reaches either the destination or an intermediate node with a valid route to the destination, an RREP is created and sent back along the reverse path toward the source. This reply packet has the information of the route from the destination to the source.
- **Route setup and maintenance:** Upon receiving the RREP, the source node chooses the best route, prioritizing the path with the fewest hops, and records it in its routing table. If multiple paths exist, the one with the fewest hops is selected. Once the route is set, data transfer occurs along this path.
- **Route maintenance and error handling:** During transmission, if a node along the path becomes unreachable due to failure or disconnection, the preceding node detects the break and sends a Route Error (RERR) message back to the source.

Table 7.1 Packet structure in LoRa Mesh

Field	Size (bytes)	Function
Type	1	Packet type (RREQ, RREP, Data)
Source ID	2	Source node identifier
Destination ID	2	Destination node identifier
Sequence number	2	Avoids routing loops
Hop count	1	Number of hops
Payload	≤20	Data (e.g., rainfall)
CRC	2	Error checking

Upon receiving this message, the source node restarts the route discovery process to find a new valid route to the destination.

Table 7.1 defines the structure of a data packet used in the AODV-based LoRa Mesh network for rainfall monitoring applications. Each field in the packet serves a specific role in ensuring accurate routing and reliable data transmission.

- Type (1byte): Specifies the type of packet being transmitted. Common types include RREQ, RREP, and Data (e.g., rainfall measurements). This field enables the receiving node to correctly interpret and process the packet.
- Source ID (2bytes): Uniquely identifies the node that originated the packet. This information is essential for route tracing, acknowledgments, and routing table updates.
- Destination ID (2 bytes): Indicates the intended recipient of the packet. Nodes use this field to determine whether to process, forward, or discard the packet.
- Sequence number (2 bytes): Prevents routing loops and ensures freshness of routing information. Each new route discovery or data transmission is tagged with a unique sequence number.
- Hop count (1 byte): Represents the number of nodes the packet has traversed. It is incremented at each hop and is used to limit packet propagation and control routing efficiency.
- Payload (≤20 bytes): Contains the actual data being transmitted, such as rainfall measurements. This field is limited in size to maintain low transmission power and latency.
- Cyclic redundancy check (CRC) (2 bytes): CRC used for error detection. It helps verify the integrity of the packet to ensure that no data has been corrupted during transmission.

The integration of LoRa Mesh networks with the AODV routing algorithm offers several advantages, particularly in enhancing coverage, improving energy efficiency, supporting dynamic adaptability, and leveraging existing LoRa infrastructure. Through multi-hop routing, data can be relayed from one node to another, effectively extending network coverage across large geographical areas without the need for extensive deployment of base stations or gateways. This feature is especially valuable in rainfall monitoring systems deployed at hydroelectric power plants, where sensors are often installed in remote and hard-to-reach locations. The ability to transmit rainfall data from distant nodes to the central gateway ensures comprehensive monitoring with minimal infrastructure requirements.

The on-demand operational nature of AODV further contributes to energy conservation. Routes are established and maintained only when data transmission is required, thereby eliminating unnecessary overhead associated with continuous route advertisements and maintenance. This reduction in control message exchanges significantly lowers the energy consumption of sensor nodes, ultimately extending their operational lifetime, a critical factor in battery-powered IoT systems deployed in remote environments.

In addition, the dynamic adaptability of the LoRa Mesh network with AODV enables the system to autonomously respond to topology changes. When a new node is added or an existing node fails, the network can rapidly reconfigure and establish alternative routes to maintain uninterrupted data transmission. This self-healing capability enhances network reliability and ensures resilience against node failures or environmental disruptions, making the system well-suited for real-world deployments in dynamic and harsh conditions.

7.4.3 AODV algorithm in WSN for rainfall monitoring

In hydroelectric power plants, rainfall measurement plays a critical role in water regulation. Accurate collection and analysis of rainfall data enable timely decisions regarding flood discharge, ensuring the safety of infrastructure and downstream areas, while also optimizing operational efficiency to maximize power generation output.

In conventional wireless rainfall sensing systems without LoRa Mesh, as shown in Figure 7.2, sensor nodes are limited to direct communication within their transmission range. As a result, if a sensor is installed in a remote or inaccessible area where the signal is too weak to reach the gateway, its data cannot be delivered to the central processing unit. This limitation significantly reduces the monitoring capability of the system and prevents timely detection and response to environmental changes occurring beyond the direct coverage of the gateway.

As shown in Figure 7.3, the integration of LoRa Mesh technology addresses these challenges by enabling the establishment of multi-hop communication paths between sensor nodes. With the assistance of intermediate nodes, data from remote

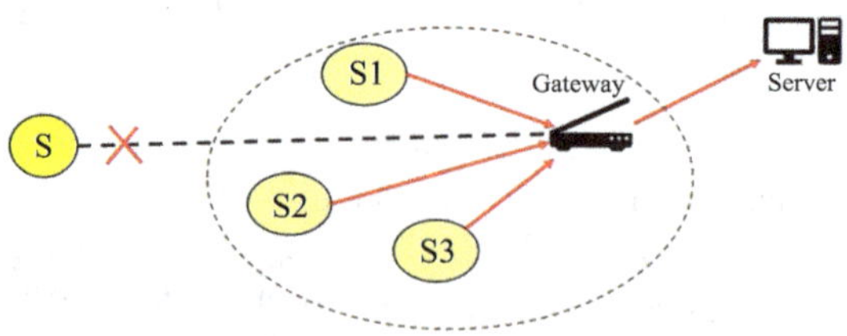

Figure 7.2 WSN without LoRa Mesh

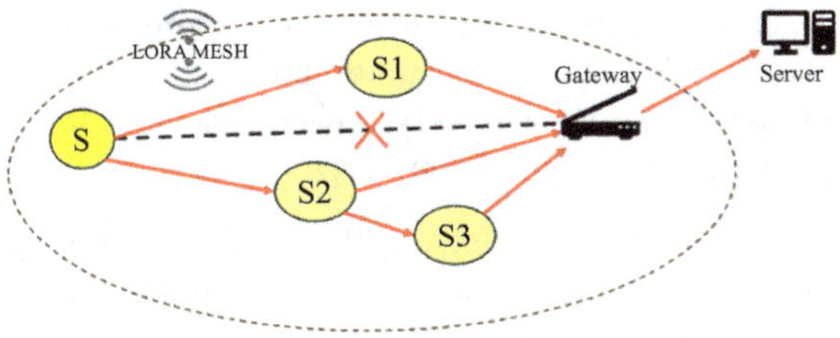

Figure 7.3 *WSN with LoRa Mesh*

sensors can be relayed to the gateway, even when a direct link is not feasible. The use of the AODV routing algorithm facilitates the dynamic discovery of optimal routes based on network demand, allowing each sensor to efficiently transmit data regardless of its proximity to the gateway.

Consequently, implementing AODV in LoRa Mesh networks significantly enhances the communication robustness and coverage of rainfall sensing systems. It ensures that data from geographically isolated or distant sensors is reliably collected and processed in real time, contributing to improved monitoring performance and operational decision-making in hydroelectric power plants.

7.4.4 Energy optimization

To optimize energy harvesting and storage for the rainfall sensors (equipped with STM32WLE5CC MCU and 3,500 mAh Lithium-ion 18650 battery), the system incorporates a Maximum Power Point Tracking (MPPT) solar charging circuit and a dedicated power management module. The MPPT circuit maximizes solar panel efficiency by dynamically adjusting the operating voltage to capture the maximum power output under varying light conditions. This ensures optimal energy transfer to the battery, even during cloudy weather prevalent in Vietnam's Central region. The power management module regulates charging and discharging of the 18650 battery, incorporating overcharge, over-discharge, and short-circuit protection to enhance battery longevity and safety. It also interfaces with the STM32WLE5CC MCU to monitor battery status, enabling intelligent power allocation during sleep and active modes. This setup supports the system's hybrid power strategy, seamlessly switching between solar and battery sources to maintain uninterrupted operation in remote settings.

To optimize energy consumption in the system, a **sleep/wake mechanism** should be implemented for nodes within the LoRa Mesh network. Each node operates in a duty cycle where it remains in **sleep mode for 9 minutes and 30 seconds**, and **wakes up for only 30 seconds** to transmit and receive data. This significantly reduces energy usage, as the node primarily operates in low-power sleep mode.

Prior to joining the network, the gateway synchronizes the current time with each node. Upon receiving this timestamp, the node aligns its internal clock with the system and begins its data transmission cycle in sync with the network.

The sleep/wake functionality enables the system to **conserve energy effectively**, as nodes are only active during necessary communication intervals. This feature extends the **operational lifetime** of battery-powered devices, making the network more sustainable for long-term deployment in remote monitoring scenarios.

- Sleep Mode: Nodes sleep for 9 minutes 30 seconds, consuming only 6.25 μW.
- Wake Mode: Nodes wake for 30 seconds to transmit/receive data, consuming 444 mW (TX) and 39.96 mW (RX) at SF = 12. The sleep/wake cycle is synchronized via the gateway, ensuring efficient energy use while maintaining data transmission frequency (every 10 minutes).

7.5 Experiments

7.5.1 Experimental setup

The system was deployed at an irrigation reservoir in a remote area, using RAK3172 CANOPUS hardware for sensor nodes and Watec gateway. The network consisted of 5 nodes, with the gateway located at the dam. Experiments were conducted to evaluate:

- Packet DSR and PLR at SF = 7 and SF = 12.
- Transmission Range: With and without LoRa Mesh.
- Energy Consumption and Battery Life: Using a 3,500 mAh, 3.7 V battery, with a transmission frequency of 10 minutes (9 minutes 30 seconds sleep, 30 seconds wake).

Sensor and relay nodes: The sensor and relay nodes, as shown in Figure 7.4, function as rainfall data acquisition stations, transmitting information to each other along pre-established paths toward the central gateway. These nodes consist of both rain sensors and intermediate relay nodes, which serve as signal repeaters to extend the LoRa coverage range and ensure reliable data delivery to the gateway, particularly in large or obstructed deployment areas.

RAK3172 CANOPUS technical specifications:

- LoRa Module: RAK3172
- LoRa Frequency: 923 MHz
- Interfaces Supported: UART / I2C / SPI / ADC / GPIO
- Operating Temperature Range: –40 °C to 85 °C
- Power Supply Voltage: 3.7 VDC to 4.2 VDC
- Ultra-low Power Consumption: 1.69 μA in sleep mode
- Backup Battery: Lithium 18650

Gateway: The gateway, as shown in Figure 7.5, is in charge of receiving rainfall data from sensor nodes and transmitting it to the central server. It supports wireless communication via LoRa and HTTP protocol. Additionally, the gateway plays a crucial role in synchronizing time across all nodes in the network and can directly communicate with them to maintain network coordination.

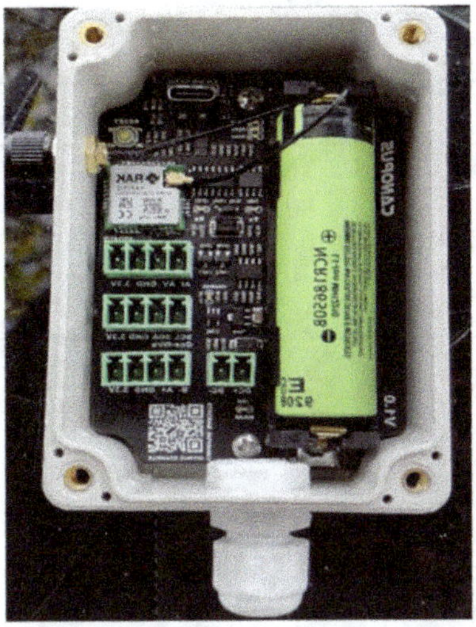

Figure 7.4 Sensor/relay node

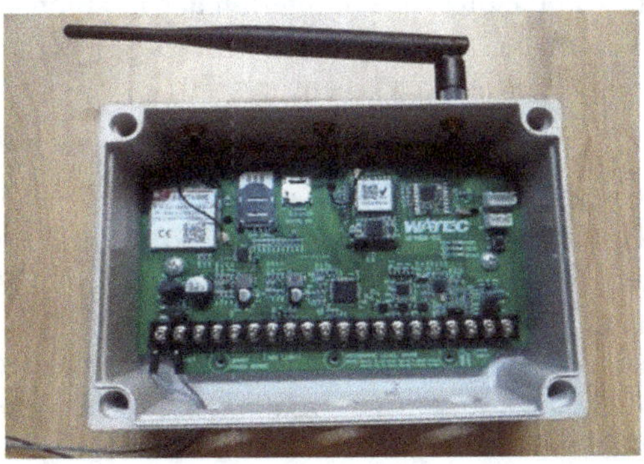

Figure 7.5 Gateway

Technical specifications:

- CPU: 32-bit ARM Cortex-M3 processor
- Processing Speed: 32 MHz
- Digital Output: 1 channel, 0–12 V
- Power Consumption: 72 mW
- Data Transmission Methods: SMS/GPRS/3G/4G/LoRa
- Storage: 32 GB SD card
- Operating Temperature: 20 °C to 70 °C
- Power Supply: 12 VDC input

Table 7.2 Reliability comparison (SF = 7)

Metric	Star topology	LoRa Mesh with AODV
Packet loss rate (PLR)	5%–20%	1%–5%
Delivery success rate (DSR)	80%–95%	95%–99%
Latency (first transmission)	0.6 s	2.7 s
Recovery capability	~0%	~90%
Energy per transmission	145.19 mJ	435.57 mJ

7.5.2 Results

7.5.2.1 Reliability

Table 7.2 compares the reliability of the proposed LoRa Mesh with AODV against a traditional star topology network at SF = 7.

Delivery success rate:

- **Star Topology:** High (~95%) when all nodes are within a 1.5 km range, but drops significantly in the presence of interference or at longer distances.
- **Mesh Network with AODV:** Highest (~95%–99%) due to acknowledgment (ACK) mechanisms and fast route recovery.

Packet loss rate:
With the AODV algorithm, the PLR is reduced to around **5%**, whereas in a star topology (without AODV), the packet loss can reach up to **20%**. Additionally, the coverage range of LoRa in star topology is smaller compared to the LoRa Mesh network using AODV.

The LoRa Mesh with AODV significantly reduces PLR to 1%–5% and achieves a DSR of 95%–99%, thanks to multi-hop routing and ACK mechanisms. However, the latency is higher (2.7 s) due to route discovery, which is acceptable for non-time-critical applications like rainfall monitoring.

7.5.2.2 Transmission range

Without LoRa Mesh, nodes beyond 2 km from the gateway failed to transmit data directly. With LoRa Mesh and AODV, intermediate nodes enabled successful data transmission over distances up to 3.34 km (3 hops), as shown in Figure 7.6.

7.5.2.3 Energy consumption and battery life

This section presents the energy consumption calculation and performance evaluation for a LoRa Mesh-based WSN utilizing the AODV routing algorithm and RAK3172 hardware. The use case involves a rainfall sensor node transmitting data to the gateway every 10 minutes. In each cycle, the sensor remains in **sleep mode for 9 minutes and 30 seconds (570 seconds)** and **wakes up for 30 seconds** to transmit data (see Tables 7.3 and 7.4).

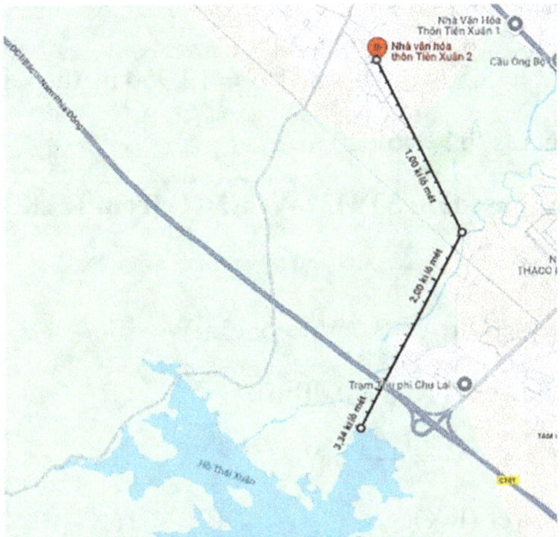

Figure 7.6 Distance from sensing node to the gateway

Table 7.3 Transmission energy per packet with different spreading factors

Spreading factor	Packet duration (s)	Transmit energy (mJ)	Receive energy (mJ)
12	1	444	39.96
7	0.3	133.2	11.99

Table 7.4 Total active energy

Spreading factor	Processing energy (mJ)	Total active energy (mJ)
12	1,036	1,519.96
7	1,087.8	1,232.99

The analysis focuses on estimating the total energy consumed during both sleep and active phases, considering the power consumption characteristics of the RAK3172 module in sleep, transmit (TX), and receive (RX) modes. The evaluation provides insight into the system's energy efficiency and its suitability for long-term deployment in low-power rainfall monitoring applications.

Backup battery specification:

- Battery type: 18650 Li-ion
- Capacity: 3,500 mAh
- Nominal voltage: 3.7 V

Total energy capacity:

$$E_{\text{battery}} = Q * U = 3{,}500 \text{ mAh} * 3.7 \text{ V} = 12{,}950 \text{ mAh} = 12.95 \text{ Wh}$$

$$12.95 \text{ Wh} * 3{,}600 \text{ s/h} = 46{,}620 \text{ mJ}$$

Power consumption (based on STM32WLE5CC from RAK3172):

- Sleep Mode Power:

$$P_{\text{sleep}} = I * U = 1.69\ \mu\text{A} * 3.7 \text{ V} = 6.25\ \mu\text{W}$$

- Transmit Mode Power (TX @ 20 dBm):

$$P_{TX} = 120 \text{ mA} * 3.7 \text{ V} = 444 \text{ mV}$$

- Receive Mode Power (RX):

$$P_{RX} = 10.8 \text{ mA} * 3.7 \text{ V} = 39.96 \text{ mV}$$

Operating cycle:

- Total cycle duration: 10 minutes = 600 seconds
- Sleep time: 570 seconds
- Active time: 30 seconds (includes data transmission and ACK reception)

Energy consumption during sleep mode:

$$E_{sleep} = P_{sleep} * t_{sleep} = 0.00625 \text{ mW} * 570 \text{ s} = 3.5625 \text{ mJ}$$

Energy Consumption During Active Mode (includes TX, RX, and MCU processing at ~10 mA):

$$E_{active} = E_{TX} + E_{RX} + E_{process}$$

Daily energy consumption:

$$E_{day} = (E_{sleep} + E_{active}) * \text{cycles per day}$$

- SF=12: $E_{day} = (3.5625 + 1{,}519.96) \text{ mJ} * 144 = 219{,}387.24 \text{ mJ}$
- SF=7: $E_{day} = (3.5625 + 1{,}232.99) \text{ mJ} * 144 = 178{,}053.56 \text{ mJ}$

Battery life estimation (10-minute frequency):

- SF = 12: resulting in a battery life of 5 hours (3,500 mAh battery).
- SF = 7: extending the battery life to 6.3 hours and 38 hours (60-minute frequency).

With a higher-capacity battery (10,000 mAh), the battery life at SF = 7 extends to 18 hours (10-minute frequency) and enables up to 5 days (60-minute frequency), making the system viable for long-term deployment in hybrid renewable energy systems.

7.6 Conclusion

This chapter presented an energy-efficient WSN solution for IoT applications in hybrid renewable energy systems, using LoRa Mesh and the AODV routing algorithm. Through experiments conducted on actual hardware, including sensor nodes and gateways, the system was validated in terms of packet quality, transmission range, and energy consumption.

The experimental results demonstrated that the LoRa Mesh system with the AODV algorithm significantly improved network reliability compared to traditional systems without mesh networking. Specifically, the ability to relay data through multiple intermediate nodes extended the coverage area, enabling sensors in remote or geographically challenging areas to reliably transmit data to the gateway. When comparing systems with and without LoRa Mesh, the mesh-enabled system achieved a higher packet DSR while maintaining acceptable latency levels—even as the distance between nodes increased. Repeated tests conducted at various distances and locations further confirmed the stability of the system, thereby enhancing the reliability of rainfall data collection and transmission. The proposed system achieves a high packet DSR (95%–99%), reduces packet loss to 1%–5%, and extends battery life through a sleep/wake mechanism.

In terms of energy consumption, the sleep/wake mechanism integrated into the sensor nodes proved to be highly effective. With a cycle of 9 minutes and 30 seconds in sleep mode and 30 seconds in active mode, the nodes operate only when necessary, significantly reducing energy usage and extending the battery life of the devices. This result is particularly well-suited to the requirements of low-power WSNs and aligns with the project's objective of optimizing energy efficiency.

Real-world experiments demonstrate its effectiveness in remote areas, making it a promising solution for sustainable IoT applications in hybrid renewable generation. Future work will concentrate on improving network scalability, optimizing routing performance in high-density node environments, and incorporating advanced energy management algorithms to maximize the use of renewable energy sources in dynamic IoT deployments.

References

[1] C. Ebi, F. Schaltegger, A. Rüst, and F. Blumensaat, "Synchronous LoRa Mesh Network to Monitor Processes in Underground Infrastructure," *IEEE Access,* vol. 7, pp. 57663–57677, 2019.

[2] V. D. Pham, T. D. Dinh, and R. Kirichek, "Method for Organizing Mesh Topology based on LoRa Technology," in *2018 10th International Congress on Ultra Modern Telecommunications and Control Systems and Workshops (ICUMT)*, Moscow, 2018.

[3] J. M. Solé, R. P. Centelles, F. Freitag, and R. Meseguer, "Implementation of a LoRa Mesh Library," *IEEE Access,* vol. 10, pp. 113158–113171, 2022.

[4] D. de Farias Medeiros, M. R. Villarim, F. B. S. de Carvalho, and C. P. de Souza, "Implementation and Analysis of Routing Protocols for LoRa Wireless Mesh

Networks," in *2020 11th IEEE Annual Information Technology, Electronics and Mobile Communication Conference (IEMCON)*, Vancouver, BC, Canada, 2020.

[5] V.-T. Ngo, M.-S. Nguyen Thi, D.-N. Truong, A.-Q. Hoang, P.-N. Tran, and N.-A. Bui, "Applying IoT Platform to Design a Data Collection System for Hybrid Power System," in *2021 International Conference on System Science and Engineering (ICSSE)*, Ho Chi Minh City, Vietnam, 2021.

[6] D. Lundell, A. Hedberg, C. Nyberg, and E. Fitzgerald, "A Routing Protocol for LoRA Mesh Networks," in *2018 IEEE 19th International Symposium on "A World of Wireless, Mobile and Multimedia Networks" (WoWMoM)*, Chania, Greece, 2018.

Chapter 8
Strategies for hybrid renewable energy sources integration in smart grid and microgrid

Michael Wood[1,2], *Silvana Matrone*[1], *Sonia Leva*[1] *and Emanuele Ogliari*[1]

This chapter examines the complex integration of hybrid renewable energy sources within smart grid and microgrid frameworks, focusing on advanced forecasting and real-world application and highlighting how these modern grid structures can enhance energy sustainability, efficiency, and resilience. It begins by discussing hybrid renewable energy systems (HRESs) that combine various renewable energy sources (e.g., solar, wind, hydro) to overcome intermittency and improve reliability. The chapter then delves into the architecture and functionality of smart grids—advanced energy networks that incorporate distributed energy resources (DERs), electric transportation, and active communication systems, creating interactive, adaptive power systems. Microgrids, as modular building blocks of smart grids, are examined for their role in localized energy control. The chapter also includes a comparative analysis of real-world examples illustrating diverse applications of smart grid and microgrid technologies worldwide.

8.1 Introduction

Over the past few decades, the global energy landscape has transformed dramatically, driven by factors such as rapid industrialization, urbanization, and technological advancement. These shifts have led to an extraordinary increase in global energy demand, which traditional energy efficiency measures and conservation efforts have only partially managed to alleviate. Rising living standards, especially in urban areas, and the expansion of industrial activities have accelerated the demand for energy, especially electricity, beyond what conventional power generation can sustainably provide.

Globalization has played a crucial role in this trend, fueling economic growth and modernization in developing regions and creating new demand centers worldwide. As

[1]Department of Energy, Politecnico di Milano, Italy
[2]muGrid Analytics, Colorado, United States

a result, energy infrastructure has struggled to keep pace, leading to severe supply-demand imbalances in many countries. This issue is particularly acute in rapidly growing economies where industrial and household energy consumption has skyrocketed, placing unprecedented pressure on energy resources. Meeting this demand through conventional, centralized power generation has become increasingly challenging, as these systems rely heavily on fossil fuels—resources that are not only finite but also unevenly distributed across the globe.

Currently, fossil fuels such as coal, oil, and natural gas provide over 70% [1] of the world's electricity. However, these resources are depleting rapidly, raising concerns about long-term energy security and stability. Furthermore, the extraction, transportation, and combustion of fossil fuels contribute significantly to environmental pollution, including the emission of greenhouse gases (GHGs) like carbon dioxide (CO_2), which are major drivers of global climate change. The reliance on these carbon-intensive resources is unsustainable and is increasingly recognized as a contributor to adverse environmental impacts, such as air pollution, habitat destruction, and global warming.

The environmental consequences of fossil fuel dependence have prompted calls for a transition to cleaner, more sustainable energy sources. Addressing the supply-demand gap solely through conventional resources is neither feasible nor desirable given the negative environmental and social impacts. This situation has spurred interest in renewable energy solutions, which promise to reduce dependence on fossil fuels while addressing growing energy needs. Renewable energy sources (RES), such as wind, solar, hydro, and biomass, are being explored as viable alternatives, not only to meet the increasing energy demand but also to support a more sustainable, resilient, and decentralized power generation model. According to the International Energy Agency in Reference [2], global renewable electricity generation is forecast to climb to over 17,000 TWh (60 EJ) by 2030, representing an increase of almost 90% from 2023. By that time, renewable energy sources are expected to account for 46% of global electricity generation, with wind and solar photovoltaic (PV) contributing significantly, together making up 30% of the total.

However, the adoption of renewable energy sources to a large extent comes with its own set of concerns. RES, such as solar, wind, and hydroelectric power, are inherently intermittent, meaning their energy production can fluctuate significantly depending on environmental conditions. Solar power, for instance, is highly dependent on sunlight, which varies by time of day, season, and cloud cover; wind energy relies on wind speed and direction, which are equally unpredictable and variable. This intermittency presents a challenge for grid stability, as electricity demand must be matched by supply in real time [3,4]. Unlike traditional fossil fuel power plants, which can adjust output on demand, renewable sources cannot be "turned on" or "off" to meet immediate needs. Consequently, integrating renewables into the grid requires careful planning, including the use of energy storage systems (ESSs) (like batteries) and complementary technologies (like flexible load management and demand response) to balance supply and demand. Addressing intermittency is essential to increasing the reliability and scalability of renewable energy, ensuring that it can contribute consistently to a stable, sustainable energy grid [5].

8.1.1 Hybrid renewable energy sources

The major problem when dealing with high penetration of RES is linked to the reliability of the systems in which they are included, especially in stand-alone applications, but also in grid-connected systems when the total contribution of renewables reaches a substantial fraction of the total energy generation. Possible solutions to this feasibility problem include the combination of several renewable sources in an HRES. These systems have the principal characteristic of being composed of one renewable and one conventional energy source, or more than one renewable source with or without any other conventional energy system; they can also include ESSs [6].

Now HRESs offer versatile configurations that leverage the complementary input of various RES, enhancing the reliability and stability of power generation despite the inherent intermittency of renewables. Coordinating different renewable sources like PV, wind, biomass, and hydroelectric power allows these systems to overcome the limitations of individual technologies. For instance, combining PV and wind power creates a complementary dynamic where solar energy is abundant during sunny days, while wind energy is often stronger at night and during colder months [7]. This combination allows for a more consistent energy flow, balancing periods of low solar generation with wind output. Similarly, pairing PV with biomass energy provides a steady base load; biomass generation can continue during cloudy days or nighttime when solar output is minimal. Biomass, often available after agricultural harvests in the autumn and winter, complements the seasonal peaks of solar power in the summer [8].

More advanced HRES configurations include ESSs, such as batteries, which store excess renewable power generated during peak periods for later use [9]. Pumped hydro storage is another option, where excess solar or wind power is used to pump water to a higher elevation [10], releasing it to generate power during periods of low renewable output. This setup provides a reserve of stored energy that can be dispatched when needed. Microgrids, which integrate multiple renewable sources in a localized network, further improve the dispatchability of renewable energy by optimizing energy flow and adapting to real-time demand fluctuations. For instance, islanded microgrids in rural or remote areas may combine PV, wind, biomass, and battery storage to meet local demands independently of the main grid.

In areas where grid access is limited, PV-diesel hybrid systems are often employed. Diesel generators provide a reliable backup when solar output is insufficient, although the environmental impact and high cost of fuel transport to remote locations are significant drawbacks. Storage systems like batteries can reduce reliance on diesel by storing surplus solar energy for later use [11]. Thus, HRES configurations create resilient, flexible energy infrastructures that reduce dependency on fossil fuels, support grid stability, and enable efficient clean energy utilization. These systems represent a viable path forward for sustainable energy, particularly in areas where a single renewable source cannot reliably meet demand.

The remainder of the chapter is structured to explore various aspects of smart grid and microgrid integration. After giving a brief definition of microgrids and smart grids, it presents real-world examples of microgrids. Following, in Section 8.2 delves into the concept of the objective function, addressing value streams, and contrasting planning with operational concerns. The focus shifts across three main areas: energetic, economic, and social and environmental aspects, emphasizing the reduction of greenhouse gas emissions. Control problems related to the integration of non-dispatchable renewables, microgrid load sharing, and islanding/un-islanding procedures are explored in Section 8.3. The last section (Section 8.4) covers the design and operational integration of microgrids with smart grids, detailing energy management systems, DER management, and simulation strategies.

8.1.2 Smart grids and microgrids

Microgrids with HRES present significant challenges due to the complexity of integrating various renewable sources, such as solar, wind, and biomass, each with distinct characteristics and fluctuating outputs. Environmental factors cause these sources to be intermittent, leading to challenges in maintaining a consistent power supply. Effective management of this variability requires advanced control systems to synchronize between sources, balance supply with demand, and efficiently manage storage and energy dispatch [12]. When HRES also includes non-renewable backup options, such as diesel generators, additional coordination is necessary to reduce fossil fuel dependence while still ensuring reliable energy availability. To address these challenges, HRES are often incorporated into microgrids and smart grids, which offer sophisticated control, monitoring, and communication technologies. These frameworks provide the infrastructure needed to optimize energy flows, stabilize supply, and enable HRES to operate more efficiently within the broader energy system [13].

Smart grids are, by definition, energy networks where prosumers, electric transportation, DERs, and communications are integral and interactive components, as in the case of active distribution networks interfaced with transmission systems [14]. Microgrids, on the other hand, characterized as the "building blocks of smart grids," are perhaps the most promising novel network structure. They are localized and self-contained energy systems that can operate independently from the main power grid (we call this off-grid mode) or as a controllable entity with respect to the main power grid (on-grid mode). It consists of DERs, such as solar PV plants, wind turbines, and storage systems such as batteries and conventional generators, all integrated and controlled by advanced software tools and communication technologies. Microgrids can serve a small energy community, a building complex or even a single home, and can operate in islanded mode or in parallel with the main power grid. They are often designed to improve, increase resilience, and reduce carbon emissions [15].

The organization of microgrids is based on the control capabilities over the network operation offered by the increasing penetration of distributed generators including microgenerators, such as micro-turbines, fuel cells, and PV arrays, together with storage devices, such as flywheels, energy capacitors and batteries and controllable (flexible) loads (e.g., electric vehicles), as the distribution level (Figure 8.1).

Microgrids enable HRES to operate in a localized, autonomous manner, balancing local supply and demand while reducing the impact of intermittency on a specific area. Smart grids, with their advanced metering infrastructure, real-time data analytics, and automation, facilitate better demand-response capabilities, allowing HRES to interact seamlessly with the larger power grid. By situating HRES within

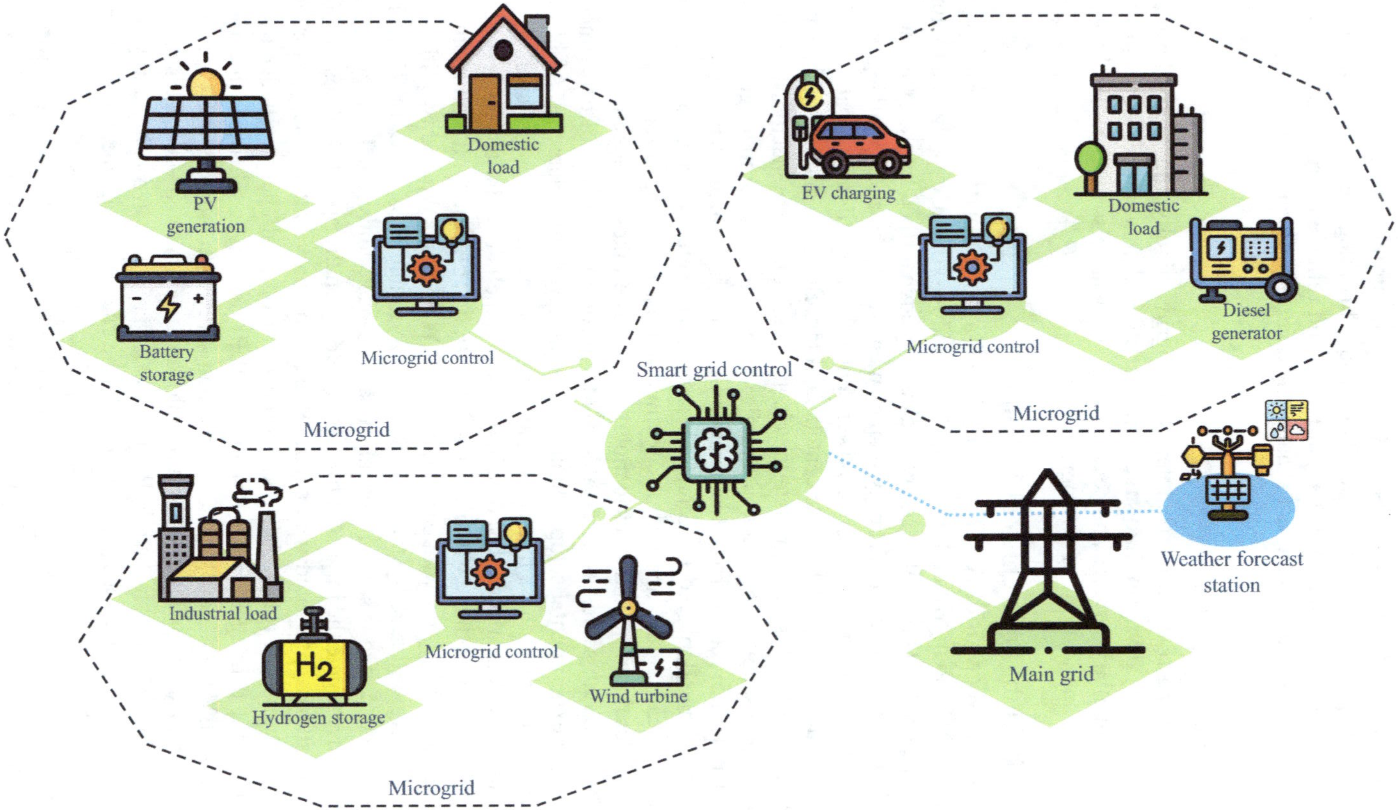

Figure 8.1 Example of a smart grid layout

these frameworks, energy operators can optimize power flows, make real-time adjustments based on generation forecasts, and even trade surplus energy with the grid when renewable production exceeds local demand. This integrated approach allows HRES to contribute to a more resilient, efficient, and reliable energy system while overcoming the inherent challenges of managing diverse renewable sources [16].

8.1.3 Examples

This section presents real-world examples of microgrids, highlighting their relevance to the integration of microgrids into smart grid systems and providing evidence of key issues such as ancillary services and peak shaving. They are also examples of both *in vivo* and *in vitro* applications of microgrids.

8.1.3.1 MG2lab (Politecnico di Milano)

The MultiGood MicroGridLAB (MG2Lab), based at the Department of Energy at Politecnico di Milano in Milan, Italy, is an advanced low-voltage experimental setup operative since 2018. This facility serves as a cutting-edge testbed for the exploration and development of innovative energy management strategies and technologies (*in vitro*). The MG2Lab consists of three power centers that function as local hubs for power distribution and control. These centers facilitate electrical and logical connectivity among MG2Lab assets, simulate the electrical effects of extended distribution lines, and enable seamless connection and disconnection from the main grid. Real-time management of the microgrid is handled by a Programmable Logic Controller (PLC) based on the deployed EMS under research. The setup supports a variety of loads that replicate future applications for microgrids, including electricity, heating, cooling, desalination, and electric mobility such as e-bikes and electric vehicles (EVs). The MG2Lab serves as a versatile platform for a wide range of experimental studies; the experimental setup is ideal for testing EMS advancements, AI-based load/renewable energy source forecasting, and innovative energy components.

The current MG2Lab layout, shown in Figure 8.2, includes a limited 100 kW grid connection, allowing both the withdrawal and injection of electricity. It features three non-dispatchable PV arrays with capacities of 23 kW and 25 kWp and two lithium-ion Battery Energy Storage Systems (BESS) with a storage capacity of 68.75 kWh. The programmable load is simulated using a back-to-back converter. This setup was employed to investigate the impact of forecast accuracy on the operational costs of an EV charging station integrated with solar PV and stationary battery energy storage [17]. This study used a rolling horizon approach for optimal energy management, leveraging a Mixed-Integer Linear Program (MILP) first-layer scheduler informed by forecasts of EV load and PV generation. Long short-term memory (LSTM) with the addition of an attention layer [18] was employed, benchmarked with a persistence model, to predict EV demand, while PV forecasts were generated using a physical hybrid artificial neural network (PHANN) [19]. The energy management system was validated on the MG2Lab. Simulations conducted across eight weeks in each season revealed that forecast accuracy significantly affects operational costs, with errors in both EV and PV forecasts increasing costs by up to 10%. An example of one week of operation is shown in Figure 8.3.

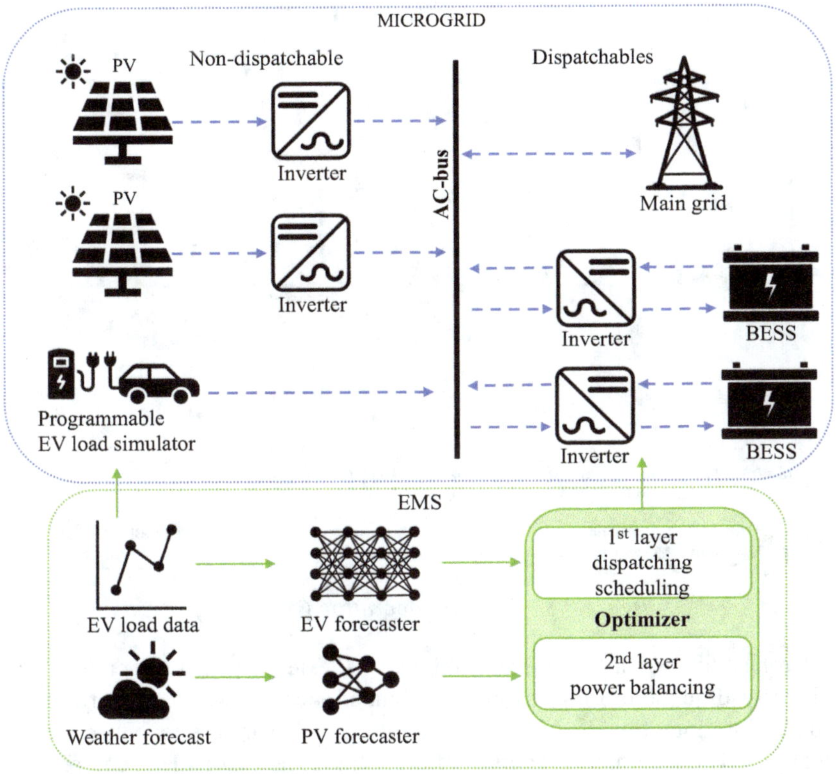

Figure 8.2 Schematic of the MG2lab

8.1.3.2 PoliGrid (Politecnico di Milano)

The PoliGrid project at the Politecnico di Milano was designed to ensure greater reliability of the electrical supply across the Leonardo-Bassini-Bonardi-Golgi campus. This advanced multi-energy system not only fulfills the campus's energy needs but also enables the delivery of flexibility services to the public distribution network. The microgrid also offers valuable opportunities for research, making it suitable as both an in *vitro* and in *vivo* case study.

The campus is composed of an internal medium-voltage grid (23 kV) that provides approximately 15 GWh of electricity annually, with peak summer demand reaching 4 MW. The grid integrates multiple solar PV installations spread across various buildings, collectively generating over 1 MW of power, as well as a 2 MW cogeneration plant. The cogeneration system efficiently produces electric and thermal energy, with the latter feeding a 2-kilometer district heating network serving the campus, allowing it to reduce the carbon footprint while meeting its energy requirements. The facility includes two 6 MW industrial boilers and an absorption chiller that delivers 1.4 MW of cooling using excess heat from the cogeneration engine. Since 2015, significant efforts have been made to modernize the campus energy infrastructure, with 11 PV systems already installed, providing a total output of 1

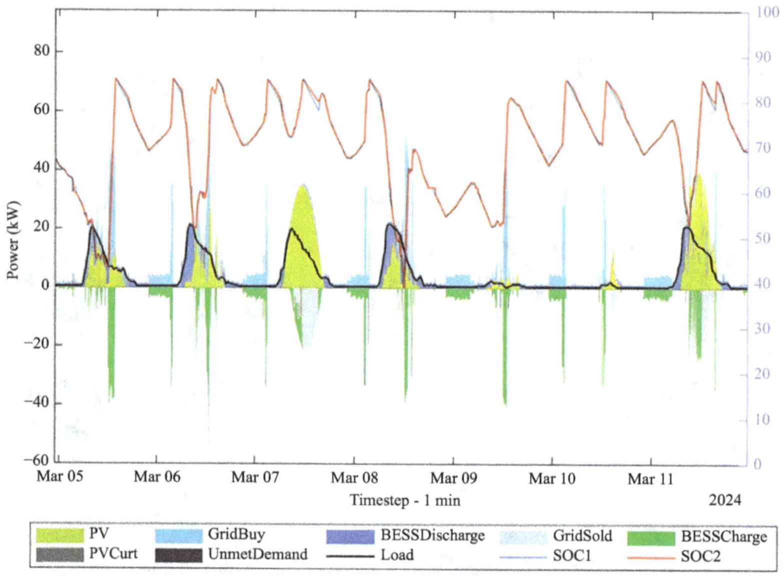

Figure 8.3 MG2lab operation (one week)

MW. A robust fiber optic communication network extends across the entire campus, enabling real-time monitoring and management of energy flows. It supports dynamic load balancing, intentional islanding capabilities, and seamless reconnection to the national grid. The schematic in Figure 8.4 shows the location of all the components of the microgrid. Since 2023, the integration of a Central Plant Controller in compliance has further enhanced the grid's flexibility. The PoliGrid thus serves as a real-world testing platform for innovative energy management solutions, supporting the integration of intermittent renewable resources while ensuring efficiency, reliability, and alignment with decarbonization goals.

The advanced load management solution described in Reference [20], utilizing components from Schweitzer Engineering Laboratories, enhances the system's ability to address network emergencies while maintaining high service quality for the local network. This solution also enables the Campus to participate in the Italian Ancillary Services Market as a virtual unit and perform Demand Response (DR). Ancillary services refer to the additional functions and support activities that maintain the reliability, stability, and quality of an electrical power system. These services are essential for the smooth operation of the grid, ensuring that electricity is delivered consistently and efficiently to the end user. More about ancillary services is described in Section 8.2.2.3, including relevant equations.

In case of disconnection from the main grid, two relays equipped with synchro-check open the two main switches and prevent their reclosing when an external network is detected. The synchro-check equipment ensures that the system can properly detect grid disconnection and transition to islanding mode, which is a form of black-start capability. In islanding mode, the combined cooling, heat and power (CCHP)

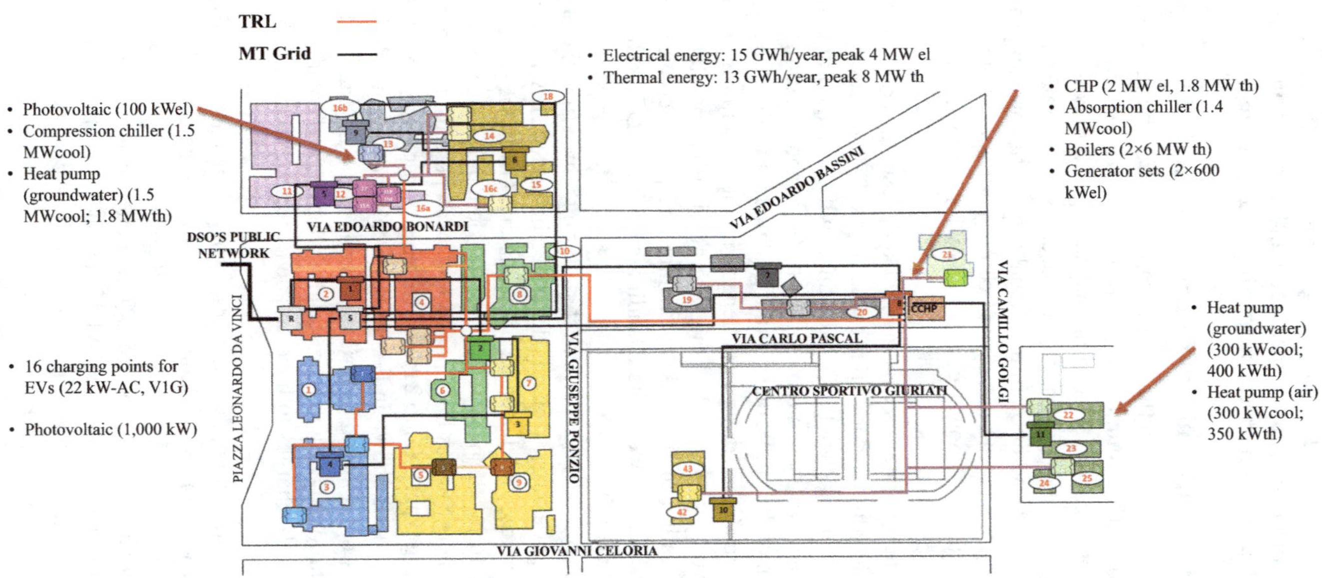

Figure 8.4 Schematic of the PoliGrid microgrid

plant controls frequency and voltage to maintain stability. If demand exceeds supply, a load shedding mechanism disconnects certain loads, and once capacity allows, these loads are gradually reconnected, supporting grid balancing. If demand exceeds supply, a load shedding mechanism disconnects certain loads, and once capacity allows, these loads are gradually reconnected, supporting grid balancing.

8.1.3.3 Bayfield County Jail and Courthouse (muGrid Analytics)

The Bayfield County Jail and Courthouse microgrid, in Washburn, Wisconsin, USA, is an innovative two-building municipal *in vivo* project that was completed in 2023. The initial design optimization and operational strategy were conceived by muGrid Analytics to meet the county's goal of resilient backup power, rate tariff savings, and reduced CO_2 emissions. In addition to municipal offices, the project provides multi-day backup power for a 911 emergency call center and data server. The schematic in Figure 8.5 shows the DER power connections with the load and grid, as well as control and communication links.

The microgrid incorporates an existing 200 kW diesel generator and an existing solar PV array, while adding more solar capacity, the Li-ion battery, the microgrid controller, and the islanding switch. The islanding switch is a medium-voltage (MV) recloser, allowing the two separate buildings to be islanded during a grid outage using a small portion of the public distribution grid under the Xcel Energy Resilience Service Pilot program. Other design challenges include operating the battery and generator in cold weather, where the daily average low temperature in January is −12°C; parallel operation of the diesel generator and power converter DERs; and robust coordination logic between the islanding switch and DERs during islanding and un-islanding procedures.

The microgrid is primarily grid-connected, during which time the objective is to reduce the retail cost of energy, which is described in Table 8.1, with the main decision variable being charge and discharge of the battery. Given the price of maximum monthly load power during the peak period, called a demand charge, and the relatively small difference in energy prices from peak to off-peak periods, the optimal strategy is mostly peak shaving from 9:00 to 21:00 local time on weekdays only. An example of two days of operation is shown in Figure 8.6. Peak shaving thresholds are calculated using muGrid's optimal and proprietary method, and the battery is dispatched to hold the microgrid load below the threshold, including any solar production. The battery opportunistically charges to have an adequate state of charge (SOC) for the next load peak.

In the case of a grid outage, the islanding switch opens and the DERs serve the load. When solar production is available and the load is below 90 kW, the battery operates in a grid-forming mode to control voltage and frequency, while the solar inverters grid-follow and simply produce as much energy as is available. Any solar production in excess of the load is curtailed. If the load rises above 90 kW, the diesel generator is turned on in grid-forming mode with its isolation switch open and is then switched into the microgrid while the battery quickly switches to grid following mode and matches the phases of the generator's voltage signal.

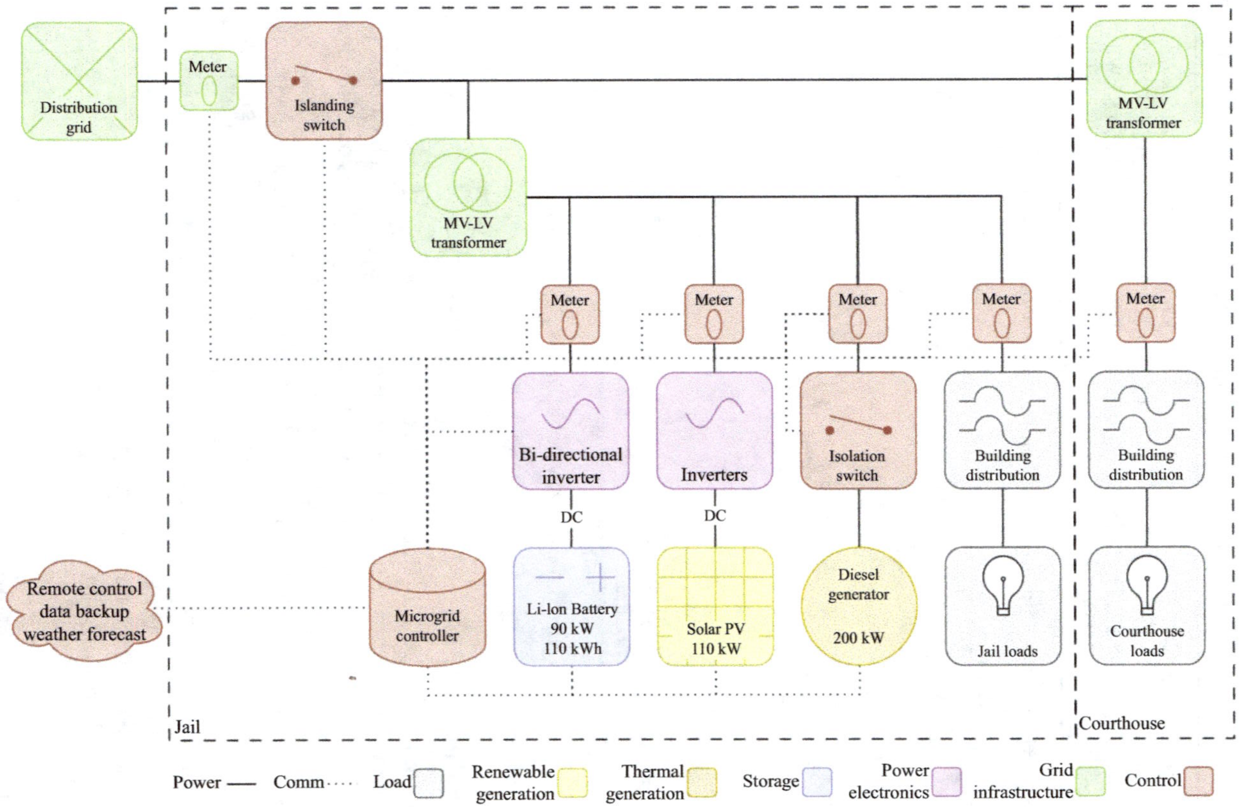

Figure 8.5 Schematic of the Bayfield County Jail-Courthouse microgrid

Table 8.1 Bayfield County Jail-Courthouse electric tariff

TOU period	Hours	Prices summer (Jun 1 to Sep 30)	Prices winter (Oct 1 to May 30)
Peak	9:00-20:59[1]	0.075 \$/kWh 13 \$/kW[3]	0.070 \$/kWh 10 \$/kW[3]
Off Peak	0:00-8:59, 21:00-23:59, 0:00-23:59[2]	0.056 \$/kWh	0.056 \$/kWh

[1]Weekdays, [2]weekends, [3]max 15-minute interval microgrid load power each month

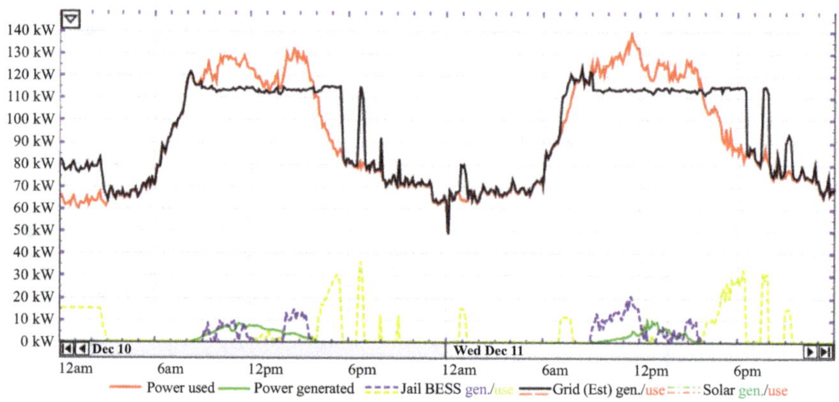

Figure 8.6 Battery peak shaving operation

Microgrid definition

A **microgrid** is a group of interconnected loads and DERs that acts as a single controllable entity with respect to the grid. It can connect and disconnect from the grid to operate in grid-connected or island mode. Microgrids can improve customer reliability and resilience to grid disturbances [21].

Advanced microgrids enable local power generation assets—including traditional generators, renewables, and storage—to keep the local grid running even when the larger grid experiences interruptions or, for remote areas, where there is no connection to the larger grid. In addition, advanced microgrids allow local assets to work together to save costs, extend the duration of energy supplies, and produce revenue via market participation [21].

8.2 Objective function

The objective, or cost, function is possibly the most important component of deciding on a strategy for cooperation between smart grids and microgrids. It essentially answers the questions: what is the goal of the microgrid and how can it be accomplished?

The general goal of every electric power system is to provide reliable and cost-effective electricity [22]. Given a market perspective, attention must be paid to what good or service is being provided, at what time and location, and what the value of that good or service is. This is especially true if energy is not necessarily exchanged (e.g., capacity markets), making the accounting less clear. Therefore, the microgrid objective function begins with the stakeholder perspective, the consumer, but can be expanded to include multiple consumers, entire distribution networks, or even an entire regional or national region of the power system.

Costs can be purely energetic, monetary, CO_2-equivalent, biodiversity-related, or societal and can have more than one aspect as in (8.1). In some cases, a purely energetic perspective can be appropriate; however, the perspective of this work is techno-economic and also includes operating and capital costs. Most holistic solutions also consider a relevant environmental or social aspect as in 8.1.

$$Cost_{total} = Cost_{monetary} + Cost_{social-environmental} \tag{8.1}$$

Defining the cost function begins with any explicit costs or revenues incurred by the microgrid. Revenue, for example, from selling renewable production to the grid, is usually treated as a cost reduction and the therefore cost can be negative. Then implicit costs, such as battery cycle degradation, do not appear as specific transactions but are nonetheless a cost of doing business due to a particular action ((8.2)). Some costs may even be ignored if they do not change with different enacted decisions, for example, a fixed monthly electric service charge that is not impacted by any customer decision besides ending the contract.

$$Cost_{monetary} = Cost_{explicit} + Cost_{implicit} \tag{8.2}$$

Finally, the objective function needs a direction of optimality, maximization or minimization, and will likely have some constraints. Equation (8.3) is a general form of the objective function, where the decision variables are contained in the vector x.

$$\begin{aligned} \min[Cost_{\$} &= Cost_{[\$],monetary}(x) + Cost_{[\$],social-environmental}(x)] \\ &\text{Subject to:} \\ &x \leq Constraints \end{aligned} \tag{8.3}$$

The cost function is often a simplification of the problem, potentially missing certain important aspects that are difficult to quantify. This is especially true of complex or subjective benefits such as the cultural value of green areas to local communities instead of a solar PV array. The cost function should therefore be considered as a guide but not a rule, especially during the microgrid design phase.

8.2.1 Energetic

It's possible to take solely an energic approach, simply minimizing the energy exchange with the grid as shown in (8.4) and (8.5). This is also called "maximizing self-consumption." Then, for each traditional consumer who becomes a prosumer, the energetic reliance on the larger grid decreases and each microgrid is said to be somewhat independent, freeing up resources for use elsewhere.

$$Microgrid^{(t)}_{[kW]} = Load^{(t)}_{[kW]} - Generation^{(t)}_{[kW]} - Storage^{(t)}_{[kW]} \tag{8.4}$$

$$min(Microgrid_{[kWh]} = \Delta t \sum_{t=0}^{T} Microgrid^{(t)}_{[kW]}) \tag{8.5}$$

One shortcoming of solely maximizing self-consumption is that peaks in net load—load less renewable generation—tend to be ignored. In Reference [23], batteries operated to maximize self-consumption reduced peaks by only 5.7%, whereas even simple heuristic peak shaving methods were capable of 22% peak reduction. So while the microgrid requires less energy from grid generators, it is almost just as dependent on the grid to meet its peaks, requiring nearly the same amount of grid infrastructure compared to without the microgrid. With a reduced load factor (average load divided by maximum load), the nearby distribution grid becomes underutilized. In economic terms, this problem manifests itself as grid companies needing to pay for the same resources with less revenue, if revenue comes mostly from kWh passing through the system. Furthermore, by simply minimizing the energy exchanged with the grid, there's no possibility for the microgrid to support the macrogrid during peak price or load events with downward flexibility.

8.2.2 Economic

Successful approaches to managing a microgrid should include the economic cost, with or without components of a larger, smart grid present. In the 20th century, electric power systems around the world experienced an increase in reliability and a decrease in cost, attributable to the competition and the liberalization of energy production and transmission [22]. Ignoring cost and forcing dependence on locally produced energy would be a step backward and will ultimately hinder the clean energy transition. Rather, an approach of optimized exchange between smart grids and microgrids is beneficial to consumers, producers, and even policy makers.

8.2.2.1 Retail electric tariff

A practical but common economic cost function is reducing a consumer's retail electric cost. This is simply because it is the one that is available in the absence of any other programs or contracts with the electricity provider, grid operator, or other consumers. To protect consumers from volatile energy prices, traditional retail electric tariffs were relatively flat per-kWh prices, which only varied seasonally or slowly over time. Dynamic or time-of-use (TOU) pricing schemes are relatively new, although in the US interest in them arose during the 1970s energy crisis [24]. In these tariffs, a few or several different energy prices vary depending on the hour of the day and day of the week, usually with not more than two to five different periods per day. Table 8.2 shows an example of TOU prices.

Since a typical billing period is one month, we can define the retail energy cost for two TOU periods with (8.6a), (8.6b), and (8.6c). This of course can be generalized

Table 8.2 Example of retail TOU energy and power prices

Property	Off peak	Peak
Period	0:00–15:59[1], 21:00–23:59	16:00–20:59
Energy price	\$0.20/kWh	\$0.30/kWh
Power price	\$0/kWh	\$10/kWh

[1]Period ends at exactly the beginning of the next hour.

for any number of TOU periods, including overlapping periods. It's assumed that the microgrid is entirely behind one electric meter.

$$Microgrid^{(OffPeak)}_{[kWh]} = \Delta t \sum_{t=0}^{T} Microgrid^{(t)}_{[kW]} \quad \forall t \in [0:00 - 15:59,\ 21:00 - 23:59] \tag{8.6a}$$

$$Microgrid^{(Peak)}_{[kWh]} = \Delta t \sum_{t=0}^{T} Microgrid^{(t)}_{[kW]} \forall t \in [16:00 - 20:59] \tag{8.6b}$$

$$Cost_{[\$]} = Price^{(Off\ Peak)}_{[\$/kWh]} \times Microgrid^{(Off\ Peak)}_{[kWh]} + Price^{(Peak)}_{[\$/kWh]} \times Microgrid^{(Peak)}_{[kWh]} \tag{8.6c}$$

In some regulatory areas such as the US and Germany [25], tariffs can also include a price on peak power, also called a demand charge. There are many implementations of this, but the following example in (8.7) can be generalized. Peak power is calculated as the maximum of the load power time series in a given billing period (e.g., 1 month) inside a certain TOU period (e.g., 9:00–21:00). Lastly, the load power time series is usually the average power over a given interval (e.g., 15 minutes).

$$\begin{aligned} \mathrm{Cost}_{[\$]} = {} & \mathrm{EnergyPrice}^{(\mathrm{OffPeak})}_{[\$/\mathrm{kWh}]} \times \mathrm{Microgrid}^{(\mathrm{OffPeak})}_{[\mathrm{kWh}]} \\ & + \mathrm{EnergyPrice}^{(\mathrm{Peak})}_{[\$/\mathrm{kWh}]} \times \mathrm{Microgrid}^{(\mathrm{Peak})}_{[\mathrm{kWh}]} \\ & + \mathrm{PowerPrice}^{(\mathrm{Peak})}_{[\$/\mathrm{kW}]} \times \max\left(\mathrm{Microgrid}^{(\mathrm{Peak})}_{[\mathrm{kW}]}\right) \end{aligned} \tag{8.7}$$

Similar to high retail energy prices, a substantial power price incentivizes flattening the load, or increasing the load factor, during that period. However, since the peak power component is a maximum, it better mimics how utilities need to invest in infrastructure to always have capacity for the maximum power.

8.2.2.2 Excess renewable energy production

So far the economic components of the cost function have only included those related to consumption. However, most prosumers will occasionally export to the grid. This may be involuntary because the load is less than the generation and the site has no

Table 8.3 Example of retail TOU energy and power prices with selling

Property	Off peak	Peak
Period	0:00–15:59[1], 21:00–23:59	16:00–20:59
Energy price	$0.20/kWh	$0.30/kWh
Energy sell price	$0.10/kWh	$0.10/kWh
Power price	$0/kWh	$10/kWh

[1]Period ends at exactly the beginning of the next hour.

available storage, or voluntary because of an economic advantage. The latter of course depends on a contractual revenue source; an example of this occurrence is presented in Table 8.3.

Unlike consumer tariffs that do not vary as much as market prices to protect the consumer, the valuation of export to the grid can even be equal to a local nodal or zonal price in every hour, or a calculation based on those prices such as is the case with the Italian exchange-on-the-spot ("scambio sul posto").

8.2.2.3 Non-tariff based incentives

While optimizing within the existing consumer tariff structure is practical and usually accepted without a further contract with the grid operator, the incentive is often missing for the microgrid to support the grid when needed, and therefore an opportunity is missed.

Demand response

Demand response was historically one of the first opportunities for consumers to provide a service to the grid, after the simple tariff-based, "passive" demand management programs. This required more integration since industry loads were usually controllable, but now the need for negative flexibility would be dispatched by a grid operator. This was originally and often still is a phone call to large single loads. However, with protocols like OpenADR, substation automation, and 3rd-party aggregators, DR can be treated like a dispatchable generation asset. This was an important step for smart grids and distributed resources, so much so, in fact that the legality of DR was finally upheld in the US Supreme Court case "*FERC v. Electric Power Supply Association*" of 2016, a case summary is available in Reference [26].

During a specific time of need determined by the grid operator, DR activation interrupts or decreases specific loads such as electric arc furnaces or lighting or can even disconnect entire buildings or microgrids. Often with DR, the magnitude of downward flexibility is somewhat stochastic since it depends on how the load would have naturally operated in the DR period. Participants in DR are usually compensated per kWh but can also be penalized heavily for not dispatching as contracted.

Flexibility

As building automation and DERs propagate into the building stock, there are more options for prosumers to somewhat increase or decrease their load without completely disconnecting from the grid. This is known as flexibility, positive in the case of increasing load and negative in the case of decreasing. Negative flexibility may even reach negative values, exporting to the grid according to our sign convention in 8.4. The exact payment structure and signal or schedule depends on the specific program and may interface directly with grid operators or via third-party aggregators. In any case, the microgrid or prosumer receives a financial benefit that typically reflects prices in ancillary services markets rather than spot market prices, in References [27], [28], and [29].

Frequency regulation
The grid frequency is maintained close to the nominal value by many tools, one of which is typically called primary or automatic frequency control. In the EU generators over 10 MW are obligated to participate by installing a frequency controller that makes small adjustments to a generator's real power output based on the frequency deviation: $\Delta f = f_{actual} - f_{nominal}$ [30]. However, in some regulatory areas, smaller generators and prosumers can also participate, more often those connected to the transmission networks rather than distribution. The remuneration is fundamentally different from the aforementioned components of the cost function because automatic frequency regulation is a continuous service and each energy asset exchanges relatively little energy compared to programs such as DR. This is why large generators can both participate in markets as well as frequency regulation.

8.2.2.4 Value of lost load

Thus far the cost function components are all explicit: actual energy exchange or financial transactions. The value of lost load—the negative economic impact of a blackout on a single consumer–is instead usually a combination of explicit and implicit costs and is difficult to estimate. Industrial or commercial consumers where a product or service is a direct result of using electricity may be able to calculate the actual monetary damage due to outages occurring at different times and lasting different durations. However, many consumers instead suffer a general interruption of life and business, where some activities are genuinely degraded and others are simply delayed until the outage ends [31]. The important consideration is to estimate, even qualitatively if necessary, the importance to the consumer of serving the load. In this way a microgrid controller can make the best decision for the consumer in emergency situations or possibly even when it's economically advantageous to tolerate loss of load, for instance by full load reduction without adequate local DERs. Also, this information is crucial during the optimal sizing of assets during the design phase of resilient microgrids.

8.2.3 Social and environmental

The clean energy transformation is being driven by a lack of valorization of natural resources, and therefore solutions to the transition should not make the same mistakes of the past.

And there are many complex and difficult to quantify benefits of green spaces in communities, biodiversity elements such as habitat for pollinators, and the ability of permeable surfaces—where rain water can easily enter the ground—to reduce urban flooding. A holistic and community-oriented microgrid project should consider many of these non-energetic and non-monetary costs.

8.2.3.1 Greenhouse gas emissions

Many communities and policy makers value the power of renewable energy and smarter, more efficient grids to reduce CO_2-equivalent emissions. This, combined with a life-cycle analysis of any impact on the local environment, can give decision makers the information to truly minimize emissions. In some cases, trees sequester carbon from the atmosphere at a lower unit cost than the reduction attributed to solar PV [32]. In some cases, the economic value of a unit of emissions reduction is relatively easy to calculate and could be included in a cost function in monetary terms, especially where the alternative is buying carbon credits on an open market. However, in other cases, a unit of CO_2-equivalent reduction may have a high and non-monetary value due to ethical considerations or the legal consequences of a state or country not meeting its pledged emissions reduction target. Often the latter is the case, and emissions reduction may need to be an independent variable in a multi-objective minimization problem.

8.2.3.2 Local environment

Lastly is a conglomerate category of local environmental benefits such as clean air and water, green areas for urban temperature management and public enjoyment, pollinator habitat, erosion control, and general biodiversity. Unfortunately the current state of research can be scarce, for instance, the impact of solar PV arrays on biodiversity [33]. Usually social and environmental costs are formulated in regulatory constraints on new construction projects, especially near wetlands or sites of historical significance. But strong environmental regulation may not exist in some contexts. The most powerful tool in this case is a life cycle assessment of each relevant aspect of the local environment and a multi-objective design optimization exercise. Attention should be paid to the impacts of different technologies which will vary widely, for example. solar PV tends to consume soil compared to micro-hydroelectric that impacts downstream aquatic habitats with reduced flow for the length of the penstock.

8.3 Control problems

Modern buildings are complex electrical, mechanical, thermal, and information technology (IT) machines that may have existing control challenges before a microgrid is introduced. Integrating a microgrid likely complicates the situation, adding a new control agent, new protocols, and new traffic on the local communication network. Furthermore, without proper and advanced understanding of the existing system, capabilities of the added equipment, and impacts of the added equipment, complications such as grounding and incompatible control algorithms may only be uncovered during the construction phase.

The focus of this chapter is a series of power control challenges and economic dispatch opportunities; however, the more practical problems are worth mentioning. They are organized by phase of the project development process:

- **Feasibility study:** Early assumptions or design decisions needed to lock in financing and secure site control can be skewed by out of date or absent building architectural drawings or environmental maps, irregularities in the supply chain, or understanding of grid operator programs and interconnection policy.
- **Optimal asset sizing:** Optimal asset sizing is heavily dependent on the asset operation, which is heavily dependent on the operation strategy and capabilities of the asset, which may not be totally known before testing the asset in situ.
- **Engineered design:** The engineered design requires full knowledge of applicable code and zoning requirements, some of which is expected to be resolved during construction, though some aspects such as fire safety can substantially change aspects such as the location of batteries.
- **Financing, procurement, and construction:** The use of public grants, rebates, and loan guarantees in distributed energy projects can be quickly complicated by third-party ownership structures or a requirements of domestically produced materials, possibly both delaying and putting a deadline on the construction process.
- **Testing and commissioning:** Discovering moderate engineering oversights during testing is expected, but can also compound quickly for more complex projects or in the difficult to navigate domain edges where neither party may fully understand or take responsibility for the interface.
- **Monitoring and validation:** Any forced modifications to the optimal design can result in decreased performance, possibly without notifying the stakeholders.

8.3.1 High current injection into distribution networks

Renewable energy injection to distribution grids often reaches a technical limit. The maximum allowable renewable injection, and therefore the maximum capacity of renewable generation given the load and other technical parameters, is called hosting capacity. The technical limit that determines hosting capacity is generally either the high and low voltage limit on the circuit or congestion at a transformer or the line itself. The exact technical limit is determined by many factors such as the allowable voltage range allowed by the regulator, grid topology, and physical parameters such as line conductance, load, and generation at different intervals along the circuit, and the presence of any grid automation. Figure 8.7 illustrates an MV circuit with multiple loads withdrawing current and one controllable solar PV and storage plant injecting current. This example also depicts the qualitative voltage profile along the circuit, helping visualize how the voltage levels vary across the network.

Since the R/X ratio of MV distributions networks is relatively high we know that real and reactive power are mostly decoupled [34]. Therefore, the voltage on the feeder is affected mostly by ohmic voltage drop, and the DC power flow model is appropriate at least for initial estimates of the voltage profile. The voltage drop can be estimated using (8.8):

$$\Delta V_k \approx \frac{R_k P_k + X_k Q_k - R_k P_{Gk} - X_k Q_{Gk}}{V_k} \tag{8.8}$$

where

- V_k is the voltage magnitude at node k [V]
- ΔV_k is the voltage drop at node k [kW]
- R_k, X_k are the resistance and reactance of the line segment to node k [Ω]
- P_k, Q_k are the active and reactive power demands at node k [kW]
- P_{Gk}, Q_{Gk} are the active and reactive power generated at node k [kW]

For the example in Figure 8.7 the voltage regulator is an on-load tap changing transformer, although often there can also be voltage regulators along the line. Current withdrawal at any node in the circuit causes the voltage at the node to decrease relative to the primary substation MV bus. Rather, net injection causes the voltage to increase relative to the substation bus. Therefore, the circuit voltage profile, the voltage as measured along the whole length of the conductors, will be lowest when net withdrawal is highest and highest when net injection is highest [34].

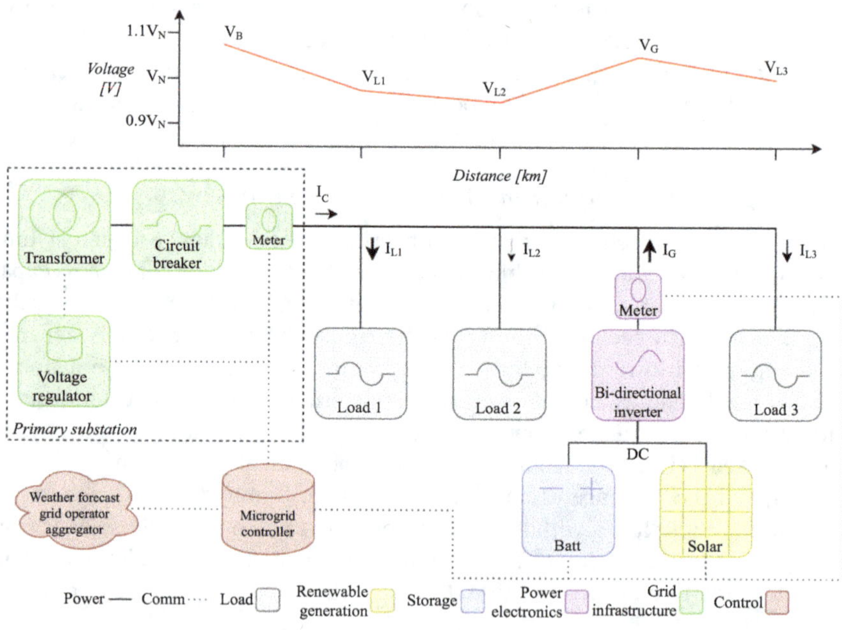

Figure 8.7 Schematic and qualitative voltage profile of example MV circuit with three loads withdrawing current and one controllable solar PV and storage plant injecting current

Voltage profile analysis therefore estimates the worst-case conditions of load and generation at each node, preferably in what is sometimes referred to as quasi-dynamic grid modeling, by changing the current values at approximately 15-minute intervals for an appropriately long period of time to have confidence that the worst case scenarios have been captured.

Similarly, current can be estimated in each conductor and going into the MV bus, also considering the thermal dynamics of conductors and transformers in worst-case ambient conditions.

8.3.2 Parallel operation of multiple microgrid sources

A microgrid with multiple energy sources must choose which source to dispatch, at what power, and when. This is referred to as the unit commitment problem, where the goal is to minimize the cost of energy produced [22]. Some sources, especially backup generators, have a control or emissions limitation preventing parallel operation with the grid, then the problem is constrained to the sources that can. There may be other constraints as well, in particular that non-dispatchable sources (e.g., solar) to be accompanied by dispatchable sources (e.g., battery) to operate. The constraints and capabilities change depending on if the microgrid is islanded, and one or more sources are grid-forming, or if the microgrid is connected to the larger grid and all sources are grid-following.

8.3.2.1 Power electronic converter sources

Primary energy sources, such as solar PV, wind, chemical battery storage, and fuel cells, require a DC-AC inverter to interface with most AC building electrical systems and the wider grid. In addition, there may be AC-DC rectifier functionality and DC-DC conversion to change voltages. The power output at the AC interface will always be limited by the primary source, for instance, the sunlight irradiation or wind speed, and efficiencies along the way. But the final AC power is also limited by the power converter capabilities and control algorithm. This can affect the minimum and maximum dispatchable real power thresholds, as well as reactive power injection or withdrawal capability. Certain power electronic converter controllers can become unstable in the presence of noise or other converters operating in parallel and filtering may be required to reduce the noise [35]. In some cases, incompatible converter technologies cannot be paralleled on the same bus at the same time, affecting the design topology or unit commitment solution. Not all power converter devices can grid-form and grid-follow, and not all can quickly switch between the two without powering down the converter.

8.3.2.2 Rotating machine sources

Primary energy sources such as internal combustion engines and hydraulic pressure instead usually interface with the AC grid with a synchronous or asynchronous rotating machine. Similarly, the output power is limited by the prime mover, although there is a small amount of energy stored in the rotational inertia of the machine. Internal combustion-type sources usually have a minimum power threshold due to limitations of the engine, and large engines often require a mandatory warm-up and cool-down period. Reactive power injection and withdrawal is not typically controllable to such a large degree as power converters [36]. Not all rotating machine generators can grid-form and grid-follow, and not all can quickly switch between the two without first de-energizing the alternator.

8.3.2.3 Grid-connected operation

While grid-connected, each source is typically in grid-following mode: the power electronic converter or combustion engine controller receives a voltage signal from the grid and injects current according to its own control algorithm and capabilities. Stability is generally maintained by the grid, which is to say, generators participating in frequency regulation. However, stability is not guaranteed, such as in the case of parallel power converters [35].

8.3.2.4 Islanded operation

While islanded, sum of the power available from each energy source must be at least as large as the total microgrid load. This is very much true on a non-transient time scale, such as the average 1-minute load and generation power, where the loads could be said to be changing from one steady state to another. At a faster time scale, such as 100 ms to 10 seconds, transient effects due to second-order RLC circuit behavior and the time constant of motors can create temporary situations where the load is quite high relative to the 1-minute average [34]. These high loads will cause voltage and possibly frequency to collapse without an appropriate response. Rotational machine sources have mechanical inertia, which can supply all or some of the transient energy according to the swing equation (8.10). Battery storage sources can quickly ramp up (or down) current to supply the energy, up to chemical or circuit breaker limits, even in excess of their steady-state design capacities (for short periods). Inverters also have a small amount of this overcurrent capability, with the limitation usually being thermal overload of the IGBT or other switching components [37].

8.3.3 Microgrid islanding and un-islanding

For various motivations, such as a grid blackout or DR, the microgrid may open its interconnection switch and create an electrical island separate from the grid. In this case typically one source needs to switch from grid-following to grid-forming, and any others remain in grid-following. This process can be quick enough for the loads to not experience an interruption in power, which is referred to as a closed transition because the islanding switch is still closed for a brief instant while one of the energy sources transitions to grid-forming. Rather if certain controllers require de-energization of the microgrid to begin grid-forming, then the process is referred to as open transition: the islanding switch opens, the microgrid de-energizes, and then one energy source transitions to grid-forming.

Several difficulties arise during islanding. Not all the energy sources capable of a closed transition may have sufficient power capacity to serve the maximum microgrid load, requiring close coordination during the transition. Furthermore, an open transition requires a black start of the microgrid: completely re-energizing the loads.

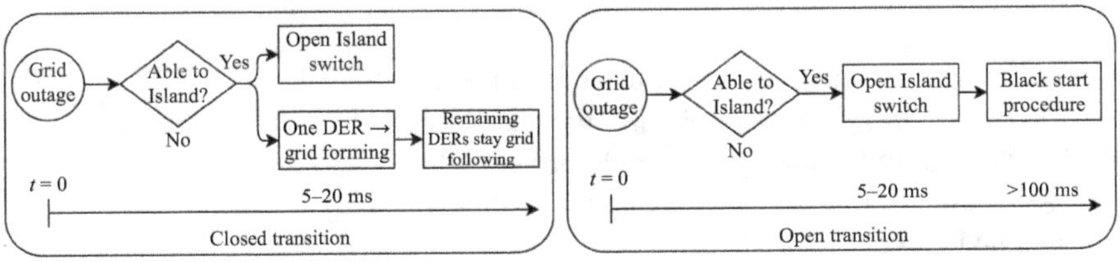

Figure 8.8 Closed and open transition sequences

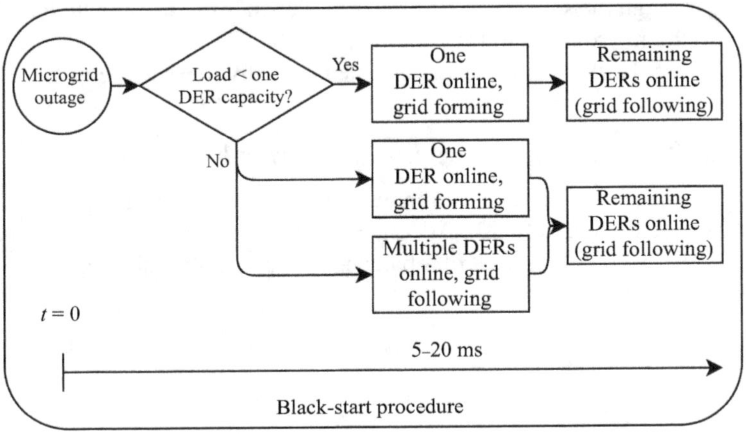

Figure 8.9 Black-start procedure

For this technical reason microgrid-capable DERs sometimes must be sized much larger than would otherwise be preferred economically.

8.3.4 Microgrid black start

Re-energizing the microgrid usually results in a transient inrush current which can be several times larger than the steady-state load current, and this current must be supplied by only the local energy sources [38]. Worse, the current may often need to be supplied entirely by the one source creating the grid-forming island due to the control problem of parallel microgrid energization from two different DERs. Figure 8.9 illustrates a black-start procedure.

The transient experienced during a black start depends on the specific microgrid infrastructure and loads but also on the amount of time the microgrid was de-energized. Transformers experience a magnetization current, as do DC-link capacitors that may be present in large power converters. Each can be estimated a priori, knowing the parameters of each device. The inrush current $I_{inrush}(t)$ for a transformer can be approximated as in the following equation:

$$I_{inrush}(t) = \frac{V_{peak}}{Z} \cdot \sin(2\pi ft + \phi) \tag{8.9a}$$

where

- $I_{inrush}(t)$ is the instantaneous inrush current [A]
- V_{peak} is the peak voltage applied to the transformer [V]
- Z is the total impedance of the transformer [Ω]
- f is the angular frequency of the applied voltage [Hz]
- t is time [s]
- ϕ is the phase angle between the voltage and current [radians]

However, load pickup, the phenomenon of many simultaneously re-energized loads going through the same start-up sequence, depends on the specific load devices connected to the microgrid and the programming and sizes of each. Therefore, load alignment is very hard to estimate and is instead usually measured experimentally.

8.3.5 Fault current and behavior of inverter-based sources

When faults occur in an AC distribution network, the fault current is provided by the sources connected to the network. Adequate fault current is necessary to locate and isolate the fault. The fault current of a balanced, three phase fault to ground is given as follows in (8.9b):

$$I_F = \frac{V_{LL}}{\sqrt{3} \cdot Z_F} \tag{8.9b}$$

where

- I_F: Fault current [A]
- V_{LL}: Line-to-line voltage of the system [V]
- Z_F: Total fault impedance [Ω]

The total fault impedance is given by (8.9c):

$$Z_F = Z_S + Z_T + Z_L \tag{8.9c}$$

- Z_S: Source impedance
- Z_T: Transformer impedance
- Z_L: Line impedance to the fault location

In most normal distribution networks, the grid source is a very low impedance source capable of high fault current for a short amount of time. The fault current is seen by protection devices, which can open to isolate the faulted circuit or even isolate in the circuit and re-energize some sections if the distribution network is so equipped. Figure 8.10 illustrates a balanced three-phase fault to ground where the fault current is supplied by the macrogrid. During the fault, power electronic converters will not increase output current by very much relative to the grid, and after some milliseconds, they may even reduce current to the nominal value [31].

If the microgrid is islanded and the macrogrid is not present, the fault current will be much lower, and typical circuit breaker or fuse devices may not engage, as shown in Figure 8.11. This creates a dangerous situation where the microgrid can be damaging equipment and putting a lethal voltage on grounded metal surfaces. Since

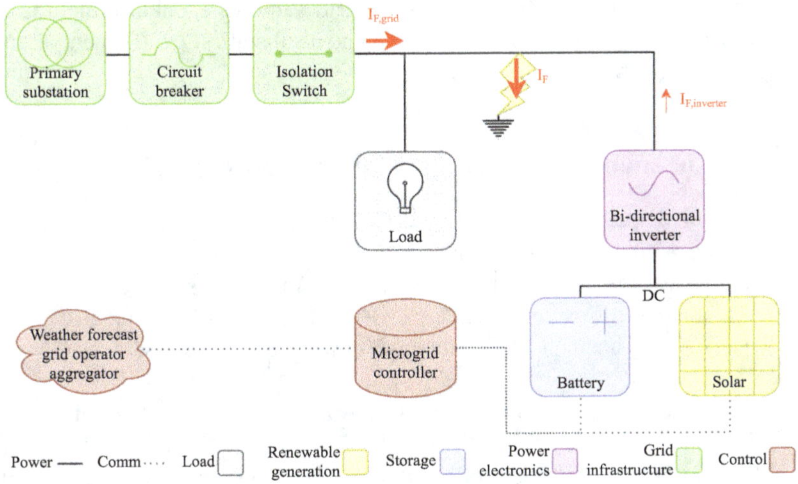

Figure 8.10 Balanced three phase to ground fault, current supplied by grid

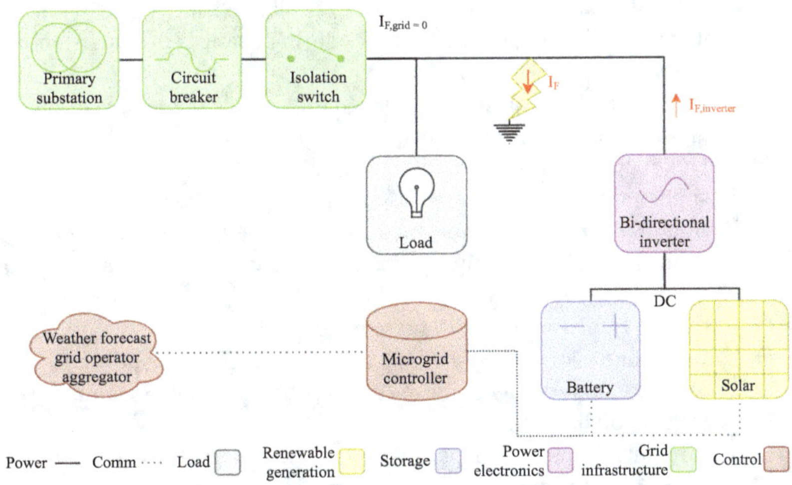

Figure 8.11 Balanced three phase to ground fault, current supplied by inverter

the power electronic sources may simply not have the prime energy source available to provide an adequate fault current, the situation could last indefinitely.

8.4 Design integration with smart grids and microgrids

The previous two sections have detailed the possible benefits of integrating microgrids with smart grids as well as some of the technical challenges. This section focuses on solutions that can mitigate the problems while retaining some or all of the benefit. The following table outlines five microgrid dispatch and control tasks that primarily take the perspective of the microgrid's interests, but also include cooperative behavior between adjacent microgrids and a substation-centric perspective on ground fault management.

Summary of smart grid microgrid integration strategies

1. **Optimal dispatch:** Monetary and CO_2 dual-objective unit commitment problem with month- and year-horizon decision variables
   (a) Real-time power balancing layer (1 s interval) with Autoregression Integrated Moving Average (ARIMA) forecast (+15 min)
   (b) Rolling horizon predict-then-optimize cost of energy layer (15 min interval)
   (c) Rolling horizon predict-then-optimize CO_2-equivalent of energy layer (15 min interval)
   (d) Calculating two-objective pareto front choosing lowest cost point above a pre-determined $\frac{\$}{tCO_2}$ threshold based on carbon market prices
2. **Power stability and adequacy:** Stabilize islanded microgrid with high percentage non-dispatchable renewables during changes in load, islanding, and black start
   (a) Maximum 1 s load power study based on available measurements and load equipment parameters
   (b) DC-couple all non-controllable sources with Li-ion battery and/or super capacitor
   (c) Synthetic inertia via non-linear dual-input $P(f, V)$ droop curve at each dispatchable source for collaborative frequency and voltage support
   (d) Closed transition islanding
3. **Optimal design:** Monetary, resilience, 100-year Global Warming Potential (GWP100), and biodiversity multi-objective optimal asset sizing with 25-year decision horizon
   (a) Simulated power balancing layer (1 s interval) with regression forecast (+15 min)
   (b) Simulated rolling horizon predict-then-optimize cost of energy layer (15 min interval)
   (c) Simulated horizon predict-then-optimize CO_2-equivalent of energy layer (15 min interval)
   (d) Pareto analysis of: lowest cost, resilience, GWP100, locally relevant social and ecological metrics
4. **Fault recovery:** Microgrid and smart grid fault detection and reaction
   (a) Monitor zero sequence currents at each microgrid source and disconnect after ride-through period (if required)
   (b) Grounding reactor at substation to detect ground faults

5. **Operational technical limits:** Manage microgrid injection to grid during high voltage or current event
   (a) Interconnection relay monitors grid voltage and curtails microgrid injection collaboratively with nearby microgrids
   (b) IEC61000 or DNP3 communication of high current event to microgrid relay which curtails injection

The following sections go into detail about most of the above strategies. A couple, such as the grounding reactor are specifically a smart grid and not a microgrid perspective, and would have very little impact on the microgrid's techno-economic operation aside from being a function the microgrid does not need to carry out. Others do not require extensive explanation, such as performing closed transitions or DC-coupling non-dispatchable renewable sources with storage.

8.4.1 Maximum load power

Microgrid optimal design must begin with adequate design, or some of the candidate solutions may be completely infeasible. This is especially true given the incentive of the optimizer to reduce the size of energy sources, whereas a black-start condition may require some amount of current too large for the reduced sources as described in Section 8.3.4.

Maximum load power of each microgrid must be estimated. Load measurements are almost obligatory, preferably 9 months (at least) of hourly (or less) interval data to understand relationships between load and exogenous variables like temperature. Furthermore, some 1 second data should be collected at different times of different days to estimate the load variation around the hourly mean. In the best case a black-start test is done, using the grid as a low impedance source to achieve a conservatively high inrush current value. In lieu of the black-start test, a list of the major transformers and motors and their design parameters should be obtained for estimating inrush current [34].

With the information gathered, the ratio between 1 second and 1 hour interval data is calculated and correlation or regression analysis determines whether the ratio has some relationship to the magnitude of the 1 hour data or to temperature. If not, the 95th or 99th percentile of the ratio's cumulative density function can be used to estimate the true 1 second peak value from the maximum measured 1 hour interval peak value. If there is a relationship between ratio and load magnitude or temperature, which is to say the higher the load magnitude the higher the ratio of 1 second peak to 1 hour average, then that relationship must be determined and then also used to estimate the true 1 second peak value. In several experiments no significant relationship is discovered.

8.4.2 Operational integration with smart grids and microgrids

The optimal dispatch of energy sources depends on making a decision in each time step for each decision variable, subject to constraints, which minimizes the cost function. The number of decision variables is, roughly speaking, the number of dispatchable resources. The decision variable in the case of battery storage is whether to charge or discharge, and at what power value. Instead, there often is not a decision variable for non-dispatchable sources like a solar PV plant, unless the power needs to be curtailed for technical reasons.

8.4.2.1 Energy Management System (EMS)

For a given set of dispatchable and non-dispatchable energy sources, this is referred to as the unit commitment problem [22] and is usually performed by the EMS. It develops a plan for future decisions, usually at a somewhat coarse time resolution such as the average hour or 15-minute interval power and SOC values. For this reason it can also be referred to as the dispatch or operation scheduler.

The EMS usually must also include an algorithm respects technical limits such as maximum grid import, balances load between multiple similar sources such as batteries, and reacts to unexpected changes to load or generation inside of the 15-minute intervals. This is often a heuristic method because of it's high speed nature: reading data, making decisions, and performing actions, every 1 second approximately.

Successful approaches vary, but there are computational advantages to delineating between (1) finding the optimal schedule of decision variables and (2) heuristic power control around that point for practical. These separate tasks are here referred to as Layer 1 and Layer 2 control, respectively. While Layer 2 control may slightly change the decision variables and therefore negatively impact the objective function, the change is usually small and outweighs the complexity of incorporating all Layer 2 objectives inside the Layer 1 optimization problem [17]. For example, small changes in battery charge or discharge power schedules are needed to balance the charge in parallel modules. Together these two layers comprise the EMS that is the main decision and control agent in a microgrid.

Layer 1: Optimal scheduling

Generally speaking, the optimization problem needs to be formulated to accurately model how changes in the decision variable affect the cost function. In the case of microgrids, this often means modeling SOC dynamics, power flow through the microgrid, how power exchange with the grid affects the monetary and CO_2 costs, and potential impacts on the main services provided by electricity, such as building heating and cooling or timely EV charging [39]. Depending on the cost function, the optimization problem may be solved only using current and past information, or may require some forecasting or assumptions about cost and performance patterns in the future. Lastly, the problem can be formulated and solved with a central agent that directly controls the energy sources, or by including distributed but cooperative agents that do not have full information but solve a local problem and act accordingly [40].

- **Downward flexibility** (reduction in net microgrid load)
  - Storage discharging
  - Combustion generator
  - Load re-scheduled to later time
  - Load curtailed (lost, not to be re-scheduled)

- **Upward flexibility** (increase in net microgrid load)
  - Storage charging
  - Curtailment of renewable production
  - Load re-scheduled to current time
  - Load opportunity (increased or added load, not anticipating future scheduled load)

The optimization problem can be approached in many different ways, each having some benefits and shortcomings. The mathematical considerations are an approach's performance in finding the global minimum, the performance given non-linear and non-convex problems, and convergence stability [41]. The practical considerations include choosing the right approach for the objective, considering also development time, cost and complexity of the required control hardware, and uncertainty of the true operational performance of the microgrid. The following are some common heuristic, rule-based methods, which may be appropriate for simple cost functions or small microgrids requiring very low equipment cost.

- **Demand response:** Downward load flexibility where dispatch is signaled from an external decision maker such as a grid operator or aggregator
- **Maximize self-consumption:** Reduce energy exchanged with the grid, often with a mix of renewable generation and storage
- **Peak shaving**: Downward flexibility during high load or high price events, often with a fixed threshold that the microgrid load (8.4) should be maintained below

The heuristic methods can often be improved by formal optimization to help make better decisions within the method's rules, for instance, the threshold value in peak shaving [42]. However, the methods are still heavily constrained by the relatively simple rules.

More powerful approaches typically involve formal optimization or machine learning methods. These, given the non-idealities of the physical and economic models, can sometimes be guaranteed to converge to the global minimum or have some statistical likelihood of doing so. However, they can also be difficult to construct and modify if the cost function changes, take minutes or hours to reach a solution, and require special and occasionally expensive computational solvers to implement. The following are some common approaches, including optimization and prediction:

- **Non-predictive optimization:** Minimize the cost function in the current time step using currently available measurements
- **Fixed horizon predict-then-optimize:** Minimize the cost function for time steps starting from the current one (t_0) to some chosen future horizon H time step ($t_0 + H$), using currently available measurements and future predicted ones, then act on the optimal decisions in each future $H - 1$ time steps
- **Rolling horizon predict-then-optimize:** Perform *fixed horizon predict-then-optimize* on the current time step t_0 and horizon H, then repeat *fixed horizon predict-then-optimize* method as time advances to the next time step t_1 using any new measurements obtained since t_0, repeating for each subsequent time step

The optimizer and predictor can take many forms and are often separate algorithms. Successful optimizers have been used such as MILP [17], Fuzzy Logic [43], and Genetic Algorithms [44]. Prediction methods vary depending on the quantity to be forecasted—such as load, solar production, or prices—but some common methods are exponential smoothing, random forest, Seasonal Autoregression Integrated Moving Average with eXogenous variable (SARIMAX) [45] and its variants, neural networks (and variants) [46], recurrent neural networks (and variants) [47], and recently transformers [48].

For some microgrids a third layer may be appropriate, often called the aggregator or Distributed Energy Resource Management System (DERMS) to coordinate between multiple microgrids or non-microgrid DER. This is particularly useful for integrating ancillary services or similar markets, where one single microgrid or DER could not be a market participant, but together as a Virtual Power Plant (VPP), the aggregator can use many DERs to provide upward and downward flexibility similar to larger plants. The layered structure is shown in Figure 8.12, while an example of EMS functioning is presented in Figure 8.2.

Layer 2: Heuristic power control

The second layer of the EMS should make only small adjustments to the power flows, based mostly on technical requirements. The second layer can help to somewhat abstract the unit commitment problem solved by the first layer, or to simplify the optimization problem and speed up its solution. Its rule-based algorithms and local-only communications allow it to work on a fast program loop speed, secondly or even faster. And if implemented on a separate computer or controller from the first layer, the second layer can even add resiliency by having a backup operational program in case of operating system or similar failure at the first layer. The second layer could be required to, for example:

- Balance SOC between parallel storage modules or cells
- Curtail renewable generation if total microgrid export exceeds a fixed value
- Curtail renewable generation dynamically based on a voltage or current signal from the distribution system

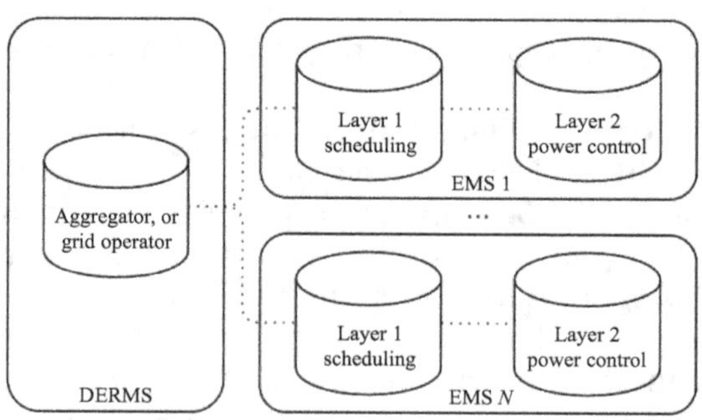

Figure 8.12 Layers of the Energy Management System and Distributed Energy Resource Management System hierarchy

- Limit maximum power due to high temperature
- Dynamically adjust downward flexibility as renewable generation varies to meet a precise average microgrid import threshold

Since the tasks required of the second layer depend on the distributed energy technology and microgrid architecture, there are few generic second-layer algorithms, and in fact the second layer may be composed of several different algorithms running on separate controllers, developed by the manufacturer of a particular energy asset. In particular second-layer functions related to the chemical battery storage may be handled by the Battery Management System (BMS).

8.4.2.2 Distributed Energy Resource Management System

Sometimes microgrids will interact with an aggregation layer, possibly alongside many other non-microgrid energy sources. The DERMS is both an information and control system to jointly operate many dispatchable energy sources in a coordinated manner. In this way, many small energy sources in different locations can have a large enough power and energy capacity to warrant participation in certain programs or markets. The geographical aggregation depends on the program or market. Often, ancillary services markets require each bid to be referenced to a single grid node or zone [22]. The aggregation may also depend on the flexibility and technology of the energy source. While energy storage is capable of both upward and downward flexibility, controllable EV charging is often only capable of downward flexibility. When EVs are capable of Vehicle-to-Grid (V2G) discharging, the flexibility options change greatly.

Although the DERMS and EMS solve similar economic problems, there may be a conflict of interest since a common objective to both may be downward flexibility during grid congestion or high-price events. If the DERMS and EMS have the same operator, the solution can be simply coordinating which system takes control in different scenarios. If the operators are different, the coordination must also be contractual in order to protect each party. Since the DERMS aggregates over many non-located sources, it is not relevant when a microgrid is islanded.

8.4.2.3 Microgrid islanding and protection devices

For microgrids that can disconnect and reconnect to the local grid, the distributed sources must have some coordination with the islanding switch or switches. As the point of technical interconnection with the grid, the islanding switch must abide by the local grid standards, and therefore is a very different instrument in different countries and regulatory zones. However, it fundamentally must be able to detect a transmission or distribution grid outage while filtering out disturbances, electrically disconnect and reconnect the microgrid from the local grid, and communicate with the energy sources that need to change operating modes. Typically a disconnect is triggered by a grid blackout, which the islanding switch controller or similar instrument must detect as well as reconnect when the grid voltage is again stable. However, microgrids may also disconnect opportunistically as a form of demand reduction or in the case of non-blackout fault or instability, such as a single-phase fault. Since both the islanding switch and EMS could initiate a disconnection, there must be communication and robust logic between the two agents. The following are a few examples:

- The islanding controller may block an EMS request to connect due to unstable voltage on the grid side, or due to a phase difference between the microgrid and grid voltages.
- The islanding controller may block an EMS request to disconnect due to ride-through requirements, where certain energy sources are required to support the grid during low-frequency events.
- The EMS may block the islanding controller request to reconnect until the grid-forming source is ready to switch to grid-following.

Protection devices are found in almost all microgrids and activate in the case of a fault to protect people and equipment from harm [34]. They may be integrated with energy sources, the islanding switch, or on various points of the distribution system. Many protection devices are not automatically resettable, such as fuses and some circuit breakers, meaning a fault can de-energize a portion or the entirety of the microgrid. In the case of single building, or smaller low-voltage distribution microgrids, a robust design will focus the energy sources near the point of interconnection with the grid so that an activated protection device in the microgrid distribution will disconnect the minimum amount of load possible, and not unnecessarily disconnect any energy sources. This is particularly true if the protection coordination is effectively designed, so that the most proximal protection device to the fault will activate [31]. Energy sources should also have an adequate capacity for the max load such that nuisance activation of their circuit breakers is minimal.

However, smart grids and microgrids often have automation to detect and isolate faults. This is most applicable to distribution networks, where overhead and underground lines are vulnerable to human- and naturally-caused short circuits, and distribution companies are required by regulation to minimize grid outages to consumers. For microgrids occupying part of a distribution network, integration of the distribution automation and the microgrid EMS and Supervisory Control and Data Acquisition (SCADA) can be quite difficult and subject to grid standards. For this reason the solutions are usually quite unique, involving substation and distribution circuit automation, which can be unique to the distribution company.

8.4.3 Simulation of operational strategy

Every microgrid experienced a design or planning process in which the capacity of each energy source was chosen. Capacity typically refers to the maximum power capacity of a source but also the maximum energy capacity of sources with storage such as *kWh* of battery. In some cases, these sizes may need to be fixed by regulation or practical consideration, but often there is a range of acceptable capacities and certain discrete values can be taken, as it is shown in Table 8.4. This discretization is due to practical and market considerations such as the power of a single solar PV module or the power and energy capacity of a single battery module.

Choosing the best energy source capacities can be formulated as a constrained techno-economic optimization. The objective function could be the same as the one

Table 8.4 Common Discretization of DER capacity

DER	Power [kW]	Energy [kWh]
Solar PV	0.2–0.4	–
Battery (Li-Ion)	5 (module), 70–100 (rack)	5[1] (module), 70–100[1] (rack)
Wind	1–5	–

[1]Assuming 1 C maximum charge and discharge rate.

for the operational problem. However, the solution is more likely to be trivial; for example, to minimize energy exchange with the grid, the optimal battery capacity is simply the largest within the technical constraints. Therefore, the objective function is usually re-framed as an efficiency statement, such as reduction of energy exchange per *kWh* of battery energy capacity or simple payback period, capital expenditure divided by annual revenue. Economic metrics such as internal rate of return and net present value are also common [22].

Evaluating the optimal capacity objective function depends on the operational strategy. For the same-size storage capacity, a heuristic strategy will often perform worse than a predictive and adaptive one, changing, for example, the simple payback period [17]. The operational strategy is intended to be implemented on a controller and run in real time with real measurements. But by simulating the operational strategy on historical measurements, such as building load and energy prices, we can infer how the strategy will perform on future data and choose the optimal capacities [45].

For a single objective capacity optimization problem, such as maximizing internal rate of return, a manual search of the solution space may be adequate. This is because simulating the EMS first and second layers on 1–5 years of 15-minute interval historical data is not an immense computational task, given that the EMS loop speed should be on the order of seconds or single-digit minutes. And while energy sources such as solar PV have very low capacity discretization capacity, around 0.4 kW, lithium-ion battery storage discretization can range from 10 to 100 kWh and wind turbines can be as high as 1 MW. Therefore, the capacity search space tends not to be exceptionally large for microgrids involving two or three differences energy sources.

8.4.4 Injection management

During the microgrid or grid infrastructure planning phase, curtailment of the renewable generator or operation of controllable DERs like the battery can be considered to increase the hosting capacity at a particular node. However, violating high voltage or current in the feeder could both damage other consumer devices and grid equipment, so charging the battery in Figure 8.7 to reduce grid injection may not be a reliable enough solution, for instance, if the battery is already at maximum SOC. The microgrid energy sources instead need to be directly managed, especially renewable generation, which the microgrid controller will usually have an incentive to export to the grid.

The microgrid interconnection relay, or similar device, will need to monitor all the distribution phase voltages and have pre-programmed thresholds relative to the technical voltage limits. For example, upon reaching 90% of the voltage limit, the relay will command the microgrid controller to reduce the grid export to below 80%, the hysteresis preventing unwanted control action. Increasing export is prohibited for 10 minutes. At 98% of the voltage limit, the microgrid will prevent all export until the voltage is below 50% of the limit and increasing export is prohibited until technical intervention [34]. The microcontroller acts by decreasing generation in steps of 1% until the new threshold is met, and it orders the curtailed energy sources from highest to lowest marginal cost.

Managing the maximum current of transformers, or congestion, functions similarly. However, the critical relay is at the primary or secondary substation instead of at the microgrid point of interconnection. The substation relay would communicate with IEC61000, DNP3, or a similar and locally appropriate protocol.

8.4.4.1 Collaboration

In the likely case that a medium- or low-voltage circuit has multiple microgrids, the microgrids need to operate somewhat collaboratively [40]. This means that the control actions should not become unstable, drastically increasing or decreasing the voltage or current, and microgrids should equally share the burden of curtailment. Hysteresis should mostly prevent instability, although each controller will be programmed to randomly resume increasing export at a time delay between 9 and 11 minutes. To share curtailment, each 1% decrease in power is performed after a random time delay between 1 and 10 seconds [34].

8.4.5 Synthetic inertia

Electric power systems the world around were built first with rotating machines for generators, both hydroelectric or internal combustion powered synchronous generators. These machines were eventually made quite large for economies of scale, but the large rotational inertia provided another benefit: it stored a small amount of energy that could be used to maintain a more stable frequency.

The dynamics of the inertia and changes in the grid frequency are described by the swing equation (8.10) for a synchronous generator:

$$M\frac{d^2\delta}{dt^2} = T_m - T_e \tag{8.10}$$

where

- M is the moment of inertia of the rotor [kg·m²]
- δ is the angular displacement [radians]
- $\frac{d^2\delta}{dt^2}$ is the angular acceleration [radians/s²]
- T_m is the mechanical torque [Nm]
- T_e is the electrical torque [Nm]

For a sample rotating synchronous machine, resistive load, and automatic frequency regulation, the machine spins at a constant frequency $\frac{d\delta}{dt}$ when the electrical load (represented as a torque T_e on the stator) equals that of the mechanical prime mover (also represented as a torque on the rotor, T_m). If the electrical load increases slightly, T_e increases and $\frac{d\delta}{dt}$ takes a negative sign, meaning that the machine is decelerating and the frequency is decreasing. The amount of deceleration per difference in torque is governed by the moment of inertia *M*, a physical parameter.

Power converters, on the other hand, have very limited electrical energy storage and no mechanical inertia [31]. For a sample power converter and resistive load, when the load increases there is a voltage drop due to Ohm's law and the converter's internal controller compensates by increasing the voltage, and frequency may not change at all. If all generators on the grid or microgrid behave exactly this way, there is likely no problem and frequency loses its physical meaning, at least on the generation side. However, many grids have a mix of traditional rotating machines, power converter generators with varying control algorithms, and both rotating and non-rotating machine loads.

For this reason, the concept of synthetic inertia has been proposed and is well researched in the literature. There are many implementations of synthetic inertia, including ones that have no resemblance at all to mechanical inertia and may be based purely on data-driven models rather than physical ones.

$$P(\Delta f, \Delta V) = \frac{1}{1 + e^{-M\sqrt{w(\Delta f) + (\Delta V)}}} \tag{8.11}$$

where

- P is the change in output real power, positive or negative, [kW]
- $\Delta f = f - f_N$, where f is measured frequency and f_N is nominal [Hz]
- $\Delta V = V - V_N$, where V is measured voltage and V_N is nominal [V]
- M is the inertial constant, experimentally determined
- w is a weighting factor, experimentally determined

The strategy chosen here is a droop function where voltage and frequency are the independent variables, and the dependent is real power. The advantage of containing both inputs in the same droop function is that in a physical power system an increased electrical load would reduce *f* according to the swing equation, but also *V* according to Ohm's law, although both are quickly impacted also by the generator governor [49].

8.5 Conclusions

Microgrids are a co-located set of loads and DERs that can operate both independent of and integrated with distribution and transmission networks and are particularly advantageous for improving the energy supply reliability to consumers. Some microgrids also have advanced control and operation features, such as dispatch optimization and integration with distribution automation systems, allowing them to better serve the consumer's objectives and also support the larger macrogrid.

The mandate to install renewable energy in electric power systems all around the world comes with both opportunities and challenges. Clean and distributed generation will reduce local air pollution, increase grid efficiency by locating generation closer to consumption, provide communities and individuals with local backup power, and democratize the energy system by allowing more choice and investment in the energy system. However, the reliance on variable renewable sources creates a load-demand imbalance problem, reverse flow on distribution grids makes voltage control and fault detection problematic, and markets are being reshaped by renewable sources with zero marginal cost of generation.

Advanced microgrids integrated with smart grids can change the energy paradigm to reap benefits and solve problems. In the past, set-and-forget networks with centralized generators relied on overbuilding infrastructure and economies of scale to reduce costs. Instead microgrids can self-supply most of their energy and manage peak load events, with both load and generation flexibility, and therefore depend less on the grid connection. During high prices or technical events such as low frequency, microgrids can use flexibility to support markets or the grid, earning revenue for the microgrid owner. And in the case of a grid blackout, hybrid renewable microgrids have the unique characteristic of providing stochastically-available backup power for an unlimited time horizon, for instance at a community cooling center or emergency coordination location.

Renewable generation and storage are decidedly on a path to mainstream adoption in power systems, but advanced microgrids are still relatively nascent and need technological and regulatory development. The following are a few important areas of required future work:

- **Battery storage control.** Whereas solar PV modules and inverters are largely interoperable given some technical constraints, battery module and inverter visibility and control by a microgrid controller are often limited or entirely impossible. Integrated ESSs partially solve this problem with a limited-functionality microgrid controller; however, most depend on heuristic-only dispatch algorithms and suffer from poor economic performance.
- **Building Energy Management System integration.** The technical problem of maximum load power, especially during a black-start event, could be better managed or even economically optimized if large loads such as Heating Ventilation and Air Conditioning (HVAC), EV charging, pumping, and lighting could be rescheduled by the microgrid controller. In addition by knowing the cost of load lost or simply priority of loads, the microgrid controller could use downward load flexibility to extend resilient backup power durations do the highest priority loads, and even bid the downward flexibility to markets, providing much-needed elasticity in spot market demand curves.
- **EV and V2G integration.** As the transportation sector is electrified, the many battery EVs plugged in to buildings and grids will quickly create a usable storage capacity on the grid many times larger than stationary distributed storage. With optimization tools and existing integrations with the grid, such as a DR program, a microgrid controller can both economically charge and discharge the EV battery, while also being the only charging source in a long-term grid outage. Currently V2G is seldom available, and where it is, the vehicle manufacturer, responsible for the battery warranty, is often incentivized to limit V2G use.

References

[1] Electricity production – Electricity Information: Overview – Analysis. Available from: https://www.iea.org/reports/electricity-information-overview/electricity-production.

[2] Global overview – Renewables 2024 – Analysis. Available from: https://www.iea.org/reports/renewables-2024/global-overview.

[3] Yang D, Wang W, Gueymard CA, *et al.* A review of solar forecasting, its dependence on atmospheric sciences and implications for grid integration: Towards carbon neutrality. *Renewable and Sustainable Energy Reviews.* 2022;161:112348. Available from: https://www.sciencedirect.com/science/article/pii/S1364032122002593.

[4] Al-Waeli AHA, Kazem HA, Chaichan MT, *et al.* The Impact of Climatic Conditions on PV/PVT Outcomes. In: Al-Waeli AHA, Kazem HA, Chaichan MT, *et al.*, editors. Photovoltaic/Thermal (PV/T) Systems: Principles, Design, and Applications. Cham: Springer International Publishing; 2019. p. 173–222. Available from: https://doi.org/10.1007/978-3-030-27824-3_5.

[5] Zahedi A. Maximizing solar PV energy penetration using energy storage technology. *Renewable and Sustainable Energy Reviews.* 2011;15(1):866–870. Available from: https://www.sciencedirect.com/science/article/pii/S1364032110002984.

[6] Deshmukh MK and Deshmukh SS. Modeling of hybrid renewable energy systems. *Renewable and Sustainable Energy Reviews.* 2008;12(1):235–249. Available from: https://www.sciencedirect.com/science/article/pii/S1364032106001134.

[7] Santos-Alamillos FJ, Pozo-Vázquez D, Ruiz-Arias JA, *et al.* Combining wind farms with concentrating solar plants to provide stable renewable power. *Renewable Energy.* 2015;76:539–550. Available from: https://www.sciencedirect.com/science/article/pii/S096014811400785X.

[8] Hossain MS, Jahid A, Islam KZ, *et al.* Solar PV and Biomass Resources-Based Sustainable Energy Supply for Off-Grid Cellular Base Stations. *IEEE Access.* 2020;8:53817–53840. *Conference Name: IEEE Access.* Available from: https://ieeexplore.ieee.org/abstract/document/9022971.

[9] Liu Z and Du Y. Evolution towards dispatchable PV using forecasting, storage, and curtailment: A review. *Electric Power Systems Research.* 2023;223:109554. Available from: https://www.sciencedirect.com/science/article/pii/S0378779623004431.

[10] Kakoulaki G, Gonzalez Sanchez R, Gracia Amillo A, *et al.* Benefits of pairing floating solar photovoltaics with hydropower reservoirs in Europe. *Renewable and Sustainable Energy Reviews.* 2023;171:112989. Available from: https://www.sciencedirect.com/science/article/pii/S136403212200870X.

[11] Pourbehzadi M, Niknam T, Aghaei J, *et al.* Optimal operation of hybrid AC/DC microgrids under uncertainty of renewable energy resources: A comprehensive review. *International Journal of Electrical Power & Energy Systems.* 2019;109:139–159. Available from: https://www.sciencedirect.com/science/article/pii/S0142061518324347.

[12] Olatomiwa L, Mekhilef S, Ismail MS, *et al.* Energy management strategies in hybrid renewable energy systems: A review. *Renewable and Sustainable Energy Reviews*. 2016;62:821–835. Available from: https://www.sciencedirect.com/science/article/pii/S1364032116301502.

[13] Fathima AH and Palanisamy K. Optimization in microgrids with hybrid energy systems – A review. *Renewable and Sustainable Energy Reviews*. 2015;45:431–446. Available from: https://www.sciencedirect.com/science/article/pii/S1364032115000696.

[14] Moreno Escobar JJ, Morales Matamoros O, Tejeida Padilla R, *et al.* A Comprehensive Review on Smart Grids: Challenges and Opportunities. *Sensors*. 2021;21(21):6978. Available from: https://www.mdpi.com/1424-8220/21/21/6978.

[15] Lasseter RH and Paigi P. Microgrid: A conceptual solution. In: *2004 IEEE 35th Annual Power Electronics Specialists Conference (IEEE Cat. No. 04CH37551)*, vol. 6. IEEE; 2004. p. 4285–4290.

[16] Al-Falahi MDA, Jayasinghe SDG, and Enshaei H. A review on recent size optimization methodologies for standalone solar and wind hybrid renewable energy system. *Energy Conversion and Management*. 2017;143:252–274. Available from: https://www.sciencedirect.com/science/article/pii/S0196890417303230.

[17] Manzolini G, Fusco A, Gioffrè D, *et al.* Impact of PV and EV Forecasting in the Operation of a Microgrid. *Forecasting*. 2024;6(3):591–615. Available from: https://www.mdpi.com/2571-9394/6/3/32.

[18] Matrone S, Ogliari E, Nespoli A, *et al.* Electric Vehicle Supply Equipment Day-Ahead Power Forecast Based on Deep Learning and the Attention Mechanism. *IEEE Transactions on Intelligent Transportation Systems*. 2024;25(8):9563–9571.

[19] Nespoli A, Ogliari E, Leva S, *et al.* Day-Ahead Photovoltaic Forecasting: A Comparison of the Most Effective Techniques. *Energies*. 2019;12(9):1621. Available from: https://www.mdpi.com/1996-1073/12/9/1621.

[20] Delfanti M, Blaco A, Bovera F, *et al.* Grid-Tie and Off-Grid Operations for an Innovative Microgrid Realized in Leonardo Campus of Politecnico di Milano. In: *2019 IEEE 5th International forum on Research and Technology for Society and Industry (RTSI)*; 2019. p. 103–108.

[21] Kroposki B, Pink C, Basso T, *et al.* Microgrid standards and technology development. In: *2007 IEEE Power Engineering Society General Meeting*. IEEE; 2007. p. 1–4.

[22] Kirschen DS and Strbac G. *Fundamentals of Power System Economics*. John Wiley & Sons; 2018.

[23] Luthander R, Widén J, Munkhammar J, *et al.* Self-consumption enhancement and peak shaving of residential photovoltaics using storage and curtailment. *Energy*. 2016;112:221–231.

[24] Zhao L, Yang Z, and Lee WJ. The impact of time-of-use (TOU) rate structure on consumption patterns of the residential customers. *IEEE Transactions on Industry Applications*. 2017;53(6):5130–5138.

[25] Schreiber M, Wainstein ME, Hochloff P, *et al.* Flexible electricity tariffs: Power and energy price signals designed for a smarter grid. *Energy*. 2015;93:2568–2581.

[26] Williams KJ. Federal energy regulatory commission v. electric power supply association. *Public Land & Resources Law Review*. 2016;(6):30.

[27] Tina GM, Aneli S, and Gagliano A. Technical and economic analysis of the provision of ancillary services through the flexibility of HVAC system in shopping centers. *Energy*. 2022;258:124860.

[28] Biegel B, Westenholz M, Hansen LH, *et al.* Integration of flexible consumers in the ancillary service markets. *Energy*. 2014;67:479–489.

[29] Sæle H, Morch A, Degefa MZ, *et al.* Assessment of flexibility in different ancillary services for the power system. In: *2020 17th International Conference on the European Energy Market (EEM)*. IEEE; 2020. p. 1–6.

[30] Li L, Zhu D, Zou X, *et al.* Review of frequency regulation requirements for wind power plants in international grid codes. *Renewable and Sustainable Energy Reviews*. 2023;187:113731.

[31] Chauhan RK, Chauhan K, and Singh SN. Microgrids for Rural Areas: Research and case studies. *Institution of Engineering and Technology*. 2020;160:477–493.

[32] Liu WY, Chiang YH, and Lin CC. Adopting renewable energies to meet the carbon reduction target: Is forest carbon sequestration cheaper? *Energy*. 2022;246:123328.

[33] Gómez-Catasús J, Morales MB, Giralt D, *et al.* Solar photovoltaic energy development and biodiversity conservation: Current knowledge and research gaps. *Conservation Letters*. 2024:e13025.

[34] Hatziargyriou N. *Microgrids: Architectures and Control*. John Wiley & Sons; 2014.

[35] Turner R, Walton S, and Duke R. Stability and bandwidth implications of digitally controlled grid-connected parallel inverters. *IEEE Transactions on Industrial Electronics*. 2010;57(11):3685–3694.

[36] Cimino M and Pagilla PR. Reactive power control for multiple synchronous generators connected in parallel. *IEEE Transactions on Power Systems*. 2016;31(6):4371–4378.

[37] Nasiri M and Mohammadi R. Peak current limitation for grid side inverter by limited active power in PMSG-based wind turbines during different grid faults. *IEEE Transactions on Sustainable Energy*. 2016;8(1):3–12.

[38] Shahparasti M, Laaksonen H, Kauhaniemi K, *et al.* Inrush current management during medium voltage microgrid black start with battery energy storage system. *IEEE Access*. 2022;10:42287–42296.

[39] Mohamed MA and Eltamaly AM. *Modeling and Simulation of Smart Grid Integrated with Hybrid Renewable Energy Systems*. vol. 121. Springer; 2018.

[40] Xu Y, Zhang W, Hug G, *et al.* Cooperative control of distributed energy storage systems in a microgrid. *IEEE Transactions on Smart Grid*. 2014;6(1):238–248.

[41] Zheng QP, Wang J, and Liu AL. Stochastic optimization for unit commitment: A review. *IEEE Transactions on Power Systems*. 2014;30(4):1913–1924.

[42] Wood M, Ogliari E, and Leva S. The Value of Forecasting: The Effect of Building Load Forecast Errors on the Performance of an Optimal Energy Management System. In: *2024 IEEE 8th Forum on Research and Technologies for Society and Industry Innovation (RTSI)*. IEEE; 2024. p. 42–47.

[43] Zehra SS, Wood MJ, Grimaccia F, *et al.* A Cost-Effective Fuzzy-based Demand-Response Energy Management for Batteries and Photovoltaics. In: *2023 11th International Conference on Smart Grid (icSmartGrid)*. IEEE; 2023. p. 1–6.

[44] Obara S and Morel J. *Clean Energy Microgrids*. vol. 90. IET; 2017.

[45] Polimeni S, Nespoli A, Leva S, *et al.* Implementation of different PV forecast approaches in a multiGood microGrid: Modeling and experimental results. *Processes*. 2021;9(2):323.

[46] Nespoli A, Leva S, Mussetta M, *et al.* A selective ensemble approach for accuracy improvement and computational load reduction in ANN-based PV power forecasting. *IEEE Access*. 2022;10:32900–32911.

[47] Wood M, Ogliari E, Nespoli A, *et al.* Day Ahead Electric Load Forecast: A Comprehensive LSTM-EMD Methodology and Several Diverse Case Studies. *Forecasting*. 2023;5(1):297–314.

[48] Hanif M and Mi J. Harnessing AI for solar energy: Emergence of transformer models. *Applied Energy*. 2024;369:123541.

[49] Nema S, Prakash V, Pandžić H. Adaptive synthetic inertia control framework for distributed energy resources in low-inertia microgrid. *IEEE Access*. 2022;10:54969–54979.

Chapter 9

Optimal transaction strategies model in energy and reserve markets considering hybrid renewable energy and storage system

Ngoc Thien Nam Tran[1], Hong-Tzer Yang[2], Minh Quan Duong[3], Marco Mussetta[4] and Francesco Grimaccia[4]

The increasing electricity demand and the imperative to reduce CO_2 emissions have accelerated the integration of renewable energy sources into power systems. However, the inherently intermittent and non-dispatchable nature of renewable generation poses significant challenges to power system reliability and stability. To address these issues, power systems can be segmented into microgrids (MGs), interconnecting to form multiple MG networks, enhancing control and operational efficiency, and providing a robust solution for addressing power imbalances from forecasting inaccuracies or equipment failures. In such systems, decentralized operation coordinated with dispatch strategies is essential to ensure overall system stability and optimize resource utilization. This approach also helps maintain the economic benefits for all participating entities. This chapter proposes an optimal transaction strategy model for hybrid renewable energy and energy storage systems (ESSs) under multiple MGs in electricity markets. Unlike prior studies focusing solely on energy markets, this framework incorporates both energy and reserve markets to minimize MG operational costs while aligning with current and future policies. Two optimization approaches are explored: a non-cooperative game for individual MG optimization and a cooperative game for coalition-based optimization. Results show significant cost reductions for interconnected MGs, with the cooperative approach achieving lower daily costs and faster computation, proving its practical effectiveness.

9.1 Introduction

In recent years, the large penetration of renewable energy sources (RESs), such as wind power (WP) and photovoltaic (PV), has made it an irreplaceable source in the

[1]Delta Electronic Inc., Tainan City, Shanhua District, Taiwan
[2]Department of Electrical Engineering, National Cheng Kung University, Tzu-Chang Campus, Taiwan
[3]Faculty of Electrical Engineering, The University of Danang – University of Science and Technology, Vietnam
[4]Department of Energy, Politecnico di Milano, Campus Bovisa, Italia

power system [1,2]. In which RESs, deployed as distributed energy resources (DERs), present operational challenges due to their intermittent and stochastic nature, particularly as traditional consumers transition into prosumers [3]. To address these issues and enhance renewable energy utilization, the MG concept has been introduced [4]. An MG integrates energy sources, storage systems, and consumers into a single controllable entity, but its reliability and power exchange capabilities are limited by RES uncertainty [5]. As a solution, multi-microgrid (MMG) systems have emerged, interconnecting MGs to operate in islanded or grid-connected modes, offering improved stability, flexibility, and energy efficiency [6,7].

In islanded mode, prosumers engage in local or peer-to-peer (P2P) energy trading within MGs [8], while in grid-connected mode, MGs exchange power with utility grids or other MGs to enhance economic benefits through inter-trading [9]. These strategies reduce transmission losses, lower power generation demands, and optimize operational costs [10], fostering mutual benefits [11]. The electricity market is evolving toward decentralization and privatization, encompassing energy and reserve markets to ensure grid reliability and stability. Game-theoretic approaches, including non-cooperative and cooperative models, play a pivotal role in optimizing energy transactions and dispatch strategies [12,13]. By integrating these methods, MGs and MMG systems improve efficiency, reliability, and cost-effectiveness, paving the way for a competitive and sustainable energy market [14].

Previous studies have optimized MG energy management by developing energy management system (EMS) frameworks that integrate RESs and ESSs [15–18]. These include methods like adaptive dynamic programming [15], particle swarm optimization (PSO) for real-time (RT) EMS [16], and improved PSO for two-level scheduling [17], with some treating ESS as generators with equivalent fuel pricing [18]. Algorithms such as genetic algorithms (GAs) and differential evolution (DE) are also applied, mainly for single MGs [16,19]. However, EMS optimization for MMG participation remains underexplored, despite its importance for future power systems. MMGs enable competitive energy markets through bidding and trading strategies, focusing on pricing mechanisms and peer-to-peer trading [20,21]. Accurate transaction pricing for all prosumers, including ESS and EVs, is vital to maintain power balance and optimize marginal prices [22]. Energy and reserve markets are managed via day-ahead (DA) and real-time (RT) schedules to address uncertainties, with pricing mechanisms ranging from marginal pricing to profit-optimized internal prices, balancing supply, demand, and participation incentives.

Energy trading optimization in electricity markets utilizes game-theoretic algorithms, bi-level programming, and robust optimization [23], enabling MGs to interact with upper-layer operators but facing challenges like time inefficiency and convergence in large systems. Bidding functions provide flexible pricing solutions, adapting to uncertainties like RES variability and demand fluctuations [24]. Aggregators use supply-demand curves to determine market-clearing prices iteratively. Game theory optimizes energy dispatch in decentralized systems, while cooperative approaches facilitate MG alliances for data sharing, benefit allocation, and price determination [25,26]. Recent studies explore reserve market strategies, integrating DA and RT schedules to address RES forecasting errors and enhance DER and ESS efficiency

Table 9.1 Literature review for previous research

References	Trading game theory	Market considered		Schedule considered	
		Energy market	Reserve market	Day-Ahead	Real-Time
[26,27]	Cooperative	Yes	No	Yes	No
[28]	Not mention	Yes	Yes	Yes	No
[29,31]	Cooperative	Yes	No	No	Yes
[30,32]	Cooperative	Yes	No	Yes	No
[33]	Stackelberg	Yes	Yes	Yes	No
[34]	Cooperative and non-cooperative	Yes	No	No	Yes

[27]. For instance, Reference [28] proposes a unit commitment model that integrates ESSs into both energy and reserve markets to mitigate forecasting errors associated with RESs. However, the lack of clear reserve market regulations poses challenges for participants in accurately assessing potential benefits, resulting in restricted energy transactions and insufficient synchronization between DA and RT market operations.

Previous studies have yet to address the complete transaction process across both energy and reserve markets for MMG energy trading with integrated DA and RT schedules. Most research focuses solely on bidding frameworks for either DA or RT schedules using game theory. For further details, relevant literature and key items pertaining to this study are summarized in Table 9.1.

Therefore, this chapter proposes innovative transaction strategies for energy and reserve markets in MMG systems, treating each MG as a prosumer. Using non-cooperative and cooperative game theory, the framework optimizes DA and RT schedules. Non-cooperative approaches minimize individual MG costs, while cooperative methods reduce total MMG costs by optimizing energy transactions under power company rules. The comprehensive mechanism incorporates all prosumer types, including RESs, DERs, and ESSs, and integrates reserve market rules with constraints on DER and ESS capabilities. Quadratic programming (QP) ensures efficient optimization, enabling convergence to market-clearing prices during bidding processes. The framework is validated using historical data and market rules from Taiwan and Australia, reflecting global goals for sustainable power systems and minimal environmental impact.

9.2 System structure and mathematical model

9.2.1 Overall system structure

The MMG structure consists of N MGs, each including RESs, Diesel generators (DGs), ESSs, and loads, with bidirectional energy transactions balancing consumption and generation as Figure 9.1. MG aggregators, acting as control centers, manage

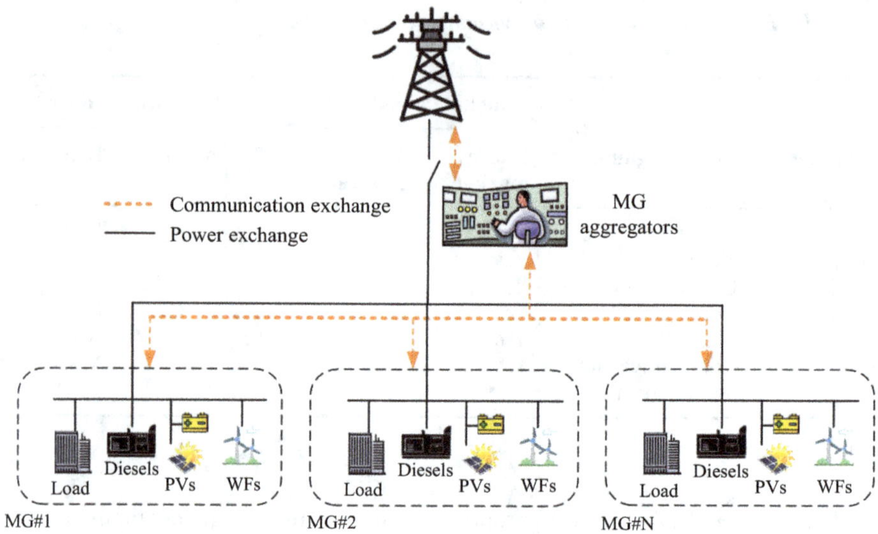

Figure 9.1 The overall system structure

transaction schedules and handle bidding and trading based on surplus or deficit energy data. They aim to minimize total energy operation costs by generating supply-demand bidding curves and exploring buying and selling price scenarios. These prices guide MG decisions on ESS charging/discharging and DG operations. Ultimately, aggregators determine net electricity import/export by coordinating energy transaction strategies through shared MG information.

Optimal transaction strategies of MMG focus on two schedules: day-ahead and real-time via the electricity market with the framework shown in Figure 9.2.

This framework integrates energy optimization strategies for MMGs in the electricity market. MG aggregators or controllers autonomously collect data from RES forecasts, ESS, and DGs to generate supply/demand curves for bidding. MGs then trade energy through the aggregator, following optimal schedules to maximize aggregator revenue or minimize operation costs with hourly resolution. The DA schedule closes when the 24-hour trading strategy and market-clearing prices for energy and reserves are announced, transitioning to RT scheduling. To manage RES forecast uncertainty, the RT schedule updates every 5 minutes, clearing errors through re-optimization using DG and ESS reserves. This approach ensures optimal DA trading prices and efficient generator operation while enabling RT adjustments.

9.2.2 Problem formulation

Energy transaction strategies aim to minimize MMG operation costs, including DG costs, ESS operation costs, and inter-trading costs, while prioritizing renewable energy. MGs must coordinate DG, ESS, and inter-trading schedules to achieve the lowest clearing price. Adaptive bidding functions replace fixed-price methods,

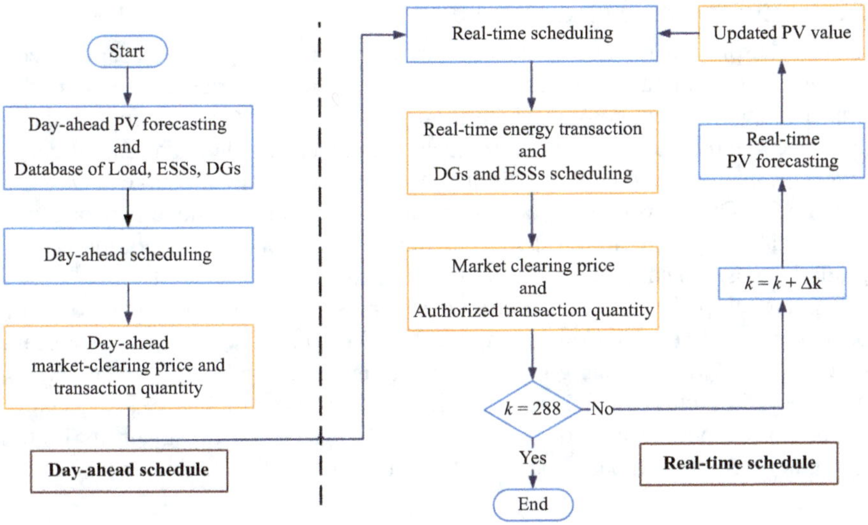

Figure 9.2 The problem framework for energy transaction strategies

allowing MG aggregators to evaluate buying and selling scenarios based on supply-demand curves. ESS charge/discharge decisions and DG operations depend on these prices, enabling operators to clear the market and optimize electricity consumption while minimizing costs.

9.2.2.1 Diesel generator cost model

In the MG, the DERs usually play a significant role in supplying power to the load demand and partly for power back up. With the development of generator technology, DGs are gradually widely used in MGs because of their flexibility. It has the ability to quickly respond to shortage power caused by RES as well as to generate stable and long-time operations like thermal generators. The operation cost of DGs is often represented by a quadratic polynomial, related to primary fuel cost and output power, which shows in (9.1):

$$C_{\mathrm{dg},i}(t) = a_i \times P_{\mathrm{dg},i}(t)^2 + b_i \times P_{\mathrm{dg},i}(t) + c_i \tag{9.1}$$

subject to

$$P_{\mathrm{dg},i,\min} \leq P_{\mathrm{dg},i}(t) \leq P_{\mathrm{dg},i,\max} \tag{9.2}$$

where $C_{\mathrm{dg},i}(t)$ is an operation cost of the ith generator at time t; $P_{\mathrm{dg},i}(t)$ is the generating power at time t; a_i, b_i and c_i are fixed cost parameters of the ith generator; $P_{\mathrm{dg},i,\min}$ and $P_{\mathrm{dg},i,\max}$ is the lower and upper limits for output power of the ith generator, respectively. When there is a maneuver of DG power from time t to $t + \Delta t$, the change in output power needs to satisfy the conditions of ramping up ($\mathrm{ramp}_{P_{\mathrm{dg},i,\mathrm{up}}}$) and ramping down ($\mathrm{ramp}_{P_{\mathrm{dg},i,\mathrm{down}}}$) limits as equation:

$$\mathrm{ramp}_{P_{\mathrm{dg},i,\mathrm{down}}} \leq P_{\mathrm{dg},i}(t + \Delta t) - P_{\mathrm{dg},i}(t) \leq \mathrm{ramp}_{P_{\mathrm{dg},i,\mathrm{up}}} \tag{9.3}$$

9.2.2.2 Energy store systems model

ESS stores surplus energy and discharges it as needed, reducing MMG operation costs but requiring consideration of installation, maintenance, and degradation expenses. Advances in battery technologies like Lithium, NaS, and Ni-Cd have improved cost-effectiveness and durability, though frequent cycling causes degradation, influenced by cycles, capacity wear, depth of discharge (DOD), and state of charge (SOC) [35]. Extreme SOC/DOD levels and cycling rates impact lifespan, but modern management systems mitigate temperature effects. This chapter models ESS operation costs as degradation costs to capture real economic impacts.

The model of the ESS degradation cost represents a direct depreciation on its capacity and lifetime [36]. Considering the *j*th ESS with capacity $E_{\mathrm{ESS},j}$, the installation cost is $C_{\mathrm{ESS},j,\mathrm{install}}$. Its charging and discharging power are $P_{\mathrm{ESS},j,\mathrm{ch}}$ and $P_{\mathrm{ESS},j,\mathrm{disch}}$, respectively, with charge efficiency $\eta_{\mathrm{ESS},j,\mathrm{ch}}$ and discharge efficiency $\eta_{\mathrm{ESS},j,\mathrm{disch}}$.

The corresponding ESS degradation cost during period Δt, given its lifetime $L_{\mathrm{ESS},j}$, can be calculated based on the total charging and discharging energy of ESS [35,36]:

$$C_{\mathrm{ESS},j}(t) = \frac{C_{\mathrm{ESS},j,\mathrm{install}}}{2 \times L_{\mathrm{ESS},j}(\Delta t, \mathrm{DOD}) \times E_{\mathrm{ESS},j}} \times \left(\frac{P_{\mathrm{ESS},j,\mathrm{ch}} \times \Delta t}{\eta_{\mathrm{ESS},j,\mathrm{ch}}} + \left| \frac{P_{\mathrm{ESS},j,\mathrm{disch}} \times \Delta t}{\eta_{\mathrm{ESS},j,\mathrm{disch}}} \right| \right) \tag{9.4}$$

The absolute value indicates that when operated in the discharged state, the ESS is still subject to degradation. *Assuming that* the charging and discharging efficiency are the same, (9.4) can be rewritten as:

$$C_{\mathrm{ESS},j}(t) = C_{\mathrm{deg},j} \times \left[P_{\mathrm{ESS},j,\mathrm{ch}}(t) + \left| P_{\mathrm{ESS},j,\mathrm{disch}}(t) \right| \right] \tag{9.5}$$

During ESS operation, it is necessary to ensure the conditions of upper and lower power limits:

$$P_{\mathrm{ESS},j,\mathrm{min}} \le P_{\mathrm{ESS},j,\mathrm{ch}}(t) + P_{\mathrm{ESS},j,\mathrm{disch}}(t) \le P_{\mathrm{ESS},j,\mathrm{max}} \tag{9.6}$$

where $P_{\mathrm{ESS},j,\mathrm{min}}$ is the lowest power limit of the battery corresponding to the rated discharge power; $P_{\mathrm{ESS},j,\mathrm{max}}$ is the maximum power limit of the battery corresponding to the rated charging power.

SOC value typically ranges from 0% to 100%. However, operators generally avoid fully charging or deeply discharging the battery to prevent rapid battery degradation. So, SOC is maintained within the upper ($SOC_{\mathrm{ESS},j,\mathrm{max}}$) and lower ($SOC_{\mathrm{ESS},j,\mathrm{min}}$) limits, as shown by (9.7) and (9.8):

$$\begin{cases} SOC_{\mathrm{ESS},j,\mathrm{min}} \le SOC_{\mathrm{ESS},j}(t) \le SOC_{\mathrm{ESS},j,\mathrm{max}} \\ SOC_{\mathrm{ESS},j,\mathrm{min}} \le SOC_{\mathrm{ESS},j}(t+\Delta t) \le SOC_{\mathrm{ESS},j,\mathrm{max}} \end{cases} \tag{9.7}$$

and

$$SOC_{ESS,j}(t+\Delta) = SOC_{ESS,j}(t) + \frac{1}{E_{ESS,j}}\left(P_{ESS,j,ch} \times \Delta t \times \eta_{ESS,j,ch} + \frac{P_{ESS,j,disch} \times \Delta t}{\eta_{ESS,j,disch}}\right) \quad (9.8)$$

9.2.2.3 Microgrid operation cost model

As mentioned, the objective function of trading strategies is to minimize MG operation costs in MMG based on the costs of DGs, ESSs, and inter-transaction among MGs in both DA and RT schedules. Since PVs and WFs are non-dispatchable sources, many countries mandate full utilization of their capacity, so this chapter assumes their generation and maintenance costs can be ignored. Therefore, that cost represents the price mechanism determined by the energy supplier to sell to consumers, mainly referring to MMG, including MG buyers and MG sellers, and contributes to incenting MMG to positively take part in energy transaction strategies. It is shown in the equation:

$$\min(C_{MG,n}) = \min\left\{\sum_{t=1}^{T}\left[\sum_{i=1}^{I} C_{dg,i}(t) + \sum_{j=1}^{J} C_{ESS,j}(t) + \lambda_{inter}(t) \times P_{inter}(t)\right]\right\} \quad (9.9)$$

and satisfy the power balance condition:

$$\sum_{i=1}^{I} P_{dg,i}(t) + \sum_{j=1}^{J} P_{ESS,j}(t) + P_{PV,n}(t) + P_{WF,n}(t) + P_{inter}(t) = P_{load,n}(t) \quad (9.10)$$

where $C_{MG,n}$ is the operation cost of nth MG; $\lambda_{inter}(t)$ and $P_{inter}(t)$ are the trading price and power exchanged among MGs, between MG and utility grid at time t, respectively; $P_{PV,n}(t)$ and $P_{WF,n}(t)$ are forecasting data of PV and WF output power of nth MG; $P_{load,n}(t)$ is the total load demand in nth MG.

The MG operation cost model as (9.9) represents energy trading strategies in Figure 9.1. Those are popular transaction strategies for MGs that have been applied in many previous studies which are mentioned. This strategy focuses on the energy market, but bidding/trading strategies on the reserve market are not explicitly mentioned. However, the reserve market also needs to be carefully studied and integrated into MG trading strategies to optimize operation costs and consider an overall problem.

9.3 The proposed approach for energy transaction strategy

9.3.1 Proposed transaction strategy

Previously, the electricity market primarily focused on the energy market with simple trading strategies among MGs as Figure 9.2. Recent studies introduced the reserve market to address energy imbalances in the RT schedule caused by RES forecasting errors but did not include it in the DA schedule. These approaches rely on RT capacity observations of DERs and ESSs for additional energy exchanges without a specific pricing mechanism or bidding process. Integrating ESSs as DERs enhances MG flexibility and stability, and their controllability, along with DGs, makes MGs well-suited for reserve market participation. Advanced forecasting and robust optimization enable MGs to manage RES and demand uncertainties effectively, allowing them to bid optimally in both energy and reserve markets for maximum profits.

This chapter proposes energy transaction strategies for MMG, optimizing operation costs in DA and RT schedules under the reserve market framework based on some power company rules, e.g., Taiwan, Australia, etc. rules [37]. The MMG aggregator manages unit commitment, reserve arrangements, and RT balancing, with marginal prices cleared in both markets. The DA schedule operates from 4–6 p.m. the previous day, while RT adjustments occur every 5 minutes as Figure 9.3.

9.3.1.1 Day-ahead schedule

First, the DA schedule is started with an hourly resolution. MGs submit all databases for aggregators to initiate the bidding process in both types of energy markets. Based on historical data such as predicted fault RESs, demand response or updated data such as forecast RESs, equipment maintenance schedules. Trading reserve capacity will be communicated to the operators and conduct the bidding, similar to the energy

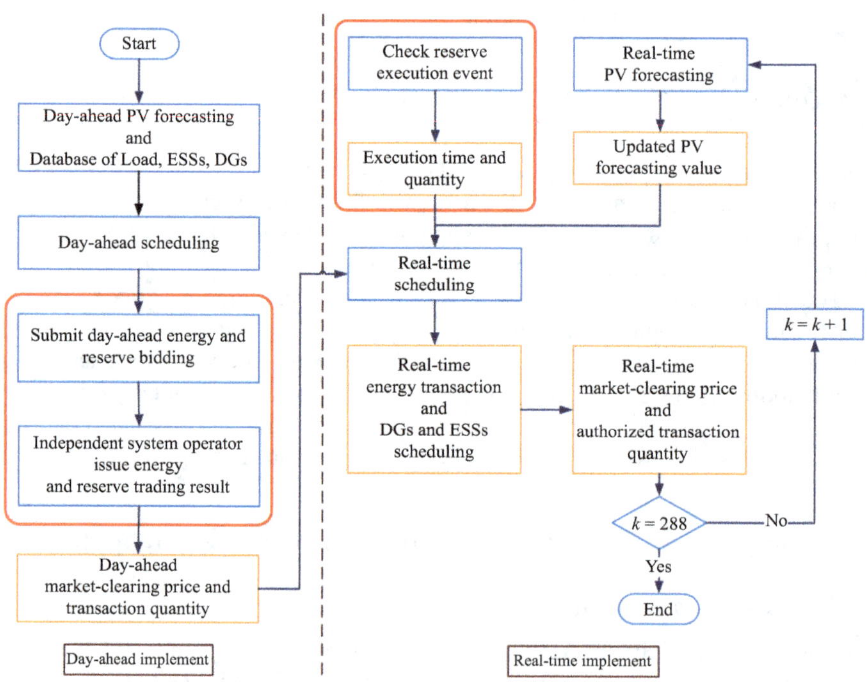

Figure 9.3 The proposed flowchart of multiple MG transaction strategies in the energy and reserve markets

market. Once the energy bidding/trading process in both markets is complete, market-clearing price and transaction quantity for the energy market are determined. MGs got the capacity payment, energy payment, and reserve capacity for the reserve market and closed the DA schedule as Figure 9.3.

In the set of MMGs containing *N* microgrid, the objective function of this process is expressed:

$$\min \sum_{t=1}^{24} \sum_{n=1}^{N} \left[C_{\text{op,DA},n}(t) - C_{r,n}(t) \right] \tag{9.11}$$

MG operating cost in DA involves the energy cost $C_{\text{op,DA},n}$ in (9.12), which is similar to (9.10), and the cost for the reserve capacity $C_{r,n}$ in (9.13). Additionally, the operation must comply with the constraints specified in (9.2), (9.3), (9.6), (9.7), (9.8), and (9.10).

$$C_{\text{op,DA},n}(t) = \sum_{i=1}^{I} C_{\text{dg},i}^{\text{DA}}(t) + \sum_{j=1}^{J} C_{\text{ESS},j}^{\text{DA}}(t) + \lambda_{\text{inter,DA}}(t) \times P_{\text{inter,DA}}(t) \tag{9.12}$$

$$C_{r,n} = \lambda_{r,\text{cap}}(t) \times P_{r,\text{cap}}(t) \tag{9.13}$$

Optimization transaction strategies observe all bidding offers of MGs, where the bidding price (or inter-trading price) for energy $\lambda_{\text{inter,DA}}(t)$ and reserve $\lambda_{r,\text{cap}}(t)$ acts as parameters in this optimization problem and changes for different prosumers over time. The power balance, e.g., power of inter-trading $P_{\text{inter}}(t)$ and reserve requirements $P_{r,\text{cap}}(t)$ need to ensure by constraints:

$$P_{\text{inter,DA}}(t) \geq P_{r,\text{cap}}(t) \tag{9.14}$$

$$P_{r,\text{cap}}(t) = \sum_{i=1}^{I} P_{\text{dg},i}(t) + \sum_{j=1}^{J} P_{\text{ESS},j}^{\text{disch}}(t) \tag{9.15}$$

Condition (9.14) states that the reserve capacity of the MG should not exceed the amount of inter-trading power exchanged with other MGs. In other words, the nature of reserve capacity is to use the surplus from reserve power of *i*th DG $P_{\text{r,dg},i}$ and *j*th ESS $P_{\text{r,ESS},j}^{\text{disch}}$ inside MG, in (9.15), to cope with power curtailment in inter-trading because faults or equipment maintenance schedules happen in other MGs. The power dispatching from reserves must satisfy the conditions (9.16), (9.17), and (9.18).

$$P_{\text{dg},i,\min} \leq P_{\text{r,dg},i}(t) + P_{\text{dg},i}(t) \leq P_{\text{dg},i,\max} \tag{9.16}$$

$$\begin{aligned} \text{ramp}_{\text{dg},i,\text{down}} \leq P_{r,\text{dg},i}(t+\Delta t) - P_{r,\text{dg},i}(t) \\ + P_{\text{dg},i}(t+\Delta t) - P_{\text{dg},i}(t) \leq \text{ramp}_{\text{dg},i,\text{up}} \end{aligned} \tag{9.17}$$

$$P_{\text{ESS},j,\min} \leq P_{\text{ESS},j}^{\text{ch}}(t) + P_{\text{ESS},j}^{\text{disch}}(t) + P_{\text{r,ESS},j}^{\text{disch}}(t) \leq P_{\text{ESS},j,\max} \tag{9.18}$$

Bidding results with price and quantities are determined via (9.11) and the conditions mentioned above. Before clearing the market, MGs will receive results to review the bidding price, and quantities should be submitted.

9.3.1.2 Real-time schedule

In the RT schedule, the MGs database is updated every 5 minutes and checks the reserve execution event provided by MG aggregators. Similarly, the objective in the RT schedule is to minimize the operation cost to balance the mismatch between RT RESs power and RES forecasting in DA. The model of this schedule is described by objective function:

$$\min \sum_{t=1}^{288} \sum_{n=1}^{N} C_{\text{op,RT},n}(t) = \min \sum_{t=1}^{288} \sum_{n=1}^{N} \left[\sum_{i=1}^{I} C_{\text{dg},i,\text{RT}}(t) + \sum_{j=1}^{J} C_{\text{ESS},j,\text{RT}}(t) + \lambda_{\text{inter,RT}}(t) \times P_{\text{inter,RT}}(t) \right] \tag{9.19}$$

Only the energy bidding process occurs during this stage, so the cost is paid for changing power generation and adjusting export or import energy to ensure balance in the system, shown in (9.20), by rescheduling DGs and ESSs and satisfy the conditions (9.2), (9.3), (9.6), (9.7), (9.8), and (9.10).

$$\sum_{i=1}^{I} P_{\text{dg},i,\text{RT}}(t) + \sum_{j=1}^{J} P_{\text{ESS},j,\text{RT}}(t) + P_{\text{PV},n,\text{RT}}(t) + P_{\text{WF},n,\text{RT}}(t) + P_{\text{inter,RT}}(t) = P_{\text{load},n}^{\text{RT}}(t) \tag{9.20}$$

This process should follow the assumed principles, built from regulations of the power companies such as Taiwan, Australia, etc. [38], shown in Figure 9.4.

The curtailment signal will be sent to MGs 10 minutes before the start of the execution event. During this time (standby period), the aggregator may issue the reserve execution instruction at any instant, and the reserve execution event should include in the RT power dispatch problem. In the standby period, from preparation time t_{cur} to

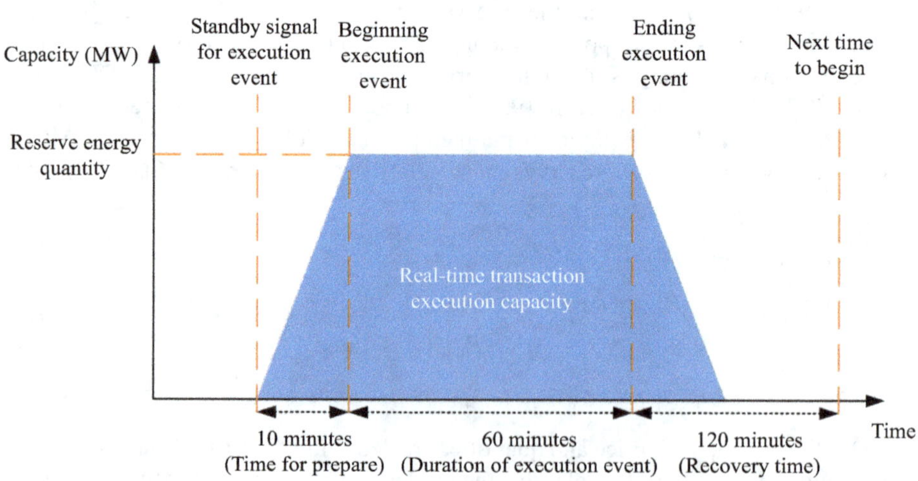

Figure 9.4 The rule for execution events in the reserve market

event start time t_{start}, $\forall t \in [t_{\text{cur}}, t_{\text{start}}]$, MG needs to secure the DGs and ESSs reserve capacity to guarantee the ability to cut down bidding capacity.

$$P_{\text{inter}}(t) \geq P_{\text{main,cap}}(t) \tag{9.21}$$

$$\sum_{i=1}^{I}\left[P_{\text{dg},i,\max} - P_{\text{dg},i}(t)\right] + \sum_{j=1}^{J}\left[P_{\text{ESS},j,\text{disch}}(t) - P_{\text{ESS},j,\min}\right] \geq P_{\text{main,cap}}(t) \tag{9.22}$$

$$SOC_{\text{ESS},j,\min} \leq \frac{1}{E_{\text{ESS}}}\left|P_{r,\text{ESS},j,\text{disch}}(t)\right| \times 1\text{ hour} \leq SOC_{\text{ESS},j,\max} \tag{9.23}$$

where $P_{main,cap}$ is the maintenance capacity at hour t of MG join in the reserve market.

When entering the execution segment, the duration is 60 minutes from the beginning event to the ending event $\forall t \in [t_{\text{cur}}, t_{\text{start}}]$. The MG checks the base capacity P_{base}, which is the trading capacity with other MGs 10 minutes before the beginning execution event, as shown in (9.24). During the event, the achievement capacity P_{achive} (or execution rate, which is calculated by P_{achive} divided by P_{base}) is the curtailment reserve capacity in RT that can be determined as (9.25). This equation states that the actual curtailment capacity will depend on the reserve market's base capacity and the actual inter-trading capacity with other MG

$$P_{\text{base}}(t) = P_{\text{inter}}(t_{\text{start}} - 10\text{ minutes}) \tag{9.24}$$

$$P_{\text{achive}}(t) = P_{\text{base}}(t) - \frac{\sum_{t=t_{\text{start}}}^{t_{\text{end}}} P_{\text{inter}}(t)}{t_{\text{end}} - t_{\text{start}}} \tag{9.25}$$

After the execution event ends, MGs will have 2 hours to recover. That also means that each MG must wait at least 2 hours to participate in the reserve market next time. To maximize revenue in the reserve market, P_{achive} needs to meet the capacity requirements. This problem will be discussed in the next section based on the rules.

9.3.1.3 Revenue of proposed reserve market mechanism

Based on the rules and notices discussed, the revenue from the reserve market is calculated by:

- **Revenue** = Capacity fees × Service quality index + Energy fees
- **Capacity fees** = Capacity payment × Bidding capacity
- **Energy fees** = Energy payment × Actual execution energy

Capacity fees can be found after the bidding process of MGs in the reserve market that calculates through capacity payment $\lambda_{r,\text{cap}}(t)$, and winning capacity. This fee refers to the service quality index (SQI) as determining the payoff for MGs to ensure that they comply with the bidding conditions for reserve capacity release. Details of the service quality index are presented in detail in Table 9.2. The second part of the revenue is the energy fee paid based on the energy payment in the reserve market, which is the DA marginal price $\lambda_{\text{inter,DA}}$, multiplies with the actual execution energy release during the event.

Table 9.2 Service quality index for execution rate in the reserve market

Execution rate	Over 95%	85% to Under 95%	70% to Under 85%	Under 70%
SQI	1	0.7	0	−240

9.3.2 Game Theory Optimization Approach

9.3.2.1 Non-cooperative game

Non-cooperative games model transactions where players maximize their benefits independently without alliances, predicting individual strategies, payoffs, and Nash equilibria. In an MMG system with NNN MGs as players, each MG optimizes import/export capacities and energy scheduling to minimize daily costs as per (9.11) and (9.19). The Nash equilibrium (NE) occurs when no MG changes its strategy. Objective functions are strictly convex, and daily costs are linear, ensuring unique NE solutions based on Theorem 1 and Theorem 3 in Reference [39]. As inter-trading capacities change, MGs adjust their strategies to minimize costs, with the Stackelberg game framework applied to determine the NE.

The Stackelberg game analyzes the trading process and the interaction between follow players and leading players. The follower in the lower layer will make multi-level decisions in reply to requests from the leader in the upper layer until mutually beneficial [40]. Energy transaction strategies are mentioned in this thesis as the Stackelberg game, with the process shown in Figure 9.5.

9.3.2.2 Cooperative game

Cooperative games prioritize maximizing collective benefits within coalitions, distributing profits based on agreed rules. Also known as coalition games, they model coalition formation and joint actions under contract law. Microgrids (MGs) adopt unified strategies to optimize objectives like power allocation and energy management, minimizing operational costs using coalition-characteristic functions (e.g., (9.11) and (9.19)). Benefits are equitably distributed, ensuring each MG earns at least as much as in independent operation.

Transaction models in cooperative games are defined as $\{N, v, \psi\}$, where N is the number of MGs, v represents coalition benefits, and ψ allocates benefits to MGs. With $2^N - 1$ possible coalitions, a four-MG system forms 15 coalitions, with each coalition's payoff analyzed through its characteristic function $v(S)$:

$$v(S) = C_S - \sum_{n \in S} C_n \tag{9.26}$$

where n is a single MG of coalition S; C_S is the total operation cost of S; $\sum C_n$ means the total operating cost of MGs in S when they operate separately.

Then, for any coalitions to be expanded or merged, they have to satisfy the condition as:

$$v(S \cup T) \geq v(S) + v(T) \tag{9.27}$$

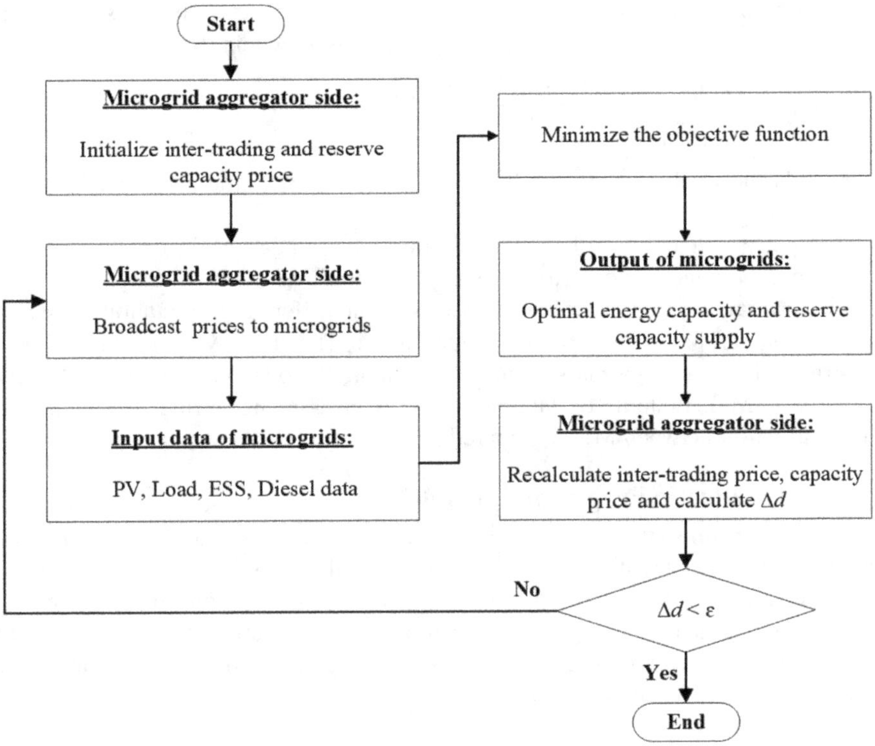

Figure 9.5 Stackelberg game flowchart

In (9.27), $v(S)$ and $v(T)$ represent two random coalitions in MMG. Therefore, the total benefit gained when the two alliances combine, shown as (9.28), is no less than the sum of the benefit when the two alliances operate without cooperation as in (9.29).

$$v(S \cup T) = C_{S \cup T} - \sum_{n \in (S \cup T)} C_n \tag{9.28}$$

$$v(S) + v(T) = \left(C_S - \sum_{n \in S} C_n \right) + \left(C_T - \sum_{n \in T} C_n \right) = C_S + C_T - \sum_{n \in (S \cup T)} C_n \tag{9.29}$$

When a new coalition is generated to guarantee the above conditions, the cooperative game is called a super additive game [41]. With a super additive game, the individual benefit of MGs in the coalition can be allocated based on the Shapley value.

The Shapley value concept, introduced in Reference [42], is an appropriate method for allocating economic benefits to multi-players in cooperative games. Due to the different contributions of each MG in the coalition, the *n*th MG's individual benefit $\psi_n(v)$ is determined via the Shapley value, as shown in (9.30):

$$\psi_n(v) = \sum_{S \subseteq \mathcal{N} \setminus \{n\}} \frac{|S|!(N - |S| - 1)!}{N!} [v(S \cup \{n\}) - v(S)] \tag{9.30}$$

where N is the total number of MGs in MMG, and $|S|$ is the number of MGs in coalition S.

The equation can be explained as follows: when the nth MG joins the coalition S at a time, the benefit it contributes is expressed by $v(S \cup \{n\}) - v(S)$. Thus, MG takes the average of that contribution over possible different permutations in which the coalition can be created as fair compensation. The strategy is approached with cooperation in the energy market and reserve markets to support the optimal transactions of MMG. Benefits of MGs are seen as cost-saving during operation with inter-transaction between MGs compared to without inter-transaction.

9.3.2.3 Quadratic Programming Solution Method

Recently, algorithms such as PSO, DE, GA, and QP have gained popularity in optimization tasks for scheduling and planning. Among these, QP is particularly effective for solving quadratic functions, optimizing multivariable functions under linear constraints as (9.31) and (9.32) [43]. Its fast processing and ability to handle complex problems make it widely applicable in signal processing, portfolio management, regression, and power plant scheduling.

$$\min f(x) = \frac{1}{2} x^T A x + a^T x \tag{9.31}$$

subject to

$$Bx = b, \quad \text{and } Cx \le c, \quad \text{and } x \ge 0 \tag{9.32}$$

In the objective function with x as the solution vector, the vector a includes all linear terms, and A contains the quadratic terms. In other words, a is the gradient of the Hessian matrix A. For constraints, $Bx = b$ and $Cx \le c$ showed all linear equality and all linear inequality constraints, respectively.

Considering the cost function of MG, it is a quadratic function due to the participation of DGs, so the general equation (9.31) represents the proposed objective functions (9.11) and (9.19). That can be solved through constraints in (9.32), which are expressed as equations (9.2), (9.3), (9.7), (9.8), (9.9), (9.10), and (9.12)–(9.22).

9.4 Simulation results and Discussion

9.4.1 Data Base and Simulation Case

Typically, DG manufacturers provide historical datasets for respective output power and fuel usage. Table 9.3 presents the parameters of the DGs, where acceleration and deceleration of power are shown by ramping level dependent on response time. The general ESSs parameter is listed in Table 9.4 [44], which uses lithium-ion batteries, one of the most popular battery technologies and a low cost. Load and RESs data

Table 9.3 Related parameters of diesel generators

Parameters	MG1			MG2		MG3		
DG No.	**1**	**2**	**3**	**1**	**2**	**1**	**2**	**3**
a (NTD/kW^2)	0.0051	0.0040	0.0039	0.0051	0.0039	0.0053	0.0045	0.0038
b (NTD/kW)	2.2034	1.9104	1.8518	2.2036	1.8545	2.5104	2.1637	1.9452
c (NTD)	15	25	40	15	40	25	35	40
$P_{\min}$ (kW)	0	0	0	0	0	0	0	0
$P_{\max}$ (kW)	150	200	300	150	450	100	200	250
Ramping (kW/hour)	150	200	300	250	350	100	200	250
Ramping (kW/10min)	75	100	150	125	175	50	100	125
Ramping (kW/5min)	37.5	50	75	62.5	87.5	25	50	62.5

Table 9.4 Related parameters of ESS

Parameters	MG 1	MG 2	MG 3
Capacity (kWh)	198.92	198.92	149.19
SOC_{Max} (%)	95	95	95
SOC_{Min} (%)	20	20	20
η_{ch} (η_{bc})	0.95	0.95	0.95
η_{disch} (η_{bd})	0.95	0.95	0.95
Cycle life (cycle)	6000	6000	6000
Cost of Battery (NTD)	6000	6000	6000
$P_{ESS,max}$ (kW)	90	90	90
$P_{ESS,min}$ (kW)	−74	−74	−74
SOC_{init} (%)	20	20	20

are collected from Kinmen Island - Taiwan, to verify the feasibility of the proposed strategies. The simulation data are scaled to fit the proposed structure represented by netload curves in Figure 9.6, including predicted and actual data of PVs and WFs.

Three different cases are performed on the same database: (i) Case 1—Transaction strategies without inter-transaction; (ii) Case 2—Transaction strategies via the non-cooperative game; (iii) Case 3—Transaction strategies via the cooperative game. The assumption bidding capacity in the reserve market is 150 kW every hour, and execution event times are set at 4:20 and 14:30.

9.4.2 Simulation results and discussions

9.4.2.1 Case 1—Transaction strategies without inter-transaction

In this case, 3 MGs operate individually and do not exchange energy with each other. That can be considered a base case indicating a necessity for transaction integration

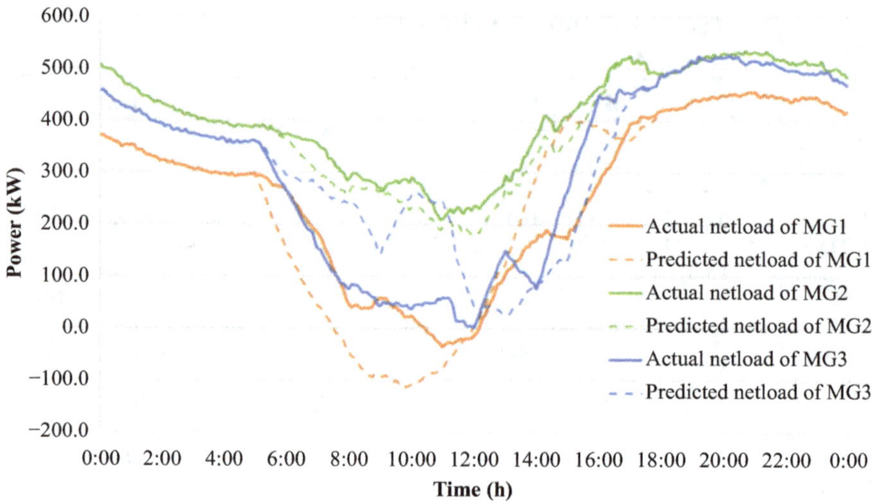

Figure 9.6 Netload curves of MMG

between MGs. Obviously, when MGs do not have inter- transactions, they cannot participate in the energy bidding processes, especially reserve capacity. In other words, MG's revenue in the reserve market is equal to zero. Results of the optimal schedule of the energy dispatching are shown in Figure 9.7.

The netload curve, representing power imbalance in each MG after RESs integration, is calculated as load curve minus PV and Wind curves. Without inter-trading, MGs rely entirely on internal DGs and ESSs to maintain supply-demand balance, leading to higher operation costs and no revenue from energy markets. This results in operation costs of 20,526.14 NTD, 34,795.38 NTD, and 22,939.30 NTD for MG1, MG2, and MG3, respectively, totaling 78,260.82 NTD for the MMG. Additionally, the absence of inter-trading prevents a uniform market-clearing price, leaving each MG with its own market-clearing price, as shown in Figure 9.8.

Without inter-trading, many negative effects are created during operation, such as low reliability, instability and increased generation costs. Because of these reasons, the simulation results of the following two trading strategies will show the efficiency when integrating energy transactions between MGs with the proposed method.

9.4.2.2 Case 2—Transaction strategies via non-cooperative game

In this approach, instead of creating several iterations for price/capacity bidding as the Stackelberg game to find the benefits for the participants, the cooperative game will be based on Shapley value to allocate profits from forming coalitions.

From Figure 9.9, the DA market-clearing price remains low between 6 a.m. and 5 p.m., as RESs reduce reliance on DGs and ESSs, lowering incremental costs (9.11). This price also serves as reserve market payment during RT events.

Reserve capacity payments are higher between 12 a.m.–2 a.m. and 9 p.m.–12 a.m., as DGs operate at high levels with minimal RES and ESS support in

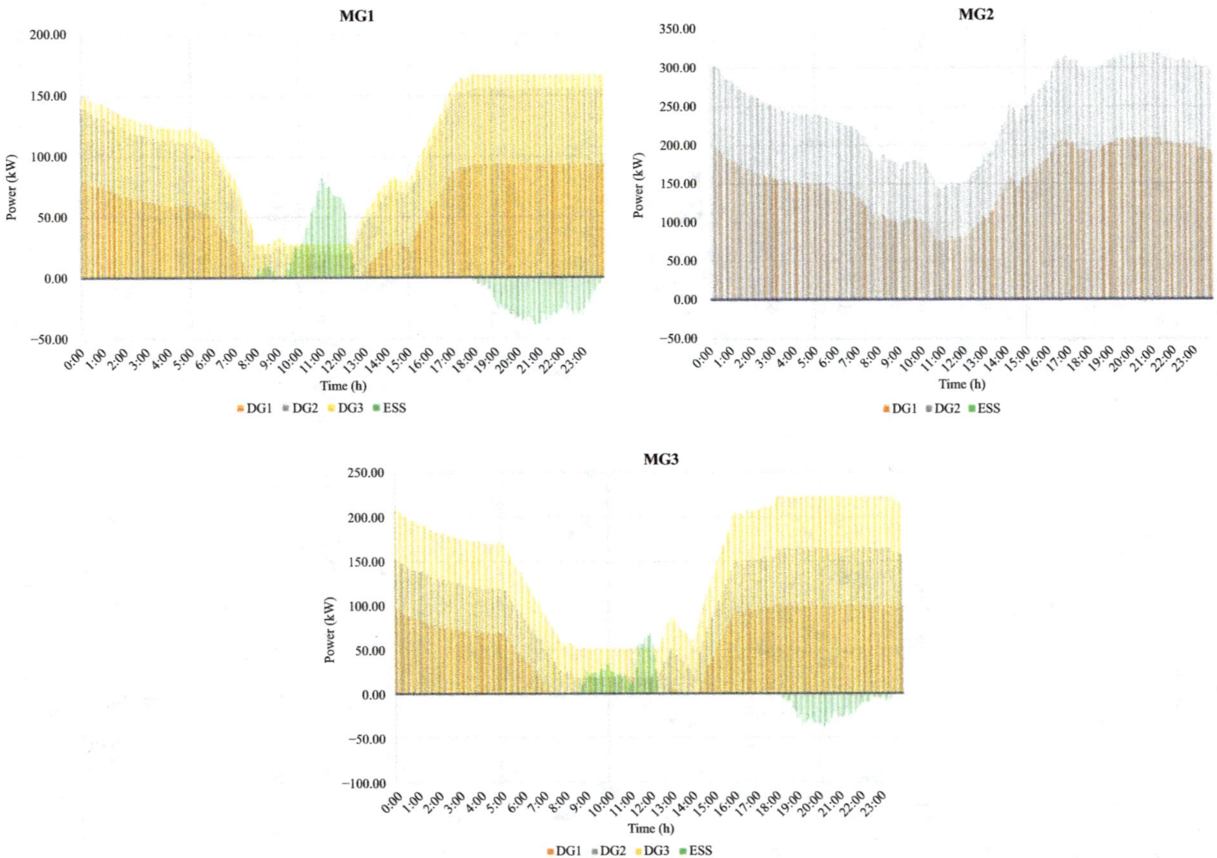

Figure 9.7 Dispatching result of Case 1

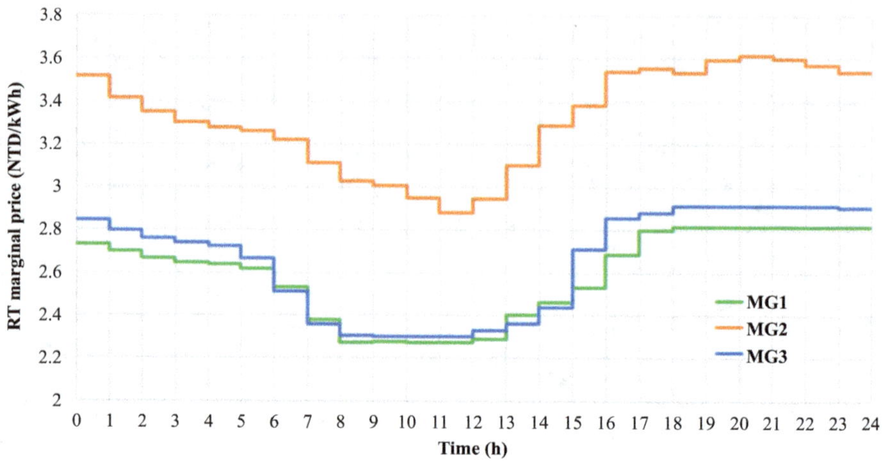

Figure 9.8 RT marginal price of each MG in Case 1

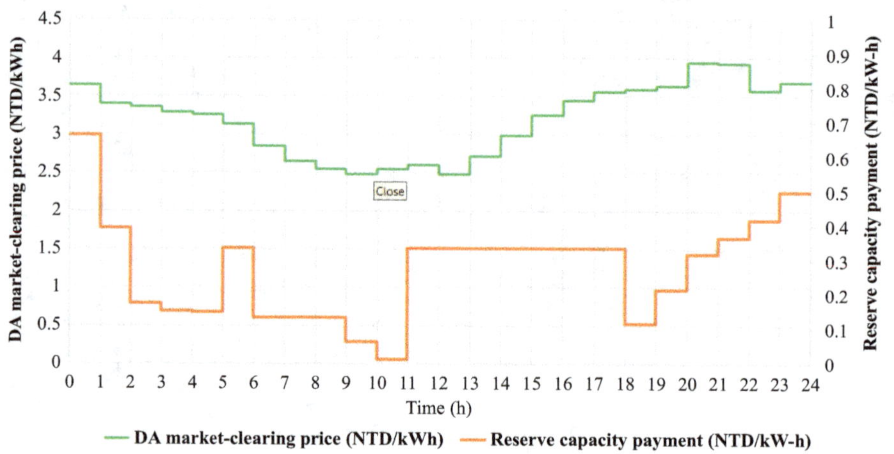

Figure 9.9 Market-clearing and reserve capacity price of DA in Case 2

Figure 9.10(a). From 9 a.m.–11 a.m., price drops to 0.06 NTD/kWh and 0.02 NTD/kWh due to high DG reserve capacity and increased RES output, reducing net-load and optimizing costs. Afterward, rising demand increases the netload curve and reserve market prices.

Based on DA bidding results, MGs determine prices and capacities for reserve market participation, as shown in Table 9.5. These capacities remain available in RT to respond to aggregator signals. Table 9.6 also outlines scenarios for RT execution events, where MGs receive a standby signal 10 minutes before execution. RT dispatching results for three MGs are illustrated in Figure 9.10(b).

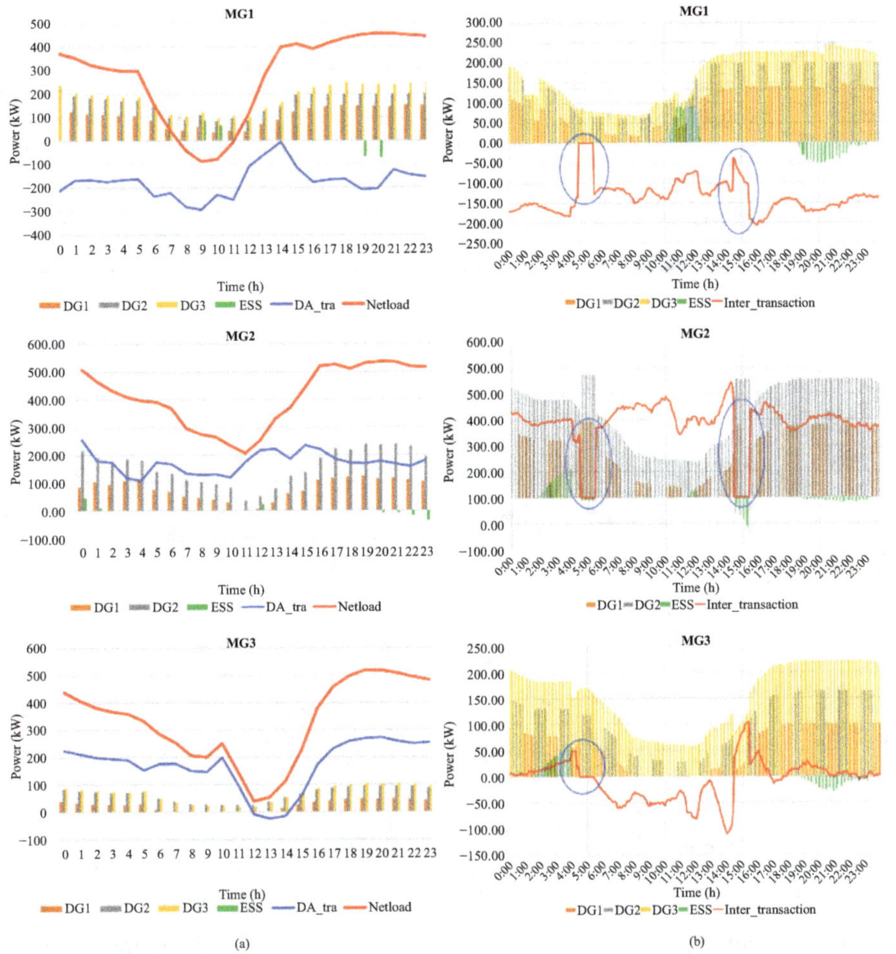

Figure 9.10 Dispatching result in Case 2: (a) day-ahead; (b) real-time

Table 9.5 Case 2: Bidding results of reserve capacity

Time (h)	Bidding quantity of MG1 (kW)	Bidding quantity of MG2 (kW)	Bidding quantity of MG3 (kW)
4:00	0	107.65	42.35
14:00	0	150.00	0.00

During reserve market execution, MG2 and MG3 halt inter-trading with MG1, reducing inter-trading power to zero at 4 a.m., while their DGs increase output to balance the shortfall. Reserve market rules restrict energy transactions during execution events, causing inter-trading among these MGs to cease. At 2 p.m., MG2 participates

Table 9.6 Case 2: Reserve execution events

Time (h)	Participating MG	Execution quantity (kW)	Standby time (h)	Execution time (h)
4:00	MG2	107.65	4:10	4:20
	MG3	42.35	4:10	4:20
14:00	MG2	150.00	14:20	14:30

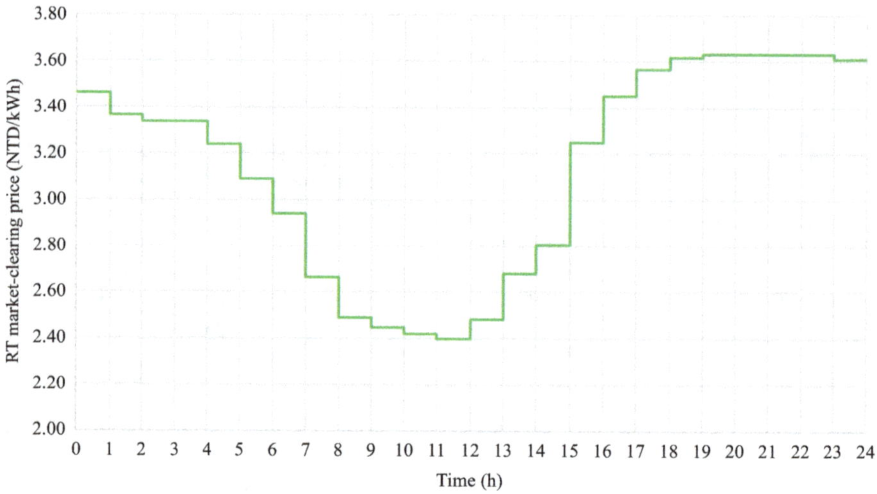

Figure 9.11 RT market-clearing of MMG in Case 2

in another execution, curtails 150 kW, and halts trading, while MG3 continues trading with others. During MG2's execution, its ESS charges as curtailment values (114.8 kW at 4:20 a.m. and 172.8 kW at 2:30 p.m.) exceeded bids. Despite minor non-compliance with Table 9.2, MGs maintain over 95% execution rates to maximize reserve market revenue, efficiently using surplus energy to charge ESSs.

After confirming the RT schedule, the hourly MMG market-clearing price is shown in Figure 9.11. Between 4:20 a.m.–5:20 a.m. and 2:30 p.m.–3:30 p.m., no market-clearing price is determined as all MGs are engaged in reserve market execution without inter-trading. However, the price is established during these periods through inter-trading between MG1 and MG3.

Finally, the operation costs of MGs and the revenue in the reserve market are summarized in Table 9.7.

9.4.2.3 Case 3—Transaction strategies via cooperative game

In this approach, instead of creating several iterations for price/capacity bidding as the Stackelberg game to find benefits for participants, the cooperative game will be

Table 9.7 Case 2: Summary of operation cost and revenue

	MG1	MG2	MG3	Total of 3 MGs
Operation cost (NTD)	20,344.61	31,652.10	22,946.52	74,943.23
Revenue in reserve market (NTD)	0	823.34	144.14	967.48

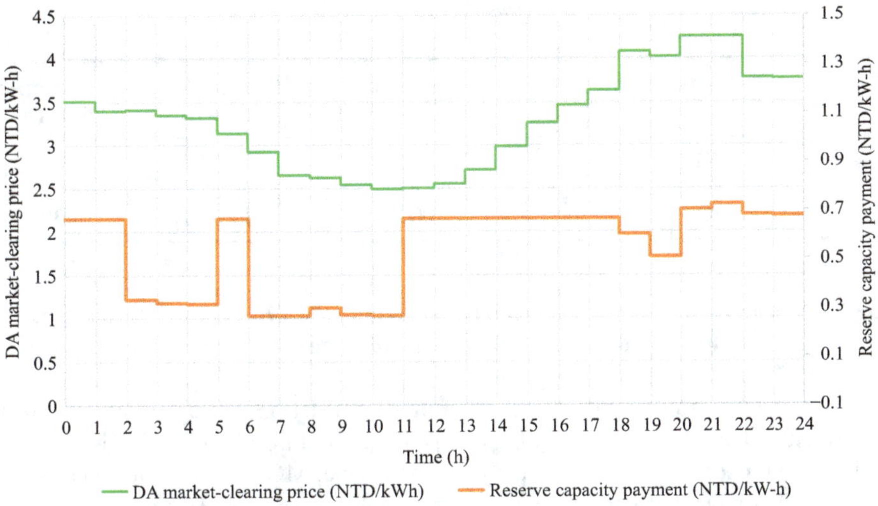

Figure 9.12 Market-clearing and reserve capacity price of DA in Case 3

based on Shapley value to allocate profits by forming coalitions. With the proposed structure, seven coalitions can be formed, including MG1, MG2, MG3, MG1, MG2, MG1, MG3,MG2, MG3,MG1, MG2, MG3.

Optimal transaction strategy results are only shown in the case of alliances formed from 3 MGs. Results of other coalitions act as provide data to allocate individual benefit of each MG.

Like non-cooperative games, energy and reserve market prices are optimized using the DA schedule objective function (9.11), as shown in Figure 9.12. Lower prices from 6–11 a.m. reflect reduced DG costs due to RESs generation, with the lowest reserve payment at 10–11 a.m. After finalizing DA schedules, the RT schedule optimizes objective function (9.19), with reserve bidding and execution details in Tables 9.8 and 9.9.

Figure 9.13(a) shows power transaction strategies of MGs in DA, when Figure 9.13(b) represents for RT transaction strategies. MGs still have inter-transaction to optimize the operation cost in whole MMG. However, until the command executes reserve capacity, as in case 2, this exchange process is interrupted. Similarly, when executing curtailment command, DGs have to generate more power to offset reserve capacity and quickly decrease after the event ends, which can be seen for MG2 and

Table 9.8 Case 3: Bidding results of reserve capacity

Time (h)	Bidding quantity of MG1 (kW)	Bidding quantity of MG2 (kW)	Bidding quantity of MG3 (kW)
4:00	0	71.17	78.83
14:00	0	150.00	0.00

Table 9.9 Case 3: Reserve execution events

Time (h)	Participating MG	Execution quantity (kW)	Standby time (h)	Execution time (h)
4:00	MG2	71.17	4:10	4:20
	MG3	78.83	4:10	4:20
14:00	MG2	150.00	14:20	14:30

MG3, because then it is cheaper to buy power from MG1 than to use the generator inside them. Finally, MGs find the RT market-clearing price, which is shown in Figure 9.14.

In addition, in cooperative game, optimal transaction strategies must be processed with all coalitions to justify benefits, and the coalition MG1, MG2, MG3 must achieve the most significant benefit. At the same time, MGs in this coalition also get the individual's profit. Specific results are presented in Table 9.10.

When the coalition consists of only 1 MG, it can be seen that MG cannot observe any benefits without inter-trading. However, when coalitions are formed with MMG, the operation cost is reduced because of bidding energy. Therefore, observed benefits in this thesis are the operating cost that each MG can save compared to when they operate separately. By applying Shapley's value, the benefits allocated to MG1, MG2, MG3 are 1,676.5 NTD, 2,003.9 NTD and 1,741.3 NTD, respectively, when 3 MGs form the coalition, and this is the largest benefit of transaction strategies. After all, MG operation costs in the coalition and the reserve market profit are shown in Table 9.10.

9.4.3 Comparison of results and discussions

From three simulation cases, comparison results of operation costs of MGs and calculation time are summarized in Table 9.11. With a non-cooperative game, the bidding process must go through several iterations to find a price/quantity that satisfies participants, which randomly causes a longer computation time. Meanwhile, the computation time in the cooperative game is proportional to the coalition's number that can be formed. The DA schedule takes 138 seconds to complete the calculation in the non-cooperative game, and the RT schedule needs 4.05 seconds per time slot (every 5 minutes). For cooperative game, the computation time is 178 seconds and 6.07 seconds for the DA and RT schedules, respectively.

Figure 9.13 Dispatching result in Case 3: (a) day-ahead; (b) real-time

Both proposed transaction strategies have shown in Table 9.12 to be effective for MMG in the energy and reserve markets. In case 2 and case 3, the market-clearing price is lower than the marginal price of MG2 in case 1 as Figure 9.15, which indicates that MG2 gets benefits from importing energy from outside, thereby reducing operation costs. Furthermore, MG1 and MG3 sell energy to MG2 at a price higher than their marginal value and achieve more profit, thereby enhancing the overall profitability of MMG.

In the non-cooperative game, the selfishness of MGs can occur because it optimizes the operation costs for each MG, then aims to optimize MMG and the individual profit in this game comes first. Transaction strategies of the cooperative game show lower daily operation costs than the non-cooperative game. This game can enhance the sustainability of MMG due to the commitment of cooperation between

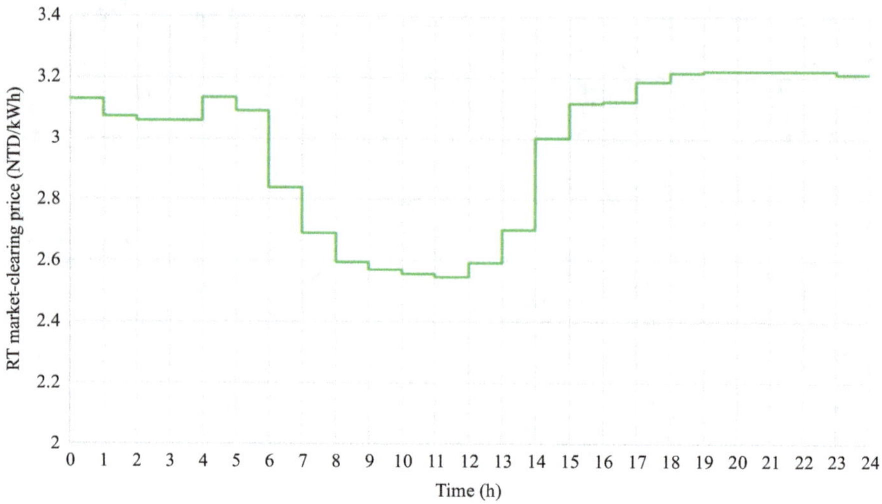

Figure 9.14 RT market-clearing of MMG in Case 3

Table 9.10 Case 3: The operation cost of the coalitions and allocated benefit

Coalitions	Operation cost (NTD)	Benefit (NTD)	Allocated benefit
{MG1}	20,526.14	0	1,794.60
{MG2}	34,795.38	0	3,263.50
{MG3}	22,939.30	0	1,163.60
{MG1, MG2}	50,577.54	4,743.98	–
{MG1, MG3}	42,921.17	544.27	–
{MG2, MG3}	54,252.70	3,481.98	–
{MG1, MG2, MG3}	72,039.03	6,221.79	–

Table 9.11 Case 3: Summary of operation cost and revenue

	MG1	MG2	MG3	Total of 3 MGs
Operation cost (NTD)	19,249.60	30,808.77	22,089.87	72,148.24
Revenue in reserve market (NTD)	0	806.64	285.36	967.48

MGs in the coalition. However, the cooperative game may be difficult to achieve due to mutual trust issues. Therefore, contracts among the MGs need to be discussed clearly to ensure profits and avoid conflicts in the coalition.

Table 9.12 Simulation comparison.

	Operation cost (NTD)			Total computation time (sec)	
	MG1	**MG2**	**MG3**	**DA schedule**	**RT schedule**
Case 1	20,926.10	32,812.67	23,831.17	30	319
Case 2	20,344.61	31,652.10	22,946.52	138	1,365
Case 3	19,249.60	30,808.77	22,089.87	178	1,782

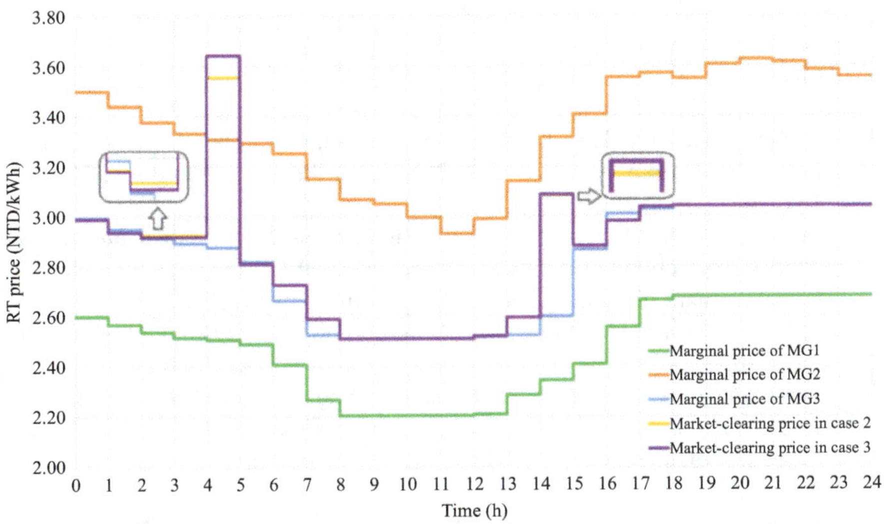

Figure 9.15 Price comparison in the RT

9.5 Conclusion

This chapter proposes optimal transaction strategies for multiple MGs in the energy and reserve markets. By employing game theory, including non-cooperative and cooperative games, the benefits of inter-trading among MGs are demonstrated. Transaction strategies are analyzed in both DA and RT schedules to coordinate energy and address forecasting errors caused by RES source uncertainties. The dispatching schedules and economic allocations align with current and future electricity market rules and contractual commitments. Results show that inter-trading strategies reduce individual operation costs and enable MGs to benefit from the reserve energy market.

Energy bidding processes in both markets are governed by strict rules for payoff and energy dispatch compliance. Transparent bidding mechanisms, utilizing game theory, determine prices and quantities while balancing individual MG benefits with the overall MMG objective. In the non-cooperative approach, MGs independently minimize their daily costs, while the cooperative approach forms coalitions to reduce total MMG operation costs. Both approaches yield similar overall results, with power exchange proving more beneficial than isolated operation. However, cooperative game theory offers advantages in lower operation costs and faster computation times, providing a superior solution for integrated energy strategies.

Optimizing energy trading strategies for MMGs is becoming increasingly important with the increasing integration of RES and storage solutions. In addition to the issue of optimal power exchange, issues of losses and voltage or frequency limitations also need to be integrated. Furthermore, integrating accurate forecast models for RES and load to enhance along with maintenance costs also needs to be considered in the model. This will enhance the reliability of the market and improve trading strategies to accommodate the rapid expansion of MGs.

References

[1] Stevens KA, Tang T, and Hittinger E. Innovation in complementary energy technologies from renewable energy policies. *Renewable Energy*. 2023;209: 431–441.

[2] Tran TH, Tran NTN, Le TMC, Le T, and Duong MQ. Assessment of influences of high photovoltaic inverter penetration on distribution automation systems: Vietnam distribution network case study. In: *Digital Technologies for Solar Photovoltaic Systems: From General to Rural and Remote Installations*. IET; 2022.

[3] Dominguez JA, Agbossou K, Henao N, Nagarsheth SH, Campillo J, and Rueda L. Distributed stochastic energy coordination for residential prosumers: Framework and implementation. *Sustainable Energy, Grids and Networks*. 2024;38:101324.

[4] Kataray T, B. Nitesh, Bharath Yarram, *et al.* Integration of smart grid with renewable energy sources: Opportunities and challenges–A comprehensive review. *Sustainable Energy Technologies and Assessments*. 2023;58:103363.

[5] Abdolmaleki L and Berardi U. Hybrid solar energy systems with hydrogen and electrical energy storage for a single house and a midrise apartment in North America. *International Journal of Hydrogen Energy*. 2024;52D: 1381–1394.

[6] Bordbari MJ and Nasiri F. Networked Microgrids: A review on configuration, operation, and control strategies. *Energies*. 2024;17(3):715.

[7] Fachini F, Bogodorova T, Vanfretti L, and Boersma S. A microgrid control scheme for islanded operation and re-synchronization utilizing Model Predictive Control. *Sustainable Energy, Grids and Networks*. 2024;39:101464.

[8] Veerasamy V, Hu Z, Qiu H, Murshid S, Gooi HB, and Nguyen HD. Blockchain-enabled peer-to-peer energy trading and resilient control of microgrids. *Applied Energy*. 2024;353:121654.

[9] Numan M, Abbas MF, Yousif M, Ghoneim SSM, Mohammed A, and Norowzi A. The role of optimal transmission switching in enhancing grid flexibility: A review. *IEEE Access*. 2023;11:143530–143548.

[10] Fu J, Yang Han, Wenhao Li, *et al.* A novel optimization strategy for line loss reduction in distribution networks with large penetration of distributed

generation. *International Journal of Electrical Power & Energy Systems*. 2023;150.

[11] Badanjak D, Pavić I, and Capuder T. Data driven approach for analyzing and correlating energy market products: Case studies of Denmark and Croatia. *CSEE Journal of Power and Energy Systems*. 2023; 11(2):503–520.

[12] Hamidi M, Liao H, and Szidarovszky F. Non-cooperative and cooperative game-theoretic models for usage-based lease contracts. *European Journal of Operational Research*. 2016;255(1):163–174.

[13] Barbato A, Capone A, Carello G, Delfanti M, Merlo M, and Zaminga A. Cooperative and non-cooperative house energy optimization in a smart grid perspective. In: *2011 IEEE International Symposium on a World of Wireless, Mobile and Multimedia Networks*. 2011:1–6.

[14] Jiang S, Li J, Zhang X, Yue H, Wu H, and Zhou Y. Secure and privacy-preserving energy trading with demand response assistance based on blockchain. *IEEE Transactions on Network Science and Engineering*. 2023;11(1):1238–1250.

[15] Joshi A, Capezza S, Alhaj A, and Chow MV. Survey on AI and machine learning techniques for microgrid energy management systems. *IEEE/CAA Journal of Automatica Sinica*. 2023;10(7):1513–1529.

[16] Billah M, Yousif M, Numan M, Saleem IU, Kazmi SAA, and Alquthami TAH. Decentralized smart energy management in hybrid microgrids: Evaluating operational modes, resources optimization, and environmental impacts. *IEEE Access*. 2023;11:143530–143548.

[17] Wongdet P, Boonraksa T, Boonraksa P, Pinthurat W, Marungsri B, and Hredzak B. Optimal capacity and cost analysis of battery energy storage system in standalone microgrid considering battery lifetime. *Batteries*. 2023;9(2):76.

[18] Rawa M, Al-Turki Y, Sedraoui K, Dadfar S, and Khaki M. Optimal operation and stochastic scheduling of renewable energy of a microgrid with optimal sizing of battery energy storage considering cost reduction. *Journal of Energy Storage*. 2023;59:106475.

[19] Jamal S, Pasupuleti J, Rahmat NA, and Tan NML. Multi-objective optimal energy management of nanogrid using improved Pelican Optimization Algorithm. *IEEE Access*. 2024;12:41954–41966:250–260.

[20] Chreim B, Esseghir M, and Merghem-Boulahia L. Recent sizing, placement, and management techniques for individual and shared battery energy storage systems in residential areas: A review. *Energy Reports*. 2024;11:250–260.

[21] Zhang D, Han R, Fu W, Ran L, and Qin J. Bi-level robust optimal energy management of a community microgrid via Stackelberg game. *IEEE Transactions on Consumer Electronics*. 2024;71(2):6889–6899.

[22] Tushar W, Chai B, Yuen C, Huang S, Smith DB, and Poor HV. Energy storage sharing in smart grid: A modified auction-based approach. *IEEE Transactions on Smart Grid*. 2016;7(3):1462–1475.

[23] Hertling R, Ross E, Jones W R, Charitopoulos V M, and Papageorgiou LG. Stochastic programming approach for optimal day-ahead market bidding curves of a microgrid. *Applied Energy*. 2023;336:120847.

[24] Heidarykiany R and Ababei C. HVAC energy cost minimization in smart grids: A cloud-based demand side management approach with game theory optimization and deep learning. *Energy and AI*. 2024;16:331–345.

[25] Ghaemi S and Anvari-Moghaddam A. Local energy communities with strategic behavior of multi-energy players for peer-to-peer trading: A techno-economic assessment. *Sustainable Energy, Grids and Networks*. 2023;34:1–21.

[26] Ali L, Muyeen SM, Bizhani H, and Ghosh A. Optimal planning of clustered microgrid using a technique of cooperative game theory. *Electric Power Systems Research*. 2020;183:10626.

[27] Deng P, Qian J, Zhang R, Wu Q, and Shi Y. Optimization scheduling for microgrid clusters: A robust game theory-based method with shared energy storage. In: *2024 7th International Conference on Energy, Electrical and Power Engineering (CEEPE)*. IEEE; 2024.

[28] Farahani M, Samimi A, and Shateri H. Robust bidding strategy of battery energy storage system (BESS) in joint active and reactive power of day-ahead and real-time markets. *Journal of Energy Storage*. 2023;59:106520.

[29] Sarfi V and Livani H. An economic-reliability security-constrained optimal dispatch for microgrids. *IEEE Transactions on Power Systems*. 2018;33(6):6777–6786.

[30] Han L, Morstyn T, and McCulloch M. Incentivizing prosumer coalitions with energy management using cooperative game theory. *IEEE Transactions on Power Systems*. 2019;34(1):303–313.

[31] Paudel K, Chaudhari C, Long C, and Gooi HB. Peer-to-peer energy trading in a prosumer-based community microgrid: A game-theoretic model. *IEEE Trans. Ind. Electron*. 2019;66(8):6087–6097.

[32] Ni M, Yang Z, and Liu H. Pricing strategy of multi-energy provider considering integrated demand response. *IEEE Access*. 2020;8:149041–149051.

[33] Yan M, Shahidehpour M, Paaso A, Zhang L, Alabdulwahab A, and Abusorrah A. Distribution network-constrained optimization of peer-to-peer transactive energy trading among multi-microgrids. *IEEE Trans. Smart Grid*. 2021;12(2):1033–1047.

[34] Liu XF, Gao B, Zhu Z, and Tang Y. Non-cooperative and cooperative optimisation of battery energy storage system for energy management in multi-microgrid. *IET Generation, Transmission & Distribution*. 2018;12(10):2369–2377.

[35] Ju P, Wang L, Goel L, and Xu Y. A two-layer energy management system for microgrids with hybrid energy storage considering degradation costs. *IEEE Transactions on Smart Grid*. 2018;9(6):6047–6057.

[36] Han S, Han S, and Aki H. A practical battery wear model for electric vehicle charging applications. *Applied Energy*. 2014;113:1100–1108.

[37] Dispatching Office. Introduction to the mechanism for non-traditional units to participate in auxiliary services. Air conditioning technology seminar. 2020/07/22. Available from: https://www.taipower.com.tw/tc/page.aspx?mid=130

[38] Taipower Company. Second Public Information Session on the Ancillary Services and Reserve Capacity Trading Pilot Platform. 2020/11/11. Available from: https://ncas.taipower.com.tw/qualification/index.html

[39] Rosen JB. Existence and uniqueness of equilibrium points for concave n-person games. *Econometrica*. 1965;33(3):520–534.

[40] Liu N, Yu X, Wang C, and Wang J. Energy sharing management for microgrids with PV prosumers: A Stackelberg game approach. *IEEE Transactions on Industrial Informatics*. 2017;13(3):1088–1098.

[41] Ni J and Ai Q. Economic power transaction using coalitional game strategy in micro-grids. *IET Generation, Transmission & Distribution*. 2016;10(1): 10–18.

[42] Shapley LS. A value for *n*-person games. *Contributions to the Theory of Games*. 1953;2(28):307–317.

[43] Nocedal J and Wright SJ. *Numerical Optimization*. 2nd ed. New York: Springer; 2016. p. 449.

[44] Delta Energy Storage System. Available from: https://www.delta-emea.com/Products/CategoryListT1.aspx?CID=18&hl=en-GB

Chapter 10

Hybrid renewable energy and smart transmission: Vietnamese approach and case studies

Thanh Viet Dinh[1], Minh Quan Duong[2], Viet Thanh Tran[3], Phuoc Khanh Ma[4], Cao Quyen Le[3], Van Phuong Vo[5], Alexander Kies[6] and Bruno Schyska[7]

The rise of various renewable energy (RE) sources creates a significant impact on transmission grid operation. In this chapter, the impacts that renewable energy source (RES) operation has on the Vietnamese power transmission grid are analyzed through practical case studies of congestion scenarios, contingency analysis, and the power system stability. New challenges to power transmission introduced by RE integration in hybrid renewable energy systems (HRES) call for smart power transmission solutions. Among them, the Battery Energy Storage System (BESS) paired with an efficient control strategy for energy storage plays a pivotal role. To achieve this objective, an Adaptive Nonlinear Droop Control strategy integrated with a State of Charge (SOC) Feedback mechanism is proposed for the effective management of Battery Energy Storage Systems (BESS) within HRES. The efficacy of the proposed control framework is validated through a case study focused on the power transmission infrastructure of the Central Highlands region in Vietnam.

10.1 Introduction

An HRES typically combines multiple renewable energy sources (RES), such as solar, wind, hydro, biomass, along with energy storage systems and conventional

[1]Department of Electrical and Electronic Engineering, Dong A University, Vietnam
[2]Faculty of Electrical Engineering, The University of Danang – University of Science and Technology, Vietnam
[3]Power Engineering Consulting Joint Stock Company 4, Nha Trang, Vietnam
[4]Central Power System Operator - Branch of National Power System and Market Operator Company, Vietnam
[5]Danang Power Company, Danang, Vietnam
[6]Department of Electrical and Computer Engineering, Aarhus University, Denmark
[7]German Aerospace Center (DLR), Institute of Networked Energy Systems, Energy Systems Analysis, Germany

power sources. The increasing integration of RES into the overall energy portfolio introduces substantial changes to power system operation. Due to the inherent variability and intermittency of RES—primarily driven by meteorological and environmental conditions—the operational dynamics of power systems are significantly affected, particularly with regard to maintaining transmission system stability and ensuring overall reliability.

Smart transmission encompasses the deployment of advanced technologies and intelligent control strategies to facilitate the effective and reliable delivery of electrical power from HRES to end-use loads. These systems are specifically designed to address critical challenges associated with the variability and intermittency inherent in RE generation, while simultaneously enhancing the overall efficiency and responsiveness of the power transmission infrastructure. They rely on real-time monitoring, control, and optimization to ensure reliable power delivery while minimizing losses and maintaining grid stability. By integrating real-time monitoring, dynamic control, and intelligent systems, smart transmission enables efficient and reliable energy delivery while addressing the challenges posed by RES.

To enable smart power transmission in HRES, an advanced infrastructure is necessary, in which the following elements are essential: Energy Storage, Advanced Communication Systems, Energy Management Systems (EMS), Dynamic Control Systems, Flexible AC and DC Transmission Systems, Artificial Intelligence and Machine Learning (predictive analytics for weather patterns, load forecasting, and maintenance, autonomous grid management through self-healing algorithms), and Cybersecurity Measures (protection of the communication and control systems against cyberattacks to ensure operational integrity).

Among the key elements mentioned above, the BESS control mechanisms are crucial for maintaining power stability and reliability in the power grid. BESS enables automated charging and discharging based on load forecasting and grid conditions, as well as real-time voltage and frequency regulation. Consequently, BESS play a pivotal role in aligning electricity supply with fluctuating demand by storing surplus energy during periods of low consumption or high RE output, and subsequently discharging the stored energy during peak load intervals or when renewable generation is insufficient. BESS also allows dynamic ramp rate control by smoothing out fluctuations caused by sudden changes in renewable generation such as cloud cover or wind gusts. BESS offer an effective solution for capturing surplus RE that might otherwise be curtailed due to transmission network limitations. By delivering stored energy during periods of elevated demand, BESS not only mitigates reliance on costly peak-load generation but also contributes to maintaining a stable and continuous power supply across the grid.

This chapter analyzes the integration of RES into the Vietnamese power system, considering the increase of renewable resources (RE) as well as electricity demand with respect to the years 2020, 2025, and 2030 (in the past and future). A case study of the Vietnamese power system with the BESS adaptive control model will also be investigated.

The locations of RES are concentrated in Southern and South Central Vietnam. The areas have tremendous potential for supplying electricity to the South in the future, as demand for electricity in the South is expected to increase significantly.

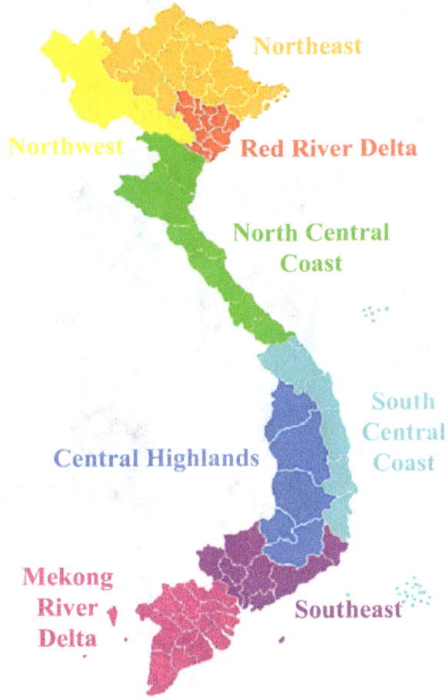

Figure 10.1 Regions in Vietnam

Over the next decade, RE development in Vietnam will primarily focus on the Southern and Central regions, particularly the Southeast, South Central Coast, and some provinces in the Mekong River Delta (Figure 10.1). While the Northern provinces also possess RE potential, their capacity remains relatively limited.

In this case study, the information on the main sources of RE for the period 2025–2030 is based on Vietnam's Power Development Plan VIII, as well as 110 kV power supply planning and power grid approved by the Ministry of Industry and Trade [1].

To evaluate the impact of RE on Vietnam's transmission grid, in the scope of this study, the power system will be analyzed under various scenarios in each period. The 24 hours during a day in Vietnam are given in simulation from 0:00 to 23:00. This study focuses on the major issues of power system operation: absorption capacity from renewable power plants, contingency analysis, and stability of the power system.

The structure of different power sources in the future years 2030, 2050 has also been indicated in Vietnam's Power Development Plan VIII [1] (Figure 10.2(a)–(b)).

10.2 Assumptions and methodology

In order to assess the impact of RES in Vietnam, the study is based on the data from actual power system operation and Vietnam's Power Development Plan VIII [1].

For the Vietnamese power system base case, it is assumed that all parallel transformers and lines are connected. Also, all the generators from conventional and renewable power plants are connected to the system.

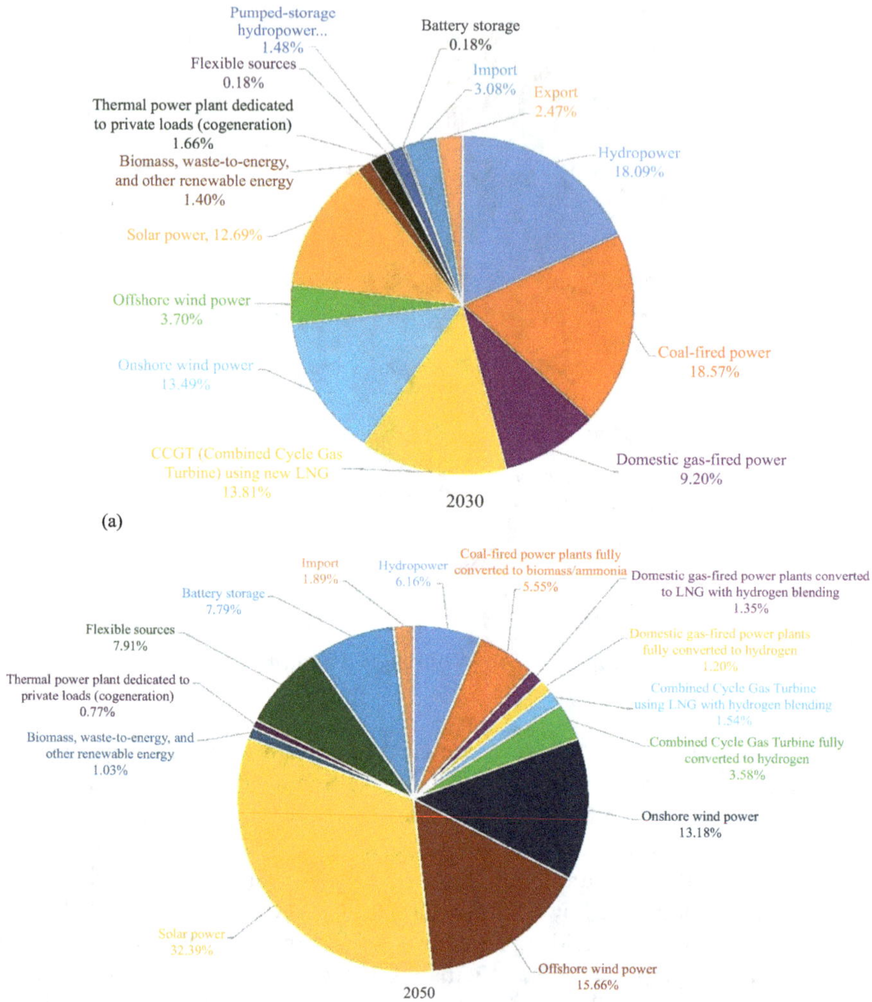

Figure 10.2 (a) Structure of power sources in Vietnamese power system in 2030. (b) Structure of power sources in Vietnamese power system in 2050.

The power transfer limits of branches are defined by the minimum value of the thermal rating, the transient stability margin, and voltage stability margin. The voltage limits must comply with the operational and planning standards applicable to the specific power system under observation and are utilized for both base-case and contingency analyses. Nonetheless, under emergency operating conditions, the system may permit broader voltage deviation ranges in order to maintain system stability and prevent cascading failures.

The permissible voltage boundaries—both upper and lower thresholds—adopted for this analysis are summarized in Table 10.1 [2]. These specified voltage constraints

Table 10.1 Defined limits for voltage levels

	Unit	500 kV		220 kV		110 kV		Generator	
		min	max	min	max	min	max	min	max
Load flow	kV	475	525	209	242	104	121		
	pu	0.95	1.05	0.95	1.10	0.95	1.10	0.95	1.05
Contingency	kV	450	550	198	242	99	121		
	pu	0.90	1.10	0.90	1.10	0.90	1.10	0.95	1.05

serve as input parameters in both steady-state load-flow simulations and contingency-based scenario evaluations.

To assess the impact of RES in different scenarios, the years 2020, 2025, and 2030 are chosen for the analysis, including selected provinces with major RE power plants. To comprehensively evaluate the integration of RES into the power system, multiple representative scenarios are selected based on different times of the day and seasonal variations throughout the year, as these temporal factors significantly influence system performance. The study considers four distinct operating scenarios, outlined as follows:

- **Scenario 1: in dry season**
  - Peak load in May
  - All generators, including conventional and renewable power plants (PPs) are connected to the system
- **Scenario 2: in dry season**
  - Medium load on Sunday
  - Dry season with the highest temperature in May
- **Scenario 3: in dry season**
  - Peak load at nighttime in May
  - All solar PPs are out of service; wind PPs produce maximum power
- **Scenario 4: in rainy season**
  - Peak load in December
  - All solar PPs are out-of-service, wind PPs produce maximum power.

The power system analysis is implemented based on PSS®E 33 power simulator.

10.3 Results of power system analysis

10.3.1 Analysis of absorption capacity for RES

10.3.1.1 Network element overload under congestions

In 2020, total RE capacity on the grid was 2683 MW, to which wind power contributed 391 MW and solar power 2292 MW. These RES are mainly located in the Southeast, South Central and Central Highlands areas. (Table 10.2) [1].

Table 10.2 Number of overloaded elements (lines/transformers) in 2020

No.	Zone/provinces	10:00		13:00		19:00		02:00	
		Lines	Transfs	Lines	Transfs	Lines	Transfs	Lines	Transfs
	Total	**18**	**10**	**13**	**3**	**3**	**3**	**0**	**3**
	North Central Coast								
1	Quang Binh	0	0	0	0	0	0	0	0
2	Quang Tri	0	0	0	0	0	0	0	0
3	Thua Thien Hue	0	0	0	0	0	0	0	0
	South Central Coast								
4	Da Nang	0	0	0	0	0	0	0	0
5	Quang Nam	0	1	0	1	0	1	0	1
6	Quang Ngai	0	0	0	0	0	0	0	0
7	Binh Dinh	0	0	0	0	0	0	0	0
8	Phu Yen	0	0	0	0	0	0	0	0
9	Khanh Hoa	0	0	0	0	0	0	0	0
	Central Highlands								
10	Kon Tum	0	0	0	0	0	0	0	0
11	Gia Lai	1	0	0	0	0	0	0	0
12	Dak Lak	0	0	0	0	0	0	0	0
13	Dak Nong	0	0	0	0	0	0	0	0
14	Lam Dong	1	0	1	0	1	0	1	0
	Southeast								
15	Ninh Thuan	4	3	4	3	4	3	4	3
16	Binh Thuan	7	1	7	1	7	1	7	1
17	Tay Ninh	2	2	2	2	2	2	2	2
18	Binh Phuoc	0	2	0	2	0	2	0	2
	Mekong River Delta								
19	An Giang	0	0	0	0	0	0	0	0
20	Long An	3	1	3	1	3	1	3	1
21	Ca Mau	0	0	0	0	0	0	0	0
22	Bac Lieu	0	0	0	0	0	0	0	0

In 2020, some transmission lines and transformers were overloaded due to the integration of renewable power plants, especially in Ninh Thuan, Binh Thuan, and Tay Ninh provinces. Besides, there were several overloaded elements in An Giang province caused by demand growth.

In 2025, many renewable power plants are to be connected to the grid with an installed capacity of over 6000 MW. They are located mainly in the Southeast, South Central Coast and Central Highlands. The 220–110 kV grid in Ninh Thuan and Binh

Thuan provinces is heavily overloaded. For example, at 13:00 in the dry season, ambient temperature is often over 30°C; the overhead conductor thermal limit is reduced to about 90% of the thermal limit at 25°C. Consequently, there are 13 overloaded components (12 overhead lines and 1 transformer) in Ninh Thuan. Similarly, 12 elements, including 11 transmission lines and 1 transformer, are overloaded in Binh Thuan province. Provinces in the Central region such as Dak Lak, Khanh Hoa, and Phu Yen will also be affected by overload with 4, 3, and 2 elements, respectively. In the Mekong River Delta area, some wind farms in Bac Lieu province cause overload on two 110 kV lines (Table 10.3).

From the power flow calculation, Ninh Thuan province's power network is heavily overloaded in the 110 kV grid. In addition, overload is also observed on a few 220 kV transmission lines and transformers in the province.

In 2030, grid congestion still occurs in 6 provinces, in which the 49 total elements (33 lines and 16 transformers) are overloaded at 13:00 daytime in the summer (dry) season. In several provinces in Southeast and South Central Coast zones, grid congestion is a major issue (Table 10.4). Especially in the Mekong River Delta, the integration of wind energy causes up to 9 overloaded 110 kV lines in Bac Lieu province.

In Ninh Thuan and Binh Thuan provinces alone, the effect of solar and wind resources leads to overload on 25 lines and transformers, specifically, 13 overloaded elements in Binh Thuan and 12 affected in Ninh Thuan province. The overloading impact of RE power plants in Khanh Hoa, Phu Yen, and Tay Ninh provinces is reduced to three to five overloaded elements at a time. The penetration of wind turbines in Bac Lieu province with a large capacity of about 200 MW results in overloading of nine 110 kV lines as well as one 220/110 kV transformer. The following Table 10.4 shows the detailed results in 2030.

10.3.1.2 Absorption capacity of RE

RES have dramatically evolved since 2019 in Vietnam. However, the national power system could not absorb all the power from distributed sources. To assess the ability to absorb RES, optimal power flow is implemented, based on constraints such as minimum cost, transmission and voltage limit.

The results show that RES cannot be fully absorbed in Ninh Thuan and Binh Thuan provinces, particularly when it comes to solar power absorption. In 2020, most of the solar power was unabsorbed, and Binh Thuan was the most affected province. The power grid could absorb only about 18% of the installed solar power. Tay Ninh and Gia Lai provinces had to shut down about 3–5% of solar power production to avoid overload (Table 10.5).

In 2025, the system receives only approximately 35% of maximum renewable electricity capacity in Ninh Thuan and about 51% in Binh Thuan province. Solar and wind power production in Dak Lak province is limited due to congestion. Only in the two provinces of Bac Lieu and Ca Mau is wind power allowed to be produced at 92% and 86% maximum capacity, respectively.

In 2030, the receiving capacity of renewable power is still limited in Ninh Thuan and Binh Thuan provinces, with only 52% and 78% of the solar power capacity in these provinces transmitted into the system.

Table 10.3 Number of overloaded elements (lines/transformers) in 2025

No.	Zone/provinces	10:00		13:00		19:00		02:00	
		Lines	Transfs	Lines	Transfs	Lines	Transfs	Lines	Transfs
	Total	**39**	**17**	**38**	**5**	**4**	**8**	**2**	**1**
	North Central Coast								
1	Quang Binh	0	7	0	2	0	6	0	0
2	Quang Tri	0	0	0	0	0	0	0	0
3	Thua Thien Hue	0	1	0	0	0	0	0	0
	South Central Coast								
4	Da Nang	0	0	0	0	0	0	0	0
5	Quang Nam	0	1	0	1	0	1	0	1
6	Quang Ngai	0	1	0	0	0	0	0	0
7	Binh Dinh	0	0	0	0	0	0	0	0
8	Phu Yen	2	0	2	0	0	0	0	0
9	Khanh Hoa	3	0	3	0	0	0	0	0
	Central Highlands								
10	Kon Tum	0	0	0	0	0	0	0	0
11	Gia Lai	2	0	2	0	0	0	0	0
12	Dak Lak	4	0	4	0	0	0	0	0
13	Dak Nong	0	1	0	0	0	0	0	0
14	Lam Dong	1	0	1	0	0	0	0	0
	Southeast								
15	Ninh Thuan	11	0	12	1	0	0	0	0
16	Binh Thuan	8	1	11	1	0	0	0	0
17	Tay Ninh	1	3	0	0	0	1	0	0
18	Binh Phuoc	0	2	0	0	0	0	0	0
	Mekong River Delta								
19	An Giang	1	0	1	0	0	0	0	0
20	Long An	3	0	0	0	1	0	0	0
21	Ca Mau	1	0	0	0	1	0	0	0
22	Bac Lieu	2	0	2	0	2	0	2	0

10.3.2 Contingency analysis (N – 1)

Results of contingency analysis ($N - 1$) for the 2020 base case for renewable power plants are presented in Figure 10.3. Insecure states in the power system associated with integrated renewable power plants are found mostly in Ninh Thuan and Binh Thuan provinces in all years: 2020, 2025, and 2030.

In 2025 and 2030, the provinces Khanh Hoa, Phu Yen, Ca Mau, and Bac Lieu can be significantly impacted by a single fault ($N - 1$) on the power grid. Many elements will be overloaded or in low voltage if a fault occurs.

Table 10.4 Number of overloaded elements (lines/transformers) in 2030

No.	Zone/provinces	10:00		13:00		19:00		02:00	
		Lines	Transfs	Lines	Transfs	Lines	Transfs	Lines	Transfs
	Total	**30**	**25**	**33**	**16**	**8**	**9**	**9**	**0**
	North Central Coast								
1	Quang Binh	0	2	0	0	0	2	0	0
2	Quang Tri	0	1	0	0	0	1	0	0
3	Thua Thien Hue	0	0	0	0	0	0	0	0
	South Central Coast								
4	Da Nang	0	0	0	0	0	0	0	0
5	Quang Nam	0	0	0	0	0	0	0	0
6	Quang Ngai	0	1	0	0	0	2	0	0
7	Binh Dinh	0	2	0	0	0	2	0	0
8	Phu Yen	0	3	1	3	0	0	0	0
9	Khanh Hoa	3	2	3	1	0	0	0	0
	Central Highlands								
10	Kon Tum	0	0	0	0	0	0	0	0
11	Gia Lai	0	0	0	0	0	0	0	0
12	Dak Lak	0	0	0	0	0	0	0	0
13	Dak Nong	0	0	0	0	0	0	0	0
14	Lam Dong	0	1	0	0	0	0	0	0
	Southeast								
15	Ninh Thuan	8	8	10	2	0	2	0	0
16	Binh Thuan	9	0	7	6	0	0	0	0
17	Tay Ninh	0	3	0	3	0	0	0	0
18	Binh Phuoc	0	1	0	1	0	0	0	0
	Mekong River Delta								
19	An Giang	4	0	3	0	0	0	0	0
20	Long An	0	0	0	0	0	0	0	0
21	Ca Mau	0	0	0	0	0	0	0	0
22	Bac Lieu	6	1	9	0	8	0	9	0

Table 10.6 shows the contingency analysis results of major RE sources penetration in six provinces Phu Yen, Khanh Hoa, Ninh Thuan, Binh Thuan, Ca Mau, and Bac Lieu.

10.3.3 Power system stability

For the power system to maintain operational security, its frequency and phasor must change within a predefined range as well as follow the national standards. If there is any disturbance, they may not deviate too far from the values for which the system

Table 10.5 Absorption capacity of renewable energy

No.	Areas/provinces	2020		2025		2030	
		Wind	Solar	Wind	Solar	Wind	Solar
	North Central Coast						
1	Quang Binh	–	100%	–	100%	–	100%
2	Quang Tri	100%	–	100%	–	100%	–
3	Thua Thien Hue	–	100%	–	100%	–	100%
	South Central Coast						
4	Da Nang	–	100%	–	100%	–	100%
5	Quang Nam	–	100%	–	100%	–	100%
6	Quang Ngai	–	100%	–	100%	–	100%
7	Binh Dinh	100%	100%	100%	100%	100%	100%
8	Phu Yen	–	100%	–	90%	–	100%
9	Khanh Hoa	–	100%	–	70%	–	86%
	Central Highlands						
10	Kon Tum	–	100%	–	100%	–	100%
11	Gia Lai	–	95%	–	100%	–	100%
12	Dak Lak	100%	100%	92%	89%	100%	100%
13	Dak Nong	–	100%	–	100%	–	100%
14	Lam Dong	–	100%	–	100%	–	100%
	Southeast						
15	Ninh Thuan	100%	88%	86%	35%	100%	52%
16	Binh Thuan	100%	82%	89%	51%	100%	78%
17	Tay Ninh	–	97%	–	92%	100%	93%
18	Binh Phuoc	–	100%	–	100%	–	100%
	Mekong River Delta						
19	An Giang	–	100%	–	100%	–	100%
20	Long An	–	100%	–	100%	–	100%
21	Ca Mau	100%	100%	86%	100%	100%	100%
22	Bac Lieu	100%	–	92%	–	76%	–

was designed. For instance, deviations from the nominal system frequency may trigger protective relays and control mechanisms, which are designed to safeguard the operational integrity and stability of the power system.

Effective integration of distributed energy resources necessitates a thorough consideration of power system stability. The incorporation of wind farms and photovoltaic power plants introduces a high degree of variability and uncertainty into power flow patterns, which can significantly influence both the steady-state and dynamic performance of the overall power grid.

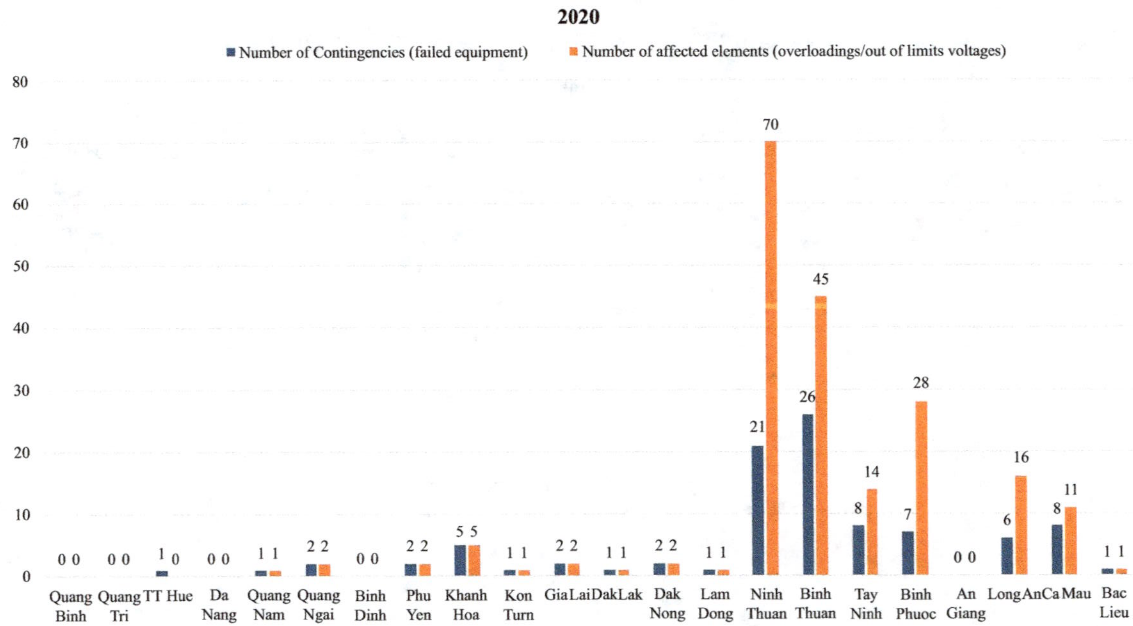

(a)

Figure 10.3 (a) Contingency analysis (N – 1) results by provinces (2020)

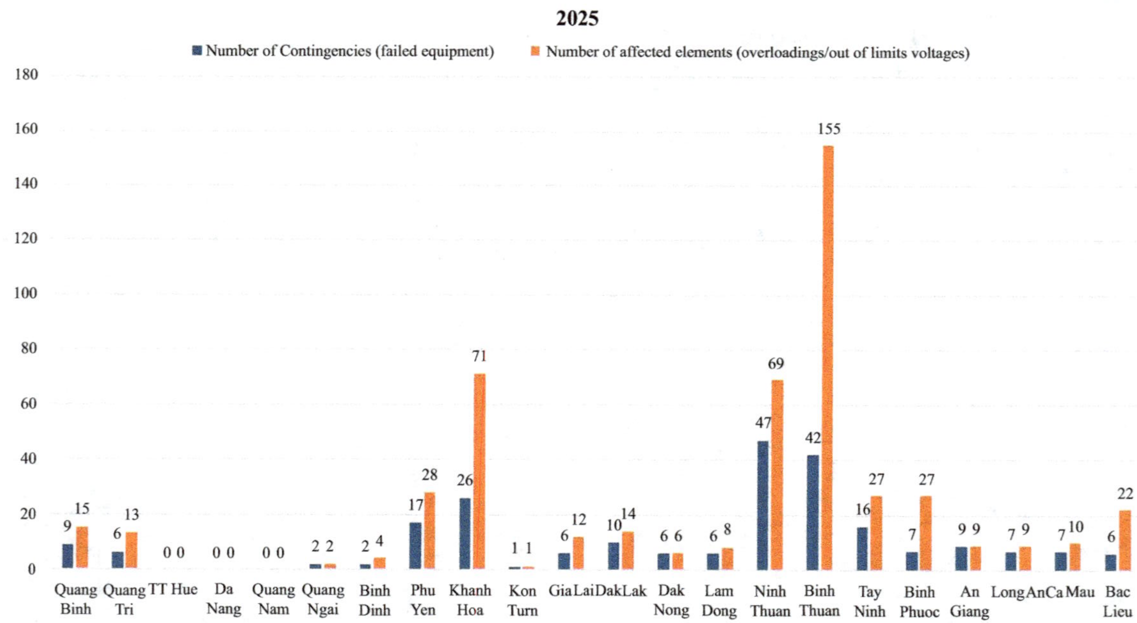

(b)

Figure 10.3 (b) Contingency analysis (N – 1) results by provinces (2025)

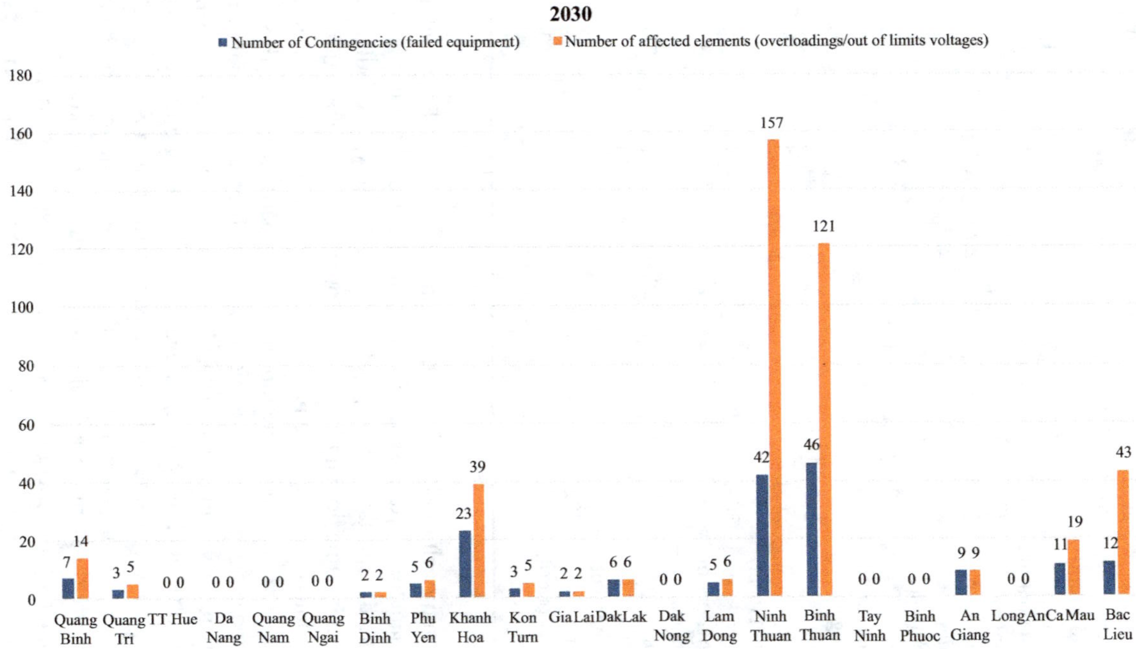

(c)

Figure 10.3 (c) Contingency analysis (N – 1) results by provinces (2030)

Table 10.6 Contingency analysis of major RE sources penetration in 6 provinces

Time and indicator			Phu Yen	Khanh Hoa	Ninh Thuan	Binh Thuan	Ca Mau	Bac Lieu
10:00	Number of Contingency (Failed equipment)	2025	15	21	47	40	10	9
		2030	3	21	41	42	3	12
	Number of overloading/ out of limits voltages	2025	28	57	154	103	13	25
		2030	4	35	140	110	3	43
13:00	Number of Contingency (Failed equipment)	2025	17	26	47	42	6	10
		2030	5	23	42	46	15	16
	Number of overloading/out of limits voltages	2025	28	71	69	155	6	29
		2030	6	39	157	121	15	41
19:00	Number of Contingency (Failed equipment)	2025	0	1	0	3	2	8
		2030	0	5	0	3	3	12
	Number of overloading/ out of limits voltages	2025	0	1	0	4	2	17
		2030	0	14	0	3	3	31
02:00	Number of Contingency (Failed equipment)	2025	0	0	5	11	7	6
		2030	0	0	0	3	11	17
	Number of overloading/ out of limits voltages	2025	0	0	7	14	10	12
		2030	0	0	0	4	19	45

Below is the approach to identifying the impact of increased RE penetration in the study of Vietnamese power system. Firstly, power system stability is evaluated with respect to inertia constant (H) or system kinetic energy. Secondly, transient stability assessment by time domain is analyzed for case of RE penetration.

In the foreseeable future, the Vietnamese power system is expected to experience a substantial increase in the penetration of RES, particularly wind, solar, and small-scale hydropower. Conversely, conventional synchronous generators may be connected less frequently. Simultaneously, the nature of electricity consumption is anticipated to evolve, with a growing prevalence of rotating electrical machines interfaced via frequency converters. These transformations are projected to significantly influence the overall system inertia, thereby necessitating comprehensive investigation to ensure the stability and reliability of power system operations.

In this research, multiple operational scenarios corresponding to the years 2020, 2025, and 2030 have been developed to evaluate the dynamic characteristics of the Vietnamese power system. These scenarios serve as the basis for estimating the system's kinetic energy, denoted as E_{sys}. The total system kinetic energy is calculated using the following expression:

$$E_{sys} = \sum S_i \bullet H_i - \sum S_{dist} \bullet H_{dist} \tag{10.1}$$

Table 10.7 Average inertia constants used to estimate inertia constants in different production scenarios

Production type (power plant)	**H (s)**
Thermal	4
Hydro conventional	3
Small hydro	1
Wind	0.42
Solar	0

where E_{sys} is the estimated system kinetic energy; $S_i \bullet H_i$ is kinetic energy of ith power plant; $S_{dist} \bullet H_{dist}$ is the kinetic energy which is removed from the system because of the disturbance.

To estimate the kinetic energy in different scenarios, some average values for inertia constants have been used [3]. These used average inertia constants are shown in Table 10.7.

When 2683 MW of renewable PPs are connected to the system, the analysis of the developed scenarios indicates that, in the year 2020, the kinetic energy of the Vietnamese power system was estimated to range from approximately 102 GWs to 158 GWs. This variation is primarily attributed to differences in the generation portfolio, particularly with respect to the type and capacity of generating units in operation. It is lower compared to the scenario of non-RE integration, with system kinetic energy between 103 GWs and 166 GWs. In 2025, the system kinetic energy is higher than in 2020, varying between 180 GWs and 284 GWs, depending on the situation. In 2030, the reduction of system kinetic energy is about 12.1% in case of integrating RE sources.

To evaluate the transient stability of the Vietnamese bulk power system, one of the most severe fault scenarios on the 500 kV grid is analyzed. Specifically, a single-phase fault is assumed to occur on the 500 kV Da Nang–Ha Tinh transmission line, with automatic reclosing failing. Such a fault in Central Vietnam impacts not only power plants within the region but also triggers dynamic responses in power plants located in both Northern and Southern Vietnam (Figure 10.4).

According to Vietnam's National Power Development Plan [1], the 500 kV backbone transmission line topology will be implemented in stages. With that, the Vietnamese power system will experience significant changes in network topology. It is estimated, in a scenario where 6000 MW of RE is replaced by 3000 MW thermal and 3000 MW hydro sources, power transfer limit from the North to Central Vietnam is reduced by 3.6% due to transient stability. When RES penetrate the power network, the power system stability may be negatively affected. In order to improve power system stability, beyond the traditional solutions such as load shedding, the spinning reserve sizing and BESS should be taken into account, especially with approximately 18,000 MW of RE to be installed in the Southern and Central regions by 2030.

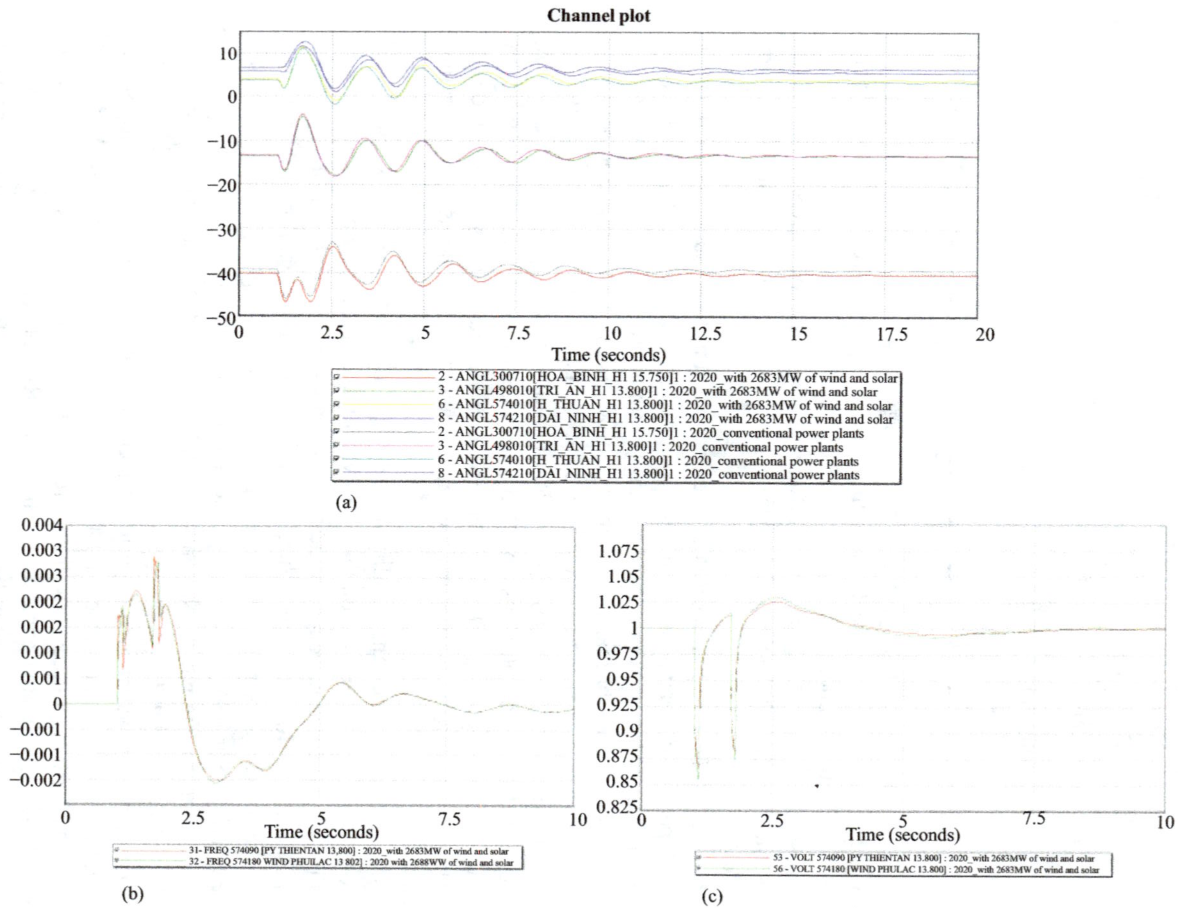

Figure 10.4 Response of power plants after disturbance

10.3.4 Summary of power system analysis result

The study presented an analysis of issues for the Vietnamese power system with the penetration of RES. There are several impacts as follows:

Firstly, with the penetration of RES, the problem of power congestion occurs on the local grid in some provinces in Vietnam (Ninh Thuan, Binh Thuan, Khanh Hoa, Phu Yen, Tay Ninh, Ca Mau, and Bac Lieu). Therefore, the power system is not able to fully absorb power from RES.

Next, the penetration of RES, especially solar, is highly concentrated in some major provinces such as Ninh Thuan, Binh Thuan, Khanh Hoa, and Phu Yen. Wide area blackouts can occur in the event of a fault on a full loading line/transformer in these provinces.

Thirdly, the stability of the power system is reduced if traditional fossil fuel-based power plants are replaced by RES.

Consequently, for a smart transmission in the hybrid Vietnamese RE, a solution for improving the absorption capacity of the grid to RE as well as power system stability should be considered.

10.4 Building a control model for BESS to enhance smart transmission and power system stability with the integration of large-scale RES

BESS have emerged as a pivotal solution for enhancing the integration of RES into modern power grids. By mitigating reliance on conventional fossil fuel-based power plants and contributing to frequency regulation, BESS play a critical role in maintaining grid stability. The accelerated expansion of RE deployment has introduced considerable operational complexities, thereby underscoring the growing global interest—both in academic research and engineering practice—in the strategic implementation of BESS. However, the practical application of BESS is still relatively new and faces numerous challenges, especially in grid frequency control when the RE penetration increases.

Globally, many studies have been conducted on BESS control models for frequency stabilization and power system stability, employing various methods and algorithms. In Vietnam, although RE has been developed significantly in recent years, earlier research has explored optimization algorithms for RE storage. For example, in research [4], author Nguyen Hong Viet Phuong applied an SOC recovery algorithm for a wind-diesel-BESS hybrid system in Phu Quy Island, maintaining battery energy within operational limits and increasing battery lifespan. In study [5], Lee *et al.* proposed a coordinated control algorithm for distributed BESS. When the SOC of a selected BESS reaches its upper or lower limit, the algorithm switches to another BESS, enabling independent voltage or frequency regulation and stabilizing the BESS system during faults. In another study [6], Shim *et al.* introduced a droop control algorithm for multiple energy storage systems based on SOC feedback, proving highly effective for grid services and SOC management. He expanded mathematical links to integrate multiple energy storage systems for load frequency regulation, enhancing the role and functionality of generation units. With the goal of controlling current flow in microgrids to address economic issues, Kaczorowska *et al.* [7] presented a current control algorithm in a microgrid integrated with energy storage systems. The system operates in charging mode during periods of low electricity prices and transitions to discharging mode when prices peak, thereby optimizing energy arbitrage and regulating power exchange between the microgrid and the main distribution network. Meanwhile, a two-layer decentralized control algorithm [8] for grid-connected microgrids was proposed to manage BESS under fluctuating RE output and load demand, achieving the desired balance between SOC and reactive power.

The above studies show that BESS applications for frequency stabilization are still in the early development stage, with frequency control remaining a complex issue dependent on various factors, such as transmission capacity dead-band range. As RE installations continue to grow, finding effective frequency control solutions remains a significant challenge.

To address these challenges, this chapter proposes a new algorithm combining two existing mathematical models: Adaptive Nonlinear Droop Control (ANDC) and SOC Feedback model. The ANDC model distributes load and power to ensure voltage and frequency stability, share power between different sources, and minimize imbalances during voltage drops. Meanwhile, the SOC Feedback model uses SOC to estimate battery status, enabling BESS devices to charge or discharge as needed to maintain system frequency in accordance with operational requirements. The combination of ANDC and SOC Feedback models inherits all the advantages mentioned above. It is not only applied in BESS systems for primary frequency control by providing reserve power during source disturbances (e.g., cloud cover, source disconnection, or power shortages), but also functions as a secondary frequency controller by absorbing surplus power to reduce grid frequency and restore charged power of the BESS system. Implementation of the algorithm is shown on the IEEE 39-bus standard grid using PowerFactory DIgSILENT 2022, adhering to technical constraints to find optimal solutions for primary/secondary frequency control. Simulation results demonstrate the algorithm's effectiveness, and are used to compare and evaluate the performance of the proposed model against the widely used CBEST system.

10.4.1 ANDC combined with SOC Feedback model

The developed model focuses on the control strategy of BESS within power systems characterized by a high penetration of RES. The control architecture and system integration framework are depicted in Figure 10.5.

The ANDC combined with SOC Feedback is a frequency regulation method that adjusts the SOC based on frequency deviations. In other words, when the system frequency changes, this method modifies the SOC of the BESS to control output power, maintain frequency stability, and ensure efficient utilization of the BESS while keeping the power system stable and operating effectively.

10.4.1.1 ANDC

The ANDC algorithm is a method that uses voltage and current signals to regulate output power through complex mathematical models. These models calculate pulse

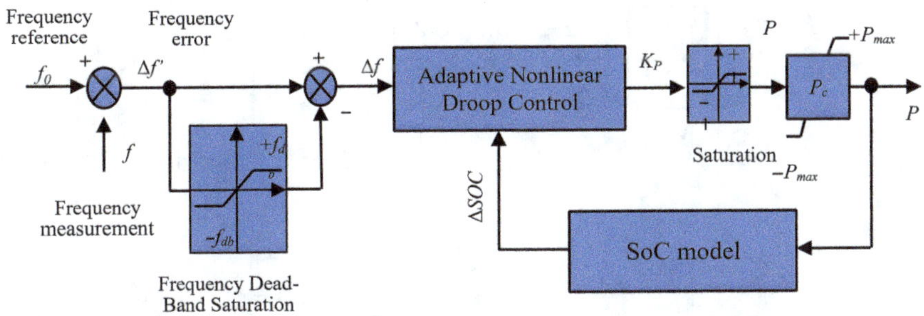

Figure 10.5 The proposed model of BESS control in power systems with high penetration of RES

width and phase angle, enabling the control of BESS devices. The effectiveness of this algorithm is demonstrated in [9] for droop control in DC microgrid systems, where it improves voltage regulation by balancing load and power distribution. Reference [10] presents a decentralized approach for active and reactive power sharing in autonomous microgrids employing voltage source inverters (VSIs), eliminating the need for physical communication infrastructure among distributed generating units. This decentralized VSI approach is applied when load fluctuations or disturbances at the power source affect the power quality.

This chapter presents an ANDC scheme designed to maintain the stability of the power grid. The algorithm uses a control method to automatically allocate loads or supply power by adjusting control parameters, which ensures flexibility, accuracy, and stability for the system. The algorithm combines two control approaches: droop control and adaptive control, addressing various power grid challenges, including:

- Balancing power between different sources in the power grid
- Ensuring voltage and frequency stability within the power grid
- Minimizing imbalances caused by changes in the power grid
- Enhancing reliability of the power grid.

Droop control

The droop frequency and voltage control is a control algorithm to distribute power between the distributed generators (DGs) [11–13]. This algorithm is used to ensure that the DG units in the system operate stably, and the output power remains balanced. Droop control simulates the droop characteristic of generators in a traditional power grid, and accordingly, regulates the output voltage and frequency of the VSI based on changes in output power. The Droop method is established by analyzing the equivalent circuit of the VSI connected to the load, as shown in Figure 10.6.

From Figure 10.6(a), we have current:

$$I = \frac{V - V_{PCC}}{R + jX} \tag{10.2}$$

$$I = \frac{(V - V_{PCC}) \cdot (R - jX)}{R^2 + X^2} \tag{10.3}$$

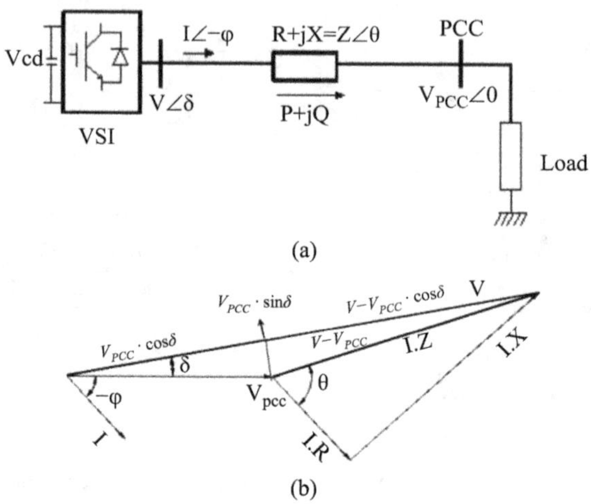

Figure 10.6 Equivalent circuit of the VSI connected to the load. (a) Power generation diagram of a voltage source inverter (VSI). (b) Current and voltage vector diagram.

$$I = \frac{R\,(V - V_{PCC})}{R^2 + X^2} - j\frac{X\,(V - V_{PCC})}{R^2 + X^2} \tag{10.4}$$

From vector diagram:

$$V - V_{PCC} = (V - V_{PCC} \cdot cos\delta) + jV_{PCC} \cdot sin\delta \tag{10.5}$$

Putting (10.5) into (10.4):

$$I = \frac{R\,(V - V_{PCC} \cdot cos\delta) + XV_{PCC} \cdot sin\delta}{R^2 + X^2} - j\frac{X\,(V - V_{PCC} \cdot cos\delta) - RV_{PCC} \cdot sin\delta}{R^2 + X^2} \tag{10.6}$$

The power after inverter in Figure 10.6(a):

$$P + jQ = V \cdot I^* \tag{10.7}$$

or

$$P + jQ = \frac{V}{R^2 + X^2} \cdot [R\,(V - V_{PCC} \cdot cos\delta) + XV_{PCC} \sin\delta] + j\frac{V}{R^2 + X^2} \cdot [X\,(V - V_{PCC} \cdot cos\delta) - RV_{PCC} \sin\delta] \tag{10.8}$$

From (10.8) we can extract P and Q as follows:

$$P = \frac{V}{R^2 + X^2}\,[R\,(V - V_{PCC} \cdot \cos\delta) + X \cdot V_{PCC} \cdot \sin\delta] \tag{10.9}$$

$$Q = \frac{V}{R^2 + X^2}\left[-R \cdot V_{PCC} \cdot \sin\delta + X \cdot (V - V_{PCC} \cdot \cos\delta)\right] \tag{10.10}$$

where

P is the active power (W);

Q is the reactive power (Var);

R, X are the resistance and the reactance of the transmission line (Ω);

I is the current flowing through the transmission line from the VSI to the Point of Common Coupling (PCC) (A);

V is the output voltage of the inverter (V);

V_{PCC} is the voltage at the PCC (V); and

δ is the phase angle between VSI output voltage and voltage at the PCC.

From (10.9) and (10.10), we have:

$$\left(R^2+X^2\right)\cdot P = V\left[R\left(V - V_{PCC}\cdot cos\delta\right) + X\cdot V_{PCC}\cdot sin\delta\right] \tag{10.11}$$

$$\left(R^2+X^2\right)\cdot Q = V\left[-R\cdot V_{PCC}\cdot sin\delta + X\cdot\left(V - V_{PCC}\cdot cos\delta\right)\right] \tag{10.12}$$

From (10.11) and (10.12) we have:

$$\left(R^2+X^2\right)\cdot P\cdot X = R\cdot X\cdot V\left(V - V_{PCC}\cdot cos\delta\right) + X^2\cdot V\cdot V_{PCC}\cdot sin\delta \tag{10.13}$$

$$\left(R^2+X^2\right)\cdot Q\cdot R = -R^2\cdot V\cdot V_{PCC}\cdot sin\delta + R\cdot X\cdot V\cdot\left(V - V_{PCC}\cdot cos\delta\right) \tag{10.14}$$

Subtract both sides of two expressions (10.13) and (10.14):

$$\left(R^2+X^2\right)\left(X\cdot P - R\cdot Q\right) = V\cdot V_{PCC}\cdot\left(R^2+X^2\right)\cdot sin\delta \tag{10.15}$$

or

$$sin\delta = \frac{X\cdot P - R\cdot Q}{V\cdot V_{PCC}} \tag{10.16}$$

Replace $sin\delta$ from (10.16) in (10.9):

$$P = \frac{R\cdot V\left(V - V_{PCC}\cdot cos\delta\right) + X\cdot V\cdot V_{PCC}\cdot\frac{X\cdot P - R\cdot Q}{V\cdot V_{PCC}}}{R^2+X^2} \tag{10.17}$$

$$P = \frac{R\cdot V\left(V - V_{PCC}\cdot cos\delta\right) + P\cdot X^2 - X\cdot Q\cdot R}{R^2+X^2} \tag{10.18}$$

$$V - V_{PCC}\cdot cos\delta = \frac{P\cdot R^2 + P\cdot X^2 - P\cdot X^2 + X\cdot Q\cdot R}{R\cdot V} \tag{10.19}$$

or

$$V - V_{PCC}\cdot cos\delta = \frac{R\cdot P + X\cdot Q}{V} \tag{10.20}$$

Assuming the transmission line has $X >> R$ and typically, the phase angle of the DGs is very small, consequently, $\sin\delta \approx \delta$ and $\cos\delta \approx 1$ [11]. Therefore, (10.16) and (10.20) can be written as:

$$\delta \approx \frac{X\cdot P}{V\cdot V_{PCC}} \tag{10.21}$$

$$V - V_{PCC} \approx \frac{X\cdot Q}{V} \tag{10.22}$$

Active power, reactive power and frequency according to the Droop control method:

$$P = \frac{V \cdot V_{PCC} \cdot \delta}{X} \tag{10.23}$$

$$Q = \frac{V^2 - V \cdot V_{PCC}}{X} \tag{10.24}$$

$$f = \frac{\omega}{2\pi} = \frac{d\delta}{dt} \tag{10.25}$$

From (10.23) and (10.24), the active power of the source is directly influenced by its phase angle, so the phase angle of the source can be adjusted by controlling its frequency (10.25). Conversely, the reactive power of the source depends on the voltage. Therefore, the output voltage of the VSI is controlled by Q, and the output frequency is controlled by P. As a result, the Droop characteristics curves P/f and Q/V (Figure 10.7) are given by the following equations:

$$m_P = \frac{\omega_0 - \omega_{\min}}{P_{\max}} \tag{10.26}$$

$$m_Q = \frac{V_0 - V_{\min}}{Q_{\max}} \tag{10.27}$$

where

m_P is the droop coefficient for P/f,

m_Q is the droop coefficient for Q/V,

V_0, ω_0 are the rated voltage and the rated angular frequency of the source and load,

$V_{\min}$, $\omega_{\min}$ are the minimum allowable voltage and angular frequency of the microgrid.

According to the EN 50160 standard, the allowable frequency deviation is 1% of the nominal frequency.

Adaptive control

Adaptive control techniques are used to adjust controllers in real-time, ensuring desired performance even when the dynamic parameters of the plant are either unknown or change over time (Figure 10.8). When the model parameters are unknown but inherently constant, the controller's structure does not depend on specific parameter values of the plant model. However, the controller's parameters cannot be precisely configured. As a solution, adaptive control algorithms implement an automatic tuning process for each controller parameter. On the other hand, a limitation of these algorithms is that the effectiveness of adaptation tends to diminish over time, requiring changes in operating conditions to restart the adaptation process [14].

The process of designing and adjusting a controller for a system can be carried out by using data collected from measurements at the system's input and output terminals under experimental conditions. The adaptive control system operates in real-time during this design and adjusting process, continuously updating the controller parameters using real-time measurements to achieve the desired performance.

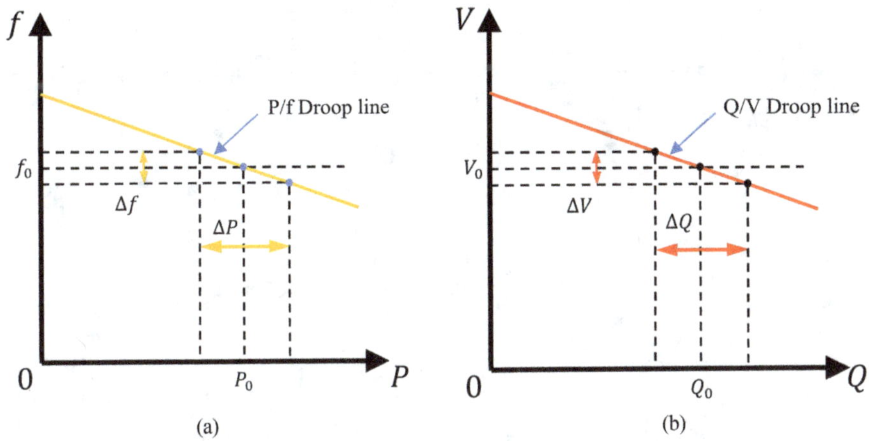

Figure 10.7 Droop characteristics diagram. (a) P/f droop characteristic. (b) Q/V droop characteristic.

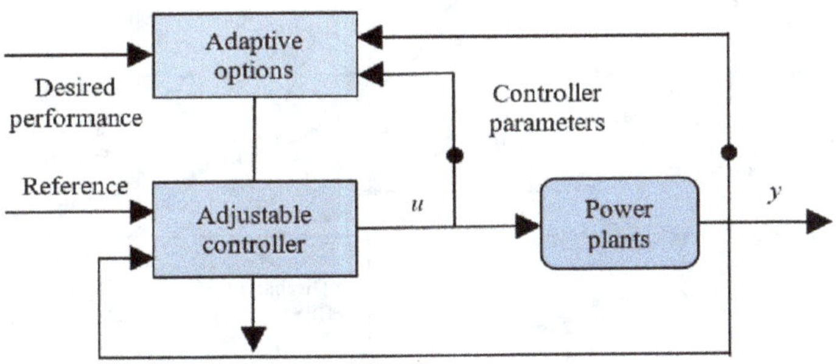

Figure 10.8 Typical Adaptive Control System diagram

In an adaptive control system, multiple different adaptive techniques can be integrated, characterized by how they process information in real-time. As shown in Figure 10.8, the adaptive control system is nonlinear because the controller parameters depend on measurements taken from the system through an adaptive loop.

The issue arising from adjusting a controller in a system with parameters that cannot be determined and change over time can be reformulated as a stochastic nonlinear control problem with complete information.

Adaptive control techniques are considered approximate methods for solving certain stochastic nonlinear control problems in controlling processes that involve undetermined parameters that change over time. This chapter focuses on the solution for frequency regulation in the system through the BESS converter. The details of the model are shown in Figure 10.9 [15].

The nonlinear droop power in the converter is calculated as follows:

$$P' = K_A\,|f_0 - f_s|\,(f_0 - f_s) = K_A|\Delta f|\Delta f = K_P\Delta f \tag{10.28}$$

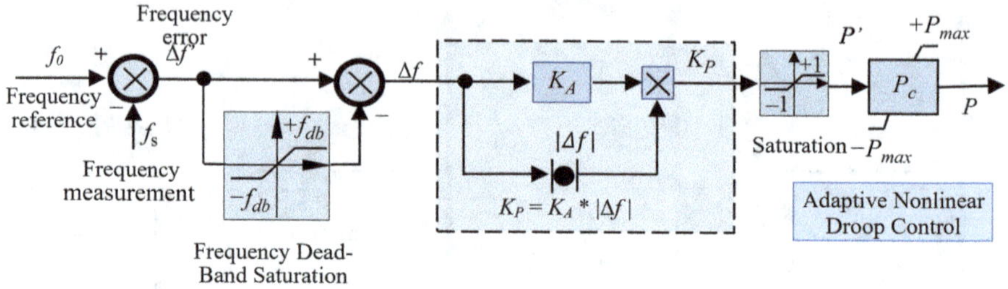

Figure 10.9 Block diagram of ANDC model

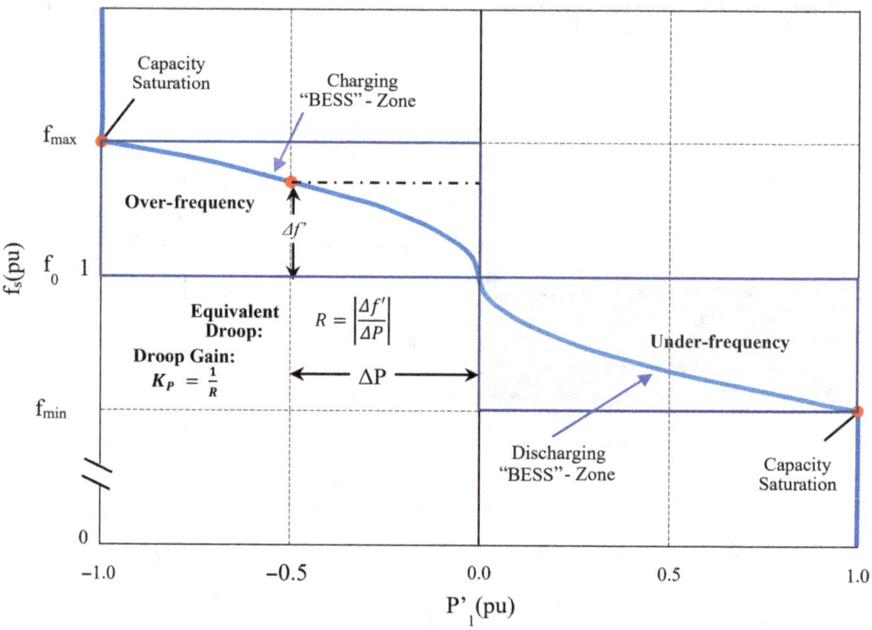

Figure 10.10 Result of the droop curve for the ANDC

where

P' is the nonlinear droop power of the converter,
K_A is the amplifier parameter,
K_P is the primary droop gain coefficient,
f_s is the estimated frequency of the phase-locked loop.

The power P' can be converted to P using the following formula:

$$P = P' \cdot P_C \tag{10.29}$$

where P_C is the rated power of BESS calculated in (pu) unit.

The droop curve of the ANDC is shown in Figure 10.10 [15].

10.4.1.2 SOC Feedback model

SOC is an indicator used to measure the level of charge in a battery or energy storage system. SOC is expressed as a percentage of the ratio between the current stored energy and the maximum energy that can be stored. The varying resistances of the transmission lines between different BESS and buses can cause an imbalance in the SOC of each battery. When imbalance occurs, it can lead to overcharging or deep discharging of the batteries, which reduces the lifespan of the BESS system. The SOC Feedback model is a control system based on a simple mathematical model and information represented as the SOC. It is used to predict the SOC, estimate the current state of the battery, and control the charge/discharge current to maintain a stable SOC.

The SOC Feedback model allows examination of the change in each SOC relative to a reference SOC (SOC_{ref}), control of the charge/discharge process according to the requirements, and balance of the load and generation, ensuring the system frequency within operational standard [15].

Figure 10.11 illustrates the block diagram of SOC Feedback method.

The deviation in the SOC is given by the equation [15]:

$$\Delta SOC_{ref} = SOC_{ref}(t) - SOC_{50\%} = \frac{(f(t) - f_n)}{K_f} = \Delta f(t)/K_f \tag{10.30}$$

where

f_n is the rated frequency,
$f(t)$ is the system frequency,
$SOC_{50\%}$ is the average value of the SOC (50% battery capacity),
SOC_{ref} is the reference value of the SOC,
ΔSOC_{ref} is the deviation of the SOC from $SOC_{50\%}$.
K_f is a coefficient, depending on the frequency deviation as in (10.31)

$$K_f = \frac{f_{\max} - f_{\min}}{SOC} \tag{10.31}$$

From Figure 10.11, when there is a frequency deviation (Δf), the input control value will be ($\Delta f - f_{ofs}$), which will result in a proportional change in power (ΔP), and from there, the State of Charge (ΔSOC) can be determined.

As a result, when a change in ΔSOC occurs, it will produce a corresponding change in f_{ofs}. This means that ($\Delta f - f_{ofs}$) will gradually approach 0 over time. In this model, when a frequency deviation occurs, the change in Δf leads to a change in the reference SOC value (ΔSOC_{ref}) as described in the equation. Consequently,

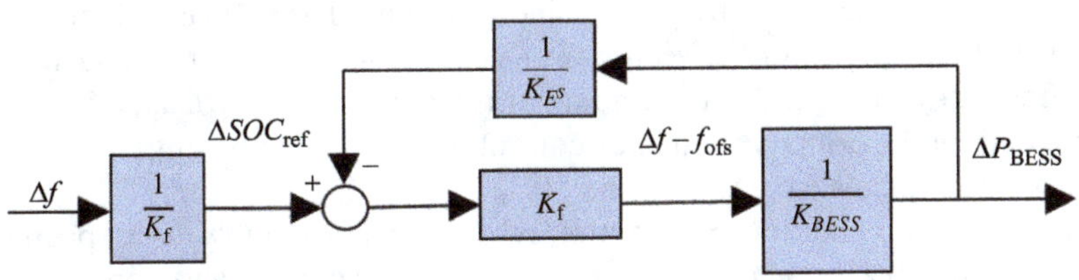

Figure 10.11 Block diagram of SOC Feedback model

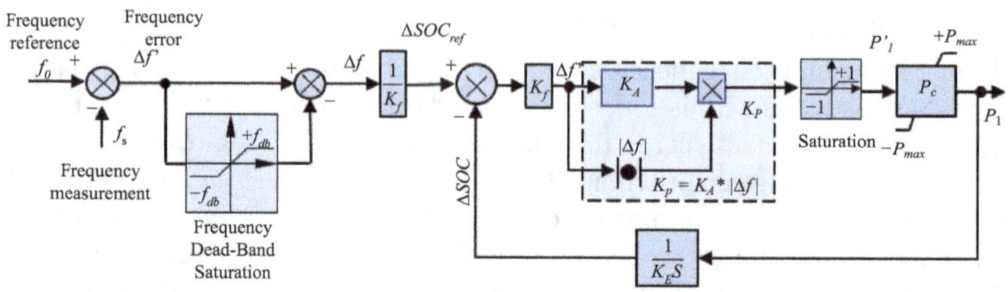

Figure 10.12 Block diagram of ANDC combined with SOC Feedback model

the change in power ΔP is based on the reference value ΔSOC_{ref} and the measured deviation ΔSOC, which in turn causes a change in the value of ΔSOC. Ultimately, $(\Delta SOC_{ref} - \Delta SOC)$ will gradually approach 0 after some time.

Figure 10.12 illustrates the block diagram of the proposed control framework, which integrates ANDC with an SOC Feedback mechanism. This configuration provides a more detailed representation and extends the control architecture previously outlined in Figure 10.5.

10.4.2 Operational response capability of the combined ANDC and SOC Feedback method

To clearly demonstrate the response capability, in this chapter, an algorithm model combining ANDC and SOC Feedback model has been developed, using the DIgSILENT Factory 2022 software. The model utilizes DSL language and other control models to assess the frequency control algorithm's response capability. Additionally, a BESS model has been integrated with an Automatic Generation Control (AGC) system, which can automatically increase or decrease power in response to changes in grid frequency.

Figure 10.13 illustrates the frequency regulation related to the BESS system compared to traditional power generation systems, operating within the same time frame [16]. The coordinated systems ensure efficient operation throughout the process.

10.4.2.1 Primary frequency control

Primary frequency regulation refers to the immediate adjustment of generator output power in response to system frequency deviations. This process is governed by the generator's static frequency-power characteristic, which is influenced by the speed control mechanism managed by the turbine governor. The primary objective is to re-establish the equilibrium between power generation and load demand when the system frequency diverges from its nominal reference value, typically 50 Hz. The time required for the primary frequency control process typically ranges from 0 to 30 seconds after an imbalance event occurs [17].

Under normal conditions, the entire system operates stably at a specific frequency (usually 50 Hz), and the balance between generation and consumption is established and maintained. However, when a disturbance occurs that affects this balance (e.g., a large generator is disconnected, a large load is connected, lightning

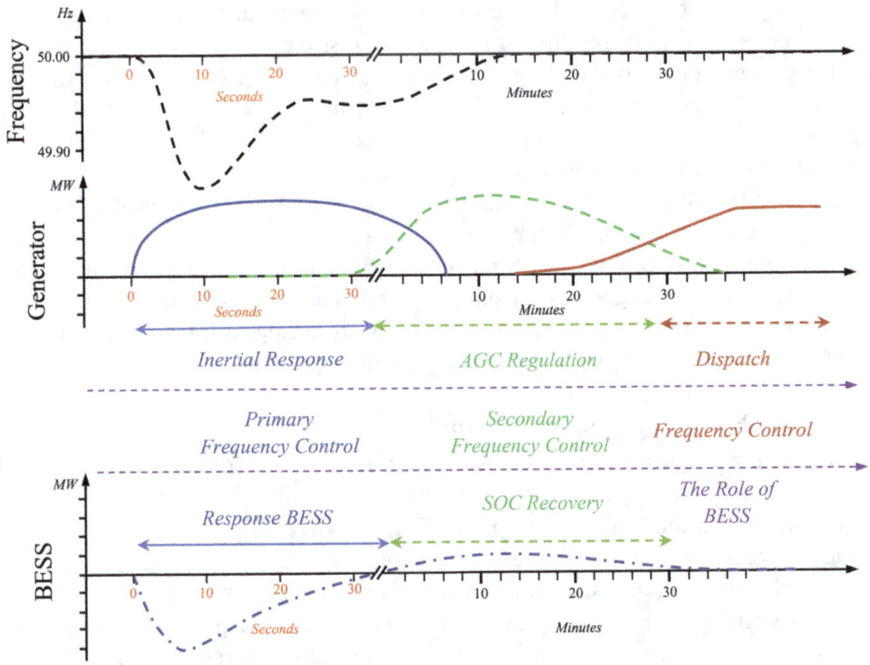

Figure 10.13 Frequency regulation by BESS system

strikes, etc.), it will immediately cause a frequency fluctuation. The grid frequency will then change rapidly, greatly deviating from the nominal value.

Following a frequency deviation, the primary control mechanisms of the generating units are activated within a few seconds to initiate corrective actions. These controllers modulate the generator output power to re-establish the equilibrium between generation and demand. Once this balance is restored, the system frequency converges to a quasi-steady-state value. However, due to the inherent droop characteristics of the generators, this stabilized frequency typically deviates slightly from the nominal setpoint.

The magnitude of the dynamic frequency deviation depends on the following factors:

- The amplitude of oscillations and the progression of faults affecting the balance of generation and consumption
- The kinetic energy of the rotating machines in the system
- The number of generating units under primary control
- The dynamic characteristics of the machines (including the controller)
- The dynamic characteristics of load consumption.

Meanwhile, the frequency deviation at a quasi-steady state is primarily determined by the amplitude of the disturbance and the stiffness of the system.

Primary frequency control requires each power generation unit to have available power reserve, allowing them to flexibly adjust their output power to respond to any frequency changes that may occur, such as increasing generation for the low frequency and decreasing generation for the high frequency. For RES (wind, solar, hydro, biomass, etc.), it is relatively easy to reduce generation when the frequency is high. However, when the frequency is low and there is a need to increase generation, it becomes quite complicated because these sources are not always available due to their dependence on weather factors [18].

The continuous growth of RES means a reduction in thermal power plants, but this also leads to challenges in frequency regulation as discussed above. This is one of the main challenges of RES in power systems. To tackle this, the use of BESS is one of the most promising solutions applied in power systems.

10.4.2.2 Secondary frequency control

When primary frequency control is used, the power reserve of each generator unit decreases, due to full or partial use. In practice, the frequency value may still be different from the nominal value after primary frequency control. In this case, secondary frequency control is employed to restore the nominal frequency value.

To accomplish this task, some power plants can perform secondary control through specific power sources reserve. The reserve level depends on the requirements of each transmission system operator and is usually a percentage of the current maximum capacity, with a minimum value set in advance to ensure independence from the maximum capacity of each generator unit.

When the system frequency falls below its nominal reference (e.g., 50 Hz), it indicates a generation shortfall relative to demand, prompting the activation of additional generation resources to restore balance. Conversely, a frequency rise above the nominal value reflects excess generation, necessitating a reduction in output from certain generating units to maintain system stability. Secondary frequency control is typically implemented automatically through a central regulator by sending signals to all generating units via a "set-point". Secondary frequency control operates minutes after primary frequency control.

10.4.3 Calculation results

This chapter presents the calculation results, and the role of the ANDC combined with SOC Feedback model in frequency regulation of the BESS system, as well as secondary frequency control using the BESS system in the Vietnamese power grid.

10.4.3.1 ANDC combined with SOC Feedback model

Models and calculation scenarios

The study develops a control model using the DIgSILENT PowerFactory 2022 software, implemented with the DIgSILENT Simulation Language (DSL). It is applied to the standard IEEE 39-bus power grid (Figure 10.14) [19], where RES contribute approximately 42% of the total power generation capacity, as shown in Table 10.8.

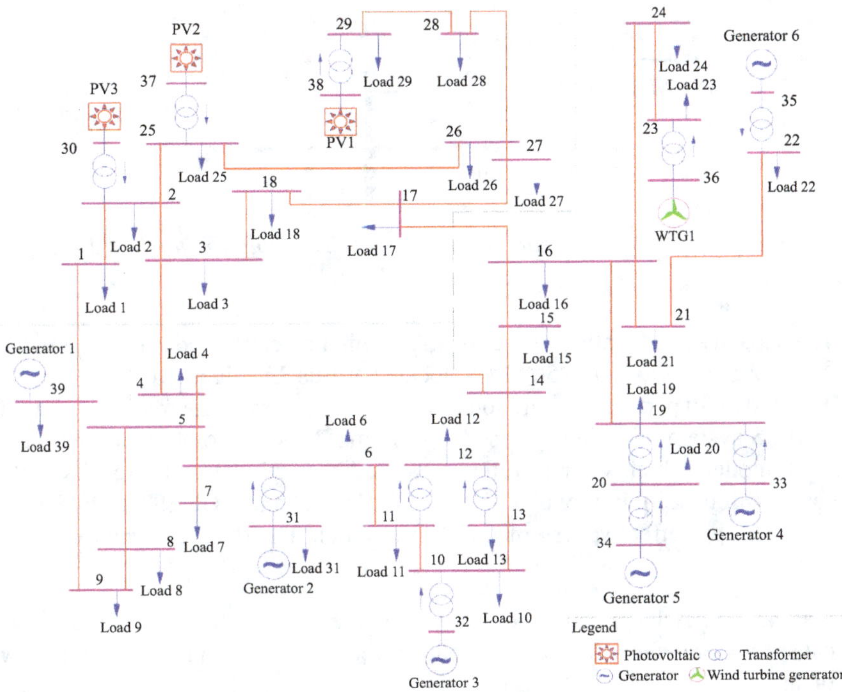

Figure 10.14 The power grid in the study—IEEE standard 39-bus system

Table 10.8 List of generators and rated capacity

	Generator	
Bus name	**Name**	***P* (MW)**
Bus 39	G1	1000
Bus 06	G2	527
Bus 10	G3	650
Bus 19	G4	632
Bus 20	G5	254
Bus 22	G6	650
Bus 23	WTG1	600
Bus 25	PV2	1200
Bus 29	PV1	600
Bus 02	PV3	250
Total		5933
Proportion of RES		42%

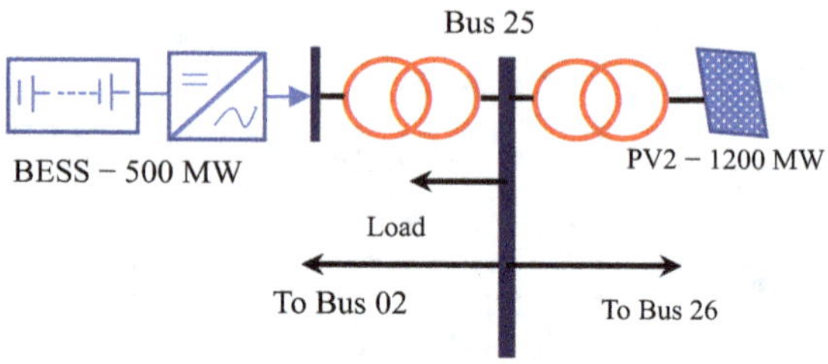

Figure 10.15 Connection diagram and BESS solution for Bus 25

The capacity and location of the BESS system are determined according to [20], with the BESS capacity set at 500 MW, located at Bus 25 (Figure 10.15).

The study will perform calculations with scenarios to evaluate frequency stability of the BESS system with the proposed algorithm ANDC in combination with SOC Feedback model, which we will refer to as the "New Model". The results will be compared with the requirements of the Grid Code [21] and a CBEST control model [22] to assess the effectiveness of this New Model. The detailed scenarios are as follows:

- **Scenario 1:** A large cloud cover changes the intensity of solar radiation and causes a significant change in output power of the PV1 and PV2 solar power plants.
- **Scenario 2:** A fault caused Generator 6 out of system.

To provide a clearer understanding of the algorithm's effectiveness in improving the operational issues (frequency amplitude, response time, etc.), we compare it with the widely used CBEST model (simulated on PSSE). The results obtained will be presented in the next section.

Calculation results

(a) Scenario 1. A worst-case scenario is considered as follows: When a large cloud cover changes the intensity of solar radiation, significantly altering the output power of PV1 and PV2 solar power plants, assuming that 1800 MW of solar power is lost. A sudden and substantial reduction in solar power output can impose considerable stress on the power system, potentially leading to dynamic instability and triggering frequency oscillations due to the abrupt imbalance between generation and demand. The cloud cover is assumed to last up to 1 minute. During the cloud cover, solar radiation rapidly decreases from 900 W/m^2 to 0 W/m^2 within 20 seconds.

After the cloud passes, solar radiation recovers and spikes to 1000 W/m^2 as the sky is clear and the power plants are stable (as shown in Figure 10.16). This calculation scenario is based on an assumption similar to the situation of a solar eclipse in the past, where Europe lost a significant amount of solar power [23], causing frequency oscillation. This simulation scenario is used to evaluate the response of the BESS system with the ANDC in combination with SOC Feedback model algorithm.

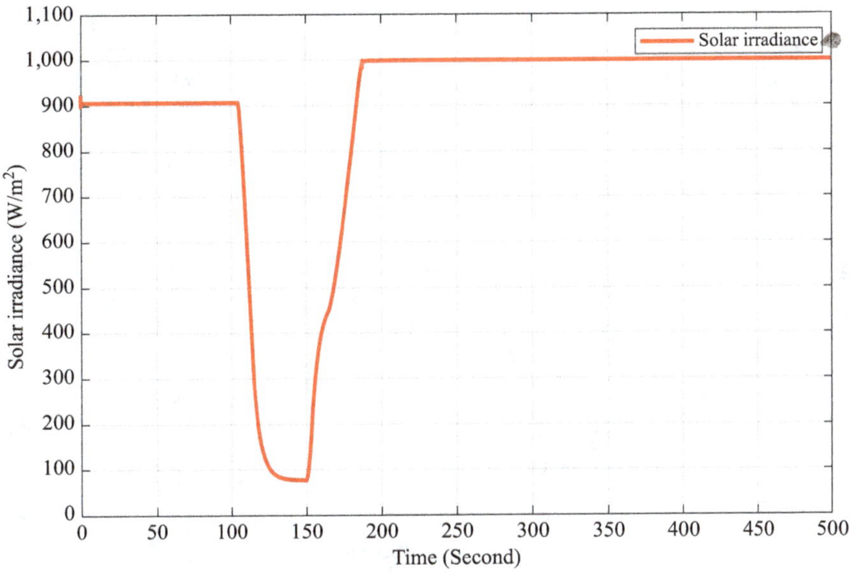

Figure 10.16 The impact of the cloud cover affects the solar irradiation on 02 solar power plants, PV1 and PV2

In Figure 10.17, it is assumed that the cloud has covered the entire area of two PV plants PV1 and PV2, causing a significant reduction in radiation intensity, and consequently, their output power fluctuates. Specifically, at the 100th second, the output power of PV1 and PV2 is 500 MW and 900 MW, respectively, which suddenly drops to 0 MW in less than 20 seconds. At the 150th second, the cloud begins to dissipate, the sky clears, and the solar radiation reaches its peak. The output power gradually increases and reaches a stable output (PV1–510 MW and PV2–1010 MW).

In this scenario, the frequency of the power system will experience significant oscillations. Frequency Ride-Through limits are shown in Figure 10.18, which represent the upper and lower boundaries within which the system frequency is allowed to oscillate. If the frequency exceeds these boundaries, the system will become unstable, and power plants must disconnect from the grid to ensure safety and prevent the worst-case scenario, grid collapse. In this situation, the role of the BESS is critical in bringing the frequency back to its initial state. In this scenario, two BESS models are compared: the New Model and CBEST.

As shown in Figures 10.18 and 10.19, when the cloud cover occurs, the system frequency drops from 50 Hz to around 49.1 Hz (close to the 49 Hz threshold). At this point, the BESS receives the signal and begins to transition from a "no power" state to generating active power to raise the frequency within approximately 30 seconds after the event. The BESS at this moment acts as a Primary Frequency Controller with a response time of about 1 minute. When the cloud dissipates, the solar radiation returns to normal. However, the output power increases because the BESS has over-generated to the grid, causing a slight increase in the system frequency (around 50.3 Hz). The

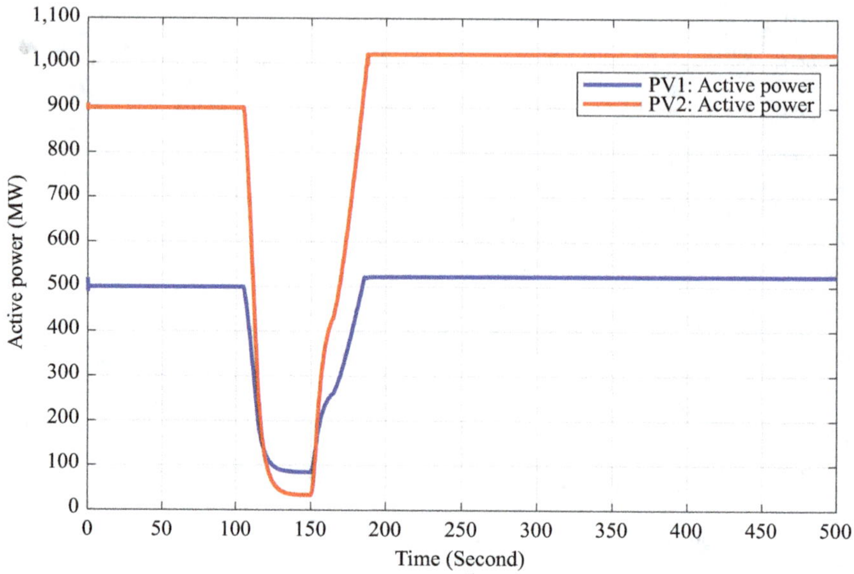

Figure 10.17 The impact of the cloud cover affects the generating capacity 02 solar power plants, PV1 and PV2

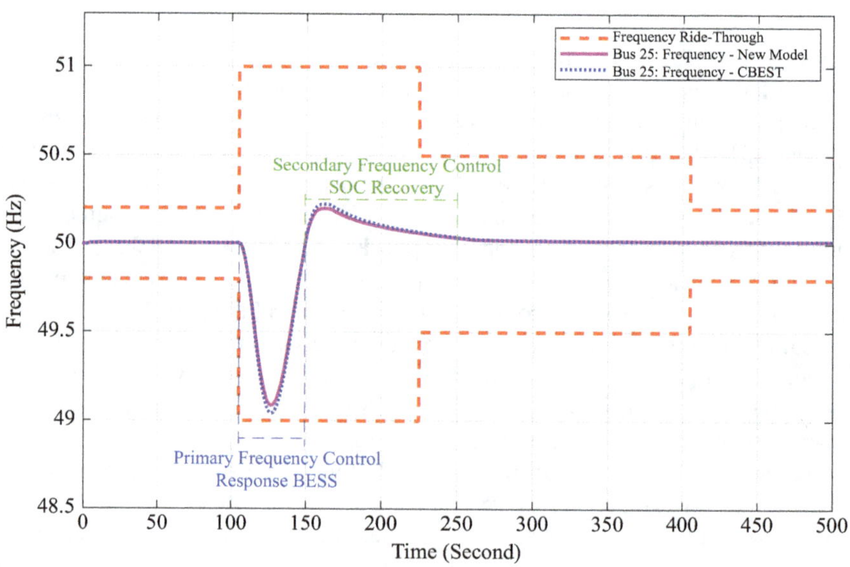

Figure 10.18 Comparison of two BESS models in power system frequency stability

BESS then absorbs active power back to charge itself, lowering the frequency back to the rated value, effectively functioning as a Secondary Frequency Controller with a response time of about 2 minutes.

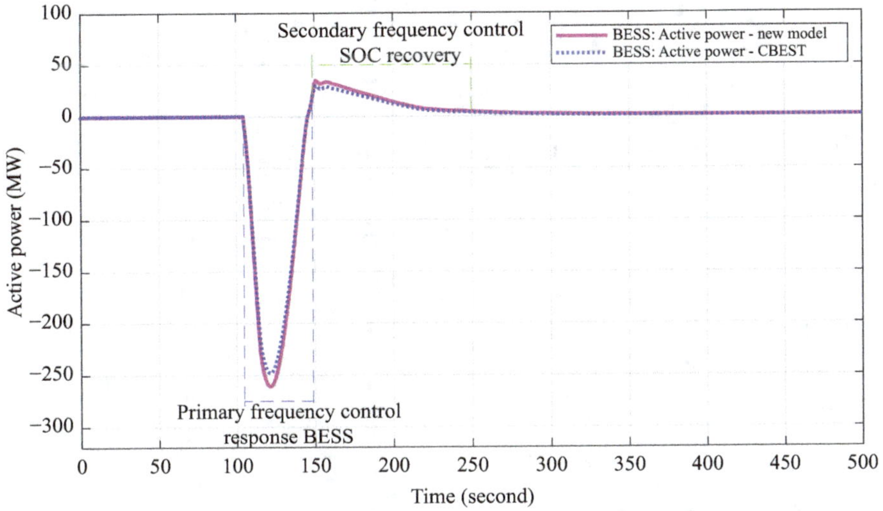

Figure 10.19 Comparison of two BESS models in power discharge

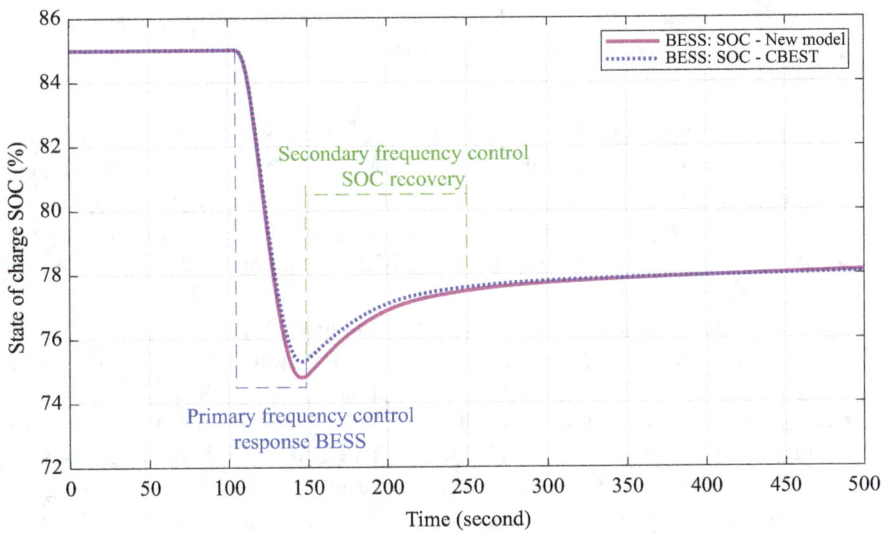

Figure 10.20 Comparison of two BESS models in SOC

The SOC of the BESS (Figure 10.20) shows a dynamic change in the reserve storage level, quickly responding to the system's needs: deeply discharging stored power to raise the grid frequency when the system is covered by clouds and rapidly recharging with surplus power to lower the frequency, while restoring the initial reserve level of the BESS system.

Through Figures 10.18–10.20, it is evident that the New Model demonstrates its advantage over CBEST, as detailed below:

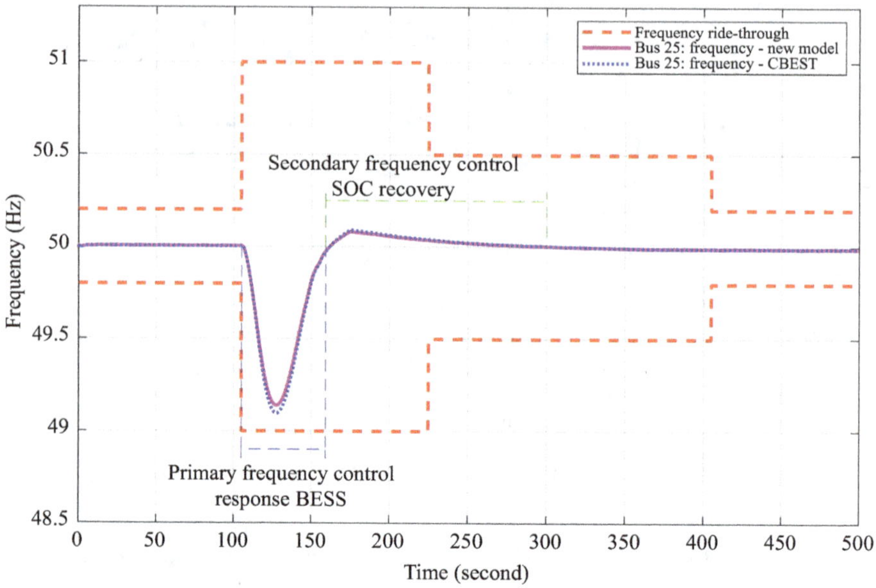

Figure 10.21 Comparison of two BESS models in power system frequency stability when Generator 6 is out of power system

- **Frequency Response:** The proposed model demonstrates a faster dynamic response by initiating primary frequency control at a threshold of 49.2 Hz, earlier than the CBEST model, which activates at 49.1 Hz. Additionally, during frequency recovery, the proposed model transitions to secondary frequency regulation more promptly compared to the CBEST benchmark, thereby enhancing overall system stability.
- **Discharge State**: The New Model discharges more deeply than CBEST within the same time frame, helping to raise the frequency more quickly. It also absorbs more surplus active power to reduce the frequency to a stable state.
- **Charging State:** The New Model has a faster response time and deeper discharge compared to the CBEST model. The enhanced performance of the proposed model can be attributed to its incorporation of SOC monitoring and the implementation of a gain factor $K_{f,}$ which defines the relationship between frequency deviation (Δf) and SOC. This approach enables a more responsive regulation strategy, minimizing latency in frequency control and optimizing the discharge behavior of the BESS.

(b) Scenario 2. In this scenario, it is assumed that Generator 6 (G6 with a capacity of 650 MW) experiences a fault and trips out of the power system, leading to a deficit in the total generation capacity of the system. After the fault is cleared, Generator 6 is restored and brought back to the grid.

In Figures 10.21 and 10.22, the system frequency oscillates significantly. At the 100th second, the frequency drops to 49.1 Hz for approximately 30 seconds as

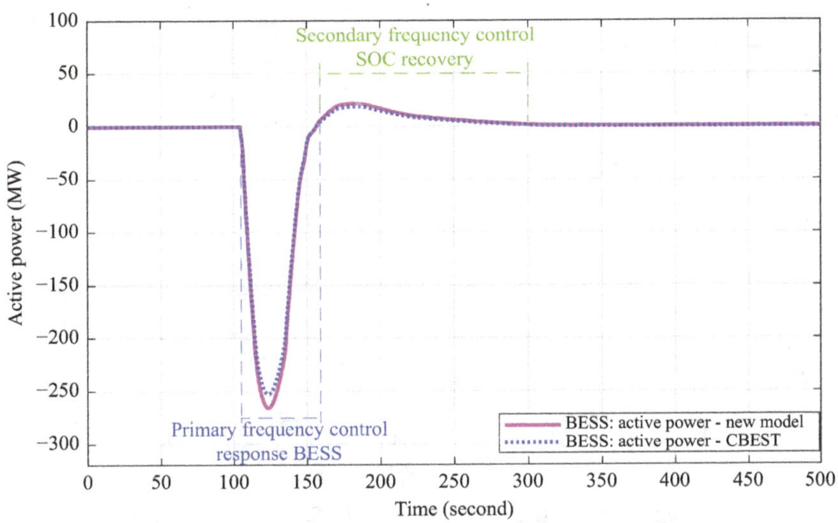

Figure 10.22 Comparison of two BESS models in power discharge when Generator 6 is out of power system

Generator 6 trips out of the grid. From the 130th second to the 165th second, the BESS is fully utilized to generate active power and restore the grid frequency (primary control). From the 165th to the 300th second, the grid frequency increases sharply due to an excess of active power as Generator 6 reconnects to the grid. During this period, the BESS absorbs the surplus power to flatten the peak-load and bring the frequency back to its stable state (secondary control).

In Figure 10.23, the SOC of the BESS operates similarly to Scenario 1: deeply discharges during primary frequency control when the grid frequency drops sharply due to the trip of Generator 6. When there is an excess of power as Generator 6 reconnects to the grid, the BESS performs secondary frequency control by recharging itself with this surplus energy.

From Figures 10.21–10.23, it can be observed that both the New Model and CBEST exhibit comparable capabilities in stabilizing the grid frequency in this scenario. However, there are slight differences in output power to the grid and the percentage of deep discharge of the energy reserve, where the New Model shows a marginal advantage over CBEST.

Considering the two scenarios, the proposed New Model not only demonstrates its ability to flexibly meet the requirements of the power system under various situations but also proves its advantage over the widely used CBEST model. With this algorithm, the BESS system not only participates in primary frequency control but also in secondary frequency control, enhancing both the stability of the power system and the efficiency of BESS utilization.

10.4.3.2 Secondary Frequency Control using BESS System in the Vietnam's power grid

Models and calculation scenarios

In 2025, more RES are planned to be allocated to the Central Highlands region of Vietnam [1]. It is important to highlight that certain utility-scale RE installations

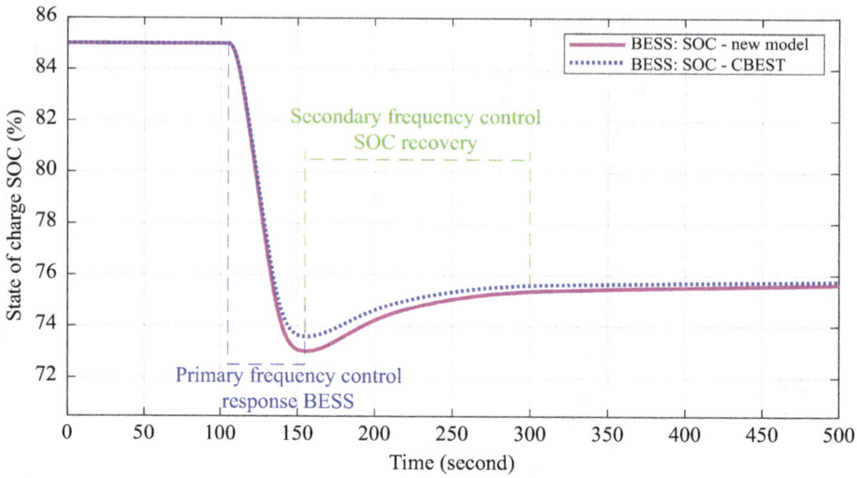

Figure 10.23 Comparison of two BESS models in SOC when Generator 6 is out of power system

are planned to be directly connected to the 500 kV transmission network, which may introduce voltage stability challenges within the power grid. To address these concerns and improve system performance, this section presents a simulation-based analysis of the Central Highlands power system, with particular emphasis on the integration of BESS as a supporting mechanism for voltage regulation and grid reliability enhancement.

Figure 10.24 presents the configuration of the power system in Vietnam's Central Highlands projected for the year 2025, featuring the integration of 500 kV/220 kV transmission infrastructure and multiple RES. This model reflects the anticipated network topology under increased penetration of renewable generation and its interaction with the high-voltage transmission system [1].

To assess the operational effectiveness of the proposed BESS control strategy, dynamic simulations will be conducted using the DIgSILENT PowerFactory software platform, which enables comprehensive modeling and analysis of power system behavior under various operating conditions. The simulation serves as a testbed to assess the effectiveness and functionality of the proposed New Model. Moreover, a new comprehensive framework for Automatic Generation Control (AGC) for BESS at the 500 kV EaNam Substation (Figure 10.25) and a new AGC method implemented in DIgSILENT software (Figure 10.26) have been applied.

The forecasted wind speed over the next 24 hours at some typical wind farms is illustrated in Figure 10.27.

Within the same area, the variation in solar radiation intensity is negligible. Therefore, we use a unique solar radiation intensity for the PV plants. Figure 10.28 illustrates the assumed solar radiation intensity over 24 hours [19].

In this chapter, a study is carried out to evaluate the power system stability with BESS integration when secondary frequency control is performed. Additionally, the

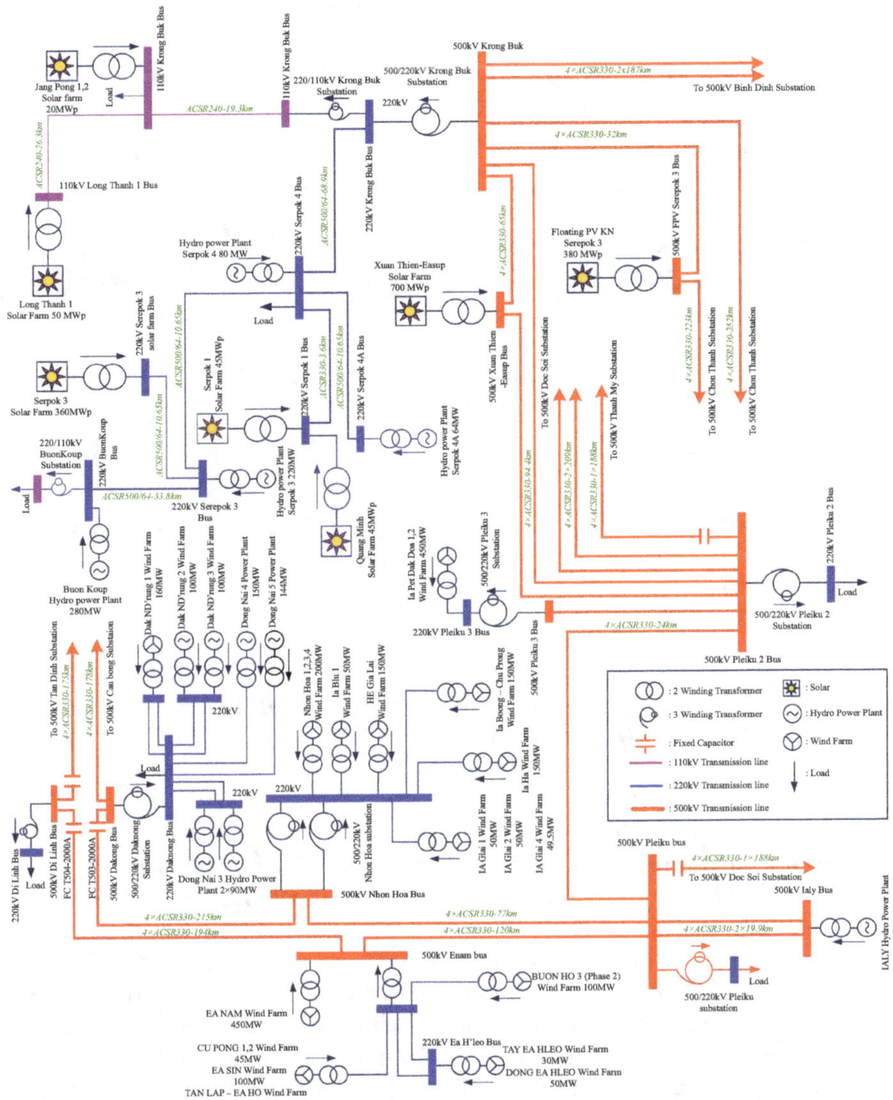

Figure 10.24 The power system of Central Highlands region in 2025

adaptive algorithm proposed above is applied to the BESS system to stabilize the voltage. The scenarios are as follows:

- **Scenario 1:** BESS system stabilizes the power grid according to the operational schedule (24 hours), maintaining the interconnection transmission limits.
- **Scenario 2**: The system experiences a loss of approximately 50 MW of Ia Ha wind power plant.

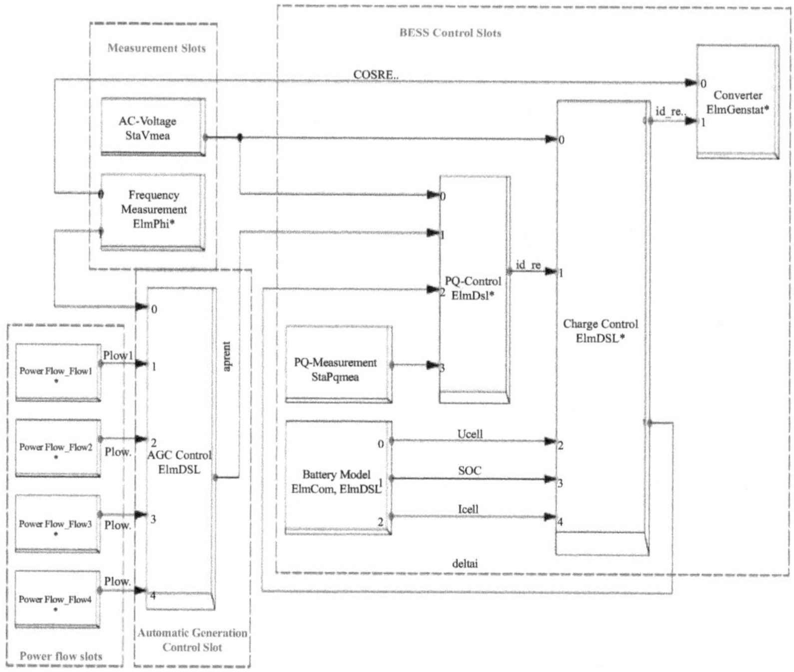

Figure 10.25 The comprehensive AGC framework for the BESS system at 500 kV EaNam in DIgSILENT software

Calculation results

The transmission grid model of the 500/220 kV in the Central Highlands region in 2025, as depicted in Figure 10.24, has been simulated in DIgSILENT to analyze the actual power generation.

(a) Scenario 1. In this scenario, the BESS control model and forecasting process are used to assess the load capacity of transmission equipment, with particular focus on series compensated capacitors. The evaluation is based on the predicted wind speeds and solar radiation over 24 hours. To identify potential overload situations for series compensated capacitors or the need to reduce the output power of RES during transmission line overload, a proactive approach has been used. By strategically regulating the charge–discharge cycles of BESS, it is possible to mitigate the risk of overloading series-compensated transmission capacitors. This regulation facilitates the modulation of power output from RES, thereby enhancing the coordination and optimal utilization of distributed energy resources. Consequently, the approach contributes to maintaining grid stability and reliability, preventing thermal overloading, and supporting the secure and efficient operation of the transmission network.

- **Operation scenario without BESS**

Based on the simulation results illustrated in Figure 10.29, the forecast indicates that between 8:00 AM and 4:00 PM, two series compensated capacitors TU 503 - DL (500 kV Di Linh Substation) and TU 502-DK (500 kV Dak Nong Substation) will experience overloading of approximately 25%, or around 2500 A, compared to their rated

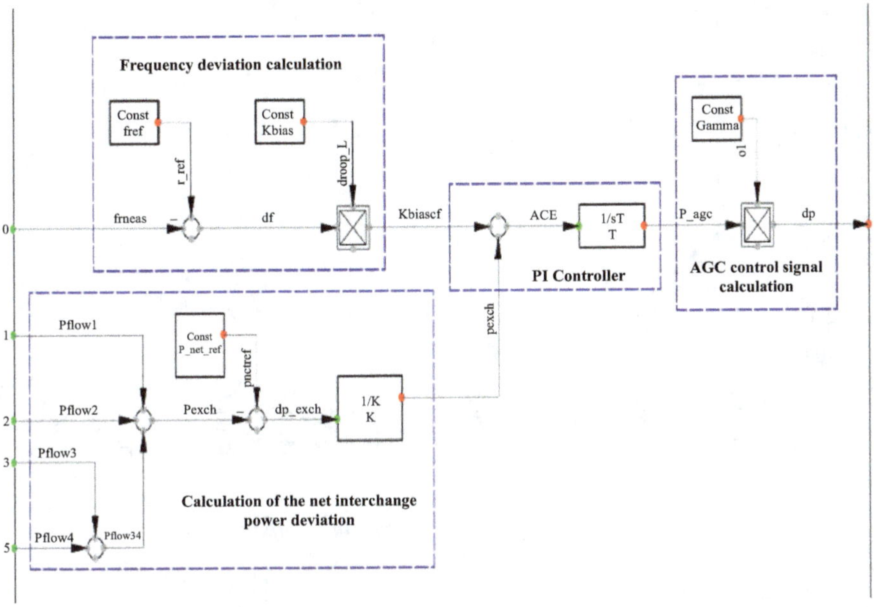

Figure 10.26 The proposed AGC method in DIgSILENT software

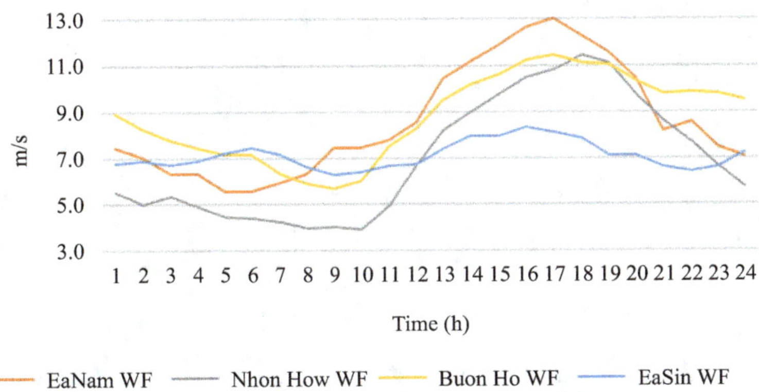

Figure 10.27 The forecasted wind speed (m/s) [19]

current (I_r), equivalent to 2000 A (100%). These overload situations could negatively affect the performance and efficiency of the power grid if not properly addressed.

To address this challenge and maintain the optimal performance of series-compensated capacitors, it is essential to implement preemptive control strategies and operational adjustments that effectively manage system dynamics under varying load and generation conditions. One of such measures is the proposed deployment of the BESS charge/discharge cycle, as discussed earlier, to manage energy flow and reduce the burden on the series compensated capacitors. By strategically using BESS, the overload problem can be minimized, ensuring smooth transmission grid operation

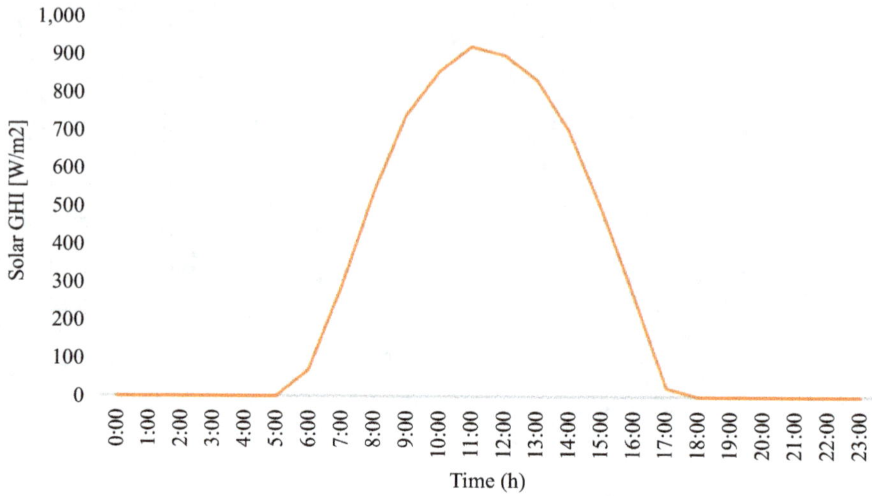

Figure 10.28 Solar radiation intensity

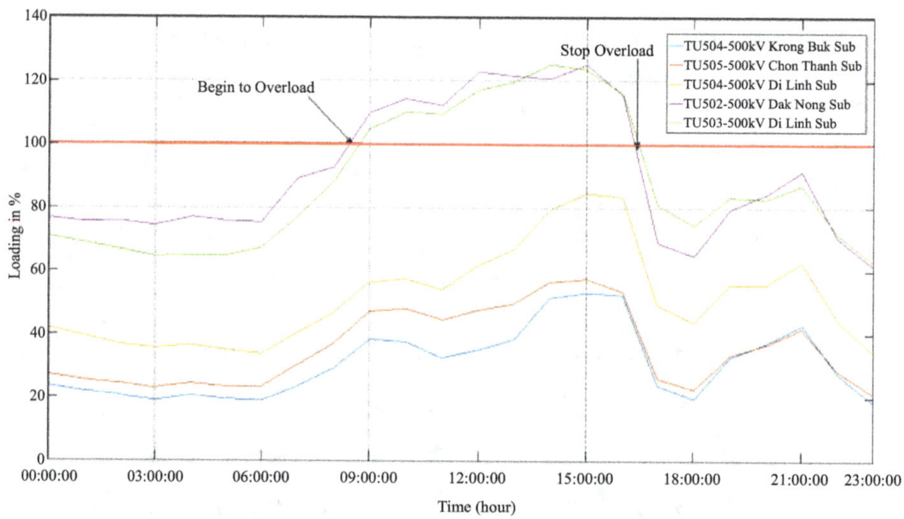

Figure 10.29 Loading status of the series compensated capacitors

and maintaining the desired loading capacity within acceptable limits for the series compensated capacitors.

- **BESS solution in secondary frequency control to ensure stable interconnection exchange on 500 kV transmission lines**

In this simulation, the BESS will be installed with a capacity of approximately 700 MW at the 500 kV EaNam Substation and 400 MW at the 500 kV Dak Nong

Substation. The main objective of BESS installation is to support the series compensated capacitors. Based on the analysis presented in Figure 10.29, it can be determined that the power grid will encounter an overload situation with the series compensated capacitors TU 503 – DL (500 kV Di Linh Substation) and TU 502-DK (500 kV Dak Nong Substation). This issue can be resolved by applying the proposed charge/discharge cycle for the BESS as described in Figure 10.30. The proposed charging cycle is expected to start at 8:00, aligned with the time when the grid begins to overload, and continue until 16:00. During this period, the BESS will be charged, absorbing and storing surplus energy very efficiently to reduce the burden on the series compensated capacitors. Following the charging period, the BESS transitions to the discharge phase, scheduled between 16:00 and 02:00 the subsequent day. During this interval, the BESS releases the previously stored electrical energy into the power grid to support system demand and enhance operational flexibility.

Upon completion of the discharge cycle, the BESS reverts to a standby mode, prepared for the subsequent charging phase. This systematic charge/discharge operation enables the BESS to dynamically contribute to load regulation, effectively alleviating grid congestion and sustaining power system stability. The integration of this operational strategy enhances the grid's capacity to respond to demand variability while maintaining the functional integrity of series-compensated transmission lines, thereby improving the overall reliability and stability of the power transmission network.

The implementation of a coordinated charge/discharge strategy for the BESS effectively mitigates the overload conditions experienced by the series-compensated

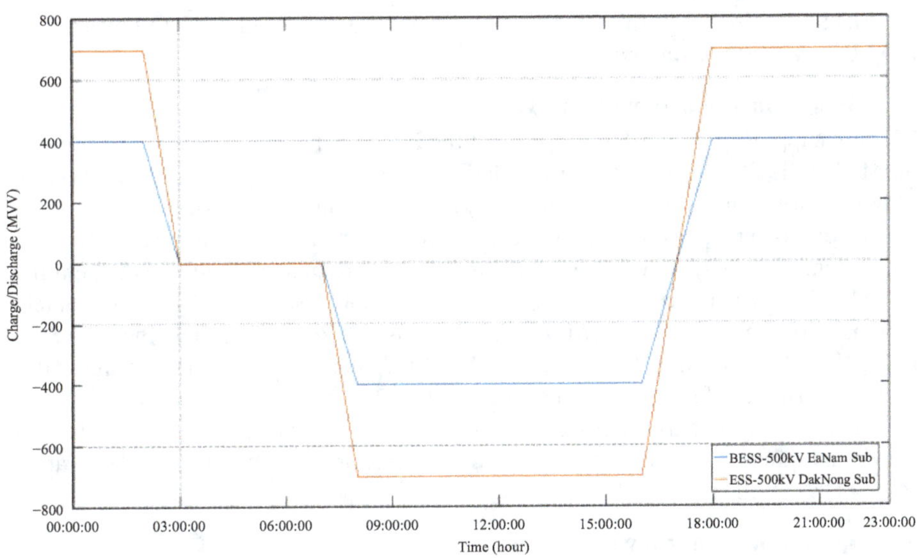

Figure 10.30 The charge/discharge cycle of the BESS

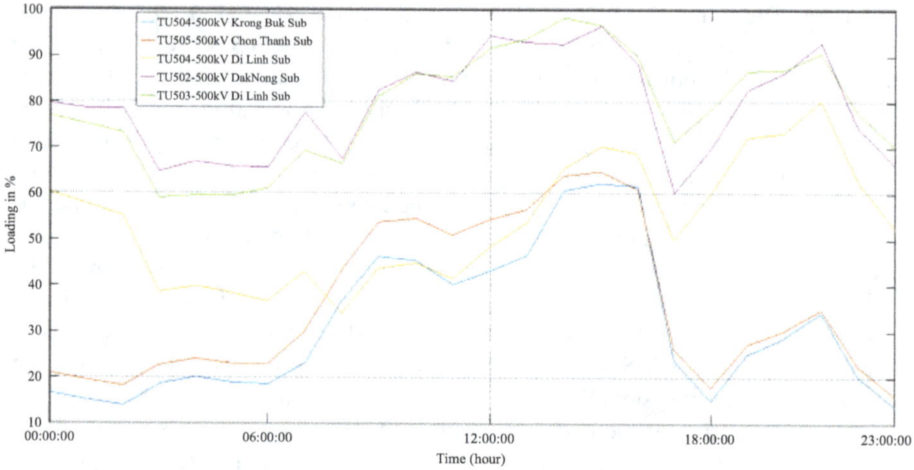

Figure 10.31 Loading status of the series compensated capacitors with BESS participation

capacitors TU 503-DL (located at the 500 kV Di Linh Substation) and TU 502-DK (located at the 500 kV Dak Nong Substation), as illustrated in Figure 10.31. Simulation outcomes validate the role of BESS in enhancing the operational performance and dynamic stability of series compensation equipment, thereby contributing to the secure and efficient operation of the high-voltage transmission network.

(b) Scenario 2. In this scenario, a system fault at the Ia Ha Wind Power Plant (50 MW) will be considered.

- **Operation scenario without BESS**

Without support of the BESS, the fault at the Ia Ha Wind Power Plant (operating at 50 MW at the 90th second of the simulation) causes the system to suddenly lose 50 MW of power. The system frequency rapidly drops from 50 Hz to 49.1 Hz.

Hydropower plants are quickly mobilized to perform frequency regulation for the system. The IALY Hydropower Plant (primary frequency control) is generating 208 MW. When a fault occurs, the system's Power Stability Control Unit immediately increases its generation to 248 MW and maintains it at 243 MW during the fault. The Dong Nai 5 Hydropower Plant, generating 50 MW, increases its output and maintains 60 MW during the fault (as shown in Figure 10.32).

After the system is stabilized, the frequency remains steady at 49.22 Hz (as shown in Figure 10.33). This value exceeds the allowable threshold for grid operation (49.22 Hz < 49.8 Hz).

- **Operation Scenario with BESS**

In this case, a BESS with a capacity of 400 MW installed at the 500 kV EaNam bus supports secondary frequency control of the system (Figure 10.34). Similar to the scenario without BESS, when a fault occurs at the Ia Ha wind power plant operating

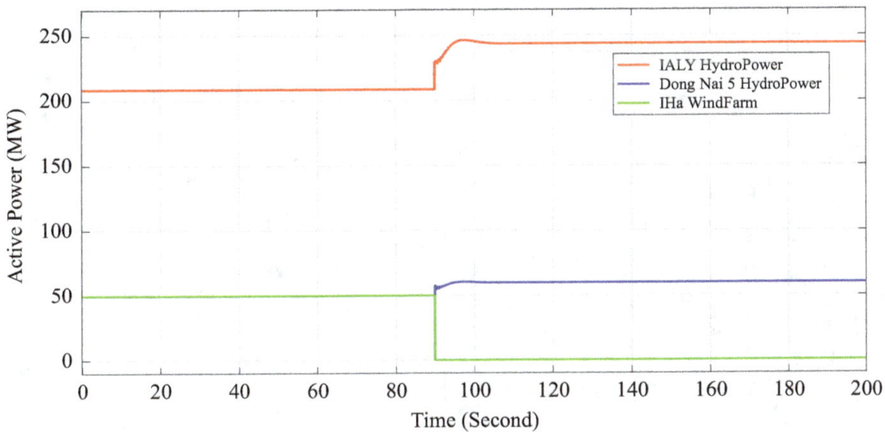

Figure 10.32 Output power of plants without BESS participation

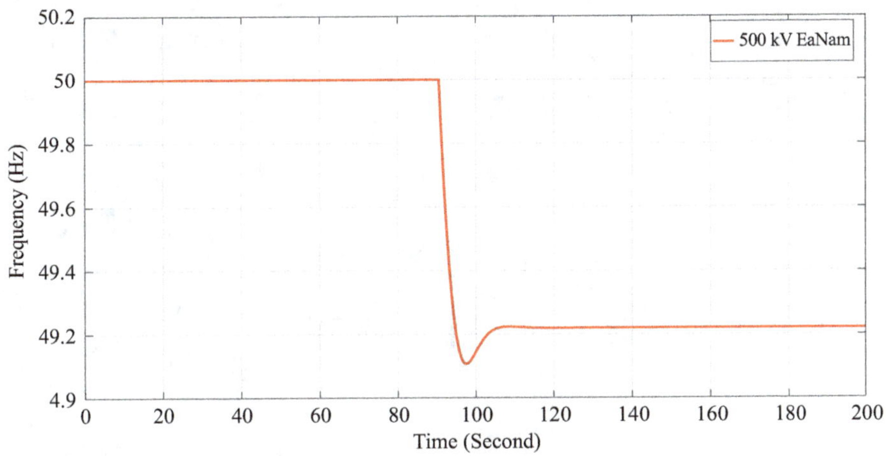

Figure 10.33 The system frequency at the 500 kV EaNam bus without BESS participation

at 50 MW at the 90th second of the simulation, the system abruptly loses 50 MW of power. The system frequency quickly drops from 50 Hz to 49.1 Hz.

To bring the frequency back to the permissible operational value, hydropower plants perform primary frequency control for the system. The IALY Hydropower Plant increases its output from 208 MW to 243 MW during the fault period, while the Dong Nai 5 Hydropower Plant increases its output from 50 MW to 60 MW. The BESS has not yet participated in frequency control, resulting in its output power gradually oscillating and decreasing to zero (Figure 10.35). At the 120th second, the BESS begins participating in secondary frequency control for the system. The EaNam BESS generates up to 40 MW of power. Consequently, the output power of the IALY and Dong Nai 5 Hydropower Plants decrease to 218 MW and 53 MW, respectively.

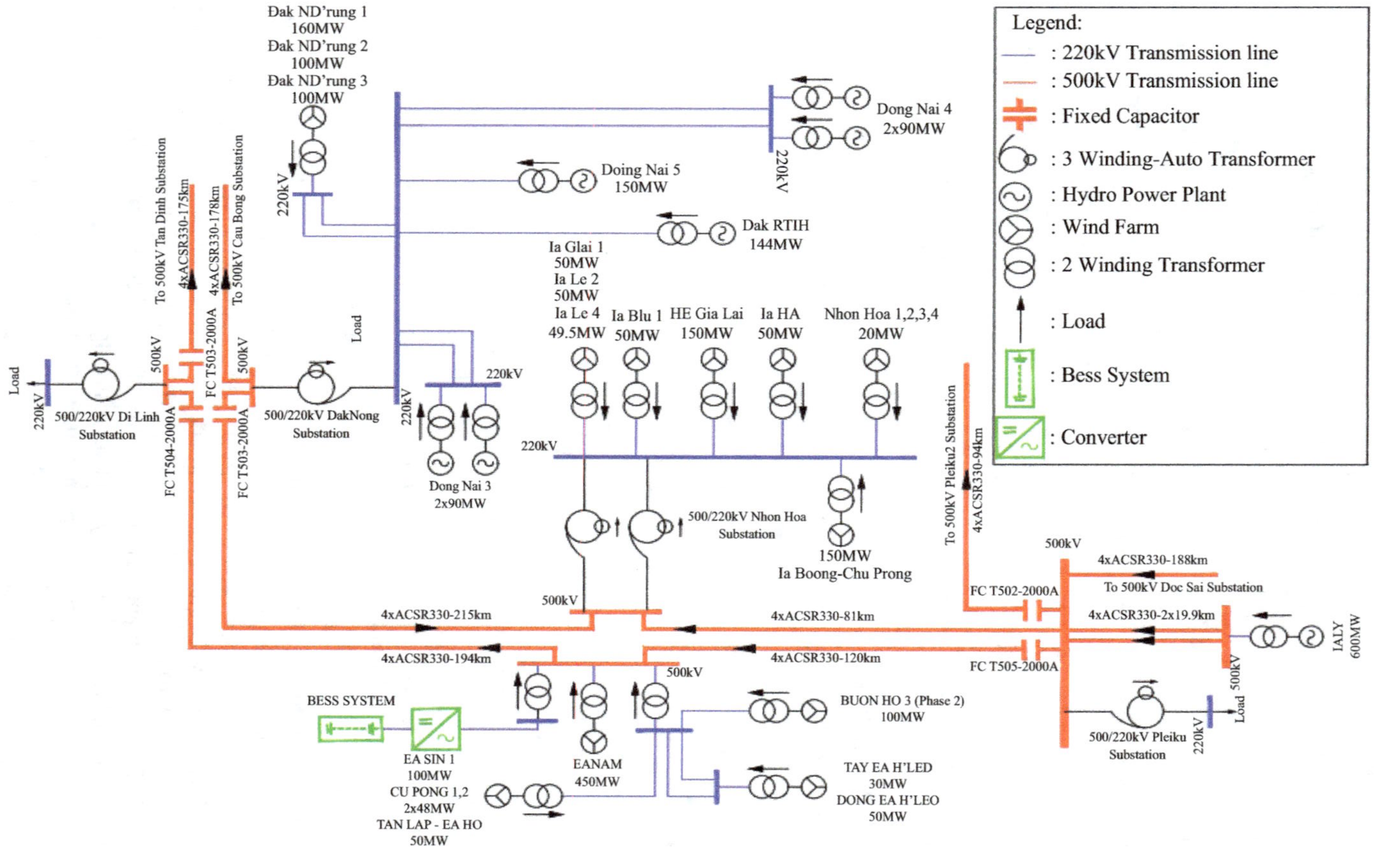

Figure 10.34 The BESS system is installed at the 500 kV EaNam Bus (400 MW)

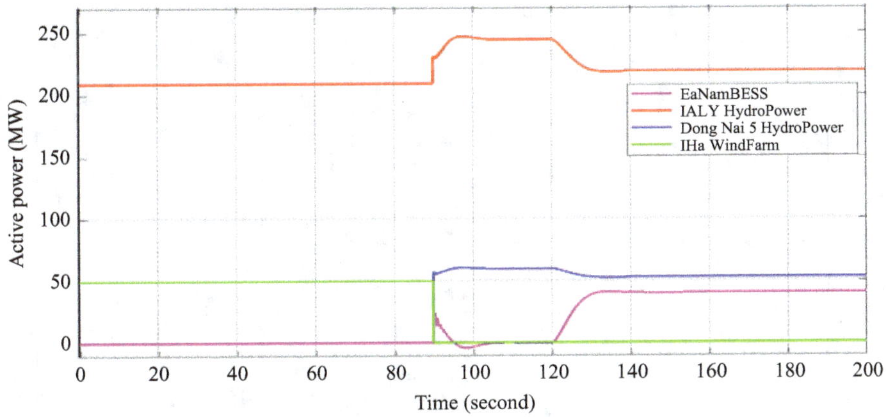

Figure 10.35 Output power of plants with BESS participation

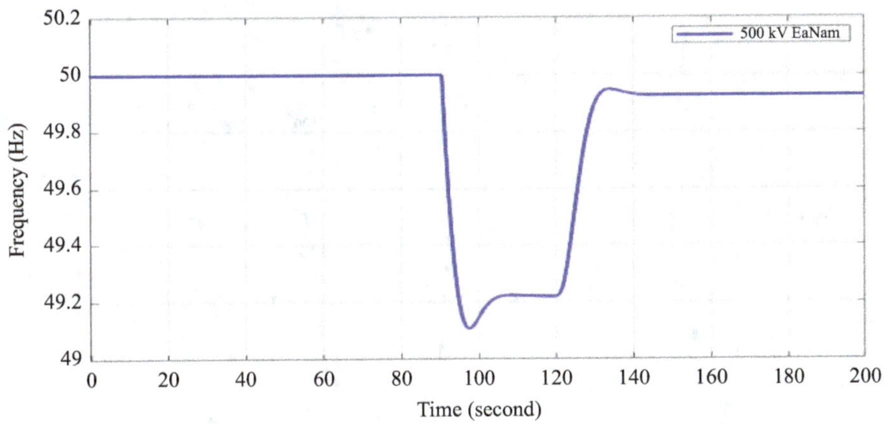

Figure 10.36 The system frequency at the 500 kV EaNam bus with BESS participation

After the system is stabilized, the system frequency remains at 49.93 Hz (Figure 10.36). This value is within the permissible range for grid operation (49.8 Hz < 49.93 Hz < 50.2 Hz).

Figure 10.37 illustrates the Area Control Error (ACE) signal, which includes frequency deviation and the total power deviation across interconnection lines. When the BESS participates in frequency control, it brings the ACE signal to zero, both the frequency and power deviations are eliminated.

As shown in Figure 10.37, at the 90th second, when the wind power plant experiences a failure, causing a change in the total system power generation and a subsequent frequency drop, the BESS does not participate in primary frequency control. As a result, the ACE value gradually returns to zero during the process where hydropower plants execute frequency regulation. At the 120th second, when

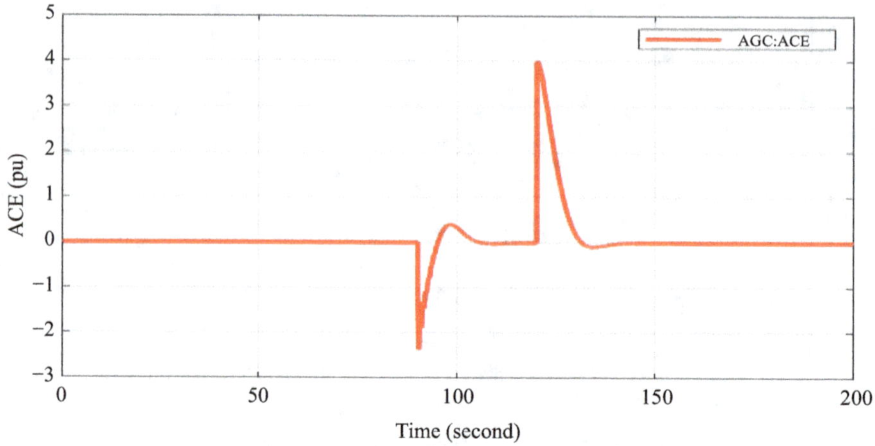

Figure 10.37 ACE signal control system of BESS

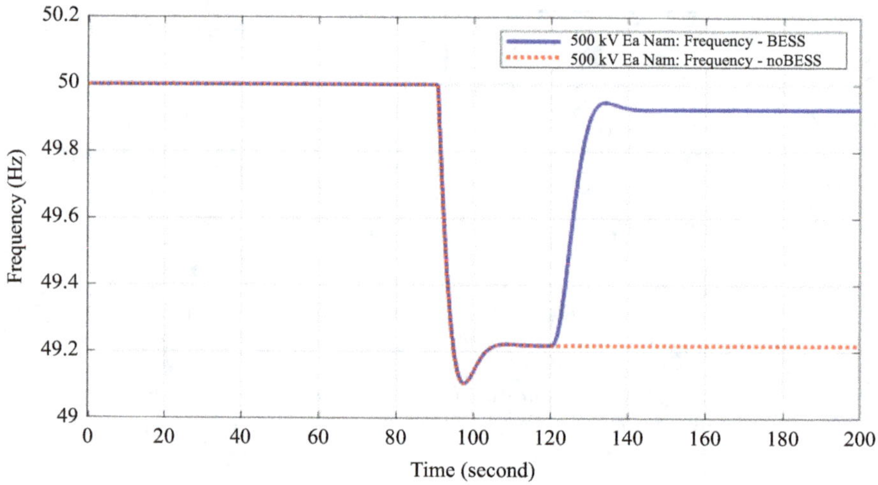

Figure 10.38 The system frequency at the 500 kV EaNam bus with/without BESS participation

the BESS engages in secondary frequency control, the deviation in system frequency from the nominal value causes the ACE signal to change. The BESS adjusts this signal to zero within the next 10 seconds, stabilizing the system frequency.

In summary, the BESS successfully fulfills its role in supporting secondary frequency control and bringing the system frequency back to the operational threshold of the grid (Figure 10.38). As a result, it fully meets the required standards and improves the overall stability of the power grid.

10.5 Conclusions

The study presented an analysis of issues for Vietnamese power system with the penetration of RES. There are several impacts as follows:

- Firstly, with the penetration of RES, the problem of power congestion occurs on the local grid in some provinces in Vietnam (Ninh Thuan, Binh Thuan, Khanh Hoa, Phu Yen, Tay Ninh, Ca Mau and Bac Lieu). Therefore, the power system is not able to fully absorb power from RES.
- Next, the penetration of RES, especially solar, is highly concentrated in some major provinces such as Ninh Thuan, Binh Thuan, Khanh Hoa, Phu Yen. Wide area blackouts can occur in the event of a fault on a full loading line/transformer in these provinces.
- Thirdly, the stability of the power system is reduced if traditional fossil fuel-based power plants are replaced by RES.

Consequently, for smart transmission in the hybrid Vietnamese RE, the solution for improving absorption capacity of the grid to RE as well as power system stability should be considered.

The ANDC algorithm has proven its effectiveness in system control. Additionally, the SOC Feedback model, which uses SOC to estimate the battery's state, controls the BESS charges/discharges as required to ensure that the system's frequency is always ready to meet operational demands. Leveraging the strengths of both methods mentioned above, the ANDC algorithm combined with the SOC Feedback model (New Model) has been proposed and implemented through DIgSILENT Factory 2022 software to build the BESS control model using DSL language. This study demonstrates the excellent support provided by the algorithm in regulating the SOC of the BESS to stabilize the system frequency during frequency fluctuation fault scenarios of power system with RE integration. Furthermore, the newly proposed model shows its superiority (fast response to faults, deep charge/discharge levels, timely frequency control, etc.) over the traditional CBEST model currently in wide use.

In addition, this chapter introduces an advanced BESS control strategy incorporating a charge/discharge cycle, designed to mitigate overload risks in series-compensated capacitors and regulate the power output of RES during transmission congestion scenarios. The proposed control framework enables BESS to act as an effective support mechanism for large-scale solar and wind power plants, facilitating the absorption of surplus energy, alleviating transmission constraints, enhancing interconnection controllability, and suppressing power fluctuations arising from the intermittency of RE generation.

It can be said that the BESS system plays a crucial role in supporting the stability of the electrical grid frequency as the share of RES continues to grow.

Acknowledgments

The former Central Region Load Dispatch Centre (Vietnam), current name: Central Power System Operator - Branch of National Power System and Market Operator

Company (Vietnam), Power Engineering Consulting Joint Stock Company 4 (Vietnam), and Deutsche Gesellschaft für Internationale Zusammenarbeit (GIZ) GmbH (Germany) supported this work technically and financially.

References

[1] Vietnam's Prime Minister, *National Electricity Development Planning Period 2021–2030, Vision to 2050*, Hanoi city, Vietnam, 2023. Available at: https://datafiles.chinhphu.vn/cpp/files/vbpq/2023/5/500-ttg.signed.pdf.

[2] Ministry of Industry and Trade, Circular No 05/VBHN-BCT, *Electricity Transmission System Regulations*, Hanoi city, Vietnam, 2015. Available at: https://thuvienphapluat.vn/van-ban/Thuong-mai/Van-ban-hop-nhat-05-VBHN-BCT-quy-dinh-he-thong-dien-truyen-tai-2015-299959.aspx.

[3] ERCOT, *Inertia Monitoring and Analysis*, Electric Reliability Council of Texas, March 2017.

[4] Phuong N.H.V., Nam N.B., Luat H.V., Hai V.Q., Cuong L.Q. and Phuoc T.D.M. 'Strategy for power control of battery storage system: PhuQuy Island study case', *Journal of Science and Technology – The University of DaNang*, Vol. 19, Iss. 4.2, pp. 12–17, 2021.

[5] Lee S.J., Kim J.H., Kim C.H., Kim S.K., Kim E.S., Kim D.U., Mehmood K.K. and Khan S.U. 'Coordinated control algorithm for distributed battery energy storage systems for mitigating voltage and frequency deviations', *IEEE Transactions on Smart Grid*, Vol. 7, Iss. 3, pp. 1–10, 2016.

[6] Shim J.W., VERBIČ G., Kim H. and Hur K. 'On droop control of energy-constrained battery energy storage systems for grid frequency regulation', *IEEE Access*, Vol. 7, pp. 166353–166364, 2019.

[7] Kaczorowska D., Janik P., Jasinski L., Rezmer J. and Suresh V. 'Power flow control algorithm in a microgrid with energy storage', *Przeglad Elektrotechniczny*, Vol. 1, Iss. 7, pp. 116–119, 2021.

[8] Kang W., Chen M., Li B., Chen F., Lai W., Lin H. and Zhao B. 'Distributed reactive power control and SOC sharing method for battery energy storage system in microgrids', *IEEE Access*, Vol. 7, pp. 60707–60720, 2019.

[9] Zhang Y., Tang M., Qu X. and Chen W. 'An improved adaptive nonlinear droop control method in DC Microgrid', in *8th Renewable Power Generation Conference (RPG)*, October 24–25, 2019, Shanghai, China.

[10] Nourollah S., Pirayesh A. and Hasanpor D.P. 'Stability preservation and power management in autonomous microgrids using adaptive nonlinear droop scheme', *Turkish Journal of Electrical Engineering & Computer Sciences*, Vol. 23, Iss. 06, pp. 1958–1980, 2015.

[11] Abdelgabir H., 'Dynamic modeling of nonlinear droop-controlled Islanded microgrids and optimum placement of renewable energy sources', PhD dissertation, The Graduate Faculty, The University of Akron, Ohio, 2018.

[12] Marin-Hurtado, A. Escobar-Mejía, A. and W. Gil-González, 'Auxiliary droop frequency and voltage control for grid-forming applied to unbalanced current

compensation', *2023 IEEE 14th International Symposium on Power Electronics for Distributed Generation Systems (PEDG)*, Shanghai, China, 2023, pp. 840–845. Doi: 10.1109/PEDG56097.2023.10215150.

[13] Panda, S. K. and Subudhi, B. 'A review on robust and adaptive control schemes for Microgrid', *Journal of Modern Power Systems and Clean Energy*, vol. 11, Iss. 4, pp. 1027–1040, 2023. Doi: 10.35833/MPCE.2021.000817.

[14] Landau L. D., Lozano R., M'Saad M. and Karimi A. *Adaptive Control Algorithms, Analysis and Applications*. London: Springer, 2011.

[15] Tan Z. 'Adaptive energy storage system control for microgrid stability enhancement', Ph.D. dissertation, Dept. Electrical and Computer Eng., Worcester Polytechnic Institute, Worcester, Massachusetts, 2018.

[16] Shim J. W., Verbic G., Kim H. and Hur K. 'On droop control of energy-constrained battery energy storage systems for grid frequency regulation', *IEEE Access*, Vol. 7, pp. 166353–166364, 2019.

[17] Akhani A.K. 'Primary secondary and tertiary frequency control', *Engineering Articles*, October 08, 2015.

[18] Tumino P. 'Frequency control in a power system', October 15, 2020. Available at: https://eepower.com/technical-articles/frequency-control-in-a-power-system/#.

[19] Bhatt G. and Affijulla S. 'Analysis of large-scale PV penetration impact on IEEE 39 - Bus power system', in *58th IEEE International Scientific Conference on Power and Electrical Engineering of Riga Technical University (RTUCON)*, October 12–13, 2017, Riga, Latvia.

[20] Thanh T.V., Viet D.T., Quyen L.C., Dai L.V. and Hieu N.T. 'A solution to enhance operational effectiveness of the renewable energy sources and power system transmission lines in Vietnam', in *6th International Conference on Green Technology and Sustainable Development (GTSD)*, July 29–30, 2022, Nha Trang City, Vietnam.

[21] Luo X, Wang J, Wojcik JD, *et al.* 'Review of voltage and frequency grid code specifications for electrical energy storage applications', *Energies*, Vol. 11, Iss. 05, pp. 1–26, 2018.

[22] Choi W.Y., Kook K.S. and Yu G.R. 'Control strategy of BESS for providing both virtual inertia and primary frequency response in the Korean power system', *Energies* (Basel), Vol. 12, Iss. 21, pp. 1–17, 2019.

[23] *The Successful Stress Test of Europe's Power Grid – More Head, Polley -15 July 2015*. Available at: https://eepublicdownloads.entsoe.eu/clean-documents/Publications/ENTSO-E%20general%20publications/entsoe_spe_pp_solar_eclipse_2015_web.pdf.

Chapter 11
Harmonizing PV- and wind-based Hybrid Renewable Energy System

Gabriela Nicoleta Tanasiev[1], Cristian Gheorghiu[1], Vladimir Tanasiev[1] and Ştefăniţă Pluteanu[1]

11.1 Introduction

Globally, transitioning to renewable energy systems and reducing reliance on fossil fuels have become essential objectives, driven by the urgent need to achieve environmental sustainability and ensure energy security [1]. Solar photovoltaic (PV) and wind power systems are among the most prevalent renewable energy technologies, offering numerous advantages such as scalability, affordability, and significant reductions in greenhouse gas emissions [2].

Nevertheless, PV and wind energy systems encounter intrinsic challenges due to their intermittent nature. Their energy output is greatly influenced by factors like weather patterns, seasonal changes, time of day, geographical conditions, equipment performance, maintenance practices, and the capacity of the electrical grid to efficiently integrate and distribute power [3]. These fluctuations pose considerable difficulties in maintaining steady and reliable energy production. Systems that rely heavily on variable renewable energy sources often struggle to deliver consistent electricity output [4].

A viable approach to mitigate these limitations is the adoption and advancement of Hybrid Renewable Energy Systems (HRES), which combine solar PV and wind power. By utilizing the complementary characteristics of these two sources, hybrid systems can balance the variability of one source with the stability of the other, thereby providing a more consistent and reliable power supply [5].

The integration of PV and wind power in hybrid systems offers several notable advantages, including improved energy security, higher generation capacity, and reduced dependence on conventional fossil fuels [6,7]. These benefits collectively contribute to a greener and more cost-effective energy landscape [8]. Moreover, hybrid systems can be tailored to the specific solar and wind conditions of various regions, ensuring optimal energy utilization while minimizing waste [9].

[1]Faculty of Energy Engineering, National University of Science and Technology, POLITEHNICA Bucharest, Romania

As the energy sector continues to decentralize, PV–wind hybrid systems are poised to play a pivotal role in meeting energy demands at both local and global levels. Their ability to provide stable power output while aligning with sustainability goals makes them an integral part of the future energy infrastructure [10].

This chapter delves into the concept of integrating PV- and wind-based HRES, highlighting their potential benefits, the obstacles they face, and the technological advancements that support their effective deployment [11]. By examining the interactions between solar and wind energy, it becomes evident that hybrid systems are essential for building a more sustainable, efficient, and resilient energy framework.

11.1.1 Overview of HRES

HRES represent an advanced and efficient approach to generating sustainable energy by combining two or more renewable energy sources. These systems often incorporate energy from solar, wind, biomass, or hydro sources, improving system reliability, enhancing efficiency, and maximizing energy production through the complementary characteristics of the resources used. For example, solar power is most effective during sunny daytime hours, whereas wind energy tends to perform better during nighttime or cloudy conditions [12].

The core components of an HRES generally consist of primary energy sources, energy storage solutions, power conversion units, and backup systems [13]. While HRES offer numerous benefits, their widespread and efficient implementation is contingent upon addressing several complex challenges. These challenges span technical, economic, environmental, and regulatory domains, with the following being among the most critical [12–17]:

- *High initial investment costs*: The integration of multiple energy sources, combined with sophisticated control mechanisms and energy storage technologies, often requires significant upfront financial investment.
- *System design complexity*: Effective integration of diverse energy sources requires advanced and customized control and management systems to ensure smooth operation and seamless coordination.
- *Accurate resource assessment*: Conducting a precise evaluation of local renewable energy resources is vital for the successful implementation of HRES. Variations in resource availability, such as fluctuating wind speeds or inconsistent solar irradiance, must be carefully considered to avoid performance issues.
- *Technological constraints*: While energy storage technologies are continuously evolving, they still face challenges related to capacity, cost, and energy density, which may impact the scalability and reliability of hybrid systems.
- *Policy and regulatory barriers*: In certain regions, the absence of supportive policies and regulatory frameworks for hybrid energy systems can limit their adoption and integration into existing power grids.

Due to their adaptable nature, HRES can serve a variety of applications across multiple sectors. Some common uses include:

- *Disaster recovery*: In regions affected by natural disasters, hybrid systems offer a rapid, reliable, and flexible energy solution, ensuring the continuity of critical infrastructure and supporting recovery efforts [14].
- *Urban energy grids*: HRES can mitigate power outages in urban areas by providing backup power, reducing the burden on centralized grids, and ensuring an uninterrupted energy supply [16].
- *Industrial power supply*: Hybrid systems can meet the energy demands of industrial operations, offering a reliable and sustainable energy solution [18].
- *Rural electrification*: In remote or off-grid areas, HRES can provide essential energy where extending the grid is unfeasible. Solar–wind combinations are particularly effective for delivering sustainable power in such settings [19].

This chapter offers an in-depth analysis of HRES, detailing their configurations, operational strategies, and the benefits they provide. Additionally, it explores the technological challenges and opportunities involved in their development, shedding light on their pivotal role in shaping the future of renewable energy systems.

11.1.2 Importance of harmonizing PV and wind in HRES

As global energy demands continue to rise and the urgency to address climate change intensifies, leveraging complementary renewable energy sources has become a critical priority. Among these sources, solar PV and wind energy stand out as two of the most advanced and widely implemented technologies. Their integration within HRES offers significant potential to address the intermittency challenges associated with stand-alone renewable energy systems. By harmonizing PV and wind energy, HRES can achieve greater reliability, efficiency, and sustainability [19,20].

11.1.2.1 Complementary nature of PV and wind energy

Solar PV and wind energy are highly compatible in many regions due to their complementary generation patterns. Solar energy output typically peaks during daylight hours when the sun is shining, while wind energy generation tends to increase at night or during cloudy and windy conditions [12]. This temporal complementarity helps to provide a more stable and dependable energy supply when both sources are combined.

For instance, in coastal areas, wind speeds are often higher during the evening and early morning hours, whereas solar radiation is most abundant throughout the day. By integrating PV and wind energy, hybrid systems can effectively balance fluctuations in energy demand across both day and night cycles [13].

11.1.2.2 Advantages of harmonizing PV and wind in HRES

Combining PV and wind energy reduces the impact of variability from each individual source. When one source generates less power, the other can help maintain a steady supply, minimizing the risk of power shortages [15].

Hybrid systems that integrate PV and wind reduce reliance on fossil fuels, lower maintenance requirements, and decrease the need for additional infrastructure, ultimately offsetting the higher initial investment with significant cost savings over time [12].

Harmonizing PV and wind energy ensures better use of available natural resources, reducing the reliance on extensive Energy Storage Systems (ESS) or backup generators. This optimization improves overall system efficiency and lowers operational costs [19].

In grid-connected systems, the complementary integration of PV and wind energy reduces the frequency and intensity of power fluctuations, ensuring a more consistent and reliable supply of electricity [17].

By providing a more consistent energy output, hybrid PV and wind systems decrease the need for fossil fuel-based backup power sources. This leads to lower greenhouse gas emissions and a significant reduction in the carbon footprint [14].

11.2 Fundamentals of PV systems

In this chapter, an in-depth presentation of PV system fundamentals will be presented, starting from the theoretical principles on which this type of distributed energy source is based and going to the particularities of actually implementing a PV system.

The basic architecture of a PV system relies on using three main components: PV modules, solar inverters, and the PV mounting structure (*it has a significant influence on the overall performance of the system*) and a couple of other ancillary equipment such as DC (Direct Current) and AC (Alternating Current) wiring and grid connection equipment.

11.2.1 PV modules

PV modules or panels represent the actual energy generators of a PV system. The working principle of a PV module is based on the PV principle, discovered by Edmond Becquerel in 1839 – called the "*Becquerel effect.*"

A PV panel is made up of the following sub-components: the aluminum framing that covers all the other components (a), the front sheet (tempered glass) (b), the top Ethylene Vinyl Acetate (EVA) encapsulant (c), the PV cell architecture (d), the bottom EVA encapsulant (e), the Back sheet (*which can be TPT / TPE / PET for one-sided PV panels and tempered glass for bifacial panels*) (f), and the junction box (g), as it can be seen in Figure 11.1.

A PV panel's energy generation capability comes from the multiple PV cells (commonly called solar cells), which are connected in series and in parallel to provide the best possible energy output (with regard to the ratio between the output voltage – *series connected cells* and the output current – *parallel connected cells*) [22] (see Figure 11.2 for details).

The PV cells are typically configured as p-n extended surface junctions made out of crystalline silicon. The semi-conductive resistive contacts are done both on the n-type part and on the p-type part of the PV cell, each cell being connected to the other cells either in series or in parallel.

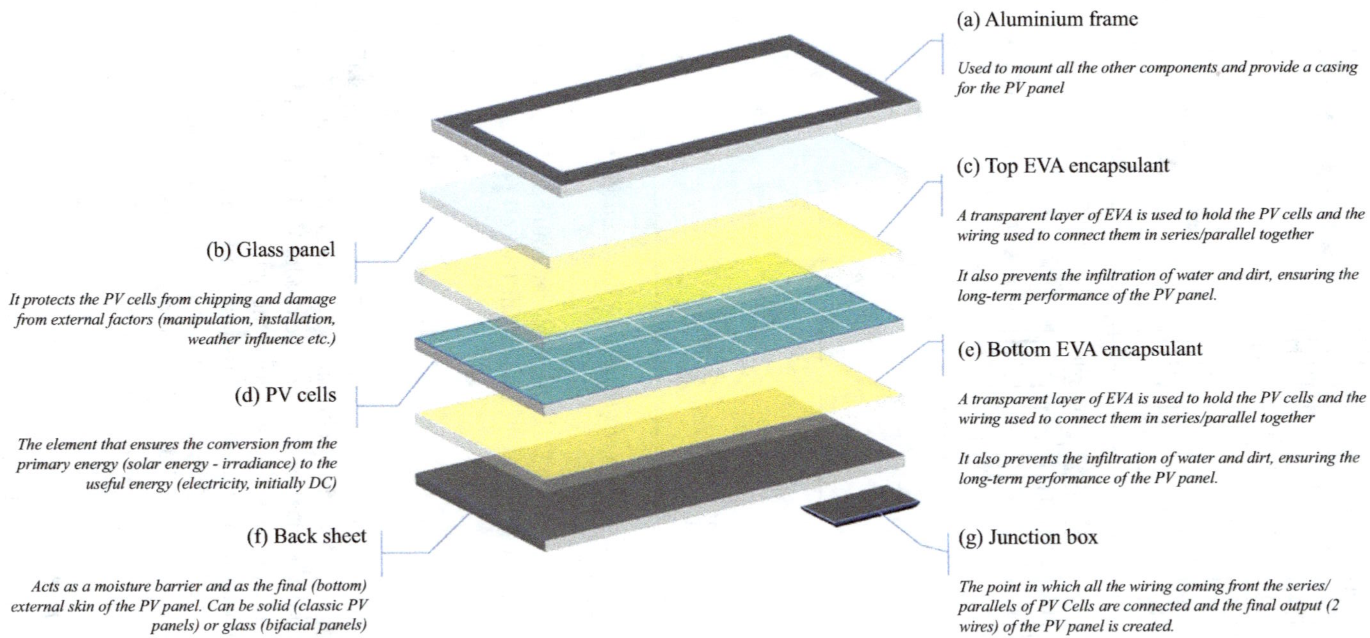

Figure 11.1 PV module architecture, adaptation after [21]

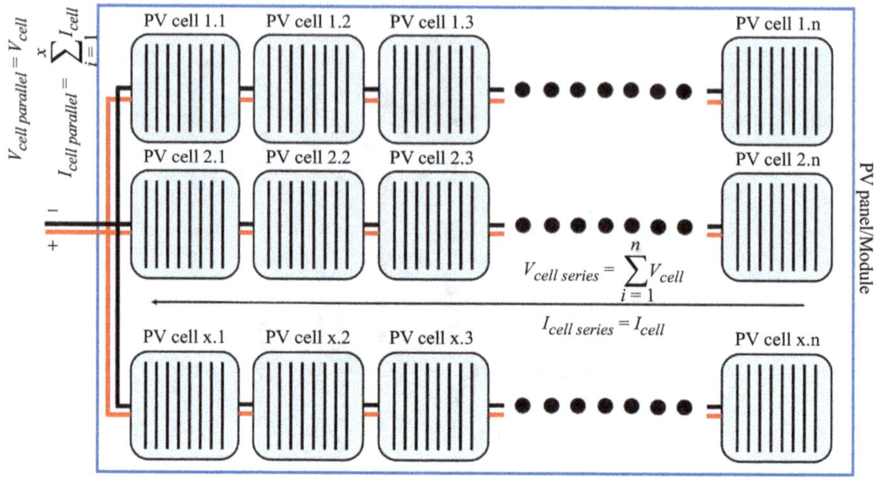

Figure 11.2 PV cells architecture

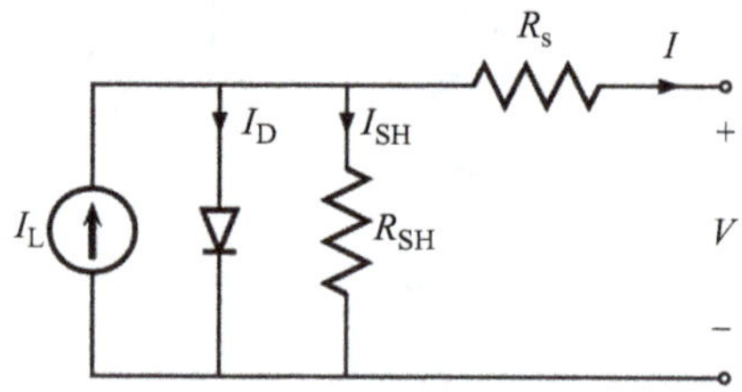

Figure 11.3 Equivalent PV cell diagram

The current generated by the PV cell can be determined by applying the Shockley diode equation and Ohm's Law to the equivalent PV cell (see Figure 11.3) with the following equation:

$$I = I_L - I_D - I_{SH} = I_L - I_S \cdot \left(e^{\frac{V_D}{n \cdot V_T}} - 1\right) - \frac{V + I \cdot R_S}{R_{SH}} \ [A] \tag{11.1}$$

where I is the output current, I_L is the current generator, I_D is the parallel connected diode (ideal case), I_{SH} is the shunt current, R_{SH} is a shunt resistor, V_D is the voltage across the diode, n is the ideality factor of the diode, V_T is the thermal voltage, V is the terminal voltage of the PV cell, and R_S is the series resistive component.

The most relevant technical characteristics of a PV panel can be classified into Electrical Characteristics (A), Mechanical Characteristics (B), and Operational Characteristics (C), as presented in Table 11.1; for exemplification, a 555 Wp PV panel was considered.

11.2.2 Solar inverters

Considering the fact that the PV panel provides DC power, a DC-to-AC conversion equipment – a solar inverter – is needed in order to be able to use the energy generated

Table 11.1 The most relevant technical characteristics of a PV Panel – exemplified for a 555-Wp model

Indicator	Value	Measuring unit	Observations
(A) Electrical characteristics			
Rated Power	555	Wp	*Watt peak – the rated power of the PV panel at STC (25°C, 1,000 W/m^2 solar irradiance and an air mass of 1.5*
Rated efficiency of the PV module	21.7	%	
Maximum power voltage, V_{mp}	42.10	V	*Values corresponding to the rated power*
Maximum power current, I_{mp}	13.19	A	
Open-circuit voltage, V_{OC}	49.95	V	*Values corresponding to the actual operational conditions of the PV panel Voltage increases as temperature lowers while current increases as temperature rises. Very important for sizing the PV strings!*
Voltage variation coefficient, $c_{V,temp}$	−0.265	%/°C	
Short-circuit current, I_{sc}	14.04	A	
Current variation coefficient, $c_{I,temp}$	0.050	%/°C	
(B) Mechanical characteristics			
Dimensions	2,256 × 1,133 × 35	mm	*Important to properly size and choose the mounting structure*
Weight of a PV panel	27.2	kg	*Very important especially for roof-mounted systems and carport systems*
PV cells			
Architecture	144 (6 × 24)	PV cells	*Six strings connected in parallel, each string having 24 cells connected in series*
Type	Half-Cut	–	*More efficient than full cells, but they have a lower rated PV cell temperature (not good for extremely high temperature climates)*
Output wiring dimensions			
Cross section	4	mm^2	*Usual sizing is 4 mm^2 and 6 mm^2*
Length	20	cm	*It usually can be customized, if needed – or extended by using MC4 connectors and additional wiring*
Maximum load on the front side	5,400	Pa	*Snow falling has to be accounted to determine the maximum on-site load*
Maximum load on the back side	2,400	Pa	*Wind speed and mounting angle have to be accounted for in order to determine the maximum on-site load*
Hailstone test	25	mm	*Usually the hailstone tests are done for a hailstone speed of 23 m/s*
(C) Operational characteristics			
Operational temperature range	−40 to + 85	°C	–
Rated PV cell temperature	45	°C	–
First year derating	2	%/year	Power loss due to Light-Induced Degradation (LID)
Rest of life-cycle derating	0.55	%/year	
Minimum loss of capacity over the life cycle	84.80	%/rated value	*Life cycle is considered to be 25 years for a PV Panel*

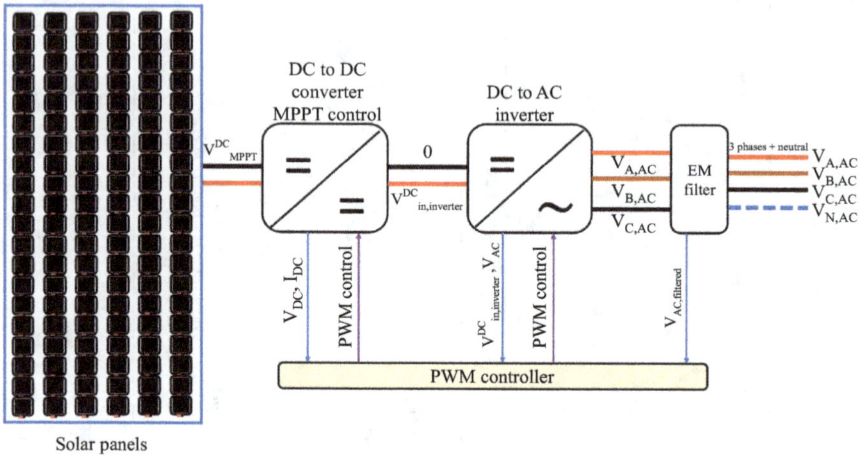

Figure 11.4 Simplified operational diagram of a solar inverter

by a PV system. The solar inverter's operation is based on a set of IGBT gate drivers that convert the initial linear DC signal (output of the PV modules) to a step signal. This step signal is then fed into a layer of filters that provides an almost sinusoidal output signal.

All modern solar inverters are also equipped with an additional DC-to-DC conversion system that acts as a MPPT (*Maximum Power Point Tracking*) Control point which is controlled via a PWM (*Pulse Width Modulation*) controller [23] (see Figure 11.4).

The most relevant technical characteristics of a PV panel can be classified into Electrical Characteristics (A), Mechanical Characteristics (B), and Operational Characteristics (C), as presented in Table 11.2; for exemplification, a 100 kW solar inverter was considered.

11.2.3 Mounting structure

Choosing the mounting structure type when developing a PV system is a crucial step in ensuring the overall performance of the project.

There are two main types of PV system projects, depending on the surface on which the PV Powerplant will be installed: (A) "Ground"-mounted PV systems and (B) Roof-mounted PV systems.

"Ground"-mounted PV systems can be further split into a number of sub-categories, as follows: (A.1) fixed ground-mounted PV systems, (A.2) one-axis tracking ground-mounted PV systems, (A.3) two-axis tracking ground-mounted PV systems, (A.4) carport-mounted PV systems, and (A.5) floating PV systems (see Figure 11.5).

Roof Mounted PV systems can be further split into a number of sub-categories, as follows: (B.1) Terrace (flat) roof mounted PV systems and (B.2) Tiled / Metal (inclined) Roof mounted PV systems (see Figure 11.6).

Table 11.2 The most relevant technical characteristics of a solar Inverter – exemplified for a 100 kW model

Indicator	Value	Measuring unit	Observations
(A) Electrical characteristics			
Output values			
Rated power	100	kW	*The rated power is the AC power provided by the inverter*
Rated European efficiency	98.4	%	*The max efficiency is not relevant in actual project development, as it is based on specific, laboratory-based tests*
Maximum AC apparent power	110	kVA	–
Maximum AC active power	110	kW	*At a power factor of 1, not recommended*
Rated output voltage	400	V	*Variable between 380 V and 480 V*
Rated AC grid frequency	50	Hz	–
Maximum output current	160.4	A	*For a 400 V grid connection*
Adjustable power factor range	0.8 leading to 0.8 lagging	–	*Can be adjusted electronically*
Maximum total harmonic distortion	<3	%	*Used to assess the quality of the filtering stage*
SS	3.5	W	*Important when assessing the needed ancillary power*
Input values			
Maximum input voltage	1,100	V	*Critical to be considered when sizing the PV panel strings*
Maximum current per MPPT	26	A	
Maximum short-circuit current per MPPT	40	A	
Start voltage	200	V	
MPPT Operating voltage range	200 - 1,000	V	
Rated input voltage	600	V	*Variable between 570 V and 720 V, depending on the rated output voltage*
Number of MPP trackers	10	–	*Critical to be considered when sizing the PV panel strings*
Max number of inputs for the inverter	20	–	*Each MPPT has 2 inputs (2 × 10 = 20 total inputs)*
(B) Mechanical characteristics			
Dimensions	1,035 × 700 × 365	mm	–
Weight	90	kg	–
(C) Operational characteristics			
Operational temperature range	−25 to + 60	Degrees Celsius	–
Max operating altitude without derating	4,000	meters	Important to consider when developing PV projects at high altitudes
Relative humidity range	0 to 100	%	Values are corelated
Protection degree	IP66	–	

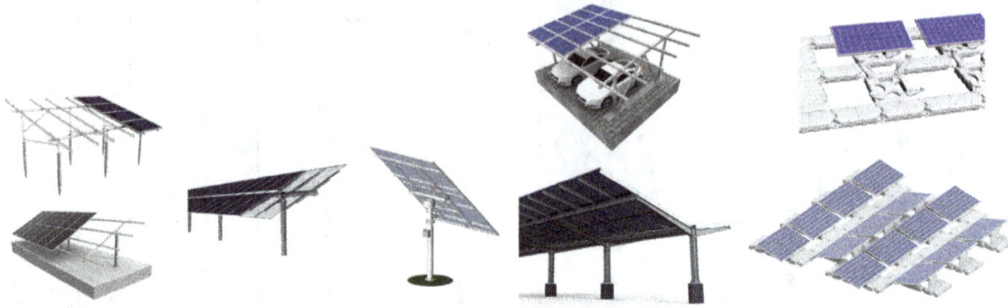

(A1) (A2) (A3) (A4) (A5)

Figure 11.5 Ground-mounted PV structure types

(B1) (B2)

Figure 11.6 Roof-mounted PV structure types

In both categories of fixed mounting structures (ground and flat roof) there are two options regarding the orientation of the Mounting structure (and, implicitly of the PV panels) – South or East-West [24].

There are a couple of factors that have to be taken into account when choosing the orientation of the mounting structure, such as:

- *The available space for installation*: For rooftop PV systems, East-West orientation maximizes the total number of installable panels as the space between the rows of panels is significantly smaller compared to the South-oriented structure (less shading). This is mainly caused by the different incline angles: South-oriented PV systems have an incline between 10° (flat roof mounted) and 35° (ground mounted) whilst East-West-oriented PV systems have an incline between 5° and 10°.
- *The desired utilization of the PV powerplant*: A South-oriented system will produce slightly more energy on a yearly basis but the average daily production curve will be steeper (more energy produced at mid-day) compared to an East-West oriented system, which will generate less yearly energy but the daily production curve will be slightly flatter (see Table 11.3 and Figure 11.7 for exemplification).

Table 11.3 *South-oriented (25° incline, 0° Azimuth) vs. East-West-oriented (10° incline) 120 kWp/100 kW$_{AC}$ PV system installed in Bucharest, Romania*

Month	South-oriented system production [kWh/year]	East-West-oriented system production [kWh/year]
January	4,963.6	3,110.8
February	6,736.9	4,781.6
March	1,057	8,466.5
April	12,226.2	10,995.1
May	13,538.3	13,100.7
June	13,745.6	13,733.1
July	14,909.4	14,636.8
August	14,384.3	13,201.1
September	11,704.7	9,717.2
October	9,134.9	6,585.6
November	5,512.9	3,541.4
December	4,516.2	2,633.3
TOTAL	**121,910**	**104,503.2**

Note: The bolded text represents the total (sum) of the monthly energy produced by the PV Powerplant in the three scenarios (fixed mounting structure vs. 1 axis tracker vs 2 axis tracker) and I've bolded it to emphasis the difference in the yearly expected energy generation and the influence that azimuth / incline angle adjustment has on it.

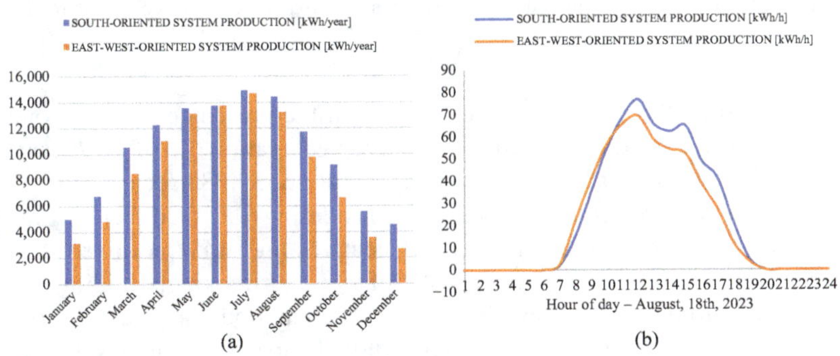

Figure 11.7 *Visual representation of the comparison between a South-oriented (25° incline, 0° Azimuth) and an East-West-oriented (10° incline) 120 kWp/100 kW$_{AC}$ PV system installed in Bucharest, Romania: (a) yearly production; (b) average daily production – simulations done with [25]*

Regarding the differences between Fixed Ground Mounting and Tracker-equipped mounting structures we can note the following:

- *Difference in performance*: The main advantage of one-axis/two-axis tracking structure is that it leads to a maximization of energy generation in most of the cases. This is mainly due to the fact that when equipped with tracking systems, the PV Powerplant will optimize the incidence angle of the solar irradiance, thus maximizing the amount of electricity generated by the PV cells. A comparison

Table 11.4 Results of simulating a Fixed mounting system, a one-axis tracking mounting system and a two-axis tracking mounting system – 120 kWp/100 kW$_{AC}$ PV system installed in Târgovişte, Romania

Month	Fixed mounting structure (kWh/month)	One-axis tracking structure (kWh/month)	Two-axis tracking structure (kWh/month)
January	5,722.17	6,137.73	8,020.89
February	9,269.94	10,276.41	12,865.18
March	12,691.72	14,280.78	16,224.89
April	12,185.31	14,087.95	15,173.02
May	15,612.09	18,843.19	20,279.52
June	16,803.56	20,559.80	22,039.16
July	18,049.33	22,464.09	24,424.11
August	17,203.04	20,701.54	22,821.54
September	15,287.16	17,681.59	20,239.44
October	12,781.33	14,212.58	17,293.36
November	7,533.97	8,103.48	10,433.92
December	58,80.64	6,265.15	8,494.4
TOTAL	**149,020.25**	**173,614.28**	**198,309.45**

Note: The bolded text represents the total (sum) of the monthly energy produced by the PV Powerplant in the three scenarios (fixed mounting structure vs. 1 axis tracker vs 2 axis tracker) and I've bolded it to emphasis the difference in the yearly expected energy generation and the influence that azimuth/incline angle adjustment has on it.

is presented in Table 11.4 and Figure 11.8. In the presented case, the one-axis tracking structure led to an increase in energy generation of 16.50%/year whilst the two-axis tracking structure led to an increase of 33.08%/year.

- *The available space for installation*: Considering the fact that the average required space for installation is 1.1–1.4 hectares for 1 MWp of fixed structure, respectively 2.2–2.4 hectares for 1 MWp of two-axis tracking, deciding upon the type of mounting structure should take into account both the available space for installation and the required / desired installed capacity of the PV Powerplant [26].
- *The difference in cost*: Another significant factor that should be taken into account is the difference in the cost of the mounting structure. If for the fixed mounting structure, the average cost is between 50 EUR/kWp and 60 EUR/kWp, the cost of one-axis tracking mounting structure exceeds 115 EUR/kWp, whilst the cost of two-axis tracking mounting structure is of about 230 EUR/kWp. More details regarding costs will be presented in Section 11.7.

11.2.4 Sizing a PV system

In order to size a PV system first the **destination** must be clear. The PV system can be used on, one hand, as a distributed energy source that provides cheap green electricity for an end-user (self-consumption) and on the other hand as core-business model (production and commercialization of energy).

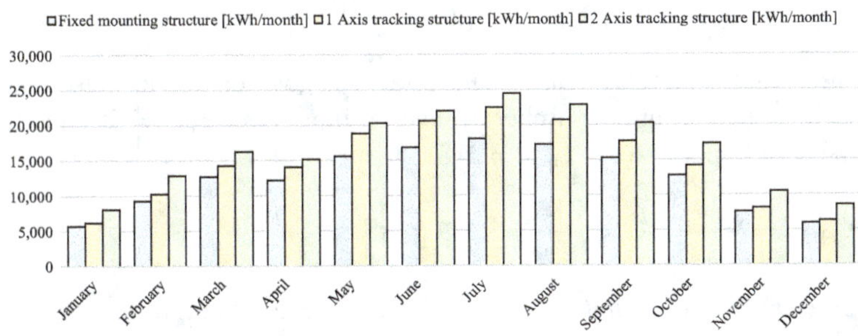

Figure 11.8 Visual representation of the comparison between fixed mounting system, a one-axis tracking mounting system, and a two-axis tracking mounting system – 120 kWp / 100 kW$_{AC}$ PV system installed in Târgovişte, Romania

If the main purpose of the PV system is to supply electricity for an existing end-user – this end-user thus becoming a *prosumer* (active end-user) – then the main considerations should be the hourly energy demand and the total available space for installing the PV system.

Next, the type of mounting structure should be selected, based on the site-specific restrictions (land layout, land surface, roof type, roof incline, etc.).

In order **to conceptualize the size** of the stand-alone PV system, an **iterative method** should be considered, in which, based on the average energy demand during the PV system production time frame (e.g., 08:00–18:00 in Romania) an initial sizing of the system is proposed. This can be done by initially considering the specific productivity of the system at approximately 1,100 kWh/kWp/year (*values are variable depending on the actual position of the system, but for a first iteration it should be sufficiently accurate*).

After determining the **approximative needed installed DC power** (PV panels), the next step is to determine the associated **installed AC power** (inverters).

For this, one should consider that the optimal **DC-to-AC ratio** to prevent oversizing and limit power losses is between **1,2** and **1,5** [27,28]. The reason is that less than 1% of the energy produced by the PV panel array throughout its life (25 years) will be at a power above 80% of the installed capacity (*see details presented in Table 11.1 regarding rated values for power in STC vs. real conditions, respectively rated voltage and current vs. the actual values depending on the outdoor temperature*).

For exemplification, let's say we decided upon installing a PV system with an installed capacity of approximately 120 kWp, by using 555 Wp panels (exemplified in Table 11.1). This means that we will need to use 218 PV panels in order to get a total installed capacity of 120.99 kWp.

Considering a DC-to-AC ratio of 1.2, the installed capacity in AC should be 100 kW – for which we will be using a 100 kW$_{AC}$ solar inverter (exemplified in Table 11.2).

The next step is to configure the strings of PV panels in order to check that the proposed PV panel and DC-to-AC ratio can be used with the selected inverter (in terms of input voltage and input current).

For this we need to determine the following:

- **String size** – PV panels connected in series (s_{size}), which will be determined with (11.2) by considering the number of inverter inputs ($n_{inv,input} = 20$) and the total number of PV panels ($n_{PV\ panels} = 218$);

$$s_{size} = \frac{n_{PV\ panels}}{n_{inv,input}} = \frac{218}{20} = 10.9\ [pcs.] \tag{11.2}$$

Considering that the number is not an integer, it means that the strings will have a different number of PV panels, in this instance: 18 inputs will have 11 panels connected in series (198 panels) and 2 inputs will have 10 panels connected in series (20 panels).

- **Maximum String voltage** (V_{string}^{max}) – in the worst-case scenario, which is at the **minimum outdoor temperature** [29] – *considered to be −13°C for the system presented in Section 11.1.3* – expected on the installation site ($t_{\min}$), determined by applying (11.3):

$$\begin{aligned} V_{string}^{max} &= \left[V_{OC} + c_{V,temp} \cdot (t_{STC} - t_{\min}) \cdot V_{OC}\right] \cdot s_{size,max} \\ &= \left[49.95 + \frac{0.265}{100} \cdot (25 - (-13)) \cdot 49.95\right] \cdot 11 \\ V_{string}^{max} &= 604.78\ [V] \end{aligned} \tag{11.3}$$

where V_{OC} is the open circuit voltage of a single PV module, $c_{V,temp}$ is the temperature coefficient of voltage, t_{STC} is the Standard Test Condition temperature, t_{min} is the minimum outdoor temperature, and $s_{size,max}$ is the maximum number of modules connected in series.

In this case the maximum string voltage is less than the maximum input voltage specified by the inverter manufacturer – 1,100 V so there is no risk of inverter failure during the cold season.

- **Minimum String voltage** (V_{string}^{min}): In the worst-case scenario, which is at the **maximum outdoor temperature** – *considered to be 48°C for the system presented in Section 11.1.3*; values are correlated ($t_{\max}$) and determined by applying (11.4):

$$\begin{aligned} V_{string}^{min} &= \left[V_{OC} - c_{V,temp} \cdot (t_{\max} - t_{STC}) \cdot V_{OC}\right] \cdot s_{size,max} \\ &= \left[49.95 - \frac{0.265}{100} \cdot (48 - 25) \cdot 49.95\right] \cdot 11 \\ V_{string}^{min} &= 515.96\ [V] \end{aligned} \tag{11.4}$$

where $t_{\max}$ is the maximum outdoor temperature.

In this case the minimum string voltage is larger than the minimum (start) input voltage specified by the inverter manufacturer – 200 V so there is no risk of inverter failure during the hot season. However, it can be seen that the minimum string voltage is slightly lower than the rated voltage of the inverter – 600 V – as such, the inverter losses during the hot season will be slightly higher than the reference values (slightly lower efficiency).

- **Maximum input current** ($I_{input,max}$): In the worst-case scenario, which is at the **maximum outdoor temperature**, determined by applying (12.5):

$$\begin{aligned} I_{input,max} &= \left[I_{SC} + c_{I,temp} \cdot (t_{STC} - t_{\min}) \cdot I_{SC}\right] \cdot n_{MPPT,input} \\ &= \left[14.04 + \frac{0.265}{100} \cdot (48 - 25) \cdot 14.04\right] \cdot 2 \\ I_{input,max} &= 29.80\ [A] \end{aligned} \tag{11.5}$$

where I_{SC} is the short-circuit current of a PV module, $c_{I,temp}$ is the temperature coefficient of current, t_{STC} is the Standard Test Condition temperature, t_{max} is the maximum outdoor temperature, and $n_{MPPT,input}$ is the number of parallel module strings connected to the inverter's MPPT input.

In this case, the maximum MPPT short-circuit current is less than the maximum short-circuit current specified by the inverter manufacturer – 40 A, so there is no risk of inverter failure during the hot season.

After configuring the PV system size, the performance of the system can be simulated by using specialized software, such as: PVSyst (*a comprehensive tool for PV system sizing, simulation, and financial analysis*), PVSol (*for projects that use less than 1,000 PV panels*), PV case (*great integration with ACAD suite perfects the layout drawing and automates the electrical diagram preparation*), HOMER (*ideal for hybrid systems, including PV and other energy sources*), HelioScope (*streamlined for layout and performance modelling*), etc.

There are also a couple of free-to-use web browser-based solutions, such as PV GIS SARAH (*European Commission-administrated platform*), RenewableNinja (*good for simulating both PV and wind projects*), etc.

11.3 Fundamentals of wind energy systems

11.3.1 Introduction

Wind energy harnesses the kinetic energy of moving air to generate electricity, making it one of the leading renewable energy technologies today. Its ability to reduce reliance on fossil fuels and lower environmental impacts has positioned it at the forefront of modern sustainable energy strategies [30]. This section delves into the principles, components, and challenges associated with wind energy systems, providing a comprehensive understanding of their functionality and future potential.

11.3.2 Operating principle of wind turbines

Wind turbines operate by converting the kinetic energy of wind into mechanical energy, which is subsequently transformed into electrical energy. This process begins when the wind interacts with the turbine blades, causing them to spin. The rotational mechanical energy is transmitted via a shaft to a generator, where it is converted into electricity [31].

The operation of wind turbines is grounded in several fundamental physical principles. One key principle is the Kinetic Energy Equation, which demonstrates that the power available in the wind is proportional to the cube of the wind velocity [32]. This relationship underscores the importance of selecting locations with optimal wind conditions for efficient energy production, expressed with (11.6):

$$P = \frac{1}{2} \cdot \rho \cdot A \cdot v^3 \tag{11.6}$$

where P represents the power output [W], ρ is the air density [kg/m_3], A is the rotor's swept area [m_2], and v is the wind velocity [m/s] [30].

Another critical concept is Betz's Limit, which establishes the theoretical maximum efficiency of a wind turbine at approximately 59.3% [31, 33]. This limit signifies the highest proportion of wind energy that can be captured and converted into mechanical energy due to the physical constraints of wind energy extraction.

Additionally, the aerodynamic design of turbine blades plays a vital role in optimizing performance. Well-designed blades maximize energy capture while minimizing turbulence and drag, ensuring effective operation [34].

The overall energy output of a wind turbine depends on several factors, including wind speed, turbine size, blade design, and the efficiency of the generator [30].

While higher wind speeds generally result in increased energy production, there are operational wind speed thresholds to ensure the safe and efficient performance of the turbine, illustrated in Figure 11.9. These thresholds include cut-in speed - the minimum wind speed required for the turbine to start generating electricity, rated wind speed - the wind speed at which the turbine achieves its maximum designed power output, cut-out speed - the upper wind speed limit at which the turbine shuts down to prevent damage to its rotor and drivetrain caused by excessive wind forces [32].

By understanding these principles and operational parameters, wind energy systems can be optimized to provide safe, reliable, and efficient energy production.

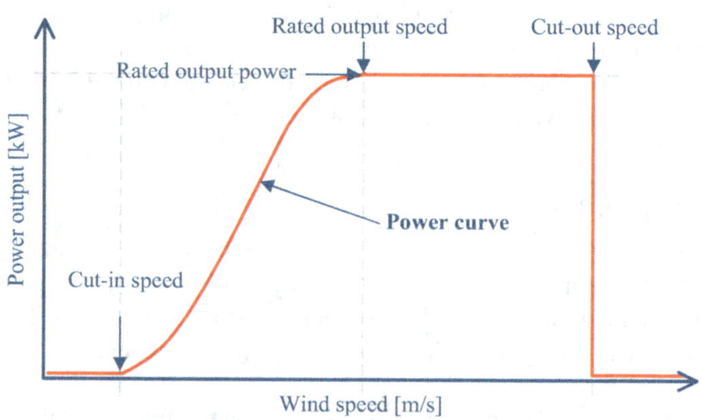

Figure 11.9 Typical power curve for a wind turbine [35]

11.3.3 Types of wind energy systems

Wind energy systems can be categorized based on various criteria, such as their size, location, and operational characteristics. The primary types include [30–32]:

- *Onshore wind turbines:* The most widely used type are installed on land in areas with suitable wind conditions. They range in size from small setups serving individuals or communities to massive utility-scale installations. Compared to offshore systems, they are simpler to construct and maintain.
- *Offshore wind turbines:* These are installed in bodies of water, usually at sea, where wind speeds tend to be stronger and more consistent. They often feature larger turbines and can produce vast amounts of energy, though installation and maintenance are more complex and expensive.
- *Small-scale wind systems:* These are designed for personal or community use, such as residential or farms. They support energy independence and can operate off-grid.

11.3.4 The main components of a wind energy system

A wind energy system relies on several essential components that work together to ensure efficient energy generation illustrated in Figure 11.10 [30–32,36]:

- *Foundation*: This provides structural support for the turbine and is engineered to withstand environmental forces such as wind pressure and seismic activity.
- *Rotor and blades:* These capture the wind's kinetic energy and transform it into rotational motion. The blades are aerodynamically optimized to enhance energy capture.
- *Nacelle*: This contains vital equipment, including the gearbox, generator, and control systems. The nacelle rotates to align with the wind for maximum efficiency.
- *Tower*: This elevates the rotor and nacelle to harness stronger, more stable winds at greater heights (Johnson, 2006).
- *Generator*: This converts the mechanical energy transmitted from the rotor into electrical energy, often using either induction or synchronous generators.
- *Control systems:* These monitor wind speed, blade angles, and overall turbine performance to ensure safety and operational efficiency. They also include braking mechanisms to protect the turbine during high wind conditions.

11.3.5 Factors affecting wind energy system efficiency

Several variables influence the efficiency and energy output of wind turbines [30–32]:

- *Wind speed*: The most critical factor, as energy production increases significantly with higher wind speeds. However, turbines are designed to operate within a defined range, from the cut-in speed (minimum operational speed) to the cut-out speed (upper safety threshold).
- *Turbine design*: Factors such as blade size, tower height, and generator efficiency directly impact a turbine's performance.

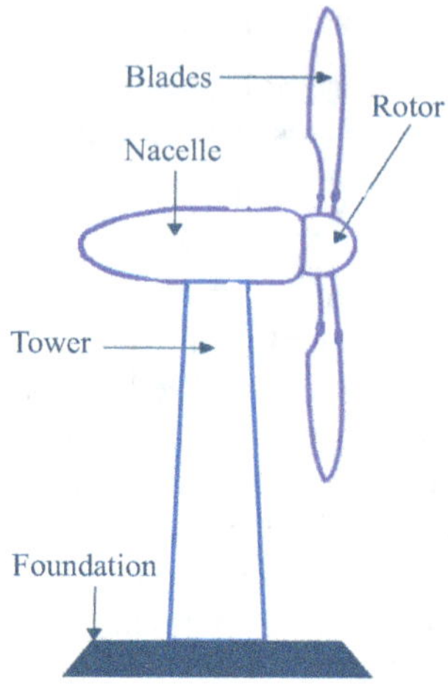

Figure 11.10 Main components of a wind turbine

- *Location*: The effectiveness of wind power plants is largely determined by their placement. Coastal regions, open plains, and mountain passes often provide the ideal conditions for consistent, high-speed winds.
- *Air density:* Denser air, influenced by temperature and altitude, allows for more energy to be extracted from the wind.

11.3.6 Advantages of wind energy

Wind energy offers several significant benefits [30–32,34]:

- *Renewable and sustainable*: As a naturally occurring and inexhaustible resource, wind energy ensures a long-term supply of clean electricity.
- *Environmentally friendly*: Wind turbines produce minimal greenhouse gas emissions, making them a cleaner alternative to fossil fuels.
- *Cost-effective*: While initial installation costs can be high, the minimal operating expenses make wind energy an economically viable option over time, especially in areas with favorable wind conditions.
- *Energy independence*: Reduces reliance on imported fuels, enhancing energy security.
- *Job creation*: The wind energy sector contributes to economic growth by generating employment in manufacturing, installation, and maintenance.

11.3.7 Challenges of wind energy

Despite its advantages, wind energy faces several challenges [30–32]:

- *Intermittency*: Wind availability fluctuates, leading to inconsistencies in energy production. This limitation often necessitates ESS or backup power sources.

- *Environmental impact:* Although wind energy is considered eco-friendly, its infrastructure can affect local wildlife, such as birds and bats, which may collide with turbine blades.
- *Noise and visual concerns*: Nearby residents sometimes report issues with noise generated by turbines and their impact on the landscape's aesthetics.
- *Land use*: Large wind farms require significant land areas, which can compete with agricultural uses or disrupt natural habitats.

11.3.8 Conclusion

Wind energy systems are a cornerstone of the global transition to renewable energy. With ongoing technological advancements and solutions to current challenges, wind energy will continue to play an integral role in ensuring a sustainable energy future. Enhanced turbine designs, improved grid integration, and the expansion of offshore wind capacities will solidify wind energy as a key contributor to a cleaner and more sustainable energy landscape [30,34].

11.4 HRES

11.4.1 Definition and concept of HRES

The HRES concept can be traced back to increased interest in energy efficiency and the use of renewable energy sources, which was an attempt to reduce dependence on fossil fuels after the 1973 oil crisis. With the evolution of digital technology and the ease of integration of various solutions in terms of energy and data, HRES quickly became a viable alternative to classic power generation solutions based on fossil fuels.

An HRES is a concept that combines two or more renewable energy sources, often complemented by an energy storage solution, to provide reliable and sustainable green energy while minimizing environmental impact.

The most used renewable energy sources in hybrid systems can vary depending on geographical location, overall technology efficiency, and costs. Nonetheless, the scientific literature includes a high number of simulations and case studies based on renewable energy sources such as solar, wind, biomass, and biofuels and less so on geothermal, hydrogen, and tidal/wave.

The main issues that HRES addresses are the intermittency of individual renewable energy sources and the unavailability of a conventional electrical grid in remote areas [37].

In terms of HRES applicability, with most of the examples found in Asia, it is mostly used in the residential sector [38]. The dominance of the residential sector highlights regional policies promoting renewable energy sources and the need for autonomous energy solutions and economic benefits, as well as scalability limitations.

When designing an HRES, economic viability plays a pivotal role in justifying initial investment and operational costs. In terms of economic assessment, the following key factors could influence the feasibility of such a solution:

- *Initial setup costs*: Acquiring the main components (solar panels, wind turbines), installation, and configuration costs.

- *Operational costs*: Maintenance of the equipment, monitoring and control, Personnel costs for regular inspection and repair, local taxes and regulations.
- *Energy prices*: With high energy prices, the economic advantages of HRES could be factored as cost per kWh in the long term.
- *The lifetime of the system*: Calculated for each piece of equipment/component (e.g., 20-25 years for solar panels, 10–15 years for inverters, and 10–15 years for Li-ion batteries with 5000-7000 cycles);
- *Access to subsidies and support policies*: Government policies, tax deductions, and subsidies will increase the economic viability of HRES.

An important aspect in an economic evaluation of HRES is the Return of Investment (ROI), which measures the profitability of an investment with respect to the initial cost of investment. For HRES, ROI typically ranges between 7 and 15 years, depending on the region [39]. In addition, to increase the economic viability of the HRES, a thorough cost optimization analysis must be considered.

Depending on the level of flexibility or initial scope, HRES can be designed as grid-connected or off-grid. Choosing the right solution is based on the analysis, which includes the following key factors: available local resources, budget, energy goals, stability of energy supply, operational costs, and ROI. Table 11.5 provides a comparative analysis for grid-connected and off-grid HRES.

When deploying an HRES, the technological challenges of integrating with existing infrastructure must be considered. Energy inputs and outputs must be adapted so that the existing infrastructure can be integrated into a diverse source of energy. In addition, from a logistical point of view, proper planning is essential for cost optimization and efficiency in terms of resource allocation.

11.4.2 Types of HRES configurations

HRES configurations, in most cases, rely on solar and wind renewable energy sources, either with or without a grid connection. Nonetheless, one of the key issues, from the practical perspective, is the integration of the existing infrastructure with HRES and capacity constraints. The HRES configurations are either based on the DC coupling or on the AC coupling, as shown in Figure 11.11.

Table 11.5 Comparative analysis of grid-connected and off-grid HRES

Feature	Grid-connected HRES	Off-grid HRES
Grid dependence	Yes	No
Energy storage	Optional, low	Mandatory
Initial investment	Moderate	High
Reliability	High	Moderate
Ideal applications	Commercial/Urban areas	Rural/Isolated areas or temporary applications

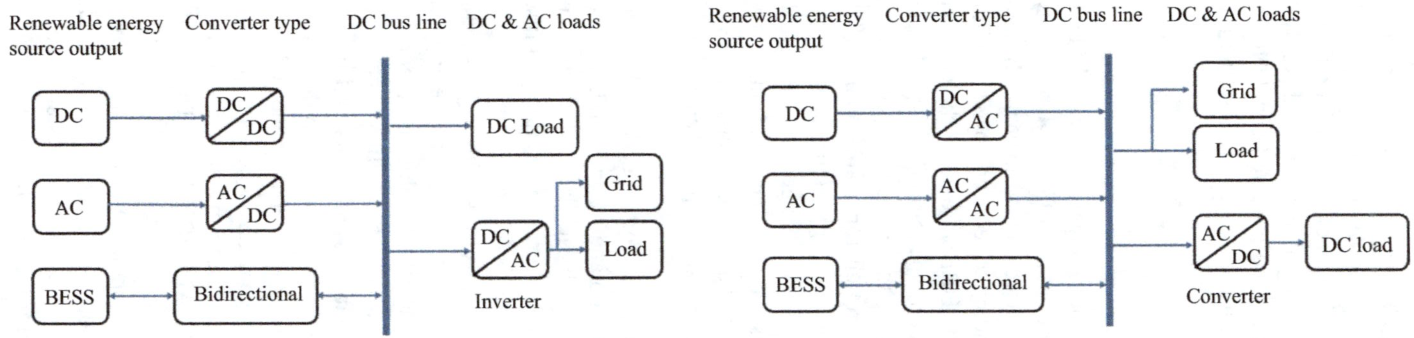

Figure 11.11 General (a) DC and (b) AC coupling of HRES

In general, DC coupling solutions are commonly used in off-grid solutions and hybrid systems that have battery storage, while AC coupling is more suitable for on-grid HRES. Some of the well-known disadvantages of DC coupling are that it is less flexible for AC loads, requiring an extra converter, limited to MPPT operational values, increased wiring, and less ideal in retrofitting AC systems. In the case of AC coupling, the main disadvantages are lower efficiency for battery charging, higher cost as an off-grid solution and it requires a grid-forming inverter when used in off-grid solutions.

11.4.2.1 On-grid HRES

Grid-connected HRES has the advantage of increased flexibility by using the grid network to cover peak consumptions and offers high reliability and economic benefits by selling the surplus of energy based on feed-in tariffs. In addition, the solution could use smaller energy storage, leading to reduced costs for initial investment. The main disadvantages of grid-connected HRES are network dependence, connectivity costs, regulations, and tariffs practiced in the region.

In an on-grid configuration, the energy supply is achieved both through the national electricity grid and through renewable sources. This solution allows the excess or insufficient production of energy from renewable energy sources to be managed through a bidirectional power exchange with the grid. The solution has proven its viability in residential, industrial, and public buildings. The biggest advantages of the on-grid systems are the optimal use of renewable energy sources and flexibility in power supply, reduced electricity costs, increased energy security, and environmental benefits.

From the practical perspective, for a successful implementation, an on-grid HRES solution requires thorough planning and compliance with regulatory requirements. For efficient integration with the grid and existing infrastructure, an HRES solution must consider grid stability, grid connection permits, metering policies, and current and future consumption needs of the customer.

Respectively, selecting solar inverters with a grid-tied connection and MPPT within operational range is important for optimized energy efficiency. Proper sizing is another important aspect of the HRES solution. In conclusion, an on-grid HRES must be designed considering a balanced solution between energy consumption, production, grid integration, reliability, and cost viability.

As an example, for residential applications, in the case of AC coupling, we can consider a solution composed of solar panels connected to an inverter. The inverter converts the DC voltage into AC voltage. Some solar inverters have two independent MPPT inputs, one AC output connection, and communication ports for monitoring and management (LAN, WLAN, RS485, and Digital I/Os). Additionally, some inverters have battery and sensor connections. For safety reasons, to prevent the inverters from supplying power when the grid is down, AC-coupled inverters shut down if the grid connection is lost due to anti-islanding mechanism protection.

In an HRES configuration, an electric generator can be integrated. The electric generator can be used in scenarios when electricity prices fluctuate throughout the day, thus avoiding peak grid tariffs; where grid connection is weak or unstable; to provide backup for critical loads such as hospitals, data centers, or factories; and during power outages. In cases where the grid has feed-in restrictions, a dump load may be required to manage the excess power.

11.4.2.2 Off-grid HRES

Off-grid HRES offers energy independence, a constant energy source, energy storage, and design flexibility. In addition, being a solution that uses exclusively renewable energy sources has a low environmental impact. However, the solution has some disadvantages, an off-grid HRES has higher operational costs and investment costs, high-complexity design, and moderate reliability.

Wind turbines and solar panels in an HRES solution are used to complement each other. In many cases, wind turbines generate energy during cloudy weather or during nighttime, while solar panels are used during the day when solar energy is available.

It typically combines at least two renewable energy sources, an energy storage and, in some cases, an electric generator to ensure constant power supply to the customers. Off-grid HRES solutions are designed primarily for remote areas without a grid connection or where extending the grid is not economically feasible. In contrast with on-grid solutions, off-grid HRES must produce, store, and manage electricity 24/7. Use cases for off-grid HRES include remote homes, industrial facilities, agricultural applications, remote emergency hospitals, remote telecom, and military operations.

As an example, in the case of a solar panel application, a typical system for producing energy consists of solar panels, multiple string MPPTs, SPD protection, a modular DC bus bar, battery packs, battery balancers, sensors, and one or more inverters, depending on the DC power (sum of the rated power of the PV panels).

In an application where a three-phase system is needed, at least three inverters are typically installed. These inverters share a common DC bus bar to charge the battery packs and supply AC loads. An optional fourth inverter may be included to optimize consumption and balance the phases. The three inverters synchronize using a master-slave configuration, where the master dictates the phase reference for the other two slaves.

In a single-phase solution, only one inverter is needed to handle AC appliances and charge the battery pack for later use. Some inverters have two AC outputs, one for critical loads and one for auxiliary loads, which can be programmed based on energy availability. Additionally, inverters include normally open and normally closed switches for power generators, as well as communication ports for remote management, serial ports for sensors (voltage, current), and an analog port for a battery temperature sensor.

An important aspect to consider when selecting an inverter is the configuration of solar panel connections, namely whether they are wired in series or parallel, to match the MPPT's operational range.

11.4.3 Advantages of hybrid systems over single source systems

Developing HRES projects should be a top priority whenever it is possible, because of the complementarity of the various renewable energy sources.

However, when choosing between an HRES and a Single Source Renewable Energy System (SSRES), one should first analyze the Renewable Energy Source Potential (RESP) of the project site, as it is not always technically feasible to develop a PV system and a wind-based system on the same site due to the difference in the primary energy source potential – for EU exemplification see Figure 11.12.

If, however, the project is located on a site that has a decent potential for an alternative energy source (e.g., primary PV + additional Wind), developing an HRES project instead of an SSRES one has the potential of generating significant advantages, both to the project's overall feasibility and to the regional power-grid.

A multi-criterial analysis was conducted and will be presented (see Table 11.6) between a stand-alone fixed mounted PV system of 60 MWp/45 MW_{AC} with a perfect South orientation and an incline angle of 25°; a stand-alone Wind Powerplant with an installed capacity of 60 MW and a hybrid system of 30 MWp / 25 MW_{AC} PV + 30 MW WP.

A number of hypotheses were considered for running the multi-criterial analysis. as follows:

- Specific investment cost – CAPEX (Capital Expenditures):
  - PV Powerplant. capacity larger than 10 MWp: 450.000 EUR/MWp [43];
  - Wind Powerplant. capacity larger than 10 MW: 1.300.000 EUR/MW [44];
- Specific operation and maintenance costs – OPEX (Operational Expenditures):
  - PV Powerplant. capacity larger than 10 MWp: 9.000 EUR/MWp/year [43];
  - Wind Powerplant. capacity larger than 10 MW: 14.000 EUR/MW/year + 0.2 EUR/MWh [44];
- Energy market structure:
  - Off-peak hours (applicable for Romania, used as an example implementation location): 11:00–15:00 [45];
  - Peak hours (applicable for Romania, used as an example implementation location): 19:00–21:00 [45];

For a clearer visualization, the yearly forecasted energy production for each type of project is presented in Figure 11.13.

From a purely multi-criterial analysis point of view, the HRES project is slightly less performant than the SSRES projects, with a total weighted score of 67.17 points.

However, when analyzing the financial performance of each project, whilst taking the Energy Market conditions into account (see Figure 11.14), it is clearly seen that the HRES will lead to an improved yearly income (YI) of 12.79 Mil. EUR/year, compared to the PV SSRES (5.28 Mil. EUR/year), but slightly lower than the WP SSRES (17.97 Mil. EUR/year).

The Simple Payback Period (SPP) of each project was determined with (11.7):

$$SPP = \frac{(CAPEX)}{YI}[years] \tag{11.7}$$

The following values were obtained:

- 60 MWp / 45 MW_{AC} PV Power Plant:

$$SPP = \frac{(CAPEX)}{YI} = \frac{27.00}{5.28} = 5.11\ [years] \tag{11.8}$$

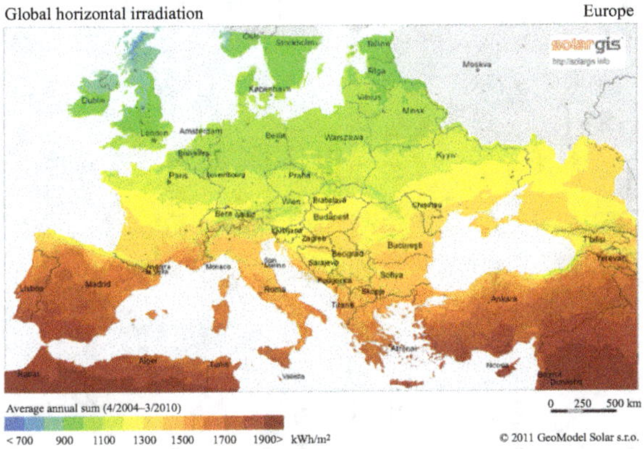

(a)

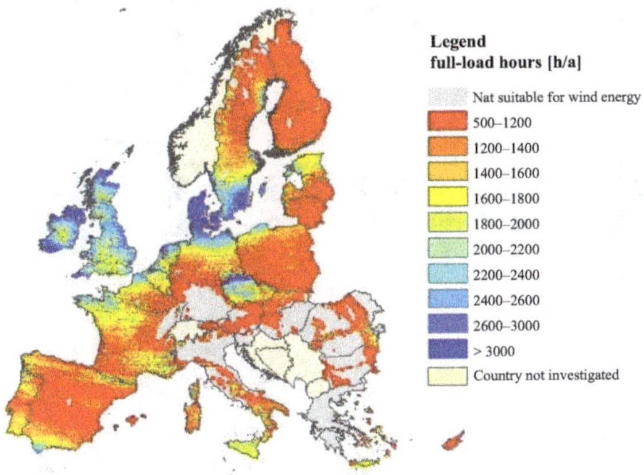

(b)

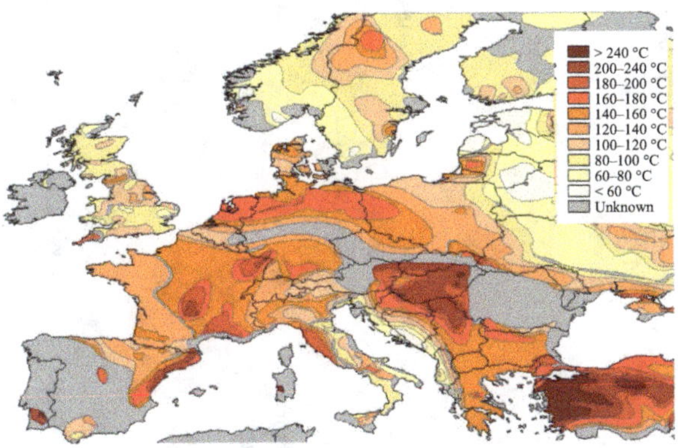

(c)

Figure 11.12 The Solar potential map [40] (a), the wind potential map [41] (b), and the geothermal potential map [42] (c) for the EU

Table 11.6 Multi-criterial analysis between an HRES and two individual SSRES projects

Selection variable	Unit of measure	Selection mode	60 MWp/45 MWAC PV	60 MW WP	30 MWp/25 MW_{AC} PV + 30 WP
Installed capacity	MW	Highest value	45.00	60.00	55.00
	Points		**75.00**	**100.00**	**91.67**
Yearly energy production	MWh/year	Highest value	121,251.05	170,505.36	71,996.75
	Points		**71.11**	**100.00**	**42.23**
Energy injected into the grid at peak-demand	MWh/year	Highest value	481.65	26,937.23	13,709.44
	Points		**1.79**	**100.00**	**50.89**
Energy injected into the grid at off-peak demand	MWh/year	Lowest value	34,351.12	25,937.72	30,144.42
	Points		**75.51**	**100.00**	**87.75**
Environmental impact reduction potential	tons of CO_2 equivalent/year	Highest value	15,398.88	21,654.18	9,143.59
	Points		**7111**	**100.00**	**42.23**
Average forecast error (*leading to Balancing market costs*)	%	Lowest value	16.00	10.00	9.00
	Points		**62.50**	**56.25**	**100.00**
Losses in power grids as a result of high-load operation [46]	%	–		3%	
Energy loss reduction in the power grid	MWh/year	Lowest value	14.45	808.12	412.28
	Points		**100.00**	**1.79**	**50.89**
CAPEX	EUR	Lowest value	27,000,000.00	78,000,000.00	52,500,000.00
	Points		**100.00**	**34.62**	**67.31**
OPEX	EUR/year	Lowest value	540,000.00	874,101.07	707,050.54
	Points		**100.00**	**61.78**	**80.89**
WEIGHTED TOTAL SCORE			**70.34**	**80.62**	**67.17**

Note: The bolded text represents points awarded for each criterion for each scenario, respectively the total points obtained by each scenario. I've bolded this text in order to make it easier to read (and follow the category/points)

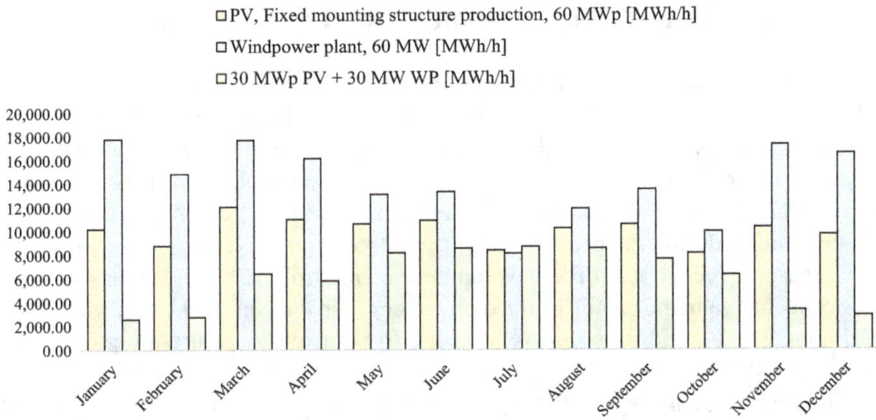

Figure 11.13 A comparison between the hourly (a) and aggregated yearly (b) energy generation of the PV and wind systems

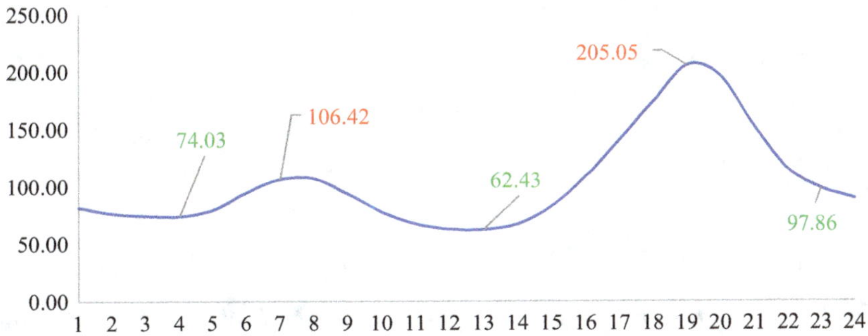

Figure 11.14 Average of day-ahead market closing prices, 2024 [EUR/MWh] [46]

- 60 MW Wind Power Plant:

$$SPP = \frac{(CAPEX)}{YI} = \frac{78.00}{17.97} = 4.34\ [years] \tag{11.9}$$

- Hybrid PV (30 MWp / 25 MW_{AC}) + WP (30 MW):

$$SPP = \frac{(CAPEX)}{YI} = \frac{52.50}{12.79} = 4.10\ [years]. \tag{11.10}$$

It can thus be concluded that the HRES project will lead to a slightly better financial profitability when compared to stand-alone projects.

The feasibility of the project can be significantly improved if the HRES is also coupled with a large-scale Battery Energy Storage Systems (BESS) that can enhance the market participation of the project, by doing peak-production shaving and thus decreasing the off-peak demand grid injection and increasing the peak demand grid injection.

11.5 Synergies and challenges in PV and wind integration

As shown in Section 11.4.3, developing hybrid PV and wind generation projects can lead to better financial performance when compared to classic, stand-alone projects. The main advantage comes from the complementary nature of the two types of energy generation technologies. PV has a bell curve generation profile, with the peak production at 12:00–14:00, whereas wind generation has a flatter generation profile, with peak production during the evening / nighttime (see Figure 11.15).

Because in recent years the development of multiple PV projects has led to a significant change in the profile of the demand curve in the National Grids worldwide (see Figure 11.16), additional measures need to be taken to ensure the proper grid stability [47,48].

One such measure is to diversify the RES power generation source in order to mitigate the primary energy source variability [50]. This can be done by developing

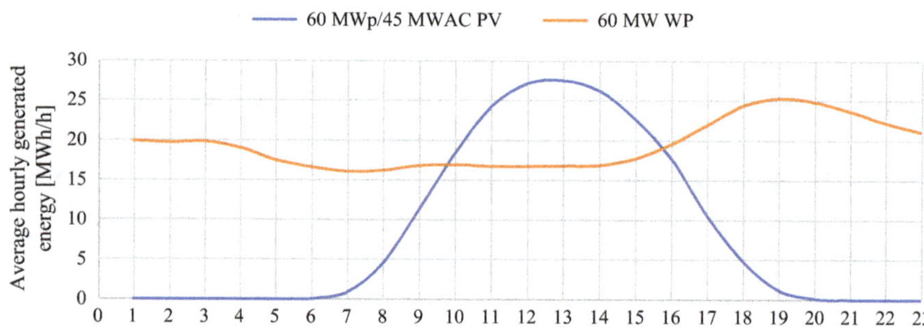

Figure 11.15 Average hourly generated energy for a 60 MWp/45 MW_{AC} PV Power Plant (blue) and a 60 MW Wind Power Plant [see Section 11.4.3 for reference]

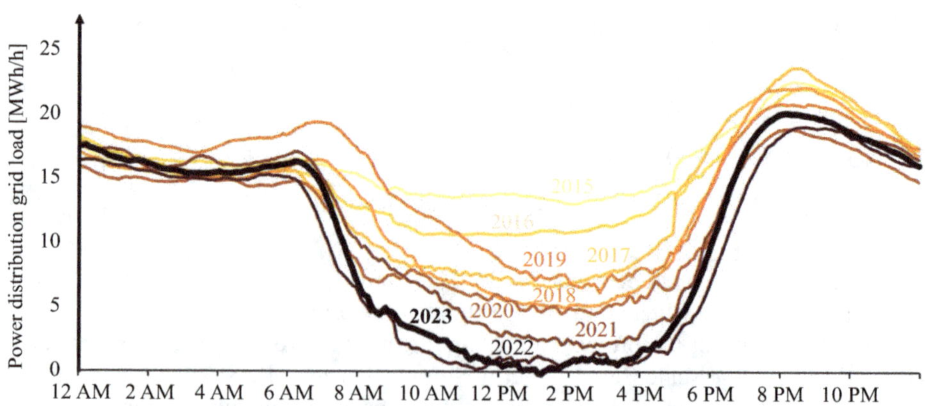

Figure 11.16 Example of the impact stand-alone PV development generates on the power-demand curve of California, USA [49]

Hybrid PV + WP projects and can further be enhanced by coupling them with large-scale ESS (that can take up additional variability in both the primary energy source and in power demand from the grid).

11.6 Challenges in harmonizing PV and wind in HRES

Solutions that use solar panels and wind turbines in an HRES must overcome several challenges related to the intermittency of renewable energy sources, grid stability and synchronization, energy storage, proper system sizing, and demand matching. Both on-grid and off-grid configurations can integrate solar panels and wind turbine mixes.

In this type of solution solar panels are used to generate energy during the day, while wind turbines could be used to generate energy during the night or cloudy days. Before installing renewable energy sources, a solar and wind map for the desired location should be consulted to determine whether the area provides sufficient wind speed and sunlight to make the project viable, both in terms of economic benefits and other objectives.

Synchronization with the local grid is also crucial for choosing the right inverter. The synchronization with the grid may vary depending on the technology used by each inverter and applicable standards. The well-known industry standards that address the synchronization with the grid, stability, and performance of the invertors in a local grid are IEEE 1547, IEC 62116, IEC 61727, IEC 62477-1, IEC 61683, IEC 62910, and EN 50549-1,2. In the technical sheets are listed reference values for Grid Synchronization Time – typically between 50 and 300 millisecond, Grid Connection Time – typically between 1 and 5 seconds, Fault Ride-Through/Recovery Time, typically between 100 and 300 milliseconds, Phase Locked Loop – between 50 and 300 milliseconds, Phase Tolerance or Phase Error Margin – between ±1° and ±5°.

In a DC-coupled solution, which is more common in off-grid solutions, solar and wind energy is used to charge the batteries before converting to AC. From the practical perspective, solar panels are connected to an MPPT that outputs a stable voltage to the DC bus bar.

In the same way, wind turbines are connected to a wind controller which outputs a stable DC voltage to DC bus bar.

11.7 Strategies for addressing integration challenges

Integrating HRES powered by PV and wind energy into power grids presents unique challenges due to the intermittent and unpredictable nature of these sources. To ensure a consistent and reliable energy supply, various strategies have been developed to address these issues. These approaches focus on improving grid stability, optimizing energy storage, enhancing predictive capabilities, and deploying advanced control systems [3,4].

11.7.1 Advanced energy storage solutions

ESS are pivotal in managing the variability of PV and wind energy generation. By storing surplus energy during periods of high production and releasing it during low production or peak demand, ESS ensures stable energy availability. Commonly used technologies include lithium-ion batteries, flow batteries, and pumped hydro storage. Hybrid storage solutions that combine different technologies enhance performance by balancing cost, capacity, and response time [4]. Additionally, smart management systems for energy storage optimize charge and discharge cycles, extending the lifespan of storage units and increasing their reliability [3].

11.7.2 Enhanced forecasting and predictive analytics

Accurate prediction of solar and wind energy output is essential for optimizing HRES. Advances in machine learning and data analytics have made it possible to model weather patterns and energy production with greater precision [50]. By leveraging historical data, real-time monitoring, and meteorological inputs, these predictive models provide valuable insights for grid operators. Enhanced forecasting minimizes the need for backup energy sources and reduces the risks of grid instability caused by abrupt energy production fluctuation [3].

11.7.3 Grid flexibility and demand-side management

Adapting power grids to accommodate the variability of HRES requires increased flexibility. Demand-side management (DSM) strategies encourage consumers to modify their energy usage to align with supply conditions. Techniques such as time-of-use (TOU) pricing, demand response programs, and integration of smart appliances enable more efficient use of renewable energy [51]. Additionally, flexible grid infrastructure, including dynamic line rating and automated reconfiguration, supports operational stability and system efficiency [4].

11.7.4 Hybrid system optimization and control

Effective control strategies are essential for maximizing the performance of hybrid systems that combine PV and wind energy. Technologies such as multi-input converters, hybrid inverters, and energy management systems (EMS) ensure optimal utilization of renewable energy while coordinating seamlessly with auxiliary power sources [52]. Adaptive algorithms and real-time monitoring allow these systems to dynamically respond to environmental changes and load variations, improving energy efficiency and reducing losses [53].

11.7.5 Policy and regulatory support

Supportive policies and regulations are critical to overcoming the challenges of integrating HRES. Governments and energy authorities can promote adoption through financial incentives, standardized interconnection protocols, and investments in research and development [54]. Policies focused on grid modernization and the deployment of smart technologies create an environment conducive to the effective integration of renewable energy sources [4].

11.7.6 Community-based microgrids

Community-based microgrids offer a localized solution to integration challenges by connecting PV and wind installations to decentralized networks. These systems enhance energy independence and resilience for communities [9]. When combined with energy storage and advanced control technologies, microgrids can transition to islanded mode during grid disruptions, ensuring uninterrupted power supply for essential services [54].

11.7.7 Hybrid system configuration and sizing

Proper system design and configuration are vital for minimizing integration challenges. By analyzing energy demand, resource availability, and load characteristics, components can be optimally sized and configured. Simulation tools and optimization algorithms are instrumental in determining the ideal combination of PV, wind, and energy storage technologies, maximizing efficiency while keeping costs in check [52].

Addressing the integration challenges of PV and wind-based HRES requires a multifaceted approach. By combining advanced technologies, enhancing grid flexibility, fostering supportive policies, and encouraging localized solutions like microgrids, these systems can achieve their full potential. This holistic strategy ensures a reliable, efficient, and sustainable energy future [4,51].

11.8 System design and optimization

This chapter presents the foundational principles and optimization strategies required for the effective design of HRES. By integrating PV arrays and wind farms, HRES can achieve a harmonious balance between reliability and sustainability while addressing the inherent variability of renewable energy sources. The chapter explores two critical aspects: the electrical collector system designs for wind farms and the integration of essential HRES components.

Figures 11.17 and 11.18 provide visual representations of the discussed concepts, ensuring clarity and deeper understanding for a PV design and a wind farm design.

As shown in Figure 11.17, the PV array design prioritizes modularity and adaptability. Each PV module represents an array of interconnected solar panels, designed to maximize energy capture under varying sunlight conditions. The figure emphasizes the importance of inverters in converting the DC power generated by the PV system into AC power for load compatibility and grid integration.

Figure 11.18 highlights a standard wind farm layout that illustrates the energy flow through an electrical collector system, which transmits electricity from individual turbines to a substation. The choice of collector system design – radial, single-sided ring, double-sided ring, or star configuration – significantly influences the performance, reliability, and cost-effectiveness of the wind farm.

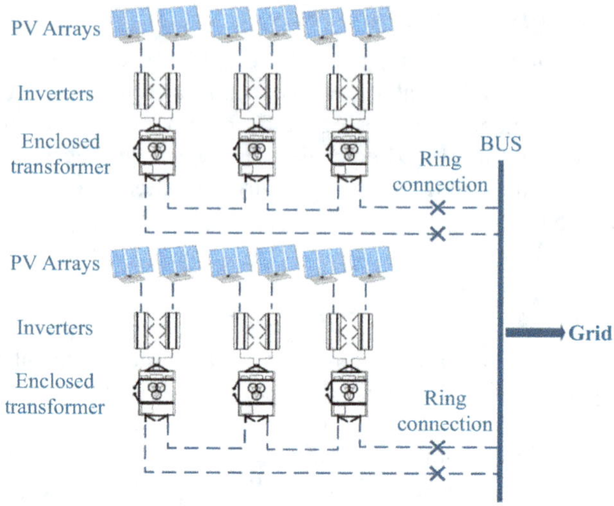

Figure 11.17 PV design

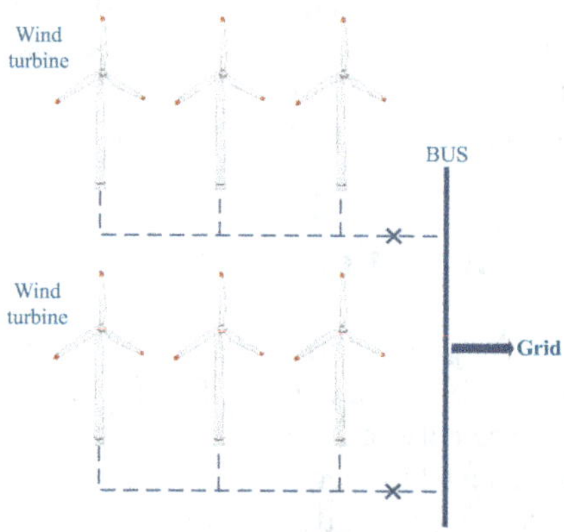

Figure 11.18 Wind farm design

11.8.1 Electrical collector system designs for wind farms

The electrical collector system is a critical component of wind farm design, as it facilitates the efficient transmission of power generated by turbines to substations or the grid. As depicted in Figure 11.18, a wind farm's layout and configuration are determined by factors such as scale, location, and desired reliability.

Radial design, referenced in Figure 11.19, is the simplest configuration, where each turbine is connected directly to the substation via a separate radial feeder. While this approach is cost-effective and straightforward to implement, it is more susceptible to energy losses over long distances and offers limited reliability.

Single-sided ring designs improve on this by creating a loop that connects turbines to the substation, providing an alternate path in the event of a fault, thus enhancing reliability. This design is represented in Figure 11.20.

Double-sided ring designs, as presented in Figure 11.21, add further redundancy, ensuring continuous operation even during equipment failures, making them ideal for

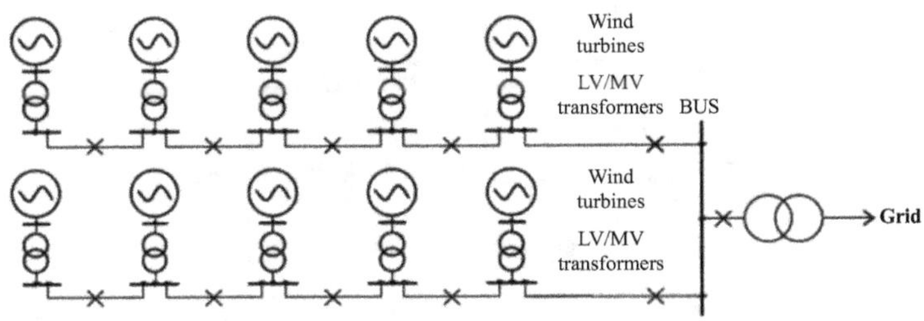

Figure 11.19 Radial design

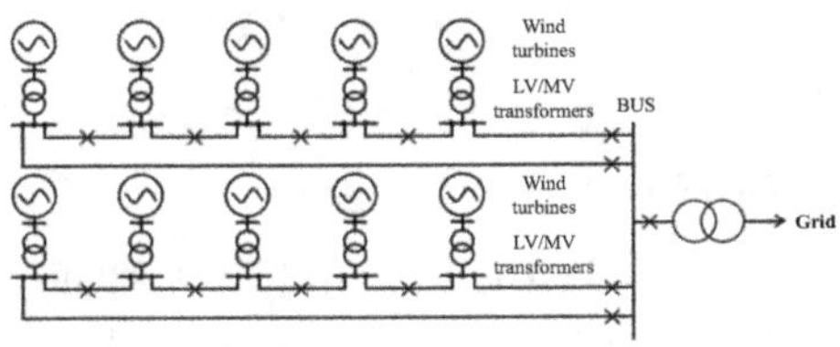

Figure 11.20 Single-sided ring design

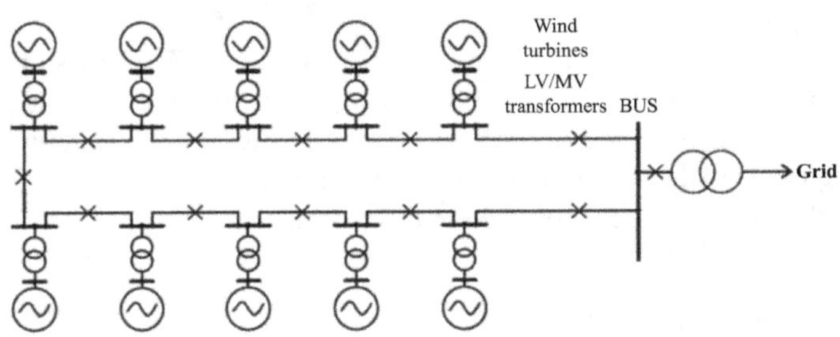

Figure 11.21 Double-sided ring design

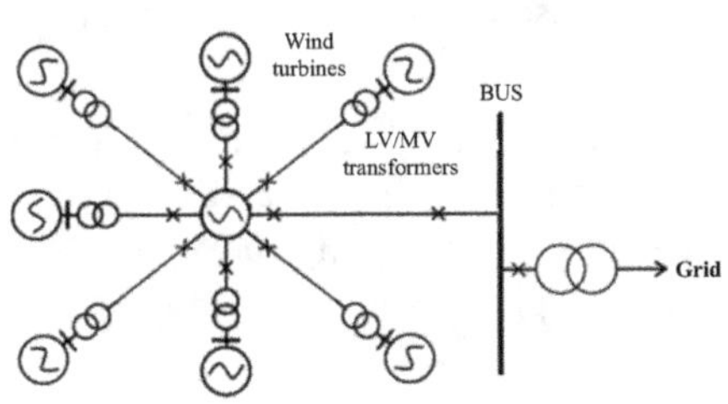

Figure 11.22 Star design

large-scale wind farms. However, the added complexity and cost must be considered during the planning stage.

The star design (see Figure 11.22) connects turbines through a central hub directly to the substation, minimizing transmission losses and distances. This layout is most suitable for compact wind farms with closely spaced turbines, offering high efficiency in smaller setups.

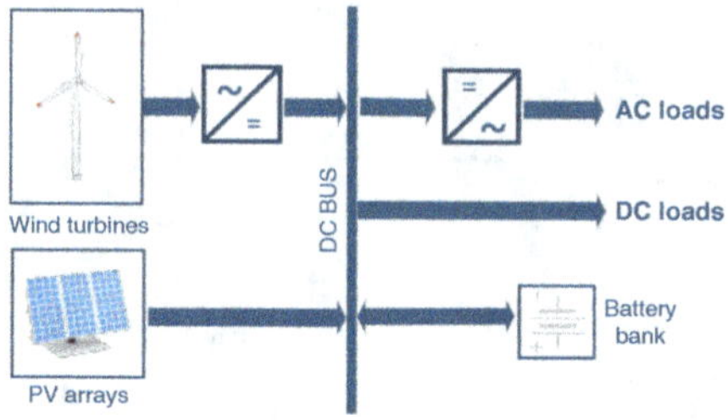

Figure 11.23 HRES design principles with key system components

11.8.2 Integration of HRES components

The integration of PV arrays, wind turbines, battery banks, and loads requires careful planning and a system-level approach. PV arrays generate electricity during daylight hours, while wind turbines often produce energy at night or during windy conditions. The natural complementarity between these sources reduces variability and ensures a more stable power output.

Battery banks (see Figure 11.23) play a critical role in HRES by storing surplus energy and providing backup during periods of low renewable generation. In DC-coupled systems, batteries are connected directly to the DC bus, allowing efficient energy storage and discharge with minimal conversion losses.

The load requirements further dictate the design:

- DC loads can connect directly to the DC bus, bypassing the need for energy conversion.
- AC loads, however, require inverters to convert DC-to-AC power, making the proper sizing of inverters essential for efficient energy delivery.

For successful integration, accurate load estimation is paramount. This ensures that components such as inverters, PV arrays, and wind turbines are appropriately sized to meet energy demands. Additionally, designing systems with scalability and modularity enables future upgrades to accommodate growing energy needs or emerging technologies.

11.9 Future trends and emerging technologies

11.9.1 Innovations in PV and wind technologies

The rapid advancements in PV and wind energy technologies have transformed the landscape of renewable energy systems. These innovations not only enhance the efficiency and scalability of HRES but also open new avenues for their application in diverse environments.

11.9.2 Innovations in PV technologies

PV technologies have undergone significant advancements in recent years, enhancing energy conversion efficiency and reducing production costs. As the technology matures, PV systems are becoming more affordable and widely adopted across various sectors. The primary PV technologies include crystalline silicon, cadmium telluride (CdTe), and the emerging perovskite solar cells, each offering unique advantages in terms of efficiency and cost-effectiveness [55].

Crystalline silicon solar modules currently dominate the global PV market, accounting for approximately 96% of total global shipments. These modules use silicon to absorb sunlight and generate electricity, achieving efficiencies exceeding 27% under laboratory conditions, while commercially available modules operate within the 20%–22% range [56]. Crystalline silicon's long lifespan and reliability make it the most commonly deployed PV technology worldwide, with continuous improvements driving down costs.

CdTe thin-film technology represents around 4% of the global PV market and offers advantages such as lower production costs and improved performance in low-light conditions. CdTe cells, which have more than doubled in efficiency over the past 15 years, leverage their superior absorption properties to generate electricity with thinner layers compared to silicon-based modules [57].

Perovskite solar cells have rapidly evolved as a promising alternative, with efficiencies increasing from 3% in 2009 to over 26% today in small-area devices [55,58]. Furthermore, perovskite-silicon tandem cells have achieved efficiencies nearing 34%. Despite their potential, large-scale commercialization remains a challenge due to stability and scalability concerns. However, ongoing research and development efforts are expected to improve durability and lower production costs, making perovskite technology a competitive contender in the PV market.

With continued advancements, the future of PV technology is set to provide more efficient and cost-effective solutions, contributing to a sustainable global energy transition.

11.9.3 Innovations in wind technologies

Wind energy technologies have experienced significant advancements in recent years, improving efficiency, capacity, and reliability while reducing costs. These innovations have enabled wind power to play a crucial role in global renewable energy expansion. Modern wind turbines are now larger, more efficient, and capable of harnessing wind energy in diverse environments, including offshore and low-wind-speed areas.

Globally, wind power capacity has grown exponentially, driven by the development of **offshore wind turbines**, which now reach capacities exceeding 15 MW per unit [59]. These turbines incorporate advanced aerodynamics and lightweight composite materials, improving power generation while reducing mechanical stress and maintenance requirements. Offshore installations benefit from stronger and more consistent wind speeds, enhancing overall energy production efficiency.

Vertical-axis wind turbines (VAWTs) have emerged as a viable option for urban and small-scale applications. Unlike traditional horizontal-axis turbines, VAWTs capture wind from multiple directions, making them suitable for decentralized energy systems in areas with fluctuating wind patterns.

Another significant innovation is the integration of **energy storage solutions**, such as batteries, to stabilize wind power output and provide consistent electricity during periods of low wind. These storage solutions enhance grid reliability and allow wind energy to contribute effectively to demand response strategies.

These innovations collectively enhance the performance, reliability, and cost-effectiveness of PV and wind systems, positioning them as critical components of modern energy systems.

11.9.4 Smart grid integration and demand response

This subsection presents an in-depth overview of smart grid integration and demand response mechanisms, focusing on their pivotal role in the effective operation of HRES. Smart grids, characterized by their ability to monitor and manage energy flow in real-time, are indispensable in balancing the variability of PV and wind energy generation with consumer demand.

Demand response (DR) in the context of HRES refers to strategies and mechanisms designed to optimize the balance between energy supply and demand by adjusting or shifting consumption patterns in response to generation availability, grid conditions, or economic incentives. DR is crucial for integrating intermittent renewable energy sources, such as wind and solar, by ensuring efficient energy use, minimizing grid stress, and enhancing system resilience.

11.9.5 HRES topologies: AC and DC coupling

The configuration of HRES significantly affects its performance, especially in terms of energy conversion efficiency and scalability. AC coupling (see Figure 11.24) integrates the outputs of PV arrays and wind turbines at an AC bus. This configuration is particularly suited for grid-connected systems and large-scale applications, as it allows direct delivery to AC loads and seamless integration with existing grid infrastructure. However, AC coupling introduces energy losses due to multiple DC-to-AC conversions and requires synchronization equipment to manage power quality and load sharing.

In contrast, DC coupling (see Figure 11.25) combines the DC outputs of PV arrays and wind turbines at a DC bus. A centralized inverter converts the DC power into AC for distribution to loads or grid connection. This configuration minimizes energy losses and simplifies battery integration, making it ideal for off-grid systems or installations with significant energy storage requirements. By connecting batteries directly to the DC bus, the system achieves efficient storage and discharge, reducing dependency on external power sources.

By leveraging advanced tools and methodologies, HRES can achieve robust and scalable designs that address the growing demand for sustainable energy solutions.

A brief comparative analysis between AC and DC coupling design will be presented in Table 11.7.

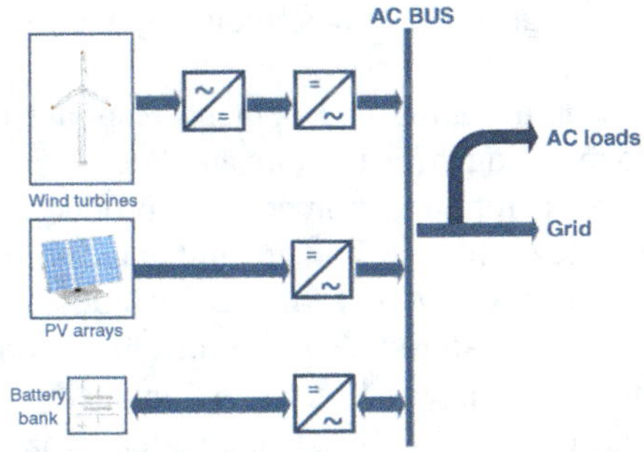

Figure 11.24 HRES design – AC-coupling topology

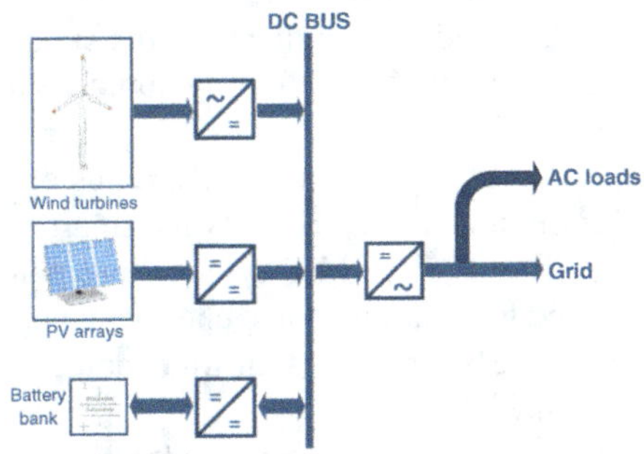

Figure 11.25 HRES design – DC-coupling topology

Table 11.7 Comparative analysis of AC and DC coupling design

Feature	AC coupling	DC coupling
Application	Grid-connected, large-scale systems	Off-grid, storage systems
Energy Losses	Higher due to multiple conversions	Lower, especially with energy storage
Complexity	Requires synchronization and control	Simplified for storage integration
Scalability	Easy to scale with additional inverters	Limited by centralized inverter capacity
Energy Storage	Less efficient battery integration	Highly efficient battery management

11.9.6 Demand response strategies

Demand response plays an essential role in balancing the variability of renewable energy sources. By aligning energy consumption with generation, HRES can achieve greater efficiency and stability. TOU pricing encourages consumers to use energy during periods of high renewable generation, such as daylight hours for PV systems. Automated dynamic load management adjusts energy consumption in real-time based on resource availability and grid conditions, reducing energy waste and improving reliability.

The role of smart grids in facilitating demand response cannot be overstated. Equipped with advanced metering infrastructure and Internet of Things (IoT) devices, smart grids enable real-time communication between producers and consumers. This ensures that energy flows are optimized, and any imbalances are addressed promptly. Peak shaving further enhances system performance by reducing demand during high-usage periods through the use of stored energy or incentivized load reduction.

Figure 11.26 illustrates the integration of IoT-enabled smart meters within an HRES, comprising PV arrays, wind turbines, and battery storage. The bi-directional smart meter shown in the figure acts as a crucial interface between the building's electrical loads and the grid. This device is responsible for measuring energy flows in both directions, ensuring precise monitoring and management of energy consumption and production. Through IoT-based communication, the smart meter can relay real-time energy data to the grid operator, enabling dynamic demand response actions based on grid conditions.

The successful integration of IoT in smart grids relies on several key technologies, including advanced communication protocols and smart metering infrastructure. **Smart meters** are equipped with embedded sensors and communication modules that transmit consumption and generation data to energy management systems. These meters support demand response by enabling real-time monitoring and remote control of loads based on grid demands.

The integration of these technologies ensures effective energy management by enabling real-time control, predictive analytics, and automated load adjustments.

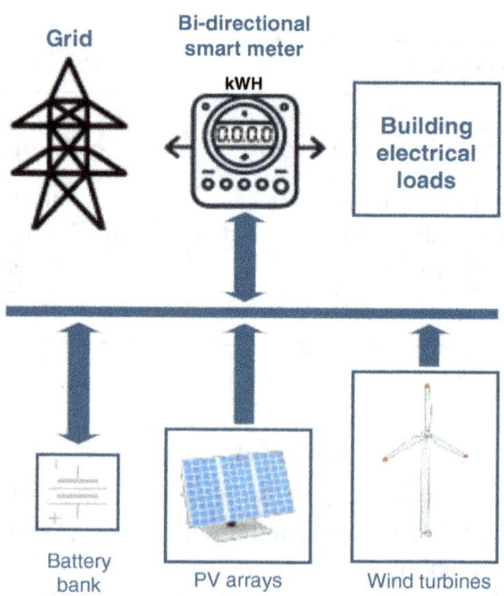

Figure 11.26 Demand response architecture

11.9.6.1 Residential and tertiary buildings HRES with TOU pricing

In both residential and tertiary (commercial and institutional) buildings, HRES with PV panels and battery storage can be leveraged to implement demand response strategies based on a TOU pricing model. This model allows consumers to adjust their energy consumption patterns, shifting high-energy activities to periods when renewable energy generation is at its peak or electricity prices are lower, resulting in substantial cost savings and enhanced energy efficiency.

For residential consumers, high-energy appliances such as washing machines, dishwashers, and electric vehicle (EV) charging can be scheduled during off-peak hours, aligning with the periods of abundant solar or wind energy generation. Tertiary buildings, such as offices, schools, and hospitals, can optimize HVAC (heating, ventilation, and air conditioning) systems, lighting schedules, and computing infrastructure to align with favorable TOU rates, reducing operational costs while maintaining service quality.

To quantify the economic benefits of demand response in HRES under TOU pricing, (11.11) can be used to calculate cost savings achieved by shifting energy consumption:

$$\text{Cost saving} = (E_{\text{shifted}} \times P_{\text{peak}}) - (E_{\text{shifted}} \times P_{\text{off-peak}}) \tag{11.11}$$

where
E_{shifted} – Energy consumption shifted to off-peak hours [kWh],
P_{peak} – Peak hour price [€/kWh],
$P_{\text{off-peak}}$ – Off-peak price [€/kWh].

11.9.6.2 Industrial HRES with demand response programs

Industrial facilities equipped with HRES, incorporating PV panels, wind turbines, and energy storage, can actively participate in demand response programs offered by utilities. These programs provide financial incentives to industrial consumers who agree to reduce or shift their power consumption during periods of high grid demand. Demand response strategies in industrial settings can be implemented through various means, including the use of backup generators, ESS, or production rescheduling to optimize energy use and reduce operational costs, using (11.12).

$$\text{Incentive} = \Delta P_{\text{reduced}} \times I_{\text{rate}} \tag{11.12}$$

where ΔPreduced is the load reduction [kW] and I_{rate} is the incentive rate [€/kW].

11.9.7 Role of energy storage in enhancing system stability

Renewable energy technologies have developed rapidly in recent years thanks to steep cost reductions, innovation, and political support. The ongoing transformation of the energy sector introduces new challenges that require changes in the way policy makers, regulators, and utilities plan, manage, and operate the energy system. The rapid expansion of renewable electricity requires a more flexible energy system to ensure that an energy system with a high share of Variable Renewable Energy (VRE) can be operated reliably and cost-effectively (see Figure 11.27).

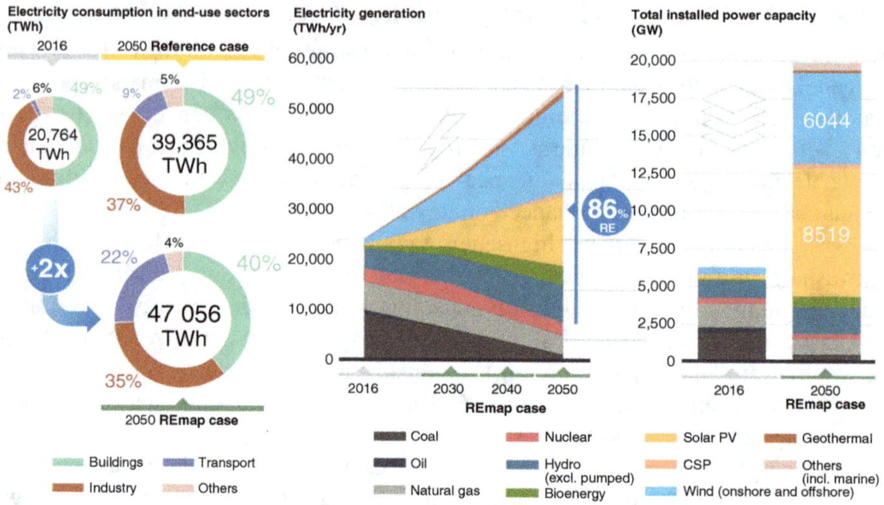

Figure 11.27 Electricity generation mix and installed capacity, categorized by fuel-type remap case (2016–2050) [60]

With its unique capabilities to absorb, store, and then re-inject electricity, electricity storage is seen as an important solution to address a number of technical and economic challenges related to the integration of renewable energy sources. Electricity storage can provide a wide range of services to support the integration of solar and wind energy and address some of the new challenges, variability, and uncertainty that solar and wind energy introduce into the energy system.

In a market setting, when allowed to participate in the wholesale market, storage can consume or supply electricity in response to price signals, in particular by increasing demand when prices are very low or even negative. While negative prices could be a sign of system inflexibility, storage can prevent such phenomena by consuming electricity and being paid for it.

When coupled with solar PV, storage can prevent the cannibalization of revenues in the middle hours of the day, increasing the profitability of solar PV and consequently creating the opportunity to install more solar power. Solar PV power production can exceed electricity demand in the middle of the day; storage can absorb some of this electricity and re-inject it later, effectively reducing restrictions due to overproduction or grid constraints. This can be done both in the market and in a vertically integrated framework.

In addition, electricity storage can participate in capacity and ancillary services markets by providing grid services such as the provision of primary and secondary reserves as well as firm capacity.

Indirectly, storage can help reduce costs by postponing the need for generation and transmission capacity by reducing the need for peaking plants and decongesting lines. When connected behind the meter, electricity storage can support the integration of distributed renewables and the active participation of prosumers in demand-side management, leading to reductions in average household electricity bills. **A key element is that storage can efficiently deliver multiple services simultaneously, thus aggregating revenues for greater cost-effectiveness**.

Overall, electricity storage could play a key role in facilitating the next stage of the energy transition, enabling an increased share of renewable electricity in energy systems and accelerating electrification in general, including in the transport sector.

However, the system value of storage is often insufficiently taken into account in electricity markets, leading to the so-called "cash crunch," where market revenues for investors are insufficient to make projects viable, leading to sub-optimal deployment of electricity storage. However, in vertically integrated environments, the same entity can realize the full value of storage, including savings realized in both production costs and investment, provided there are adequate incentives to explore the potential of storage to reduce the cost of supply [60].

The deployment of batteries – both stationary and in EVs – is currently on the rise; they are expected to play a key role in increasing flexibility in the power sector, although in energy terms, pumped hydropower remains by far the largest source of electricity storage [61].

In addition to the benefits for grid stability, battery energy storage coupled with variable renewable electricity generation capacity, typically generated by solar PVs, is an attractive business case in cities (see Figure 11.28).

BESS are the leading candidate to address the flexibility and resilience of power grids through applications such as **peak-shaving,** as shown in Figure 11.29. Batteries are of particular interest at small and medium scales due to their relatively high energy density, lack of geographical restrictions, minimal noise, and low maintenance requirements.

BESS play a key role between renewable energy supply and electricity demand response. Energy supplied from renewable sources or from the electricity grid is available for instantaneous consumption, and many factors, such as the variation of solar panels or electricity market demand, have a significant impact on the cost of electricity.

Locating energy storage in the same place as demand, such as in factories, has several important advantages. These advantages include peak shaving for both grid imports and exports from embedded renewables.

BESS provide back-up power, delay infrastructure reinforcement, improve power quality, and increase self-consumption of embedded renewables.

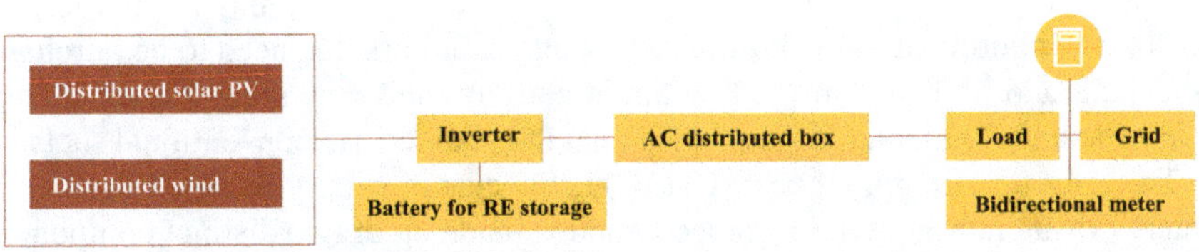

Figure 11.28 Hybrid RES projects and energy storage capacities general architecture

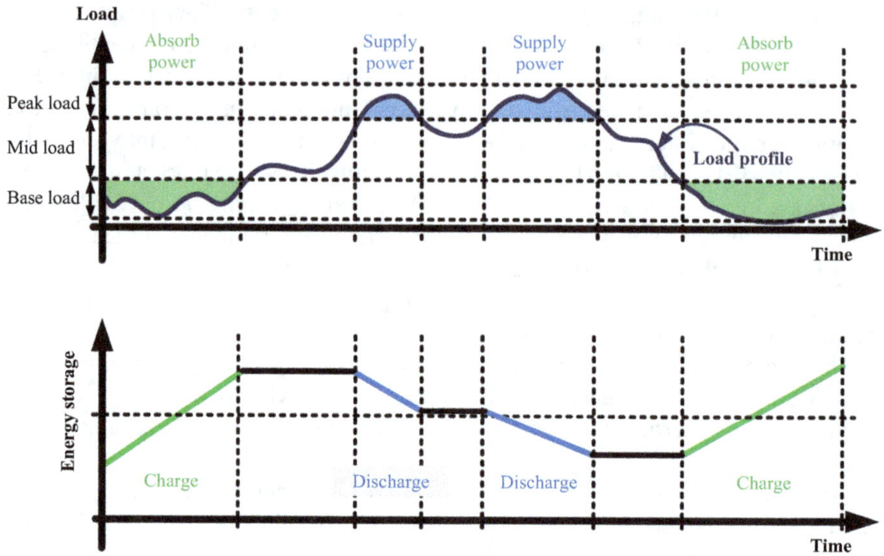

Figure 11.29 Example of utilizing BESS to ensure a Peak shaving load-curve flattening procedure [62]

This would benefit the Distribution System Operator, and subsequently other electricity consumers in the distribution area, through reduced future distribution charges resulting from limited/delayed investment in the distribution system.

It's worth noting that a renewable energy generation project is not mandatory in order to reap the benefits of energy storage. Peak load reduction and emergency backup are examples of applications that work without solar panels, and customers can still benefit from the cost savings, stability, and resilience that energy storage offers.

The peak of the daily load curve, although short in time, generates a major impact on the energy and operational efficiency of power grids. The conventional approach for minimizing the impact of this phenomenon is to increase the power generation capacity during the peak period [63].

However, this option is not always economically feasible, may be inefficient in terms of genset utilization (as it implies having generators that need to be purchased, maintained, and used for only a few hours per day) and may significantly increase the specific environmental impact when peaking sources use conventional fuels.

The long-term marginal cost (LTMC) is the cost of supplying an additional unit of energy using new generating capacity and is made up of two distinct components: the additional cost of adding new capacity and the additional cost of fuel and variable maintenance and operating expenses incurred in providing the additional energy and is therefore high.

The constant increase in peak loads on electricity grids increases the likelihood of damage due to undelivered energy and increases the marginal cost of energy supply. In this context, balancing the power generation-transmission-distribution capacity with the electricity demand in real time has become a major issue for power companies [64–66].

As peak electricity generation is needed for a very short period of the day, existing power plants (Short-Term Marginal Cost – STMC) are often used, fully depreciated in financial terms, with virtually no CAPEX – their modernization generating negligible development CAPEX compared to the initial CAPEX.

The transition to environmental neutrality leads to an increase in CAPEX for peak load generation from new, energy-efficient, and environmentally neutral plants.

These two approaches lead to a dramatic increase in the price of peak-load electricity from the point of view of its generation [67] to guarantee the payback of the investment and annual costs, respectively, over the life cycle of the peaking power plants in the context of their use for a limited number of hours per year.

One of the biggest difficulties of the STOs is to ensure the balance between electricity generation and electricity demand [68], as an imbalance between these two components can lead to instability, voltage variations, and even blackouts.

Imbalances between power generation and power demand materialize in additional demands on the generating sets and in low-value PQIs (Power Quality Indices), such as frequency drops outside the permissible quality limits. The most effective solution to minimize these imbalances is to obtain a load curve that is characterized by a small difference between the peak and the dip [69–75].

Replacing power transformers and power lines is a very costly operation, and other solutions to increase the power transfer capacity are being sought. Also, obtaining permits and agreements for new power line routes and land for transformer stations is a difficult and lengthy process.

As such, **peak shifting** by using ESS is a readily available and cost-effective way to mitigate these issues whilst ensuring an optimal LCOE (Levelized cost of electricity) level (see Figure 11.30).

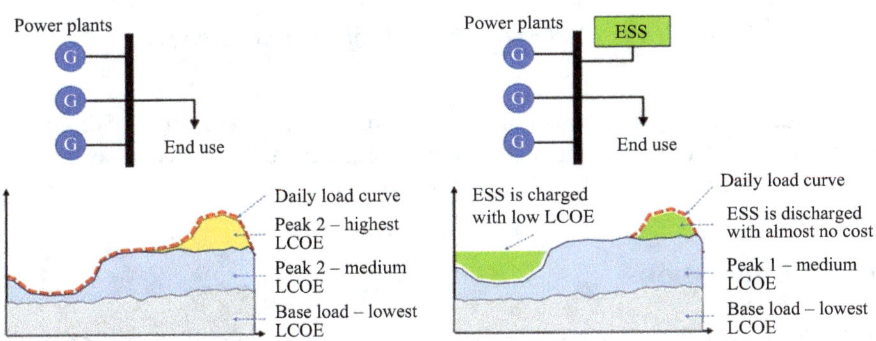

Figure 11.30 Peak shifting by using Energy Storage Systems (ESS), adaptation after [76]

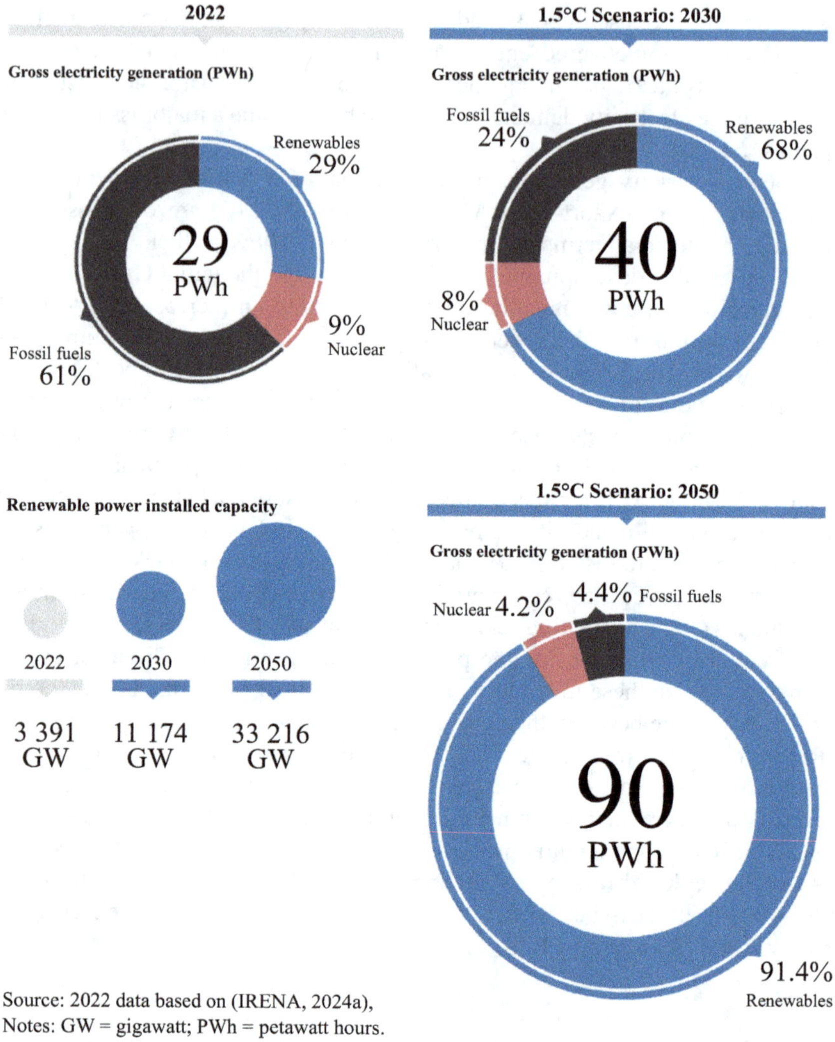

Figure 11.31 RES weight in total energy generation [77]

In order to ensure the transition to 100% renewable energy by 2050, an estimated 1,465 GWh of electrochemical battery storage capacity is needed, as it can be observed in Figure 11.31.

11.10 Conclusions

This chapter has offered a detailed examination of HRES, highlighting the synergy between PV and wind energy systems to maximize their potential as a combined system. Beginning with the essentials of PV and wind energy systems, the chapter emphasizes their complementary nature. PV systems generate power during sunny periods, while wind systems often operate during different weather conditions or at night, making their combination a dependable solution to ensure a stable energy supply.

The discussion covered the definition and setup of HRES, demonstrating their advantages over single-source systems in terms of performance, versatility, and better use of renewable resources. The chapter tackled significant challenges in integrating PV and wind systems, such as production inconsistencies, grid stability concerns, and the added system complexity. Strategies like incorporating ESS, improved forecasting techniques, and deploying smart controllers were explored to overcome these obstacles.

Design and optimization strategies were reviewed, highlighting the need for context-specific approaches that account for location, economic, and energy demand differences. The combination of ESS, smart grid technologies, and demand response measures was emphasized as essential to maintain a stable and reliable energy delivery.

The chapter also explored the economic and environmental impacts of HRES, focusing on their potential for cost reduction and reduced carbon emissions. Furthermore, the chapter highlighted future trends and innovations, including advancements in PV and wind technologies, the application of artificial intelligence for optimization, and the growing role of HRES in smart grid systems.

In summary, HRES present a promising pathway for advancing sustainable energy targets by merging the strengths of PV and wind energy. With the support of energy storage, smart grid systems, and ongoing innovations, these systems can overcome existing challenges to provide cost-effective, dependable, and eco-friendly energy solutions. As technology evolves, HRES will increasingly play a central role in the global transition toward a cleaner and more resilient energy future.

References

[1] International Renewable Energy Agency (IRENA). Renewable capacity statistics 2020 [Internet]. Abu Dhabi: International Renewable Energy Agency; 2020 [cited 2025 Jan 28]. Available from: https://www.irena.org.

[2] Santos A, Silva M, and Pereira A. Advances in renewable energy technologies and their integration. *Energy Reports*. 2019;5:823–32.

[3] Zhang L and Xu L. Predictive analytics for wind and solar hybrid systems in grid integration. *Energy Reports*. 2019;5:467–78.

[4] Liu Q, Song H, and Zhang T. Integration challenges of hybrid renewable energy systems: Advances and perspectives. *Renew Sustain Energy Rev*. 2017;77:452–65.

[5] Fadaei H, Haghifam MR, and Ghasemi H. Operation and control of hybrid renewable energy systems. *Renew Energy*. 2020;146:2323–42.

[6] Almeida L, Pereira L, and Barbosa J. Integration of solar and wind energy in hybrid systems: A review. *Renew Sustain Energy Rev*. 2021;129:109948.

[7] Khalid R, Syed HH, and Khan MA. Optimizing hybrid renewable energy systems for a sustainable future. *Sustain Energy Technol Assess.* 2020;38:100706.

[8] Bai Y, Song H, and Zhang L. Renewable energy hybrid systems: Benefits, challenges, and prospects. *Energy Reports*. 2019;5:1292–304.

[9] Tao X, Zhang X, and Chen H. Community-based microgrids for renewable energy integration. *Energy Reports*. 2020;6:1425–36.

[10] International Energy Agency (IEA). World energy outlook 2021: Energy transitions and global emissions [Internet]. Paris: International Energy Agency; 2021 [cited 2025 Jan 28]. Available from: https://www.iea.org.

[11] Yang Y, Li Y, and Liu B. Wind-solar hybrid systems for sustainable energy integration: Review and prospects. *Appl Energy*. 2021;287:116460.

[12] Gokhan A and Kaya Y. Challenges in hybrid renewable energy systems integration: Overview and opportunities. *J Energy Eng*. 2021;147(3):04021036.

[13] Cruz RA and Silva PA, Fernandes TA. Optimal design and integration of hybrid renewable energy systems. *J Renew Sustain Energy*. 2020;12(2):025301.

[14] Liu B, Wang S, and Zhang X. Integration of hybrid renewable energy systems for enhanced grid stability. *Energy Reports*. 2020;6:1573–81.

[15] Yilmaz A, Bayindir R, and Cebeci Y. Control and management strategies in hybrid renewable energy systems: A review. *Renew Sustain Energy Rev.* 2019;115:109370.

[16] Zhao X and Lu D. Energy storage technologies for hybrid renewable energy systems: Current status and future perspectives. *Energy Storage*. 2019;1(1):34–48.

[17] Zhang L, Li Y, and Zhang J. Policy frameworks and regulatory barriers in the integration of hybrid renewable energy systems. *Renew Energy*. 2020;144:1534–45.

[18] Guo J, Peng J, Luo Y, Zou B, and Luo Z. Study on the hybrid energy storage for industrial park energy systems: Advantages, current status, and challenges. *Nat Sci Open*. 2024;3(3):20230051. Available from: https://doi.org/10.1360/nso/2023005.

[19] Cheng X, Yang L, and Zhao M. Hybrid renewable energy systems for rural electrification: A review. *Renew Sustain Energy Rev*. 2018;90:1024–36.

[20] Qusay H, Sameer A, Aws ZS, Hayder MS, and Marek J. A review of hybrid renewable energy systems: Solar and wind-powered solutions: Challenges, opportunities, and policy implications. *Results Eng*. 2023;20:101621. Available from: https://doi.org/10.1016/j.rineng.2023.101621.

[21] Gong J, Xie L, Li Y, Ni Z, Wei Q, Wu Y, and Cheng H. Analysis of the impact resistance of photovoltaic panels based on the effective thickness method. *J Renew Mater*. 2021;9:33–51. Available from: https://doi.org/10.32604/jrm.2021.016262.

[22] Ji D, Zhang C, Lv M, Ma Y, and Guan N. Photovoltaic array fault detection by automatic reconfiguration. *Energies*. 2017;10:699. Available from: https://doi.org/10.3390/en10050699.

[23] Katche ML, Makokha AB, Zachary SO, and Adaramola MS. A comprehensive review of maximum power point tracking (MPPT) techniques used in solar PV systems. *Energies*. 2023;16:2206. Available from: https://doi.org/10.3390/en16052206.

[24] Adua EA, Sanda B, Sadiq A, and Musa J. Comparisons of east/west and south oriented photovoltaic (PV) systems in terms of performance and land utilization with the best optimum tilt angles. *Int Res J Modern Eng Technol Sci*. 2023;5:2582–5208.

[25] PVsyst [homepage on the Internet]. Software solution. Available from: https://www.pvsyst.com/https://www.pvsyst.com.

[26] van de Ven DJ, Capellan-Peréz I, Arto I, *et al.* The potential land requirements and related land use change emissions of solar energy. *Sci Rep*. 2021;11:2907. Available from: https://doi.org/10.1038/s41598-021-82042-5.

[27] Díaz-Bello D, Vargas-Salgado C, Águila-León J, and Lara-Vargas F. Methodology to estimate the impact of the DC to AC power ratio, azimuth, and slope on clipping losses of solar photovoltaic inverters: Application to a PV system located in Valencia Spain. *Sustainability*. 2023;15:2797. https://doi.org/10.3390/su15032797.

[28] İşin E and Çetinkaya N. Determination of optimum DC/AC ratio for PV power plants using artificial intelligence. *3rd International Word Energy Conference*, pp. 606–613, 2023, ISBN: 978-625-367-567-7.

[29] Praveen GB, Satyendra Kumar Shet M, and Latha Shenoy K. (2024). Review and Study of Solar String Inverters for a PV System. In: Kajampady S, Revankar ST. (eds) *Advances in Renewable Energy & Electric Vehicles. AREEV 2022. Lecture Notes in Electrical Engineering*, vol 1083. Springer, Singapore. https://doi.org/10.1007/978-981-99-6151-1_1.

[30] Burton T, Sharpe D, Jenkins N, and Bossanyi E. *Wind Energy Handbook*. Wiley; 2012.

[31] Manwell JF, McGowan JG, and Rogers AL. *Wind Energy Explained: Theory, Design and Application*. Wiley; 2010.

[32] Sathyajith M. *Wind Energy: Fundamentals, Resource Analysis, and Economics*. Springer; 2006.

[33] Betz A. The maximum of the theoretically possible exploitation of wind by means of a wind motor. *Wind Energy System*. 2013;37(4):441–46.

[34] Johnson GL. Wind Energy Systems. Manhattan, KS: Prentice-Hall; 2006.

[35] The Roundup, "Wind Turbine Power Curve," The Roundup, 2024. [Online]. Available from: https://theroundup.org/wind-turbine-power-curve/. Accessed on 29.01.2025.

[36] Prateek J and Gokhale-Welch C. Fundamentals of wind energy. National Renewable Energy Laboratory. 2022. Available from: https://www.nrel.gov/docs/fy23osti/84501.pdf.

[37] Hashem Nehrir M, Strunz K, Aki H, and Wang C. A review of hybrid renewable/alternative energy systems for electric power generation: Configurations, control, and applications, *IEEE Trans Sustaina Energy*, 2011;2(4):392–403, Available from: https://doi.org/10.1109/TSTE.2012.2157540

[38] León Gómez JC, De León Aldaco SE, and Aguayo Alquicira J. A review of hybrid renewable energy systems: Architectures, battery systems, and optimization techniques. *Eng.* 2023;4:1446–67. Available from: https://doi.org/10.3390/eng4020084

[39] Udoka Eze VH, Mwenyi JS, Ukagwu KJ, Eze MC, Eze CS, and Okafor WO. Design analysis of a sustainable techno-economic hybrid renewable energy system: Application of solar and wind in Sigulu Island, Uganda, *Sci Afr.*, Volume 26, 2024, e02454, Available from: https://doi.org/10.1016/j.sciaf.2024.e02454

[40] Solar radiation database. Available at: https://solargis.info/. [Accessed on: 20.12.2024]

[41] Matthias H, Desislava D, and Thomas H. Integration of wind and solar power in Europe: Assessment of flexibility requirements. *Energy.* 2014;69. Available from: https://doi.org/10.1016/j.energy.2014.02.109.

[42] German Aerospace Center (DLR), Institute of Technical Thermodynamics, Section Systems Analysis and Technology Assessment, Concentrating Solar Power for the Mediterranean Region, MED-CSP Study Report commissioned by Federal Ministry for the Environment, Nature Conservation and Nuclear Safety Germany, Stuttgart, 16 April 2005, http://large.stanford.edu/courses/2012/ph240/aljamaan2/docs/algerien_med_csp.pdf.

[43] IRENA (2024). Renewable power generation costs in 2023. International Renewable Energy Agency. Abu Dhabi.

[44] IEA (2023). Wind onshore LCOE index based on average annual input costs. 2018-2024. IEA. Paris https://www.iea.org/data-and-statistics/charts/wind-onshore-lcoe-index-based-on-average-annual-input-costs-2018-2024.

[45] Romanian Electricity Market Operator database. Available at: https://www.opcom.ro/acasa/ro. [Accessed on: 20.12.2024]

[46] Wang J, Wan Y, Xiong Z, Xiong X, and Ouyang J. Cooperative overload control strategy of power grid-transformer considering dynamic security margin of transformer in emergencies. *Int J Electr Power Energy Syst.* 2022;140:108098.

[47] Iwabuchi K, Watari D, Zhao D, Taniguchi I, Catthoor F, and Onoye T. Enhancing grid stability in PV systems: A novel ramp rate control method utilizing PV cooling technology. *Appl Energy.* 2025;378(Part A):124737. Available from: https://doi.org/10.1016/j.apenergy.2024.124737.

[48] Patil GB, Raghuwanshi SS, and Arya LD. Stability Analysis of Grid-Integrated PV Systems. *Int J Eng Trends Technol.* 2024;72(4):51–66. Available from: https://doi.org/10.14445/22315381/IJETT-V72I4P106.

[49] California ISO, Transport System Operator data base, Available at: https://www.caiso.com. [Accessed on: 20.12.2024].

[50] Nwagu CN, Ujah CO, Kallon D, and Aigbodion VS. Integrating solar and wind energy into the electricity grid for improved power accessibility. *Unconvent Res.* 2025;5:100129. https://doi.org/10.1016/j.uncres.2024.100129.

[51] Gao W, Liu X, and Zhang Y. Advanced techniques for hybrid renewable energy system forecasting. *Renew Energy.* 2017;118:15–23.

[52] Khan MJ, Sarker MAR, and Rahman A. Optimization of hybrid systems for renewable energy integration. *IEEE Trans Power Electron*. 2020;35(4): 3256–69.

[53] Karapistoli E, Makridis N, and Papadopoulos D. Demand-side management for hybrid renewable energy systems. *Renew Sustain Energy Rev*. 2019;105: 145–54.

[54] Hossain MS, Alam MM, and Rahman S. Policy frameworks for hybrid renewable energy systems integration. *Energy Policy*. 2020;137:1127–35.

[55] The National Renewable Energy Laboratory of the U.S. Department of Energy. Cell efficiency. https://www.nrel.gov/pv/cell-efficiency.html.

[56] U.S. Department of Energy, Solar Energy Technologies Office. Crystalline silicon photovoltaics research. https://www.energy.gov/eere/solar/crystalline-silicon-photovoltaics-research.

[57] U.S. Department of Energy, Solar Energy Technologies Office. Cadmium telluride. https://www.energy.gov/eere/solar/cadmium-telluride.

[58] U.S. Department of Energy, Solar Energy Technologies Office. Perovskite solar cells. https://www.energy.gov/eere/solar/perovskite-solar-cells.

[59] U.S. Department of Energy, Wind Energy Technologies Office. Wind energy technologies office. https://www.energy.gov/eere/wind/wind-energy-technologies-office.

[60] IRENA (2020), Electricity Storage Valuation Framework: Assessing system value and ensuring project viability, International Renewable Energy Agency, Abu Dhabi.

[61] Kim T, Song W, Dae-Yong Son D-Y, Ono LK, and Qi Y, Lithium-ion batteries: outlook on present, future, and hybridized technologies, *J Mater Chem A*, 2019.

[62] Rana MM, Atef M, Sarkar MR, Uddin M, and Shafiullah G. A review on peak load shaving in microgrid-potential benefits, challenges, and future trend. *Energies*. 2022;15:2278. Available from: https://doi.org/10.3390/en15062278.

[63] Uddin M, Romlie MF, Abdullah MF, Halim SA, Bakar AHA, and Kwang TC, A review on peak load shaving strategies, *Renew Sustain Energy Rev.*, Available from: http://dx.doi.org/10.1016/j.rser.2017.10.056, Vol. 82(P3), pg. 3323–3332, 2017.

[64] Rahimi A, Zarghami M, Vaziri M, and Vadhva S. A simple and effective approach for peak load shaving using battery storage systems, *Proceedings of the North American Power Symposium*, IEEE, pg. 1–5, 2013.

[65] Chua KH, Lim YS, and Morris S. Energy storage system for peak shaving, *Int J Energy Sector Manag*, 2016;10:3–18.

[66] Sinha S and Chandel S. Review of software tools for hybrid renewable energy systems, *Renew Sustain Energy Rev.*, 2014;32:192–205.

[67] Van den Bergh K and Delarue E. Cycling of conventional power plants: technical limits and actual costs, *Energy Conversion Manage.*, 2015;97:70–7.

[68] Joshi KA and Pindoriya NM, Day-ahead dispatch of Battery Energy Storage System for peak load shaving and load leveling in low voltage unbalance distribution networks, *Proceedings of the IEEE Power & Energy Society General Meeting*, pg. 1–5. 2015.

[69] Thompson CC, Oikonomou PK, Etemadi AH, and Sorger VJ, Optimization of data center battery storage investments for microgrid cost savings, emissions reduction, and reliability enhancement, *IEEE Trans Ind Appl.*, 2016;52: 2053–60.

[70] Yan X, Zhang X, Chen H, Xu Y, and Tan C. Techno-economic and social analysis of energy storage for commercial buildings, *Energy Convers Manag.*, 2014;78:125–36.

[71] Luo X, Wang J, Dooner M, and Clarke J. Overview of current development in electrical energy storage technologies and the application potential in power system operation, *Appl Energy*, 2015;137:511–36.

[72] Reihani E, Motalleb M, Ghorbani R, and Saoud LS. Load peak shaving and power smoothing of a distribution grid with high renewable energy penetration, *Renew Energy*, 2016;86:1372–79.

[73] Son S and Song H. Real-time peak shaving algorithm using fuzzy wind power generation curves for large-scale battery energy storage systems, *Int J Fuzzy Log Intell Syst.*, 2014;14:305–12.

[74] Lavrova O, Cheng F, Abdollahy S, *et al.* Analysis of battery storage utilization for load shifting and peak smoothing on a distribution feeder in New Mexico, *Innovative Smart Grid Technologies (ISGT)*, pg. 1–6, 2012.

[75] Telaretti E and Dusonchet L. Battery storage systems for peak load shaving applications. Part 2: Economic feasibility and sensitivity analysis, *Proceedings of the 16th International Conference on Environment and Electrical Engineering*, pg. 1–6, 2016.

[76] Setiawan A, Arifin Z, Jufri F, Sudiarto B, Garniwa I, and Setiabudy R. Economic analysis of cost-based load shifting implementation on large utility systems using battery energy storage system. *Int J Sustain Energy*. 2023;42:1182–201. Available from: https://doi.org/10.1080/14786451.2023.2248503.

[77] International Renewable Energy Agency, World Energy Transition Outlook 2024, IRENA, Abu Dhabi.

Chapter 12

An integrated Whale Optimization combined with Perturb and Observe Algorithm for efficient MPPT of PV system under partial shading

Zhi-Kai Fan[1], *Abhishek Tarikallu Thippesh*[1], *Kuo Lung Lian*[1] *and Pham Anh Hao*[1]

The power output of photovoltaic (PV) systems is often reduced by partial shading conditions (PSCs), which may be caused by vegetation, cloud cover, dirt buildup, or sudden changes in the weather. It becomes more challenging to determine the actual global maximum power point (GMPP) in these situations, as they create multiple peaks in the power-voltage characteristic curves (P-VCCs). This study addresses this issue by combining the Whale Optimization Algorithm (WOA) with the Perturb and Observe (P&O) method to present a novel Maximum Power Point Tracking (MPPT) approach. While the P&O method fine-tunes the tracking process and ensures accurate convergence to the true GMPP, the WOA effectively explores complex multi-peak power curves. Under PSCs, the combined strategy enhances energy efficiency, minimizes power oscillations, and accelerates tracking. According to simulation results, the proposed hybrid approach outperforms traditional metaheuristic algorithms (MHAs) regarding tracking time and accuracy. These findings illustrate the effectiveness of the WOA-P&O integration in overcoming the limitations of conventional MPPT techniques, thereby improving the overall performance of partially shaded systems.

12.1 Introduction

The worldwide demand for electrical energy has been consistently increasing due to economic expansion and population development. The substantial reliance on fossil fuels for electricity generation has resulted in significant environmental issues, notably climate change, caused by greenhouse gas emissions. To mitigate these issues, a shift toward renewable and sustainable energy sources, including solar, wind, and hydroelectric power, has become imperative [1]. Among these, solar energy captured by PV systems is distinguished as one of the most promising methods for extensive energy generation. Solar energy is a dependable, clean, and non-polluting

[1]Department of Electrical Engineering, National Taiwan University of Science and Technology, Taiwan

power source capable of fulfilling future electricity demands without causing environmental harms [2]. Nevertheless, PV systems encounter considerable obstacles, especially because of PSCs. Obstacles like buildings, trees, clouds, or dirt can result in uneven solar irradiation, causing power loss and the possible production of hot spots, which may inflict severe damage to PV modules [3]. Bypass diodes are frequently incorporated into PV systems to mitigate these effects; nevertheless, they alter the power–voltage (*P–V*) and current–voltage (*I–V*) properties, resulting in a *P–V* curve with several peaks [4]. This phenomenon complicates the extraction of optimal power, since local maximum power points (LMPPs) arise.

The efficiency of a solar PV system relies on the appropriate impedance matching between the PV module and its load. If the system fails to function at the GMPP, the produced power stays beneath its maximum capacity, and we utilize MPPT methodologies to enhance power extraction and address this challenge [5]. The effectiveness of an MPPT algorithm primarily depends on its tracking speed and accuracy, both of which significantly influence the total efficiency of PV systems under varying weather conditions [6]. In recent years, the extensive use of solar energy has been driven by declining prices related to PV panel production and installation. The proliferation of rooftop solar installations and large-scale solar farms worldwide underscores the need for efficient MPPT techniques [7]. Conventional MPPT techniques, including P&O, golden selection search (GSS), and incremental conductance (INC) [8], have been extensively utilized. However, current methodologies inadequately monitor the GMPP in the presence of several LMPPs, necessitating the development of more sophisticated and adaptable control techniques [9].

To address the challenges posed by PSCs, two primary methodologies have been examined. The initial aspect pertains to hardware modifications, which include adaptive reconfigurations, power electronics equalizers, multilayer DC-link inverters, and distributed MPPT topologies [10,11]. These solutions are generally costly and difficult to implement, which impedes their widespread adoption, despite their efficiency. The second method prioritizes software-based MPPT approaches that employ sophisticated control algorithms, thereby eradicating the necessity for supplementary hardware and improving cost-effectiveness [12]. Artificial intelligence (AI)-based methodologies, MHAs, and deterministic approaches (DAs) are the three primary categories into which software-based MPPT solutions are classified [13,14]. DAs, such as hill-climbing and P&O methods, frequently produce unsatisfactory results when subjected to PSCs [15,16]. Artificial neural networks (ANNs) and other AI methodologies have shown the potential to improve the precision of tracking [17]. However, they require substantial computational resources and training data [18]. Generally, MHAs can be categorized into four types, which are evolutionary algorithms, swarm intelligence, physics-based, and human-based algorithms, as shown in Figure 12.1.

MHAs have garnered considerable interest owing to their capacity to navigate intricate search spaces and evade entrapment in local optima [19]. MHAs offer several advantages that make them highly suitable for solving complex optimization problems. One of their key strengths is problem independence—they can be applied to a wide range of problems without requiring detailed problem-specific knowledge or derivative information. Their stochastic nature enables them to escape from

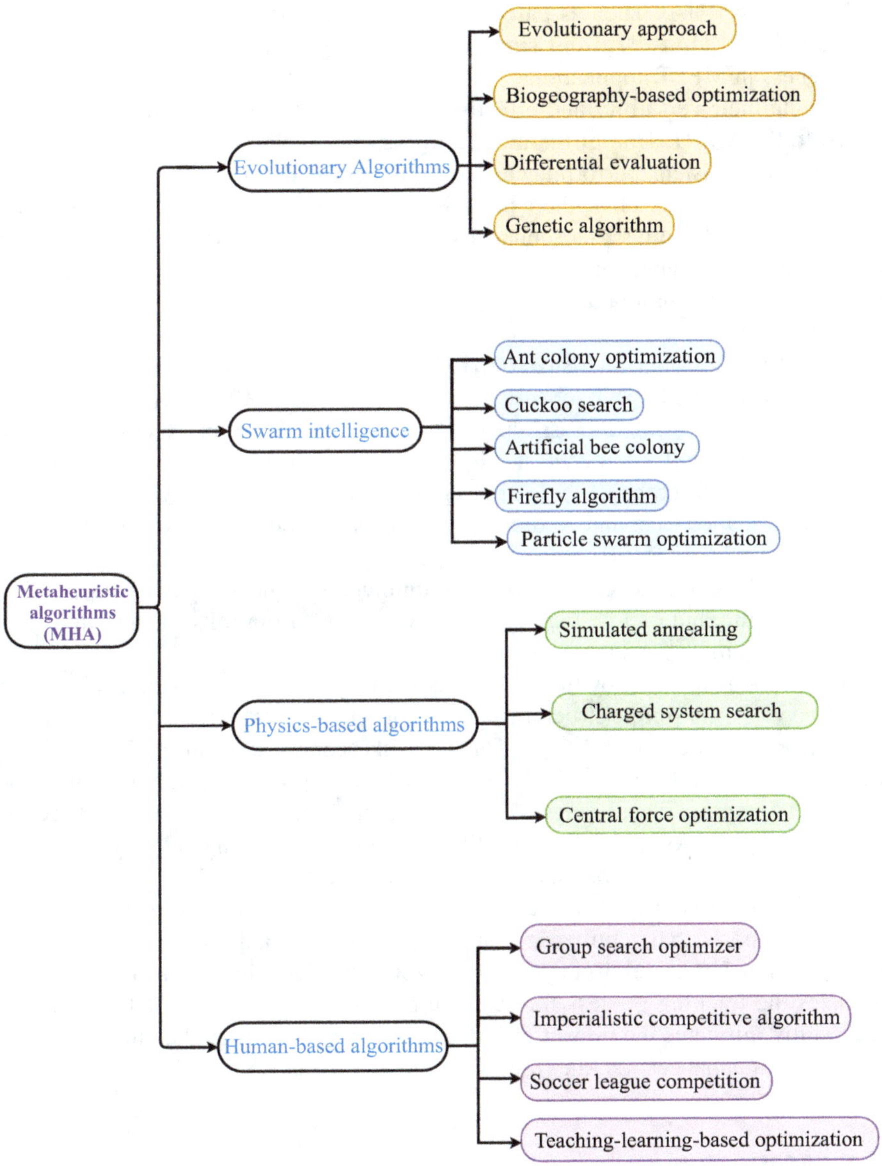

Figure 12.1 Some of the classification of metaheuristic algorithms

local optima, improving the likelihood of finding global or near-global solutions. Meta-heuristics are also flexible and adaptable, allowing for easy customization or hybridization to fit specific applications. They are generally simple to implement, requiring minimal mathematical background, and they offer strong global search capabilities, making them effective in navigating complex, nonlinear, and multi-modal search spaces, such as in the case of MPPT.

Numerous MHAs, such as particle swarm optimization (PSO) [20], grasshopper algorithm (GOA) [21], and simulated annealing (SA) [22], etc., have shown effectiveness in MPPT applications.

Despite their versatility and effectiveness, MHAs have some limitations when applied to the MPPT. One of the main drawbacks is their stochastic nature, which can cause inconsistent performance, requiring repeated runs or careful tuning to achieve reliable results. Thus, there has been a trend to combine MHAs with DAs for MPPT. The hybrid approach can enhance the robustness of the MPPT algorithm as the deterministic component can stabilize the search process, while the MHA maintains diversity and avoids premature convergence. Additionally, DAs can refine solutions found by MHAs, reducing computational effort and improving convergence reliability. For example, Tomar and SuwarnozLian [23] developed a hybrid Crow Search Algorithm (CSA) and P&O approach to enhance tracking efficiency. Wijaya *et al.* [24] introduced a modified Grasshopper Optimization Algorithm (GOA) integrated with the INC method. Lian *et al.* [25] presented a hybrid approach that integrates P&O with PSO. Nugraha *et al.* [26] integrated the CSA with the GSS to enhance MPPT performance Lian and Andrean [27] proposed a modified INC algorithm that combines with SA for MPPT.

Nevertheless, the need for parameter tuning for some of the MHAs is often problem-specific and can be particularly challenging to optimize if the number of the control parameters is large.

The WOA is inspired by the bubble-net hunting technique of humpback whales and is known for its simple structure and ease of implementation, with relatively few control parameters compared to other MHAs like genetic algorithm (GA) or PSO. Mirjalili *et al.* identify three primary stages of WOA: prey seeking, encircling, and bubble-net foraging. These stages facilitate the attainment of equilibrium between exploration and exploitation. This equilibrium improves tracking velocity and precision, rendering WOA a viable contender for MPPT application [28]. In this chapter, an MPPT method is proposed to combine WOA with P&O to improve MPPT efficiency. This approach first utilizes the intelligent search capabilities of WOA to locate the GMPP area. Upon identifying the GMPP zone, the algorithm shifts to P&O, which optimizes the tracking process for accurate power extraction. This hybrid solution successfully mitigates the impact of PSCs by utilizing the optimal elements of both approaches. It enhances tracking precision and optimizes the overall performance of PV systems.

12.2 Methodology

Sections 12.2.1 and 12.2.2 present a concise summary of two principal methods employed for identifying the GMPP in PV systems subjected to partial shading: the WOA and the P&O algorithm.

12.2.1 Review of Whale Optimization Algorithm

Whales are captivating organisms; they are universally acknowledged as the biggest mammals on the planet. Certain whale species may reach lengths of up to thirty meters and weigh as much as one hundred eighty tons. Seven primary species of whales exist: sei, finback, killer, minke, humpback, right, and blue whales. Excluding their considerable dimensions, whales are primarily predators. Remarkably, they never really sleep; they must surface for breath. Instead, merely one section of their brain sleeps at a time, enabling them to maintain alertness while obtaining some rest [29]. Whales have amazing intelligence and emotional depth. Research conducted by Hof and Van Der Gucht [30] indicates that the whales possess specialized brain cells termed spindle cells, analogous to those seen in humans. These cells are associated with emotions, decision-making, and social behaviors, depicting whales as one of the few mammals that possess this cognitive characteristic alongside humans. Whales possess double the total number of cells with spindles in contrast with adult humans, which enhances their intricate behavior. Research indicates that whales possess the capacity for cognition, learning, decision-making, and communication that parallels human relationships, but with a different degree of complexity [31]. Scientists have observed killer whales developing unique dialects that are exclusive to their groups.

Whales have pronounced social tendencies. Although some individuals choose isolation, most creatures exist in social groupings. Killer whales, for example, frequently remain in their familial pods for their whole lives. These whales predominantly consume krill and tiny fish. A particularly intriguing characteristic of humpback whales is their distinctive hunting method known as "bubble-net feeding." During hunting, they focus on aggregations of krill or tiny fish near the ocean's surface [30]. They generate a unique arrangement of bubbles in a circular or figure-eight configuration to ensnare their prey. Until the year 2011, the researchers could solely observe this phenomenon from the ocean's surface. Cade *et al.* [32] offered enhanced insights by affixing tag sensors to humpback whales. Utilizing these sensors, they documented more than three hundred bubble-net feeding occurrences involving nine distinct whales. Their research revealed two primary feeding strategies: "upward spirals" and "dual loops." In the upward spiral technique, whales descend beneath a school of fish and ascend circularly while emitting bubbles, so they trap their prey. The research uncovered insights into how these clever mammals synchronize their hunting techniques, emphasizing their sophisticated cognitive capabilities.

12.2.1.1 Encircling prey

Humpback whales are recognized for their proficiency at locating and encircling their prey efficiently. In optimization, this behavior is mimicked by positing the best candidate solution identified thus far, which signifies the target prey or is at least proximate to the ideal solution. Due to the uncertainty of the optimal solution's precise placement inside the search space, other search agents dynamically modify their locations, converging toward the best-known solution. This procedure enhances search and increases the likelihood of identifying the global optimum. The following is a mathematical description of this behavior:

$$\vec{Dis} = \mid \vec{C} \cdot \vec{X^{*}_{whale}}(t) - \vec{X_{whale}}(t) \mid \tag{12.1}$$

$$\vec{X_{whale}}(t+1) = \vec{X^{*}_{whale}}(t) - \vec{A} \cdot \vec{Dis} \tag{12.2}$$

whereas t illustrates the present iteration, while $\vec{C}$ and $\vec{A}$ are coefficient vectors that influence the search process. $\overrightarrow{X^*_{whale}}$ denotes the position vector of the optimal solution found accordingly and $\overrightarrow{X_{whale}}$ represents the position vector of an individual search agent. The vertical bars || indicate the absolute value, and the · symbol signifies element-wise multiplication.

The coefficient vectors $\vec{C}$ and $\vec{A}$ are determined using the following equations:

$$\vec{A} = 2\vec{a} \cdot \vec{r_1} - \vec{a} \tag{12.3}$$

$$\vec{C} = 2 \cdot \vec{r_2} \tag{12.4}$$

whereas $\vec{a}$ covers both the exploitation and exploration stages. Meanwhile, $\vec{r_1}, \vec{r_2}$ are the random vectors with values varying between 0 and 1. As discussed in the preceding part, the humpback whales assault the prey using the bubble-net method, as depicted in Figure 12.2.

12.2.1.2 Bubble-net hunting behavior: A model for exploitation in optimization

This scheme plays a crucial role in their foraging strategy, and this behavior is mathematically modeled in the WOA through two key mechanisms: **shrinking encircling** and **spiral updating positions.**

Shrinking encircling strategy

In nature, humpback whales gradually tighten their circular movement around their prey to trap them efficiently. In WOA, this behavior is replicated by gradually reducing the value of parameter $\vec{a}$ in (12.5). As the value of $\vec{a}$ decreases, the search agents (whales) move closer to the optimal solution identified so far, mimicking the real-world tightening motion of whales as they close in on their prey:

$$\overrightarrow{X_{whale}}(t+1) = \overrightarrow{Dis'} \cdot e^{bl} \cdot \cos(2\pi l) + \overrightarrow{X^*_{whale}}(t) \tag{12.5}$$

where $\overrightarrow{Dis'} = \mid \overrightarrow{X^*_{whale}} \cdot (t)\overrightarrow{X_{whale}}(t+1)\ mid$ showcases the distance between the whale (search agent) and the prey (present optimal solution), b is a constant that dictates the shape of the logarithmic spiral, and l is a randomly generated number between -1 and 1.

Spiral updating coordinate

Apart from encircling their prey, humpback whales also follow a spiral-shaped trajectory while hunting. This motion is mathematically expressed as:

$$\overrightarrow{X_{whale}}(t+1) = \begin{cases} \overrightarrow{X^*_{whale}}(t) & if\, p \geq 0.5 \\ \overrightarrow{Dis'} \cdot e^{bl} \cdot \cos(2\pi l) + \overrightarrow{X^*_{whale}}(t) & else \end{cases} \tag{12.6}$$

In addition to using the bubble-net strategy, humpback whales also search for prey in a random manner. This behavior can be mathematically represented in (12.6), where p is a randomly generated number between 0 and 1.

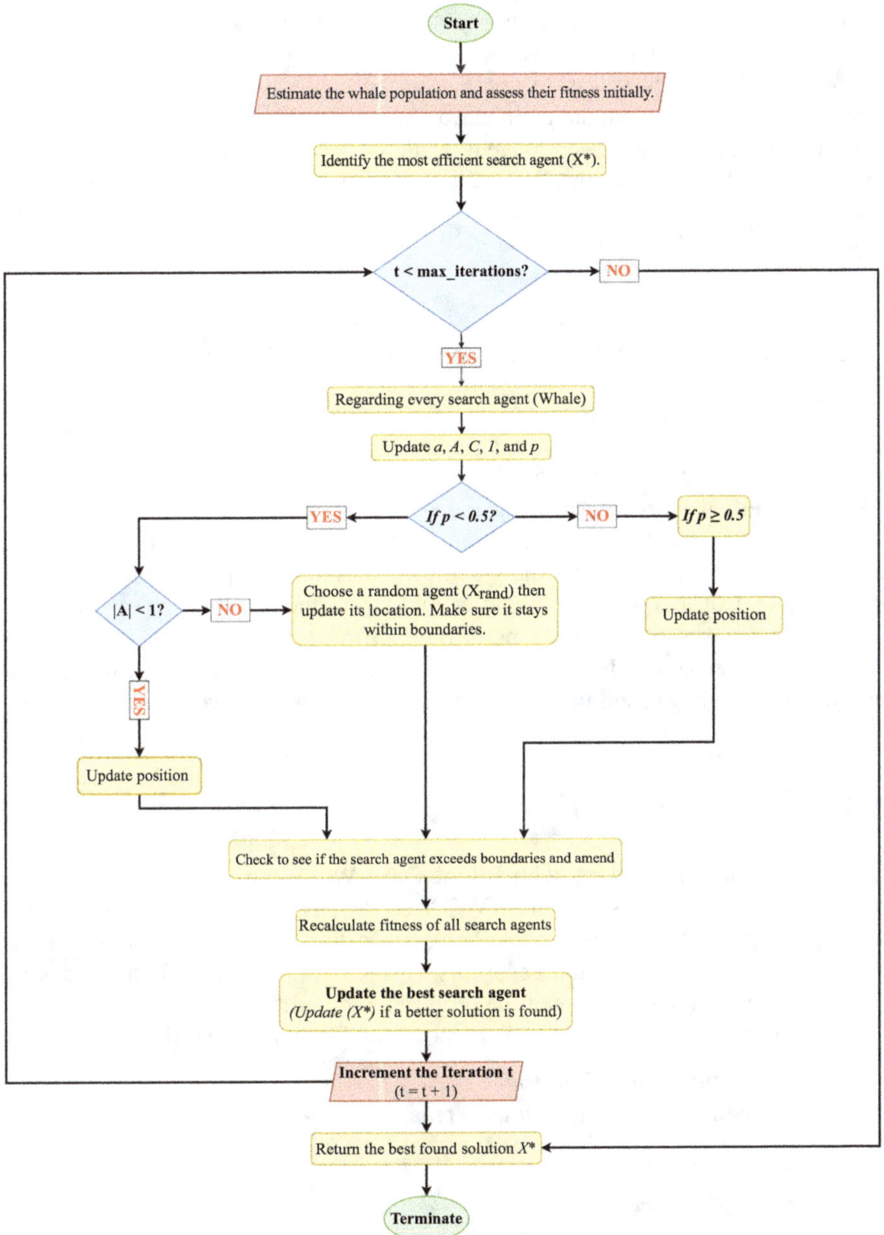

Figure 12.2 Workflow of Whale Optimization Algorithm

Prey search (exploration period)

To find prey, one may use the same technique based on the $\vec{A}$ vector variation (exploration). Humpback whales pursue objects in random order based on their relative positions. By assigning random values to $\vec{A}$ that are either higher than 1 or less than -1, we help the search agent move away from the best-known solution and explore new parts of the search space. In contrast to the exploitation phase, we employ a search agent selected at random to update a search agent's position in the exploration phase, rather than the best search agent available. This strategy emphasizes exploration, and $|\vec{A}| > 1$ allows the WOA algorithm to do a worldwide search. Here's how the mathematical model appears:

$$\vec{Dis} = |\vec{C} \cdot \vec{X}_{rand} - \vec{X_{whale}}| \tag{12.7}$$

$$\vec{X_{whale}}(t+1) = \vec{X}_{rand} - \vec{A} \cdot \vec{Dis} \tag{12.8}$$

where the random position vector (a random whale) — $\vec{X}_{rand}$ is selected from the existing population.

12.2.2 Review of P&O Algorithm

The P&O method works by altering the operating voltage of a PV solar system and observing the resulting change in output power to track the MPP [33]. In each iteration, the algorithm perturbs the system by marginally adjusting the voltage (or the duty cycle) of a converter, thereafter measuring the resultant power output. If the power increases, it continues perturbing in the same direction; if the power decreases, it reverses the direction of perturbation. This process is repeated continuously, allowing the system to oscillate around the MPP. The workflow is shown in Figure 12.3.

12.3 Proposed method

The proposed algorithm starts with WOA. Once WOA completes its optimization process, the algorithm transitions from WOA to P&O, using the best solution obtained by WOA as the initial condition for P&O. Since this solution is already near the GMPP, P&O's fast perturbation capability ensures accurate and efficient tracking. Figure 12.4 shows the workflow of the proposed method.

The WOA switches to P & O when one of the following conditions is met:

1. A whale is sufficiently close to the prey.
2. A predefined number of iterations is reached.

12.4 Simulation results

12.4.1 Testing preparation

The proposed method is initially tested and analyzed using MATLAB® ver. R2024b. In the following studies, four series-connected PV modules form a PV string to generate six *P–V* curves with different GMPPs representing various PSCs. The PSCs are simulated by varying the irradiance levels of the modules, as shown in Table 12.1. Meanwhile, Table 12.2 lists the GMPPs and voltage at the GMPP (VGMPP) of the

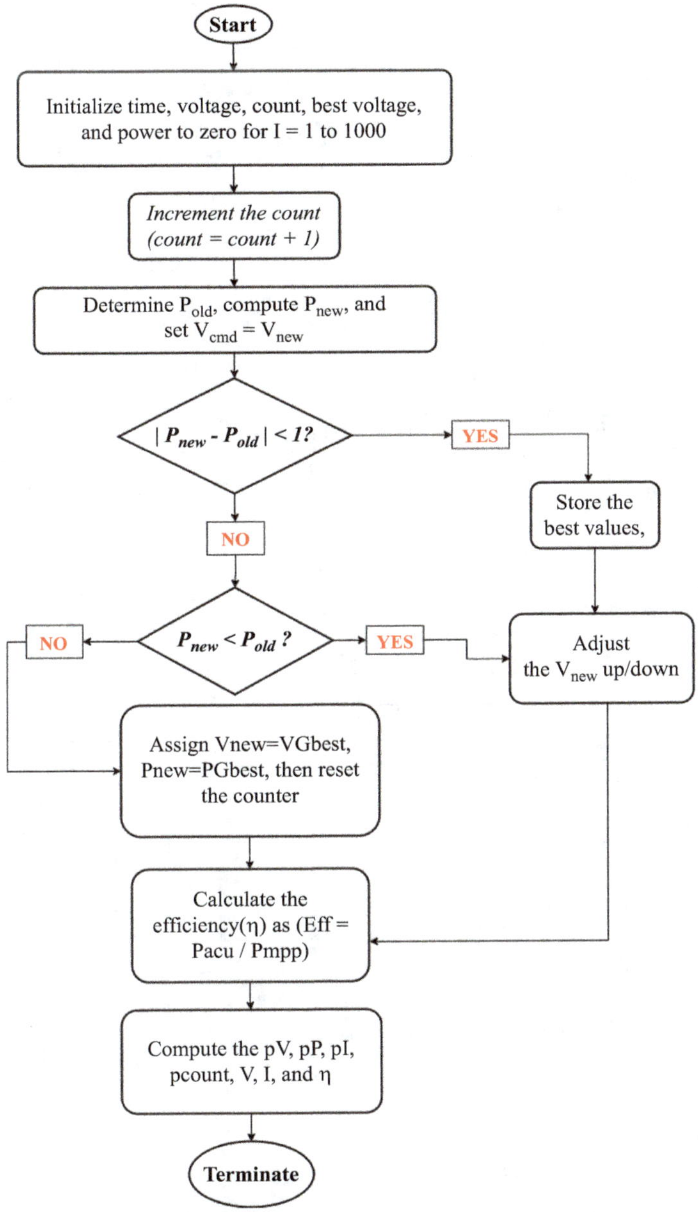

Figure 12.3 Workflow of P&O Algorithm

six *P–V* curves. Figure 12.5 illustrates the form of six *P–V* curves. These curves are subsequently utilized to assess the effectiveness of the proposed method.

For comparison purposes, the proposed method was evaluated alongside WOA, P&O, and SA in tracking the GMPP across all generated *P–V* curves. The performance of each algorithm was assessed based on tracking time and accuracy. To

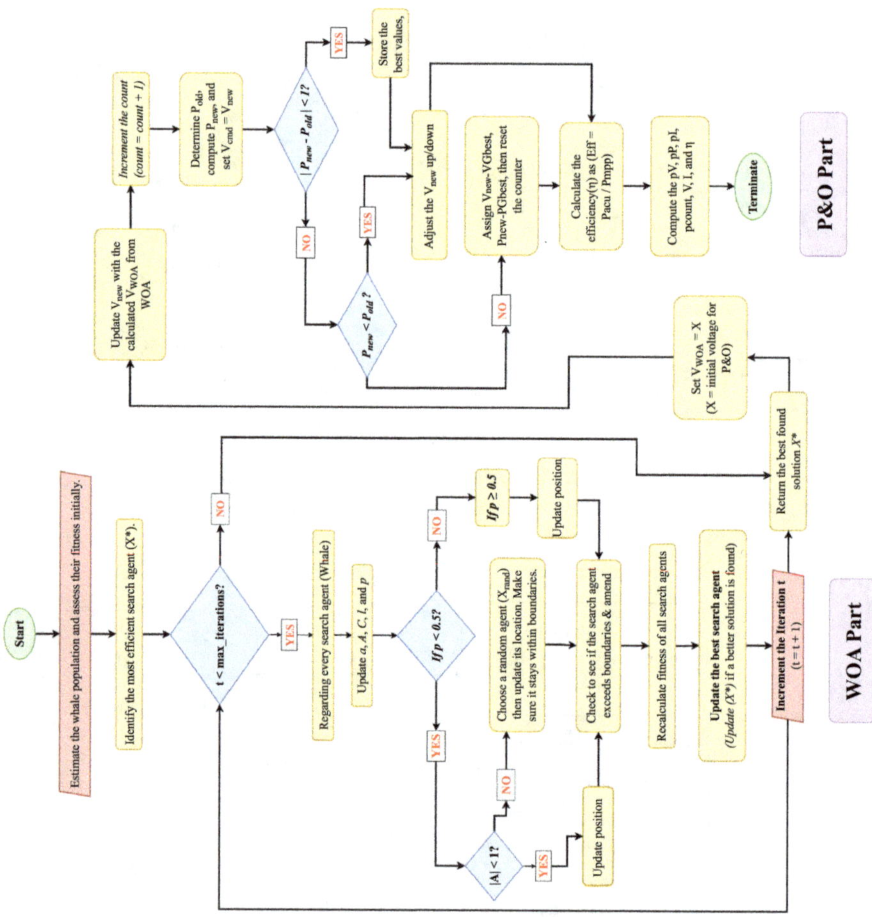

Figure 12.4 Workflow of the Proposed method

Table 12.1 Specifications of the P–V curves

	Irradiance (W/m^2)			
	Module			
Pattern	**Number 1**	**Number 2**	**Number 3**	**Number 4**
A	700	700	400	400
B	1000	600	500	400
C	1000	1000	700	400
D	1000	800	600	400
E	600	600	200	200
F	1000	950	300	200

Table 12.2 P–V curve description

Pattern	**VGMPP (V)**	**GMPP (W)**	**Description**
A	180.06	200.42	Two peaks, GMPP in the right side
B	184.52	209.84	Four peaks, GMPP in the right side
C	138.71	266.69	Three peaks, GMPP in the middle
D	138.22	229.57	Four peaks, GMPP in the middle-right side
E	86.44	136.12	Two peaks, GMPP in the left side
F	88.58	220.32	Three peaks, GMPP in the left side

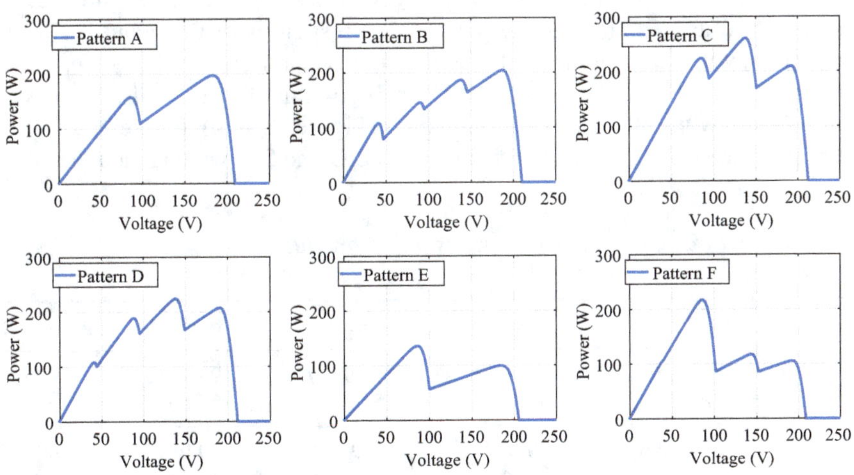

Figure 12.5 The tested P–V validation curves

ensure a fair comparison, the initial values for both the proposed method and WOA were uniformly set at [90%, 75%, 60%, 45%, 30%] times the voltage of open-circuit

(V_{oc}) during the first iteration. Meanwhile, the initial values for P&O and SA were uniformly set at 0.75 times V_{oc}.

The tracking accuracy is defined as:

$$\text{Accuracy} = \frac{P_{\text{tracked}}}{P_{\text{reference}}} \times 100\% \tag{12.9}$$

where P_{tracked} is the average power delivered by the MPPT system over a given period and $P_{\text{reference}}$ is the reference maximum power available from the system during the same period.

12.4.2 Parameter selection

WOA involves two key parameters, *a* and *b*, which must be carefully selected, as they significantly affect both convergence time and tracking accuracy. In this study, the selection of these parameters is prioritized based on accuracy to ensure effective tracking of the GMPP. Based on this criterion, extensive simulations were conducted to determine the optimal values. As a result, *a* and *b* were set to 1.5 and 0.5, respectively. Here, *a* represents the convergence factor, while *b* is the constant that defines the shape of the logarithmic spiral. The corresponding results are presented in Table 12.3.

12.4.3 Simulation results

The power tracking curve of the proposed method under Pattern A is shown in Figure 12.6(a), where the GMPP is reached in just 2.94 seconds. On the other hand, WOA requires 5.49 seconds to reach the GMPP, as depicted in Figure 12.7(a). Figures 12.8(a) and 12.9(a) display the tracking responses of P&O and SA for the same pattern, with tracking times of 1.36 and 4.19 seconds, respectively. For Pattern B, the proposed method achieves GMPP tracking within 3.03 seconds, as shown in Figure 12.6(b), while WOA takes 5.99 seconds (Figure 12.7(b)). Figures 12.8(b) and 12.9(b) illustrate the tracking performances of P&O and SA, which reach the GMPP

Table 12.3 Tracking results with various a and b values

		$a = 0.5$	$a = 1$	$a = 1.5$	$a = 2$
$b = 0.5$	Accuracy	99.9776	99.9727	99.9819	99.963
	Time	4.76	5.495	5.265	5.55
$b = 1$	Accuracy	99.9445	99.9756	99.9812	99.9676
	Time	4.485	5.51	5.245	5.66
$b = 1.5$	Accuracy	99.9665	99.9678	99.9765	99.964
	Time	4.325	5.24	5.025	5.32
$b = 2$	Accuracy	99.944	99.9204	99.9631	99.9613
	Time	4.37	5.1	4.89	5.21

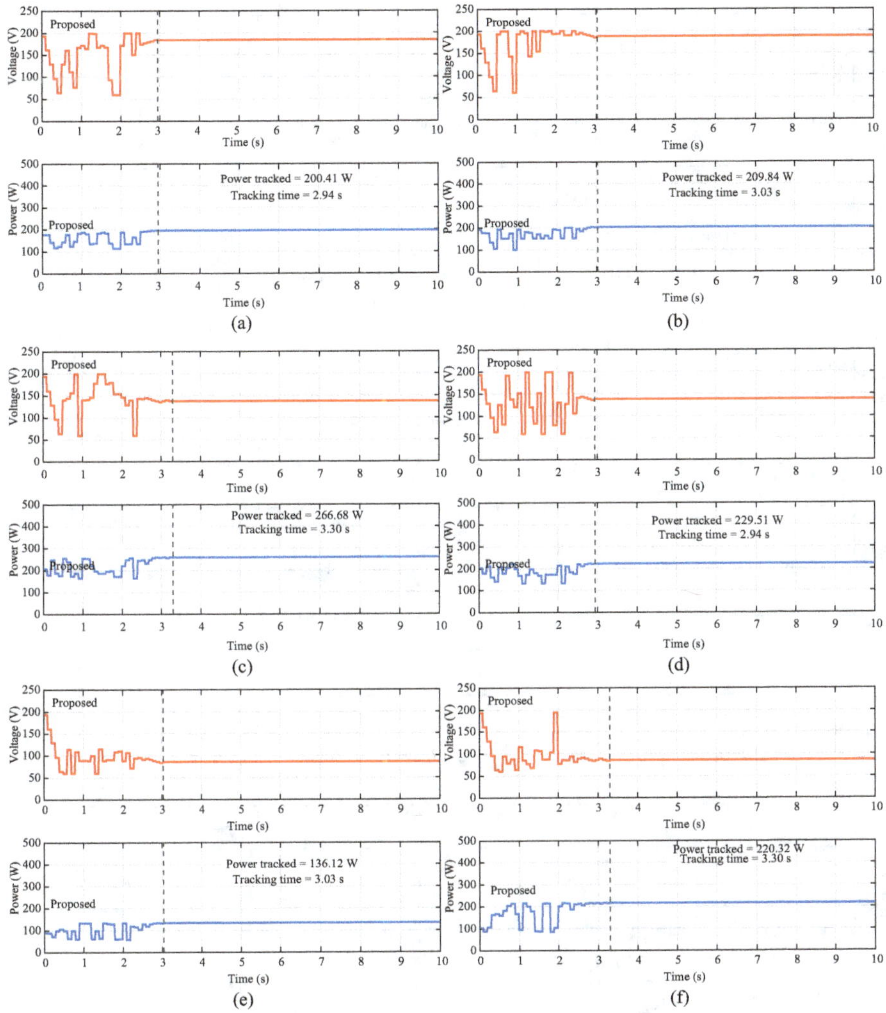

Figure 12.6 Simulation result of the proposed method under Patterns A to F

in 1.45 and 5.59 seconds, respectively. For Patterns D–F, the tracking profiles of the proposed method and WOA are generally consistent with those observed in Patterns A and B. However, P&O becomes trapped in local maxima for Patterns C–F, failing to reach the GMPP. Similarly, SA exhibits a similar issue under Pattern E. The average tracking performance of the four MPPT algorithms across all patterns is summarized in Tables 12.4 and 12.5. Overall, the proposed method delivers the highest accuracy at 97.9251%, followed by WOA (97.8262%), SA (92.8593%), and P&O (85.1608%). These results confirm the superiority of the proposed method in terms of tracking accuracy under varying irradiance conditions.

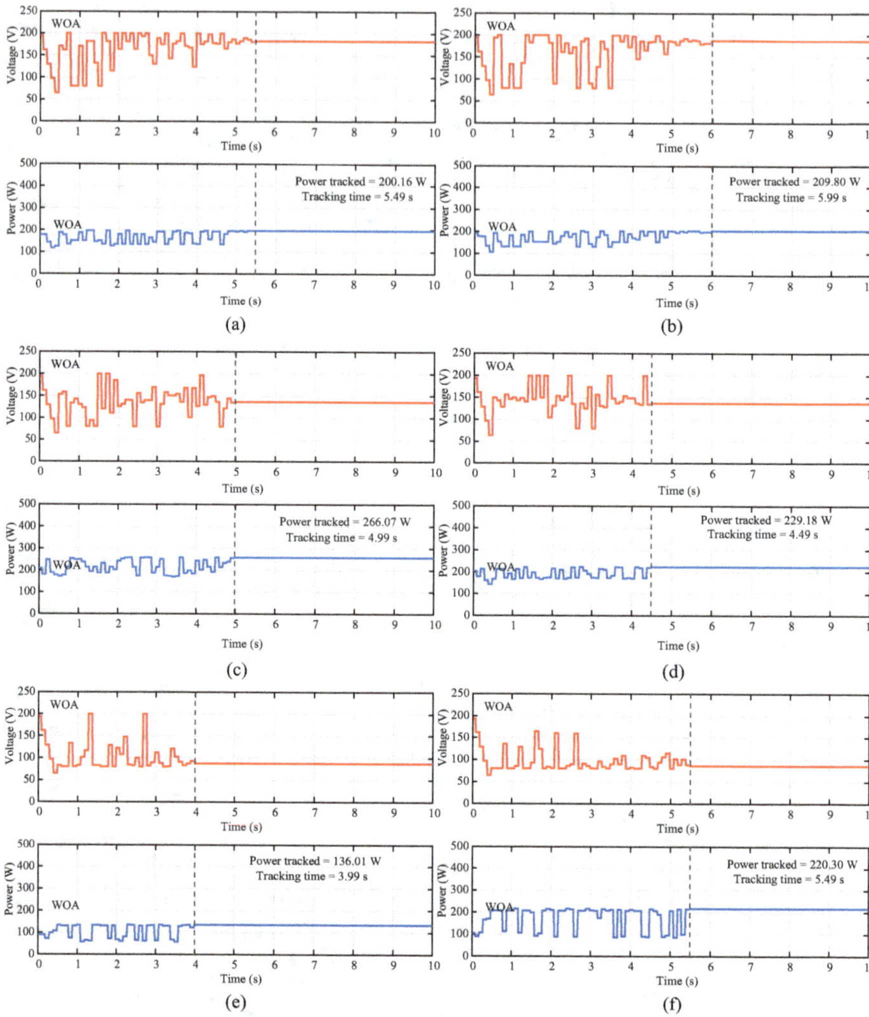

Figure 12.7 Simulation result of the WOA under Patterns A to F

12.5 Hardware experimentation analysis and results

The experimental setup in this study consists of an emulator of solar panels, a digital signal processor (DSP), a DC power supply, and an interleaved boost converter, as illustrated in Figure 12.10. The PV emulator accurately simulates PV panel characteristics under various environmental conditions, including PSCs. The MPPT control algorithm is implemented on a DSP, chosen for its ability to generate precise gating signals required for converter control. The converter used in this experiment comprises double sets of diodes, inductors, and electronic switches, offering higher

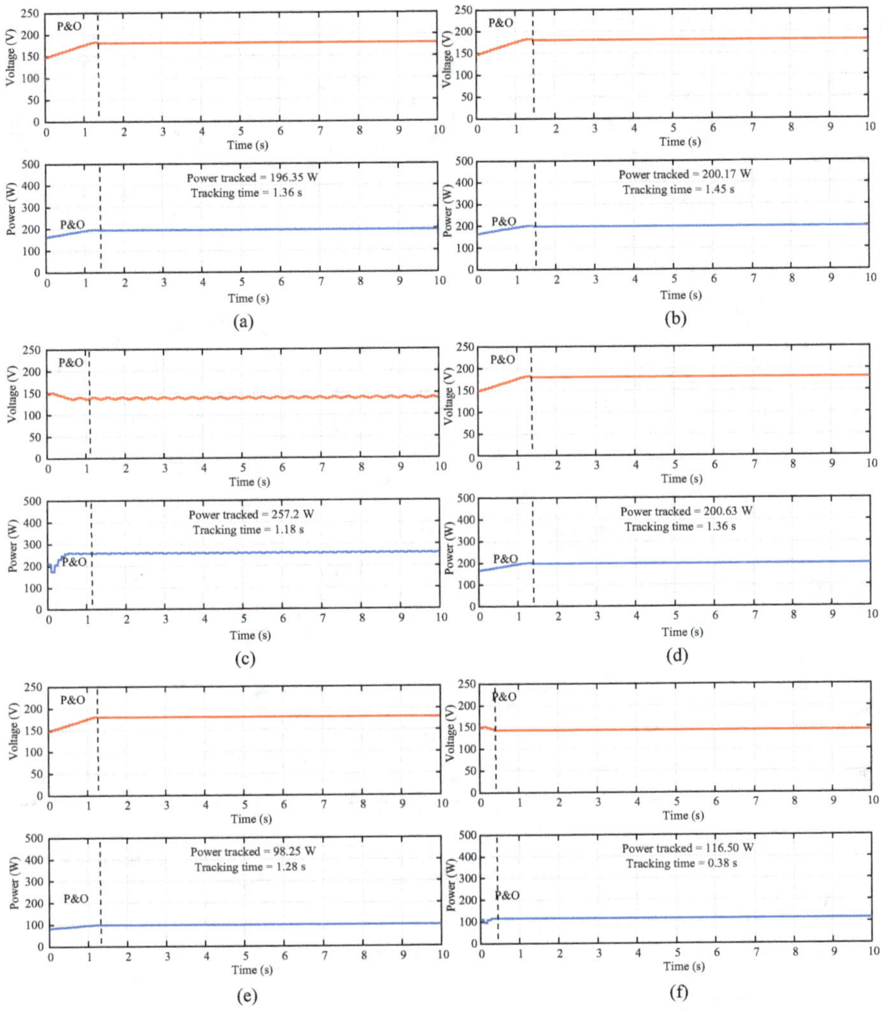

Figure 12.8 Simulation result of the P&O under Patterns A to F

efficiency compared to a standard boost converter. The DC power supply maintains a voltage of 450 V at the load terminals.

12.5.1 Results of static case analysis

Figures 12.11 to 12.14 are the results for the static case using experimental data. The tracking accuracy and tracking time are shown in Tables. 12.6 and 12.7. On average the proposed method achieved the highest accuracy at 94.654%, outperforming WOA (94.2629%), SA (92.1857%), and P&O (85.5005%). For tracking time, on average the proposed method achieves 2.30 s, WOA with 4.52 s, P&O with 1.05 s, and SA with 4.05 s. P&O exhibited the shortest average tracking time at 1.05 s. However, it

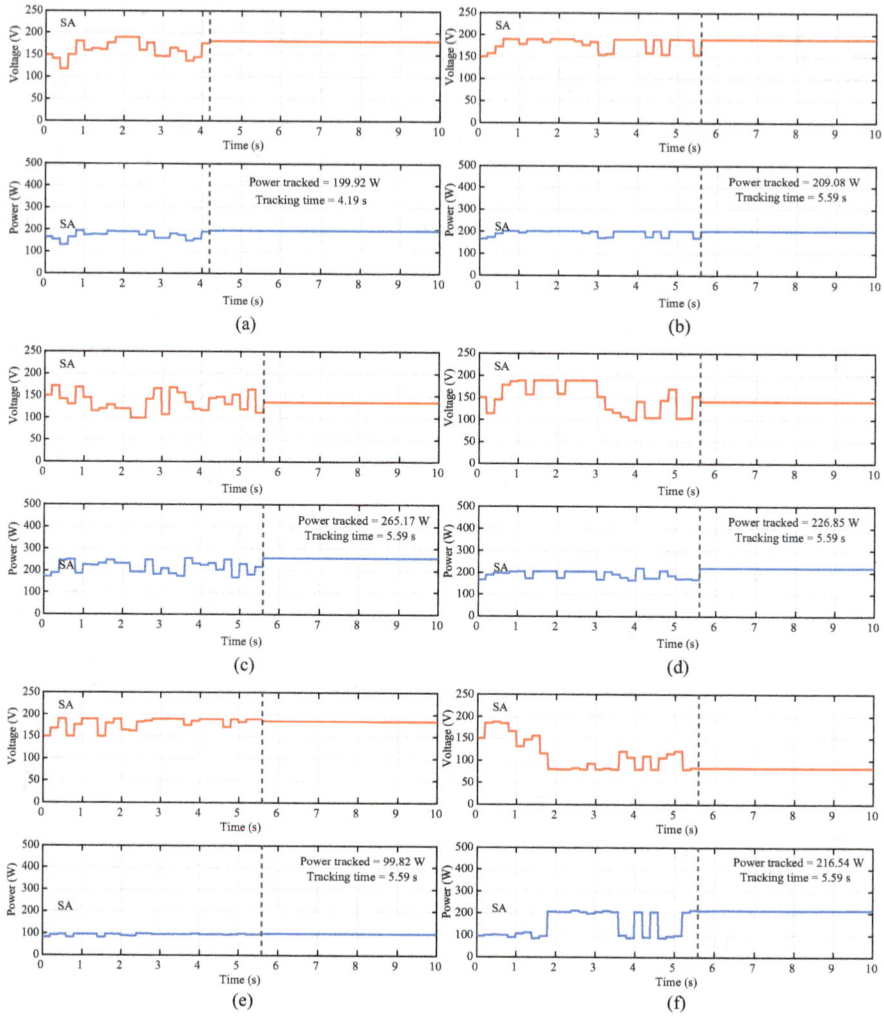

Figure 12.9 Simulation result of the SA under Patterns A to F

failed to identify the GMPP under patterns D to F due to convergence to local optima. In comparison, the proposed method maintained a higher reliability with an average tracking time of 2.30 s, resulting in a 9.15% improvement in accuracy over P&O.

12.5.2 Results of dynamic case analysis

The rate at which clouds change, thus affecting solar irradiance and MPPT, can be surprisingly rapid. The dynamics of cloud movement and formation can lead to irradiance transients (fast changes in sunlight intensity) that challenge MPPT algorithms [34,35].

Table 12.4 Tracking accuracy of four methods (Simulation)

Pattern	Method			
	Proposed	WOA	P&O	SA
A	98.0732%	97.9508%	97.9693%	97.8334%
B	97.0584%	97.0399%	95.3917%	96.7069%
C	97.5412%	97.3181%	96.4416%	96.9889%
D	97.4878%	97.3476%	96.1058%[1]	96.3579%
E	99.1294%	99.0493%	72.1790%[1]	72.6939%[1]
F	98.2604%	98.2515%	52.8776%[1]	96.5746%
Overall	97.9251%	97.8262%	85.1608%	92.8593%

[1] Trapped at local maximum.

Table 12.5 Tracking time of four methods (Simulation)

Pattern	Method			
	Proposed	WOA	P&O	SA
A	2.94 (s)	5.49 (s)	1.36 (s)	4.19 (s)
B	3.03 (s)	5.99 (s)	1.45 (s)	5.59 (s)
C	3.30 (s)	4.99 (s)	1.18 (s)	5.59 (s)
D	2.94 (s)	4.49 (s)	1.36 (s)[1]	5.59 (s)
E	3.03 (s)	3.99 (s)	1.28 (s)[1]	5.59 (s)[1]
F	3.30 (s)	5.49 (s)	0.38 (s)[1]	5.59 (s)
Overall	3.09 (s)	5.07 (s)	1.17 (s)	5.36 (s)

[1] Trapped at local maximum.

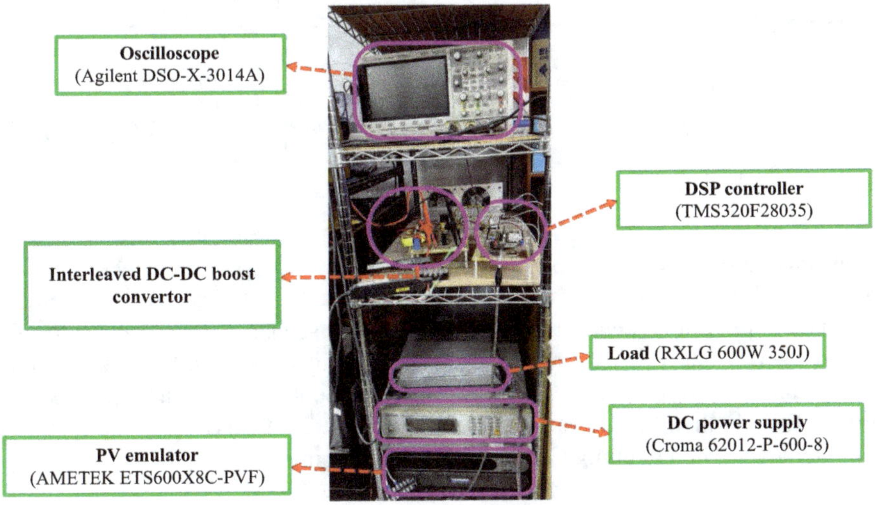

Figure 12.10 Experimental setup of PV system

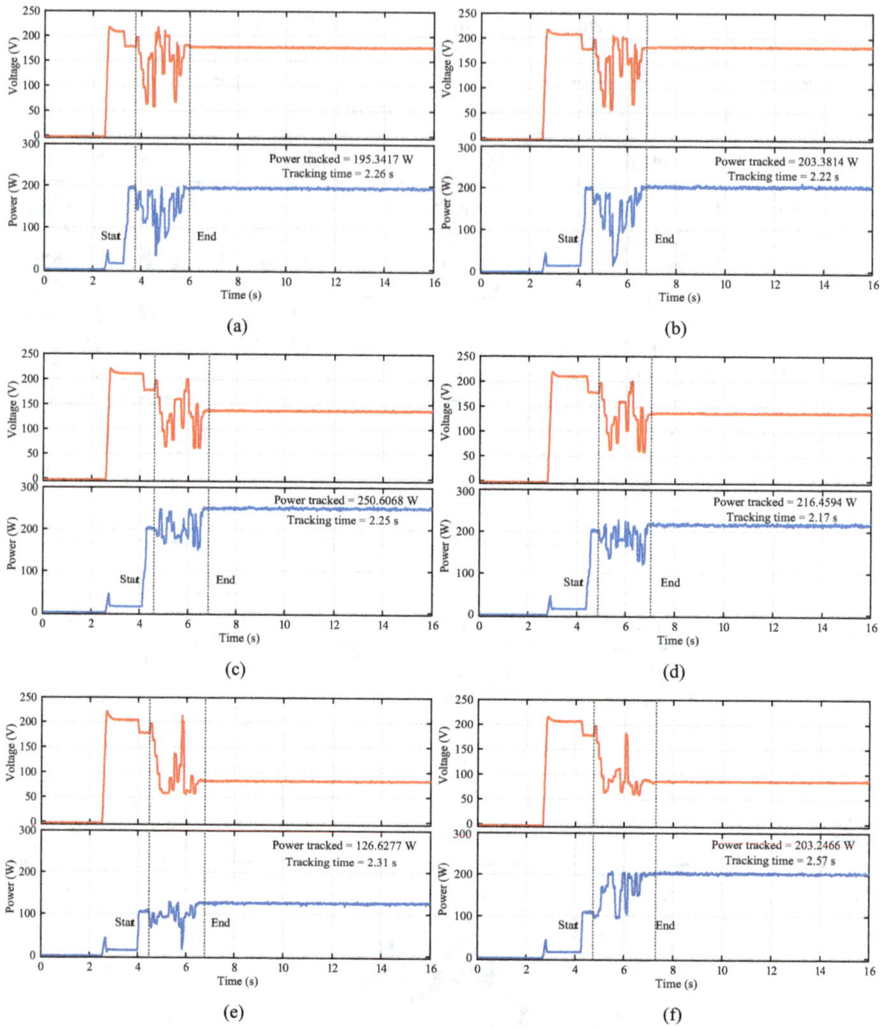

Figure 12.11 Experimental result of the proposed method under Patterns A to F

In this study, two scenarios designed to evaluate the proposed method under varying conditions. In Case 1, the transition follows the sequence from patterns C to D, then to pattern A, and finally to pattern E. This scenario is intended to test the method's performance when tracking the GMPP as it shifts from the highest to the lowest value. Conversely, Case 2 follows the reverse sequence, transitioning from patterns E to A, then to pattern D, and ending at pattern C, assessing the method's ability to adapt when tracking GMPP in the opposite order.

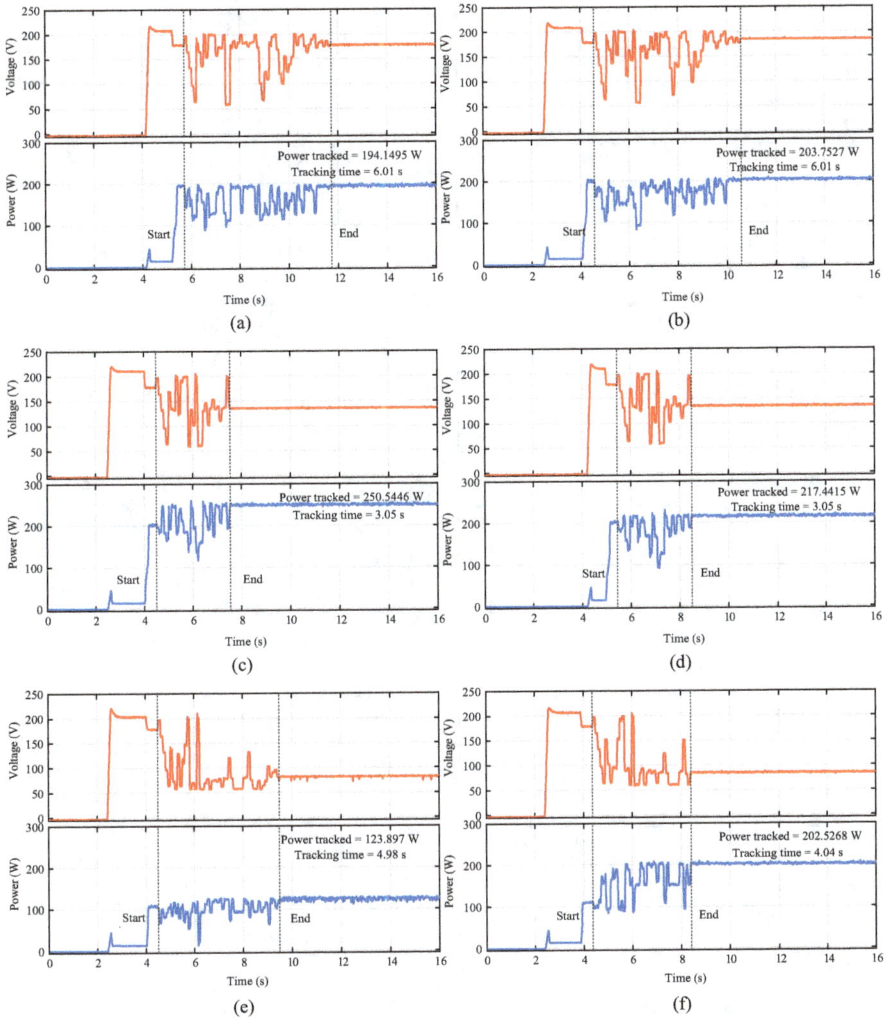

Figure 12.12 Experimental result of the WOA under Patterns A to F

In the dynamic testing, a 12-second gap was used between each irradiance condition change. Two main equations are utilized. The first equation is a criterion for determining whether a curve transition has occurred, as shown in (12.10).

$$\frac{|P_{\text{GMPP}} - P_{\text{GMPP,last}}|}{P_{\text{GMPP,last}}} \geq \Delta P(\%) \tag{12.10}$$

where P_{GMPP} specifies the GMPP power at the most recent operating point, whereas ΔP represents the threshold difference, set at 5%. If this condition is satisfied, the MPPT algorithm resets its search agents to find the new GMPP.

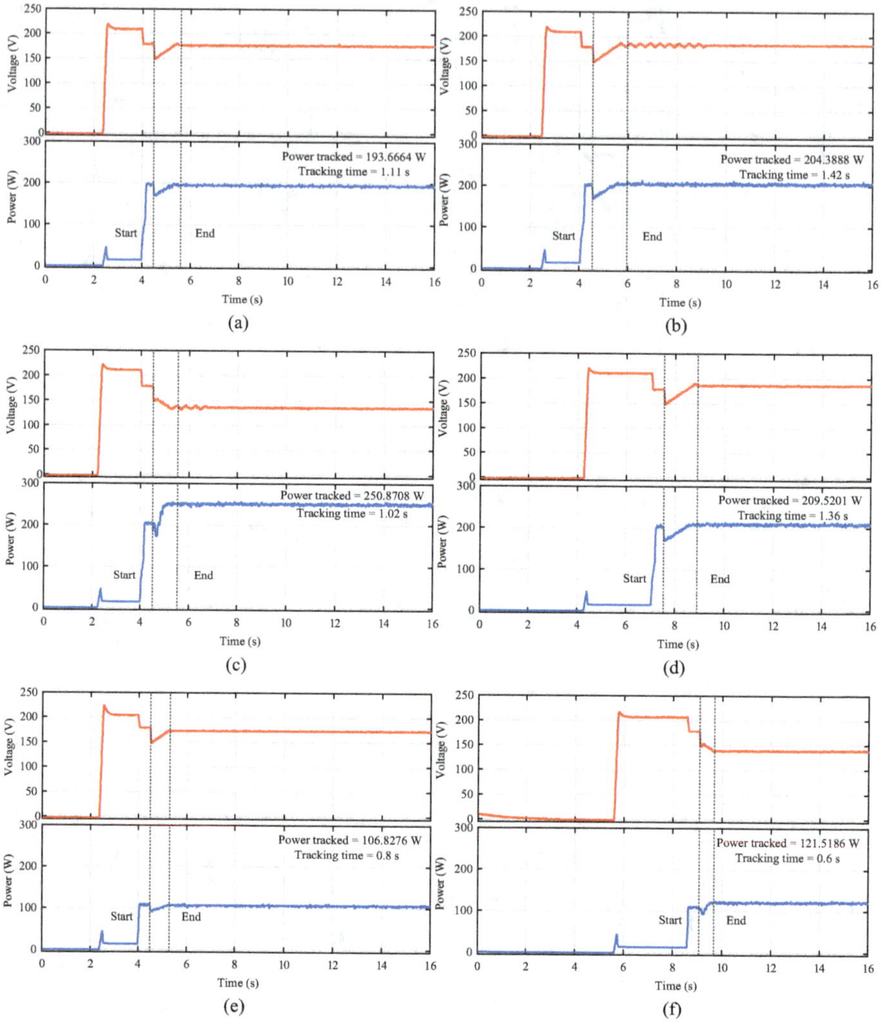

Figure 12.13 Experimental result of the P&O under Patterns A to F

The second equation is used to calculate the overall dynamic tracking efficiency, as presented in (12.11).

$$\eta_d = \frac{\sum_i V_{\text{pv},i} \cdot I_{\text{pv},i} \cdot \Delta T_i}{\sum_j \text{GMPP}_{\text{th},j} \cdot \Delta T_j} \times 100\% \tag{12.11}$$

where ΔT_j specifies the time in which the GMPP ($\text{GMPP}_{\text{th},j}$) is ideally attained, and ΔT_i denotes the sampling interval during which voltage ($V_{\text{pv},i}$) and current ($I_{\text{pv},i}$) measurements are obtained. Thus, along with low fluctuations, fast algorithm convergence to the GMPP under different PSCs has higher η_d.

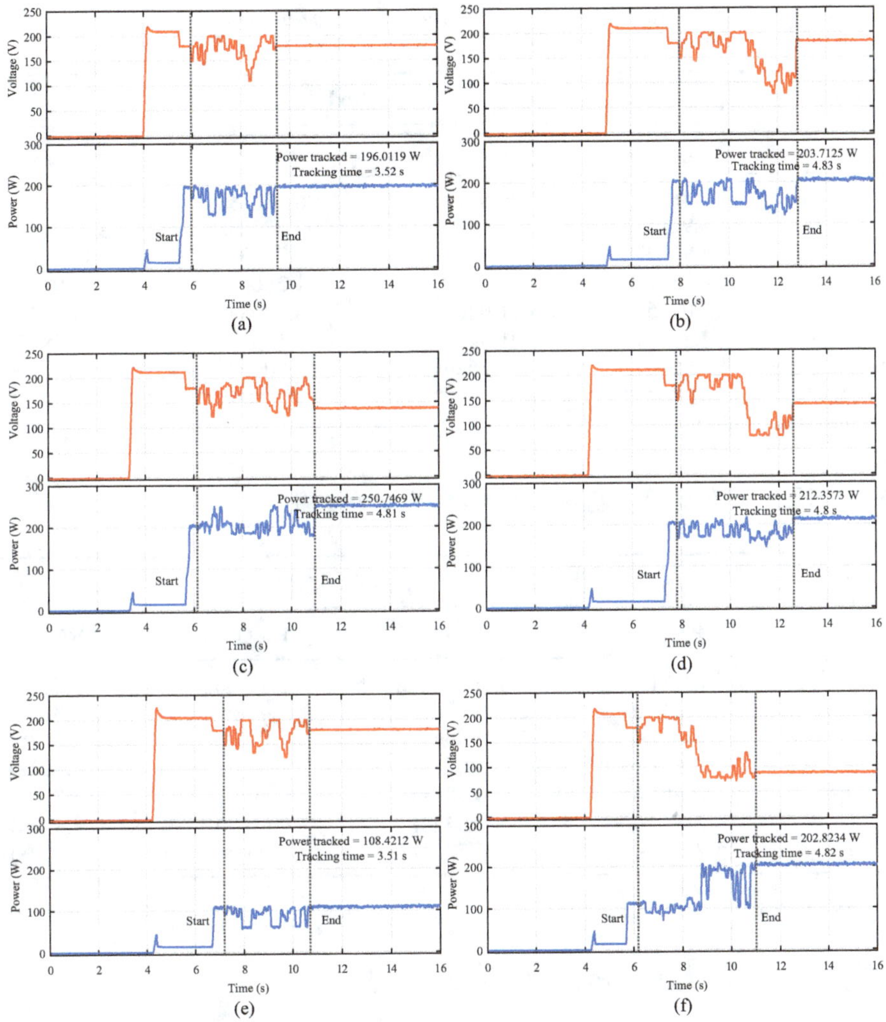

Figure 12.14 Experimental result of the SA under Patterns A to F

The results are shown in Table 12.8 and Figures 12.15 and 12.16. The proposed method achieved the highest accuracy at 91.5975%, outperforming WOA (88.8101%), SA (88.0808%), and P&O (88.4805%). For Case 2, the results are shown in Figure 12.16(a)–(d). The accuracy achieved by each algorithm is 91.1971% for the proposed method, 87.4067% for WOA, 91.3229% for P&O, and 86.8775% for SA. Overall, the proposed method achieved the highest accuracy at 91.3973%, outperforming WOA (88.1084%), SA (87.4792%), and P&O (89.9017%).

Table 12.6 Tracking accuracy of four methods (Experiment)

Pattern	**Method**			
	Proposed	**WOA**	**P&O**	**SA**
A	97.4662%	96.8713%	96.6303%	97.8006%
B	96.9221%	97.0991%	97.4022%	97.0799%
C	93.9693%	93.9460%	94.0683%	94.0219%
D	94.2891%	94.7169%	91.2663%[1]	92.5022%
E	93.0265%	91.0204%	78.4805%[1]	79.6512%[1]
F	92.2506%	91.9236%	55.1555%[1]	92.0586%
Overall	94.6540%	94.2629%	85.5005%	92.1857%

[1] Trapped at local maximum.

Table 12.7 Tracking time of four methods (Experiment)

Pattern	**Method**			
	Proposed	**WOA**	**P&O**	**SA**
A	2.26 (s)	6.01 (s)	1.11 (s)	3.52 (s)
B	2.22 (s)	6.01 (s)	1.42 (s)	4.83 (s)
C	2.25 (s)	3.05 (s)	1.02 (s)	4.81 (s)
D	2.17 (s)	3.05 (s)	1.36 (s)[1]	4.80 (s)
E	2.31 (s)	4.98 (s)	0.80 (s)[1]	3.51 (s)[1]
F	2.57 (s)	4.04 (s)	0.60 (s)[1]	4.82 (s)
Overall	2.30 (s)	4.52 (s)	1.05 (s)	4.38 (s)

[1] Trapped at local maximum.

Table 12.8 Efficiency results for dynamic cases

	Method			
Scenarios	**Proposed**	**WOA**	**P&O**	**SA**
Case 1	91.5975	88.8101	88.4805	88.0808
Case 2	91.1971	87.4067	91.3229	86.8775

12.6 Conclusion

This study introduces an innovative MPPT algorithm that combines the WOA with the P&O approach to improve tracking performance. The WOA component effectively mitigates the issue of local maxima by facilitating a global search. At the

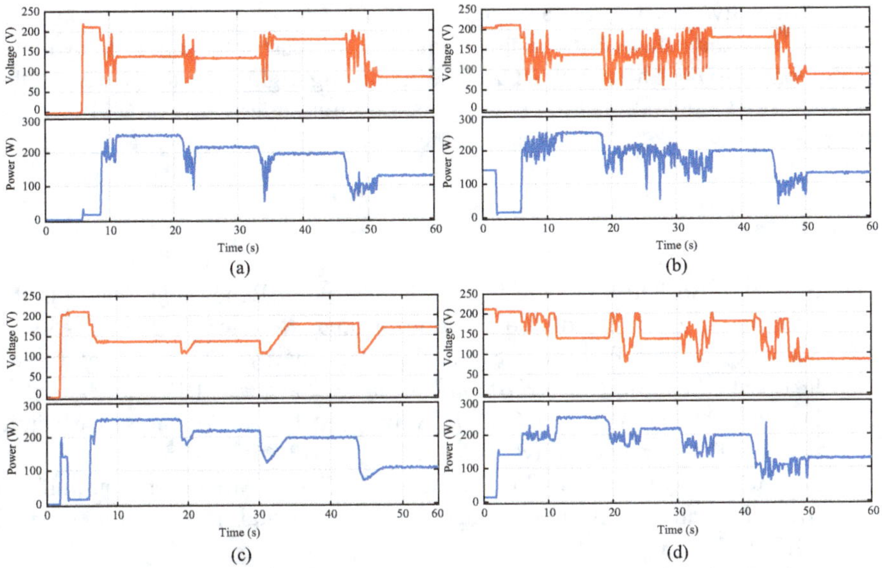

Figure 12.15 Trajectories of the dynamic Case 1: (a) Proposed; (b) WOA; (c) P&O; (d) SA

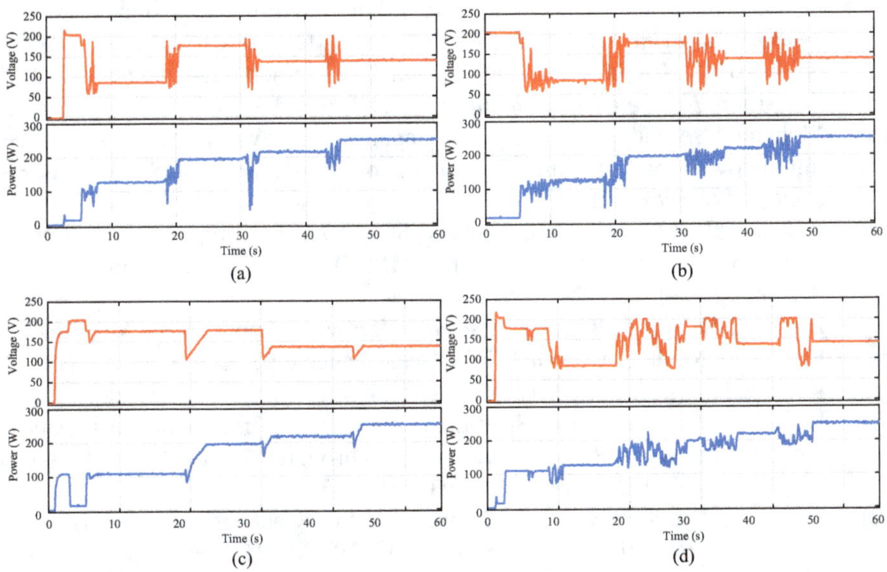

Figure 12.16 Trajectories of the dynamic Case 2: (a) Proposed; (b) WOA; (c) P&O; (d) SA

same time, a strategically designed transition mechanism enables a smooth transition to P&O for precise tracking of the GMPP. A comparative examination reveals that the proposed method not only attains more accuracy than the WOA for the specified objective function but also significantly decreases tracking time. The validation findings indicate that the proposed method decreases tracking time by about 50% compared to the traditional WOA. The results highlight the effectiveness of the proposed MPPT method, making it an optimal option under PSCs of MPPT in PV systems.

References

[1] Gielen D, Boshell F, Saygin D, *et al.* The Role of Renewable Energy in the Global Energy Transformation. *Energy Strategy Reviews*. 2019;24:38–50.
[2] Kabir E, Kumar P, Kumar S, *et al.* Solar Energy: Potential and Future Prospects. *Renewable and Sustainable Energy Reviews*. 2018;82:894–900.
[3] Deline C, Dobos A, Janzou S, *et al.* A Simplified Model of Uniform Shading in Large Photovoltaic Arrays. *Solar Energy*. 2013;96:274–282.
[4] Di Vincenzo MC and Infield D. Detailed PV Array Model for Non-uniform Irradiance and Its Validation against Experimental Data. *Solar Energy*. 2013;97:314–331.
[5] Bai J, Cao Y, Hao Y, *et al.* Characteristic Output of PV Systems under Partial Shading or Mismatch Conditions. *Solar Energy*. 2015;112:41–54.
[6] Huang L, Pan B, Wang S, *et al.* Review on Maximum Power Point Tracking Control Strategy Algorithms for Offshore Floating Photovoltaic Systems. *Journal of Marine Science and Engineering*. 2024;12(12):2121.
[7] Nijsse FJMM, Mercure JF, Ameli N, *et al.* The Momentum of the Solar Energy Transition. *Nature Communications*. 2023;14:6542.
[8] Mohamed SA and Sattar MAE. A Comparative Study of P&O and INC Maximum Power Point Tracking Techniques for Grid-Connected PV Systems. *SN Applied Sciences*. 2019;1:174.
[9] Katche ML, Makokha AB, Zachary SO, *et al.* A Comprehensive Review of Maximum Power Point Tracking (MPPT) Techniques Used in Solar PV Systems. *Energies*. 2023;16(5):2206.
[10] Bassi H, Salam Z, Ramli MZ, *et al.* Hardware Approach to Mitigate the Effects of Module Mismatch in a Grid-connected Photovoltaic System: A Review. *Energies*. 2019;12(22):4321.
[11] Raza MA, Zahra S, Raza S, *et al.* Mitigating the Impact of Partial Shading Conditions on Photovoltaic Arrays Through Modified Bridge-Linked Configuration. *Sustainability*. 2025;17(3):1263.
[12] Sagonda AF and Folly KA. A Comparative Study between Deterministic and Two Meta-Heuristic Algorithms for Solar PV MPPT Control under Partial Shading Conditions. *Systems and Soft Computing*. 2022;4:200040.
[13] Boubaker O. MPPT Techniques for Photovoltaic Systems: A Systematic Review in Current Trends and Recent Advances in Artificial Intelligence. *Discover Energy*. 2023;3(9):1–40.
[14] Seyedmahmoudian M, Horan B, Soon TK, *et al.* State of the Art Artificial Intelligence-based MPPT Techniques for Mitigating Partial Shading Effects on PV systems – A Review. *Renewable and Sustainable Energy Reviews*. 2016;64:435–455.

[15] Liu F, Kang Y, Zhang Y, *et al.* Comparison of P&O and Hill Climbing MPPT Methods for Grid-connected PV Converter. In: *Proceedings of the 3rd IEEE Conference on Industrial Electronics and Applications (ICIEA)*. IEEE; 2008. p. 804–807.

[16] Andrean V, Chang PC, and Lian KL. A Review and New Problems Discovery of Four Simple Decentralized Maximum Power Point Tracking Algorithms—Perturb and Observe, Incremental Conductance, Golden Section Search, and Newton's Quadratic Interpolation. *Energies*. 2018;11(11):2966.

[17] Villegas-Mier CG, Rodriguez-Resendiz J, Álvarez-Alvarado JM, *et al.* Artificial Neural Networks in MPPT Algorithms for Optimization of Photovoltaic Power Systems: A Review. *Micromachines*. 2021;12(10):1260.

[18] Olabi AG, Abdelkareem MA, Semeraro C, *et al.* Artificial Neural Networks Applications in Partially Shaded PV Systems. *Thermal Science and Engineering Progress*. 2023;37:101612.

[19] Millah IS, Subroto RK, Chang YW, *et al.* Investigation of Maximum Power Point Tracking of Different Kinds of Solar Panels Under Partial Shading Conditions. *IEEE Transactions on Industry Applications*. 2021;57(1):17–25.

[20] Wasim MS, Amjad M, Habib S, *et al.* A Critical Review and Performance Comparisons of Swarm-based Optimization Algorithms in Maximum Power Point Tracking of Photovoltaic Systems under Partial Shading Conditions. *Energy Reports*. 2022;8:4871–4898.

[21] Mahmoodi Tabar S, Shahnazari M, and Heshmati K. Maximum Power Point Tracking in Partially Shaded Photovoltaic Systems Using Grasshopper Optimization Algorithm. *IET Renewable Power Generation*. 2023;17(2):389–399.

[22] Lyden S and Haque ME. A Simulated Annealing Global Maximum Power Point Tracking Approach for PV Modules Under Partial Shading Conditions. *IEEE Transactions on Power Electronics*. 2016;31(6):4171–4181.

[23] Tomar Y and SuwarnozLian KL. A Novel MPPT Method Based on Crow Search Algorithm Combined with Perturb & Observe for Partial Shading Conditions. In: *2021 IEEE 12th Energy Conversion Congress & Exposition – Asia (ECCE-Asia)*; 2021. p. 1475–1480.

[24] Wijaya BH, Subroto RK, Lian KL, *et al.* A Maximum Power Point Tracking Method Based on a Modified Grasshopper Algorithm Combined with Incremental Conductance. *Energies*. 2020;13(17):4329.

[25] Lian KL, Jhang JH, and Tian IS. A Maximum Power Point Tracking Method Based on Perturb-and-Observe Combined with Particle Swarm Optimization. *IEEE Journal of Photovoltaics*. 2014;4(2):626–633.

[26] Nugraha DA, Lian KL, and Suwarno. A Novel MPPT Method Based on Cuckoo Search Algorithm and Golden Section Search Algorithm for Partially Shaded PV System. *Canadian Journal of Electrical and Computer Engineering*. 2019;42(3):173–182.

[27] Lian KL and Andrean V. A New MPPT Method for Partially Shaded PV System by Combining Modified INC and Simulated Annealing Algorithm. In: *2017 International Conference on High Voltage Engineering and Power Systems (ICHVEPS)*; 2017. p. 388–393.

[28] Mirjalili S and Lewis A. The Whale Optimization Algorithm. *Advances in Engineering Software*. 2016;95:51–67.
[29] Friedrich-Alexander-Universität Erlangen-Nürnberg. The language of whales. *FAU News*, 2020; Accessed: 2025-11-8.
[30] Hof PR and van der Gucht EV. Structure of the Cerebral Cortex of the Humpback Whale, *Megaptera novaeangliae* (Cetacea, Mysticeti, Balaenopteridae). *The Anatomical Record*. 2007;290(1):1–31.
[31] Project DE. The Evolution of Killer Whale Dialects. *CORDIS EU Research Results*. 2015; Accessed: 2025-03-18.
[32] Cade DE, Friedlaender AS, Calambokidis J, *et al.* Kinematic Diversity in Rorqual Whale Feeding Mechanisms. *Current Biology*. 2016;26(19):2617–2624.
[33] Dzulqarnain MW, Lian KL, Suwarno, *et al.* Accelerated Multiple Perturb and Observe Method for Maximum Power Point Tracking Utilizing Maximum Voltage Distribution Data. In: *2024 6th International Conference on Power Engineering and Renewable Energy (ICPERE)*; 2024. p. 1–6.
[34] Ali MH, Zakaria M, and El-Tawab S. A Comprehensive Study of Recent Maximum Power Point Tracking Techniques for Photovoltaic Systems. *Scientific Reports*. 2025;15(1):14269.
[35] Al Farisi FK, Fan ZK, and Lian KL. Comparative Study of White Shark Optimization and Combined Meta-Heuristic Algorithm for Enhanced MPPT in Photovoltaic Systems. *Energies*. 2025;18(8):2110.

Chapter 13

A quadratically constrained mixed-integer programming formulation for reconfiguration in electrical distribution networks with shunt capacitors and on-site generation

Pham Nang Van[1] *and Le Thi Minh Chau*[1]

This chapter introduces a formulation of an optimization problem that involves both integer and continuous variables with quadratic constraints (i.e., mixed-integer quadratically constrained program, MIQCP) to best ascertain the structure of a network for the minimization of the loss of electrical power in electrical distribution grids, taking into account dispersed generating (DG) and capacitors connected in parallel to the load. The suggested optimization model is developed from the original nonlinear optimization model by applying a modified linear approach to distribution power flow analysis. This optimization model can be efficiently resolved using typical commercial solvers like CPLEX. The proposed MIQCP formulation is verified on the 33-node test system and a practical 190-bus distribution system located in Vietnam during the year 2021. The impact of required electrical load levels on optimum solutions is examined comprehensively. Additionally, the outcomes of the load flow analysis derived based on the adjusted distribution power flow analysis method are juxtaposed with those obtained by resolving nonlinear load flow equations by employing the power summation technique, which provides precise answers.

13.1 Introduction

Restructuring of the network is the optimization of electrical distribution system topology by the alteration of sectionalizing and tie switch statuses.

Network reconfiguration can serve both planning purposes and real-time operational control. The goal of the grid reconfiguration challenge is to minimize power loss or balance loads to get the radial architecture of electrical distribution grids [1,2]. The distribution systems utilize a radial architecture for two primary reasons: (1) to facilitate system coordination and fault protection and (2) to reduce abnormal current flow caused by a fault [2]. Optimizing network design with reactive power sources

[1]School of Electrical and Electronic Engineering, Hanoi University of Science and Technology, Vietnam

like shunt capacitors may substantially enhance energy loss reduction and improve voltage profiles.

The expanding deployment of DG resources within distribution grids enhances voltage profiles, improves power supply reliability, and yields economic advantages, including reduced power dissipation and load balance [3]. The positioning of DG significantly influences the optimum operational topology of distribution networks [4,5].

Network reconfiguration is typically formulated as a problem based on non-linearly constrained programming with integer variables (MINLP). Approaches for tackling this optimization problem can be broadly divided into two primary categories: heuristic and exact mathematical approaches [6]. Authors in [3] introduced an advanced meta-heuristic technique referred to as the Harmony Search Algorithm (HAS) to co-optimize grid configuration and site selection for distributed generation installation, aiming to minimize power losses and enhance voltage profiles. A two-stage robust optimization model, addressed through an iterative column-and-constraint generation approach to reconfigure feeders under uncertain load conditions, was suggested in [7]. The study [6] proposed an optimization model based on a mixed-integer program with quadratic conic constraints (MISOCP) that utilized second-order conic relaxation, big-M methods, and piecewise linearization to address a combination optimization issue involving reactive power along with network architecture. Switching operation and exchange method for managing uncertainty-aware multi-period grid reconfiguration, accounting for the unpredictability in power consumption and solar production, was proposed in [8]. In [9], the authors presented a discrete evolutionary algorithm designed to concurrently optimize distribution system topology optimization and capacitors connected in parallel to the distribution lines. An enhanced particle swarm optimization method was presented in [10] to deal with the topology of the power distribution grid issue in conjunction with the reactive power regulation of distributed generating units. These methodologies, grounded on artificial intelligence algorithms, are labour-intensive and frequently fail to yield globally ideal answers. The imposition of radial structure constraints in the distribution system reconfiguration problem presents challenges related to computational efficiency were developed and validated in [2,11].

This chapter aims to introduce a mathematical formulation utilizing MIQCP for the optimal determination of network architecture, taking into account distributed generating units and parallel-connected reactive compensator devices. This MIQCP approach is constructed using a linearly expressed formulation of distribution power flow, known as the modified branch flow model (Modified DistFlow, MD) for electrical distribution grids. This chapter has made substantial contributions as outlined in the following:

- To translate the network reconfiguration problem's an optimization framework characterized by nonlinearities and integer decision variables, becomes a programming model comprising discrete and continuous optimization elements variables and quadratic constraints.
- To validate the proposed MIQCP model using an actual distribution system comprising 190 buses in Vietnam (2021).

- To examine how the demand level affects the best solutions to the network reconfiguration problem.

There are five sections in this chapter. The optimization problem's nonlinear formulation is shown in Section 13.2. Section 13.3 provides an enhanced linear formulation of the DistFlow model, while Section 13.4 presents the network reconfiguration issue model based on MIQCP. Using a 190-bus actual electrical distribution system in Vietnam and a distribution network with 33 nodes, Section 13.5 presents numerical findings and comments. Section 13.6 draws conclusions.

13.2 Nonlinearly constrained programming model with integer variables

The main goal of the network topology optimization problem addressed in this research is to achieve power loss reduction. Consequently, this target function is articulated as shown in (13.1).

$$\min_{x_{ij}, U_i, P_i, Q_i, P_{ij}, Q_{ij}} \sum_{ij \in \Phi_B} R_{ij} \frac{P_{ij}^2 + Q_{ij}^2}{U_i^2} \tag{13.1}$$

where x_{ij} is a binary decision variable indicating the status of line energization (1 if energized, 0 otherwise); U_i represents the voltage magnitude at bus i; P_i and Q_i refer to the active and reactive powers injected to node i, respectively; P_{ij} and Q_{ij} stand for the active and reactive power flows at the sending end of branch ij, respectively; R_{ij} is the series resistance associated with branch ij; and Φ_B represents the complete collection of distribution lines within the network.

The formulation of the power network topology optimization includes the model's requirements listed in the following.

13.2.1 Constraints of binary variables

The binary variable x_{ij} indicates the switching status of edge ij. It takes the value 1 if the branch ij is closed, and 0 otherwise.

$$x_{ij} = \{0, 1\}; \quad \forall ij \in \Phi_B \tag{13.2}$$

In addition, if line ij is disconnected, the active and reactive power flows across it are constrained to zero, as captured by the linear inequalities in (13.3).

$$\begin{aligned} -x_{ij}M \le P_{ij} \le x_{ij}M; \quad \forall ij \in \Phi_B \\ -x_{ij}N \le Q_{ij} \le x_{ij}N; \quad \forall ij \in \Phi_B \end{aligned} \tag{13.3}$$

where M and N are big enough positive constants.

13.2.2 Constraints of power balance

According to DistFlow [12–14], the active and reactive power balance is formulated as follows:

$$P_{hi} = \sum_{j \in \Phi_{N(i)}, j \neq h} P_{ij} + R_{hi} \frac{P_{hi}^2 + Q_{hi}^2}{U_h^2} - P_i \quad \forall i \in \Phi_N \tag{13.4}$$

$$Q_{hi} = \sum_{j \in \Phi_{N(i)}, j \neq h} Q_{ij} + X_{hi} \frac{P_{hi}^2 + Q_{hi}^2}{U_h^2} - Q_i \quad \forall i \in \Phi_N \tag{13.5}$$

$$P_i = -P_{Di} + P_{Gi}; \quad \forall i \in \Phi_N \tag{13.6}$$

$$Q_i = -Q_{Di} + Q_{Gi} + Q_{Ci}; \quad \forall i \in \Phi_N \tag{13.7}$$

where Φ_N is the complete collection of buses; $\Phi_{N(i)}$ denotes the group of neighbouring buses adjacent to node i; Q_C denotes the amount of reactive power supplied by the parallel capacitor; and P_{Gi} and Q_{Gi} denote the active and reactive powers injected by on-site generating unit at node i, respectively.

13.2.3 Constraints of nodal voltage equation

The expression for voltage drop along an energized branch in distribution systems is given by:

$$U_i - U_j = \frac{R_{ij}P_{ij} + X_{ij}Q_{ij}}{U_i}; \quad \forall ij \in \Phi_B \tag{13.8}$$

For the open branches, a big-H method is employed to incorporate the voltage constraint equations, as shown in the following equation [15]:

$$-(1 - x_{ij})H \leq U_i - U_j \leq (1 - x_{ij})H; \quad \forall ij \in \Phi_B \tag{13.9}$$

By combining (13.8) and the voltage-related constraints are formulated in (13.10) and (13.11)

$$U_i - U_j \leq (1 - x_{ij})H + \frac{R_{ij}P_{ij} + X_{ij}Q_{ij}}{U_i}; \quad \forall ij \in \Phi_B \tag{13.10}$$

$$U_i - U_j \geq (1 - x_{ij})H + \frac{R_{ij}P_{ij} + X_{ij}Q_{ij}}{U_i}; \quad \forall ij \in \Phi_B \tag{13.11}$$

This is explained as follows. If the branch ij is closed, then $x_{ij} = 1$ and $(1 - x_{ij})H = 0$. The composition of (13.10) and (13.11) is equivalent to (13.8). If the branch ij is open $(x_{ij} = 1)$ and H is large enough, then constraints (13.10) and (13.11) are unenforced.

13.2.4 Constraints of branch power flow

The bounds on real power and reactive power in edge ij are given by:

$$-P_{ij}^{\max} \leq P_{ij} \leq P_{ij}^{\max}; \quad \forall ij \in \Phi_B \tag{13.12}$$

$$-Q_{ij}^{\max} \leq Q_{ij} \leq Q_{ij}^{\max}; \quad \forall ij \in \Phi_B \tag{13.13}$$

$$-P_{ij}^2 + Q_{ij}^2 \leq (S_{ij}^{\max}); \quad \forall ij \in \Phi_B \tag{13.14}$$

where $S_{ij}^{\max}$ is the thermal flow restriction on branch ij; $P_{ij}^{\max}$ and $Q_{ij}^{\max}$ denote the allowable power flow limits on line ij, respectively. Constraints (13.12)–(13.13) can be employed to prevent reverse flow in the distribution network by constraining the minimum power value of the power flow to zero.

13.2.5 Bounds of node voltage magnitude

The magnitude of the voltage at each bus is subject to the following constraints:

$$U_i^{\min} \leq U_i \leq U_i^{\max}; \qquad \forall i \in \Phi_{\mathrm{N}} \tag{13.15}$$

where $U_i^{\min}$ and $U_i^{\max}$ represent the lower and upper voltage limits at bus i, respectively.

13.2.6 Constraints of radial structure

The following requirements are utilized to ensure the radiality configuration of electrical distribution grids [2]:

$$\sum_{i \neq j} x_{ij} = N_{\mathrm{N}} - N_{\mathrm{sub}}; \quad \forall ij \in \Phi_{\mathrm{B}} \tag{13.16}$$

$$\sum_{l|r(l)=i} k_l - \sum_{l|s(l)=i} k_l = K_i; \quad \forall i \in \Phi_{\mathrm{N}} \tag{13.17}$$

$$K_i = 1; \qquad \forall i \in \Phi_{\mathrm{G}} \tag{13.18}$$

$$K_i = 0; \qquad \forall i \notin \Phi_{\mathrm{G}} \cup \Phi_{\mathrm{sub}} \tag{13.19}$$

$$-x_{ij}N_{\mathrm{G}} \leq k_{ij} \leq x_{ij}N_{\mathrm{G}}; \qquad \forall ij \in \Phi_{\mathrm{B}} \tag{13.20}$$

where Φ_{G} and Φ_{sub} are the collection of on-site generators and all main substations, respectively; N_{G} is the overall number of all DGs; and N_{sub} is the overall number of substation buses; N_{N} is the overall number of nodes.

The general form of the optimization problem described above is formulated as a nonlinear programming model with integer variables. In Section 13.3, a linearized version of the distribution flow model is presented to transform this general problem into a program with mixed-integer variables and quadratic constraints.

13.3 Modified DistFlow model

Consider a distribution network comprising two buses with the equivalent circuit representation shown in Figure 13.1.

The phasor diagram of the voltage drop for this system is shown in Figure 13.2.

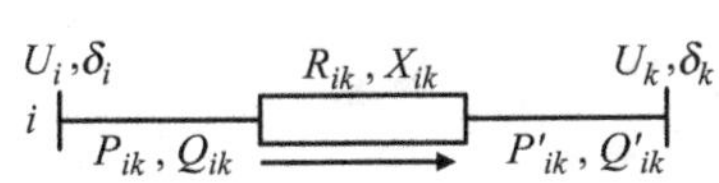

Figure 13.1 A power distribution line

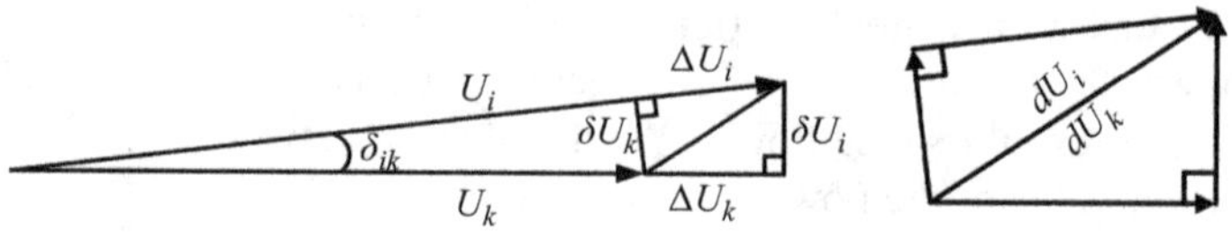

Figure 13.2 Phasor diagram illustration of voltage drop in a distribution system with two buses

The respective real and imaginary parts of the voltage drop in branch *ik* are described in the following:

$$\Delta U_{ik} = \frac{R_{ik}P_{ik} + X_{ik}Q_{ik}}{U_i}; \qquad \delta U_{ik} = \frac{X_{ik}P_{ik} - R_{ik}Q_{ik}}{U_i} \tag{13.21}$$

where bus *i* is selected as the reference for voltage phase angles.

The Modified DistFlow (MD) technique suggested in this research is based on the following assumptions.

The first assumption states that the angular displacement between phases between buses *i* and *j* is negligible. Under this assumption, the approximate expression given in can be attained.

$$\sin\delta_{ik} \approx \delta_{ik}; \quad \cos\delta_{ik} \approx 1 - \frac{1}{2}\delta_{ik}^2 \tag{13.22}$$

The element contributing to the voltage drop in the horizontal (real) direction can be calculated using equations (13.23)–(13.24) and as follows:

$$\Delta U_i = U_i - U_k\cos\delta_{ik} \approx U_i - U_k\left(1 - \frac{1}{2}\delta_{ik}^2\right) \tag{13.23}$$

$$\Delta U_k = U_i\cos\delta_{ik} - U_k \approx U_i\left(1 - \frac{1}{2}\delta_{ik}^2\right) - U_k \tag{13.24}$$

Subtracting from results in expression as in (13.25).

$$\Delta U_i - \Delta U_k = \frac{1}{2}(U_i + U_k)\delta_{ik}^2 \tag{13.25}$$

Based on the first assumption, the following approximate equation is derived.

$$\Delta U_i \approx \Delta U_k \tag{13.26}$$

Likewise, the following mathematical states can be achieved:

$$\delta U_i = U_k\delta_{ik}; \quad \delta U_k = U_i\delta_{ik} \tag{13.27}$$

By substituting (13.21) into (13.26) and (13.27), the active and reactive power at the sending terminal are connected to those at the destination end through the following expressions:

$$\frac{P_{ik}}{U_i} = \frac{P'_{ik}}{U_k} - \frac{X_{ik}}{R_{ik}^2 + X_{ik}^2}(U_i - U_k)\delta_{ik} \tag{13.28}$$

$$\frac{Q_{ik}}{U_i} = \frac{Q'_{ik}}{U_k} - \frac{R_{ik}}{R_{ik}^2 + X_{ik}^2}(U_i - U_k)\delta_{ik} \tag{13.29}$$

Leveraging the first assumption, (13.28) and (13.29) can be simplified further as expressed in (13.30).

$$\frac{P_{ik}}{U_i} \approx \frac{P'_{ik}}{U_k}; \quad \frac{Q_{ik}}{U^i} \approx \frac{Q'_{ik}}{U^k} \tag{13.30}$$

The mathematical statements (13.30) demonstrate that the proportion of active power flow to voltage magnitude at the sending end is nearly the same as that at the receiving end. A similar observation holds for the reactive power flow.

The sending-end power flow on branch ik can be calculated as follows:

$$P_{ik} = \frac{R_{ik}(U_i^2 - U_iU_k\cos\delta_{ik}) + X_{ik}U_iU_k\sin\delta_{ik}}{R_{ik}^2 + X_{ik}^2} \tag{13.31}$$

$$Q_{ik} = \frac{X_{ik}(U_i^2 - U_iU_k\cos\delta_{ik}) - R_{ik}U_iU_k\sin\delta_{ik}}{R_{ik}^2 + X_{ik}^2} \tag{13.32}$$

To be multiplied (13.31) by R_{ik} and (13.32) by X_{ik} results to the following equation:

$$U_i - U_k\cos\delta_{ik} = \frac{P_{ik}R_{ik} + Q_{ik}X_{ik}}{U_i} \tag{13.33}$$

Under the first assumption, (13.33) is rewritten in the following:

$$U_i - U_k \approx R_{ik}\frac{P_{ik}}{U_i} + X_{ik}\frac{Q_{ik}}{U_i} \tag{13.34}$$

Let

$$\hat{P}_{ik} = \frac{P_{ik}}{U_i}; \quad \hat{Q}_{ik} = \frac{Q_{ik}}{U_i}; \quad \hat{P}_i = \frac{P_i}{U_i}; \quad \hat{Q}_i = \frac{Q_i}{U_i} \tag{13.35}$$

Furthermore, the expression (13.34) is described as in (13.36).

$$U_i - U_k = R_{ik}\hat{P}_{ik} + X_{ik}\hat{Q}_{ik} \tag{13.36}$$

Using the Taylor series approximation at $U^{(0)} = 1$, the resulting mathematical expression is given in the following:

$$\begin{aligned}\frac{1}{U} &\approx \frac{1}{U^{(0)}} + \left(\frac{1}{U}\right)'\Big|_{U^{(0)}} \cdot (U - U^{(0)}) = \frac{1}{U^{(0)}} - \frac{1}{(U^{(0)})^2} \cdot (U - U^{(0)}) \\ &= 1 - (U-1) \approx 2 - U\end{aligned} \tag{13.37}$$

By combining (13.37) and (13.36), the following equation represents the voltage relationship in the two-bus distribution network:

$$\frac{1}{U_k} - \frac{1}{U_i} = R_{ik}\hat{P}_{ik} + X_{ik}\hat{Q}_{ik} \tag{13.38}$$

Let

$$W = \frac{1}{U} \tag{13.39}$$

then, the following expressions can be obtained.

$$\hat{P}_i = P_i W_i; \quad \hat{Q}_i = Q_i W_i; \quad \hat{P}_{ik} = \hat{P}'_{ik}; \quad \hat{Q}_{ik} = \hat{Q}'_{ik};$$
$$W_k - W_i = R_{ik}\hat{P}_{ik} + X_{ik}\hat{Q}_{ik} \tag{13.40}$$

The general form of the MD technique is derived through the following formulas.

$$\hat{P}_{hi} = \sum_{\substack{k\in\Omega_i \\ k\neq h}} \hat{P}_{ik} - \hat{P}_i, \quad \forall i \in \Omega_{\mathrm{N}} \tag{13.41}$$

$$\hat{Q}_{hi} = \sum_{\substack{k\in\Omega_i \\ k\neq h}} \hat{Q}_{ik} - \hat{Q}_i; \quad \forall i \in \Omega_{\mathrm{N}} \tag{13.42}$$

$$W_k - W_i = R_{ik}\hat{P}_{ik} + X_{ik}\hat{Q}_{ik}; \quad \forall ik \in \Omega_{\mathrm{B}} \tag{13.43}$$

$$\hat{P}_i = P_i W_i; \quad \hat{Q}_i = Q_i W_i \quad \forall i \in \Omega_{\mathrm{N}} \tag{13.44}$$

After the model in (13.41)–(13.44) is solved, the voltage magnitude and branch flow can be determined using (13.35) and (13.39).

13.4 Quadratically constrained mixed-integer programming-based model

Utilizing the updated DistFlow formulation introduced in Section 13.3, the original nonlinear problem of network topology optimization is reformulated into an MIQCP model as follows.

13.4.1 Objective function

The objective function (13.1) is reformulated in the following:

$$\min_{x_{ij}, W_i, U_i, \hat{P}_i, \hat{Q}_i, \hat{P}_{ij}, \hat{Q}_{ij}} \sum_{ij\in\Phi_{\mathrm{B}}} R_{ij}\left(\hat{P}_{ij}^2 + \hat{Q}_{ij}^2\right) \tag{13.45}$$

13.4.2 Constraints of binary variables

The constraints relating to binary variables are modified in the following:

$$x_{ij} = \{0, 1\}; \quad \forall ij \in \Phi_{\mathrm{B}} \tag{13.46}$$

$$-x_{ij}M \leq \hat{P}_{ij} \leq x_{ij}M; \quad \forall ij \in \Phi_{\mathrm{B}} \tag{13.47}$$

$$-x_{ij}M \leq \hat{Q}_{ij} \leq x_{ij}M; \quad \forall ij \in \Phi_{\mathrm{B}} \tag{13.48}$$

13.4.3 Constraints of power balance

The constraints relating to nodal power balance are modified in the following:

$$\hat{P}_{hi} = \sum_{j\in\Phi_{\mathrm{N}(i)}, j\neq h} \hat{P}_{ij} - \hat{P}_i; \quad \forall i \in \Phi_{\mathrm{N}} \tag{13.49}$$

$$\hat{Q}_{hi} = \sum_{j \in \Phi_{\mathrm{N}(i)}, j \neq h} \hat{Q}_{ij} - \hat{Q}_i; \quad \forall i \in \Phi_{\mathrm{N}} \tag{13.50}$$

$$\hat{P}_i = -P_{\mathrm{D}i} W_i + \hat{P}_{\mathrm{G}i}; \quad \forall i \in \Phi_{\mathrm{N}} \tag{13.51}$$

$$\hat{Q}_i = -Q_{\mathrm{D}i} W_i + \hat{Q}_{\mathrm{G}i} + \hat{Q}_{\mathrm{C}i}; \quad \forall i \in \Phi_{\mathrm{N}} \tag{13.52}$$

13.4.4 Constraints of nodal voltage equations

The constraints relating to voltage equations are amended in the following:

$$W_j - W_i \leq (1 - x_{ij})M + R_{ij}\hat{P}_{ij} + X_{ij}\hat{Q}_{ij}; \quad \forall ij \in \Phi_{\mathrm{B}} \tag{13.53}$$

$$W_j - W_i \geq -(1 - x_{ij})M + R_{ij}\hat{P}_{ij} + X_{ij}\hat{Q}_{ij}; \quad \forall ij \in \Phi_{\mathrm{B}} \tag{13.54}$$

13.4.5 Constraints of power flow on branches

The constraints of power transfer on branches are modified as follows.

$$-P_{ij}^{\max} W_i \leq \hat{P}_{ij} \leq P_{ij}^{\max} W_i; \quad \forall ij \in \Phi_{\mathrm{B}} \tag{13.55}$$

$$-Q_{ij}^{\max} W_i \leq \hat{Q}_{ij} \leq Q_{ij}^{\max} W_i; \quad \forall ij \in \Phi_{\mathrm{B}} \tag{13.56}$$

$$\hat{P}_{ij}^2 + \hat{Q}_{ij}^2 \leq \left(S_{ij}^{\max} W_i\right)^2; \quad \forall ij \in \Phi_{\mathrm{B}} \tag{13.57}$$

13.4.6 Limits of bus voltage magnitude

The limitations relating to node voltage magnitude are amended in the following:

$$2 - U_i^{\max} \leq W_i \leq 2 - U_i^{\min}; \quad \forall i \in \Phi_{\mathrm{N}} \tag{13.58}$$

13.4.7 Constraints of radial configuration

The limitations pertaining to radial topology are given in the following:

$$\sum_{i \neq j} x_{ij} = N_{\mathrm{N}} - N_{\mathrm{sub}}; \quad \forall ij \in \Phi_{\mathrm{B}} \tag{13.59}$$

$$\sum_{l|r(l)=i} k_l - \sum_{l|s(l)=i} k_l = K_i; \quad \forall i \in \Phi_{\mathrm{N}} \tag{13.60}$$

$$K_i = 1; \quad \forall i \in \Phi_{\mathrm{G}} \tag{13.61}$$

$$K_i = 0; \quad \forall i \notin \Phi_{\mathrm{G}} \cup \Phi_{\mathrm{sub}} \tag{13.62}$$

$$-x_{ij} N_{\mathrm{G}} \leq k_{ij} \leq x_{ij} N_{\mathrm{G}}; \quad \forall ij \in \Phi_{\mathrm{B}} \tag{13.63}$$

13.5 Results and discussions

In the following part, the proposed quadratic-constrained optimization framework with mixed-integer variables for determining the best topology of electricity distribution systems is validated through two case studies: a 33-node distribution network and a power distribution system consisting of 190 nodes located in Vietnam, with data from the year 2021. The proposed optimization problem is addressed using IBM CPLEX within the GAMS environment on a personal computer equipped with a 1.60 GHz Intel Core i5 processor and 4 GB of RAM. In addition, power flow analysis for the radial distribution network is conducted using the power summation method implemented in the MATPOWER toolbox [16] within MATLAB® R2018a.

13.5.1 Power distribution network with 33 nodes

We examine the 33-node electrical distribution network illustrated in Figure 13.3. This system is designed to operate at 12.66 kV nominal voltage. In the base case scenario, the total load demand across the network amounts to 3715 kW of real power and 2300 kVAr of reactive power.

Data for line branches and demands of the 33-node distribution system are presented in Appendix A. In the 33-node electrical distribution grid, the branches represented by solid lines are considered normally not open, while those depicted with dashed lines are normally open. The system includes four distributed generation units situated at nodes 18, 22, 25, and 33. Each DG unit is assumed to generate 200 kW of active power and 150 kVAr of reactive power. In addition, two shunt capacitors are installed at buses 18 and 33, with rated capacities of 400 kVAr and 600 kVAr, respectively. The nodal voltage boundaries are set between 0.95 p.u. and 1.05 p.u. The total active power loss under the base case scenario, before network reconfiguration, is 84.58 kW.

Four test scenarios are formulated and analysed for comparison, as described in the following:

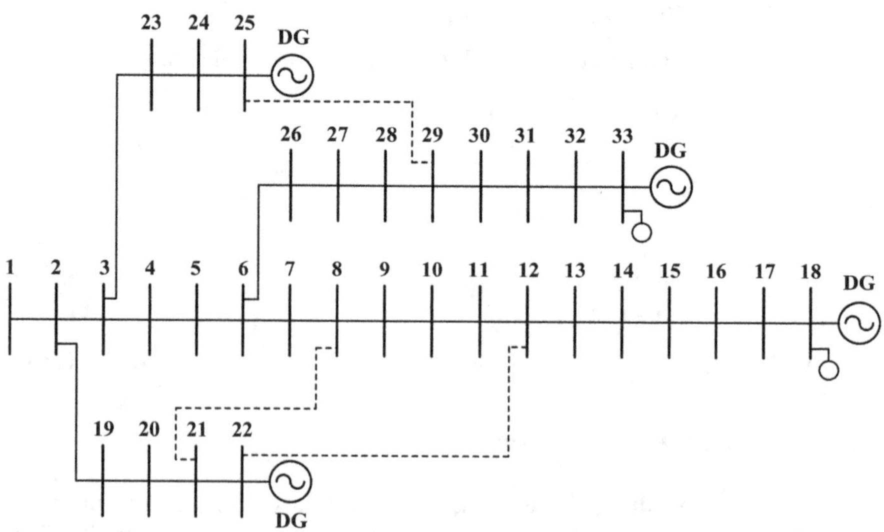

Figure 13.3 Diagram of the 33-node distribution system

- **Scenario 1:** The system operates under nominal conditions, with demand set at 100% of the baseline load.
- **Scenario 2:** The overall load demand is raised to 150% of its original value.
- **Scenario 3:** The system load is further scaled up to 200% of the baseline.
- **Scenario 4:** The load demand is raised to 220% of the baseline, representing the most stressed operating condition.

The branch configurations and corresponding computational times under all four scenarios, as determined by the MIQCP formulation described in Section 13.4, are summarized in Table 13.1.

Based on the optimal branch statuses presented in Table 13.1, electrical load distribution analysis is performed to determine network power dissipation, nodal voltage profiles, and the active power and reactive power flowing through branches. Table 13.2 presents the total system active power losses before and after applying network configuration optimization across the four scenarios.

Table 13.2 shows that the network reconfiguration considerably reduces the power loss, a decrease of 31.94% for scenario 3.

The nodal voltage profiles corresponding to the 100% and 200% loading levels are illustrated in Figures 13.4 and 13.5, respectively. As observed from these figures, network reconfiguration significantly improves the overall voltage profile. Specifically, under the 200% load condition, the smallest nodal voltage rises from 0.9270 p.u. (pre-reconfiguration) to 0.9612 p.u. (post-reconfiguration). Moreover, the post-reconfiguration pattern of voltage magnitudes exhibits a flatter and more uniform distribution compared to the pre-reconfiguration case.

Table 13.1 Branch status outcomes and computational time for 33-bus test network

Demand level (%)	Opened edges	Time (s)
100	7–8; 9–10; 28–29	0.315
150	7–8; 10–11; 28–29	0.396
200	7–8; 10–11; 28–29	2.412
220	6–7; 10–11; 28–29	1.378

Table 13.2 Real power loss outcomes for the 33-bus test system

Demand level (%)	Pre-reconfiguration power loss (kW)	Post-reconfiguration power loss (kW)	Power loss decrease (%)
100	84.58	61.14	27.71
150	238.03	165.97	30.28
200	512.38	348.72	31.94
220	663.21	467.46	29.52

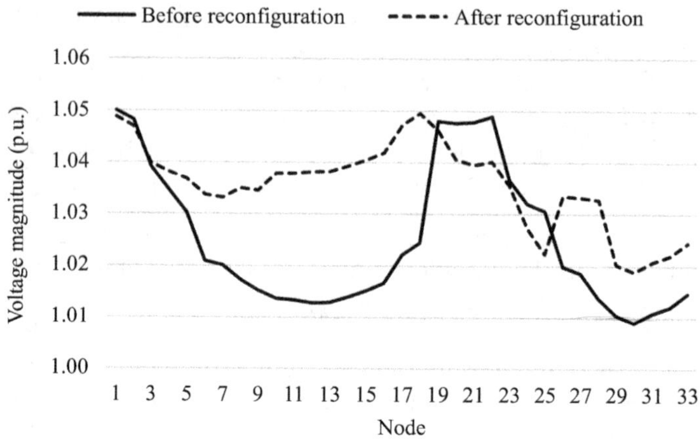

Figure 13.4 Nodal voltage pattern of the 33-node distribution system under 100% load level

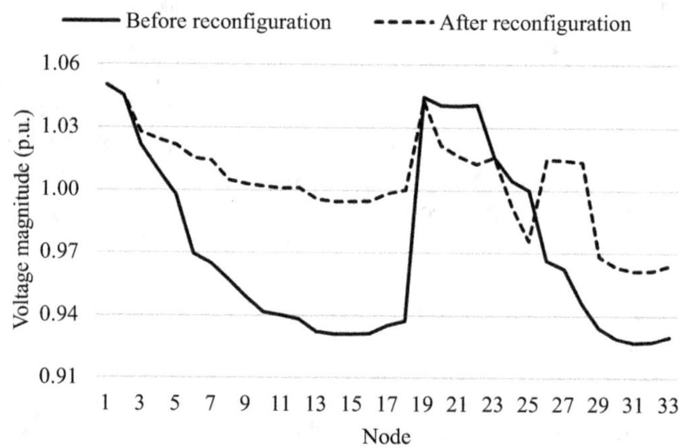

Figure 13.5 Nodal voltage pattern of the 33-node distribution system under 200% load level

Figures 13.6 and 13.7 present the smallest and mean nodal voltages of the 33-node distribution system under four different loading cases, respectively. According to Figure 13.6, the minimum voltage magnitude under the 200% loading condition improves from 0.9083 p.u. before power network reconfigured to 0.9500 p.u. after power network reconfigured. Similarly, Figure 13.7 shows that the average nodal voltage reaches 1.0038 p.u. from the initial 0.9725 p.u. for the same loading level. These results indicate that network reconfiguration enhances both the minimum and average voltage levels, particularly under high-load conditions.

Tables 13.3 and 13.4 present the voltage magnitudes and branch power flows obtained by resolving the optimization issue using the MD model based on MIQCP and by addressing the nonlinear power flow equations, for the 33-node electricity

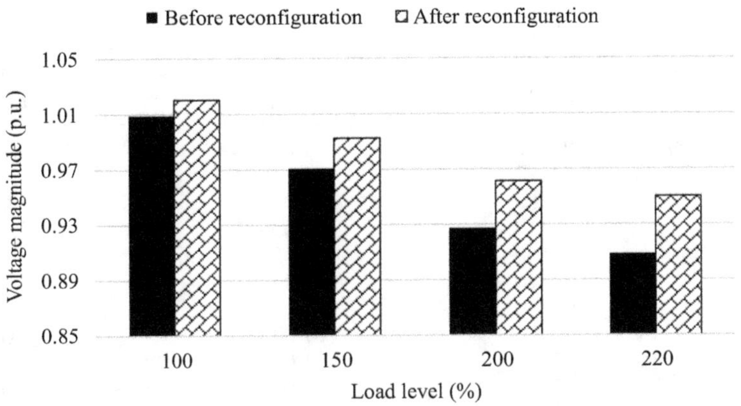

Figure 13.6 Minimum nodal voltage of the 33-node test system under different loading scenarios

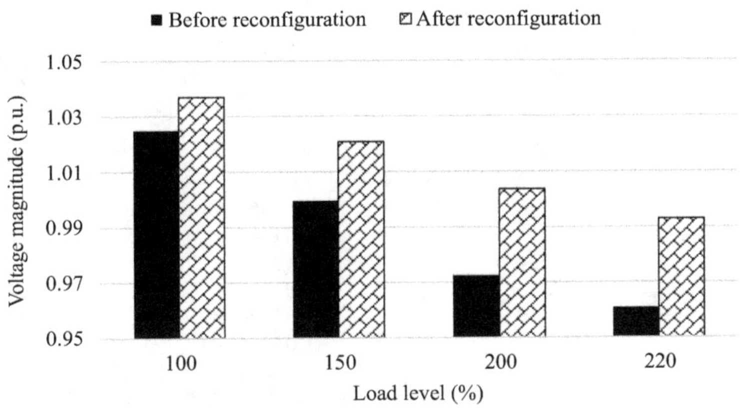

Figure 13.7 Mean voltage results for the 33-bus system

distribution system after network reconfiguration under the 100% loading condition. In addition, Table 13.5 summarizes the calculated overall real power loss using both the MIQCP-based optimization approach and the AC power flow (ACPF) method across four different load levels. The comparison indicates that the errors in nodal voltages, branch power flows, and total loss of power resulting from the approximate MD model are minimal when compared to those obtained using the exact ACPF analysis.

13.5.2 Vietnamese distribution network with 190 nodes

This subsection presents the computational outcomes of the distribution network topology optimization problem applied to the electricity distribution grid in Vietnam, based on 2021 operational data. Figure 13.8 shows the one-line diagram of the

Table 13.3 Nodal voltage comparison for 33-node test system

Bus	ACPF	MIQCP	Error (%)
1	1.04876	1.04876	0.0000
2	1.04692	1.04694	0.0017
3	1.03980	1.03984	0.0037
4	1.03810	1.03814	0.0043
5	1.03669	1.03673	0.0041
6	1.03370	1.03375	0.0049
7	1.03310	1.03315	0.0049
8	1.03496	1.03523	0.0265
9	1.03449	1.03477	0.0267
10	1.03782	1.03827	0.0436
11	1.03790	1.03835	0.0437
12	1.03817	1.03862	0.0433
13	1.03827	1.03877	0.0485
14	1.03924	1.03975	0.0492
15	1.04041	1.04094	0.0506
16	1.04188	1.04241	0.0511
17	1.04726	1.04779	0.0504
18	1.04948	1.05	0.0495
19	1.04619	1.04623	0.0036
20	1.04061	1.04083	0.0212
21	1.03954	1.03981	0.0261
22	1.04022	1.04055	0.0313
23	1.03548	1.03554	0.0062
24	1.02732	1.02742	0.0100
25	1.02234	1.02247	0.0131
26	1.03343	1.03348	0.0044
27	1.03319	1.03324	0.0049
28	1.03269	1.03274	0.0046
29	1.02026	1.02041	0.0148
30	1.01884	1.01901	0.0167
31	1.02067	1.02086	0.0183
32	1.02188	1.02207	0.0184
33	1.02446	1.02465	0.0181

Vietnamese power distribution system, comprising 190 nodes. Detailed data on the branch lengths and demands of this system can be found in Appendix A. The unit resistance and unit reactance of all branches are 0.254 Ω and 0.358 Ω, respectively.

The Vietnamese distribution system operates at the nominal voltage level of 35 kV. Two main substations are sited at nodes 1 and 2. Under the base case scenario, corresponding to 100% demand level, the total system demand amounts to 26,673.4 kW of real power and 12,916.5 kVAr of reactive power.

The Vietnamese distribution network includes eleven tie switches. Prior to network reconfiguration, five of these tie switches—located on branches 15–105, 116–162, 157–168, 87–176, and 155–189—are normally in the open position. In addition, the system is equipped with seven capacitors installed at nodes 47, 54, 66,

Table 13.4 Comparison of branch power flow (p.u) for 33-node test system

Sending end	Receiving end	MIQCP		Error (%)	
		Active power	Reactive power	Active power	Reactive power
1	2	2.96087	0.721958	0.515352742	3.43150927
2	3	2.01513	0.785786	0.13494316	1.392698456
3	4	0.622446	0.30107	0.07721467	0.321851641
4	5	0.501601	0.220693	0.062507286	0.376295737
5	6	0.441002	0.190434	0.056340773	0.394555324
6	7	0.199896	0.099948	0.079337125	0.2324505
8	9	0.0599542	0.0199847	0.116320343	0.162645138
10	11	−0.0599121	−0.0199707	0.146500082	0.146500077
11	12	−0.10485	−0.0499281	0.147189358	0.146854378
12	13	0.249825	−0.338852	2.661394911	0.110954579
13	14	0.189951	−0.373847	2.650255672	0.223091461
14	15	0.0703206	−0.454074	5.689461237	0.295384152
15	16	0.0105012	−0.464575	2.410076506	0.319992446
16	17	−0.049376	−0.485195	4.833377989	0.3193297
17	18	−0.109494	−0.507653	0.571973706	0.28714951
2	19	0.840802	−0.124953	1.479679269	11.78237777
19	20	0.750428	−0.164783	1.613974324	8.095272686
8	21	−0.259706	−0.119861	0.122304274	0.130195352
20	21	0.656705	−0.203866	1.624698168	3.42390331
12	22	−0.414614	0.253963	1.677894778	0.196391128
21	22	0.305346	−0.363992	3.196316625	1.270618338
3	23	1.28917	0.439453	0.197170751	1.932828251
23	24	1.19395	0.387697	0.242267361	1.796277962
20	21	0.656705	−0.203866	1.624698168	3.42390331
12	22	−0.414614	0.253963	1.677894778	0.196391128
21	22	0.305346	−0.363992	3.196316625	1.270618338
3	23	1.28917	0.439453	0.197170751	1.932828251
23	24	1.19395	0.387697	0.242267361	1.796277962
24	25	0.764908	0.184809	0.473860284	1.819896091
6	26	0.179905	0.069961	0.106244879	0.138876585
26	27	0.119927	0.0449712	0.104187616	0.14332707
27	28	0.0599645	0.0199882	0.100429555	0.168067817
25	29	0.541335	0.133944	0.71499332	1.250086006
29	30	0.420294	0.0637035	0.942076522	1.536193781
30	31	0.219788	−0.536167	1.762296217	0.108097625
31	32	0.070252	−0.60711	2.098169493	0.070474569
32	33	−0.139562	−0.707781	0.448857719	0.080232849

80, 95, 130, and 151, with respective reactive power ratings of 300, 225, 225, 150, 300, 150, and 450 kVAr. It is also assumed that eight on-site generating units, each rated at $300 + j150$ kVA, are connected at buses 10, 22, 64, 76, 90, 110, 148, and 174.

An analysis is performed for six scenarios with demand levels set at 100%, 125%, 150%, 175%, 200%, and 225%, respectively.

Table 13.5 Comparative results of the overall power loss (kW) for the 33-node test system

Demand level	ACPF	MIQCP	Error (%)
100%	61.14	60.55	0.962
150%	165.97	164.79	0.710
200%	348.72	346.19	0.726
220%	467.46	463.43	0.863

Table 13.6 presents the branch switching statuses and computation times for the six loading scenarios of the Vietnamese electrical distribution grid, as obtained using the MIQCP-based optimization technique proposed in Section 13.4.

Based on the optimal branch configurations presented in Table 13.6, load flow analysis is performed to evaluate the system's active power loss, nodal voltage magnitudes, and real and reactive power flows through the branches. Table 13.7 summarizes the overall system power losses in pre- and post-reconfiguration states across diverse loading scenarios. The results indicate a significant reduction in power losses due to reconfiguration. Notably, for the 200% loading condition, the overall power loss in the Vietnamese electrical distribution grid is reduced by 29.44%.

The nodal voltage profiles of the Vietnamese distribution system under 100% and 200% loading conditions are illustrated in Figures 13.9 and 13.10, respectively. Additionally, Figure 13.11 presents the average voltage magnitudes across all nodes for six different load levels.

As observed in Figures 13.9 and 13.10, network reconfiguration leads to a noticeable improvement in voltage profile quality. Specifically, at the 200% load level, the smallest nodal voltage rises from below 0.985 p.u. (pre-reconfiguration) to 1.008 p.u. (post-reconfiguration). Moreover, the post-reconfiguration voltage profile is noticeably flatter and more uniform compared to the pre-reconfiguration case, indicating improved voltage stability across the system.

Figure 13.11 shows that the average nodal voltage increases from 1.011 p.u. (pre-reconfiguration) to 1.022 p.u. (post-reconfiguration) under the 200% loading condition.

Figures 13.12 and 13.13 illustrate the errors in nodal voltage and branch power flow, respectively, for the MD model—derived by solving the optimization issue using the MIQCP formulation—compared to the results obtained from the exact ACPF method. These results pertain to the Vietnamese electrical distribution grid after topology optimization under the 100% loading condition. The maximum observed error at this demand level is 0.0074%, which is negligible.

Furthermore, Table 13.8 presents the total active power losses calculated using both the MIQCP-based MD model and the ACPF method across six load scenarios. The results confirm that the errors linked with the overall power loss in the MD formulation are very small in comparison with the exact ACPF values, thereby validating the accuracy and applicability of the proposed approach.

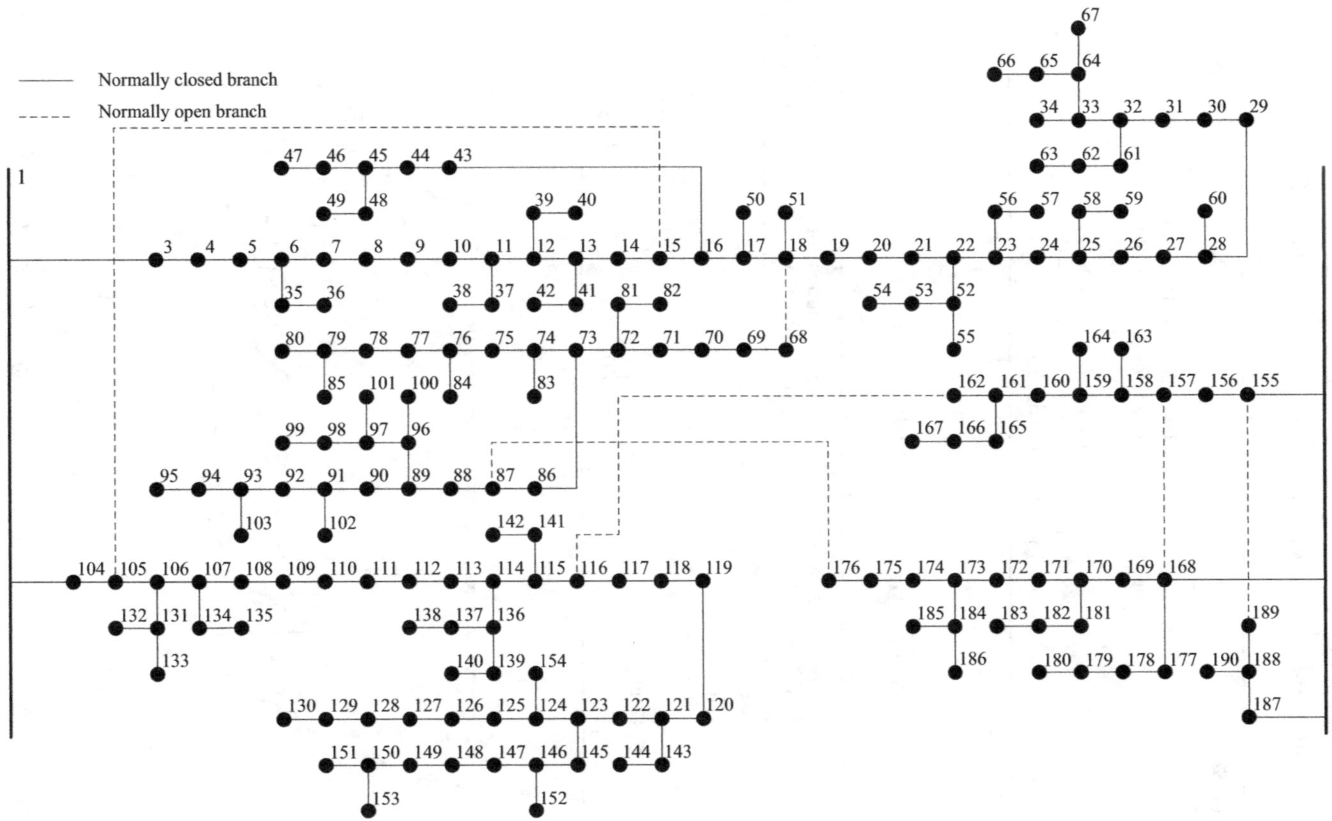

Figure 13.8 Vietnamese distribution system

Table 13.6 Branch state and computation time for the Vietnamese electrical distribution system under six loading scenarios using the MIQCP-based model

Load level (%)	Opened edges	Time (s)
100	18–68; 1–104; 15–105; 157–168; 155–189	3.783
125	18–68; 1–104; 15–105; 157–168; 155–189	2.843
150	18–68; 1–104; 15–105; 157–168; 155–189	4.455
175	18–68; 1–104; 15–105; 157–168; 155–189	2.893
200	18–68; 1–104; 15–105; 157–168; 155–189	4.121
225	18–68; 15–105; 116–162; 157–168; 155–189	1.533

Table 13.7 Power loss for Vietnamese distribution system

Demand level	Pre-reconfiguration power loss (kW)	Post-reconfiguration power loss (kW)	Power loss decrease (%)
100%	285.04	207.65	27.15
125%	480.50	346.77	27.83
150%	731.19	523.46	28.41
175%	1039.59	738.75	28.94
200%	1408.37	993.70	29.44
225%	1840.47	1408.40	23.48

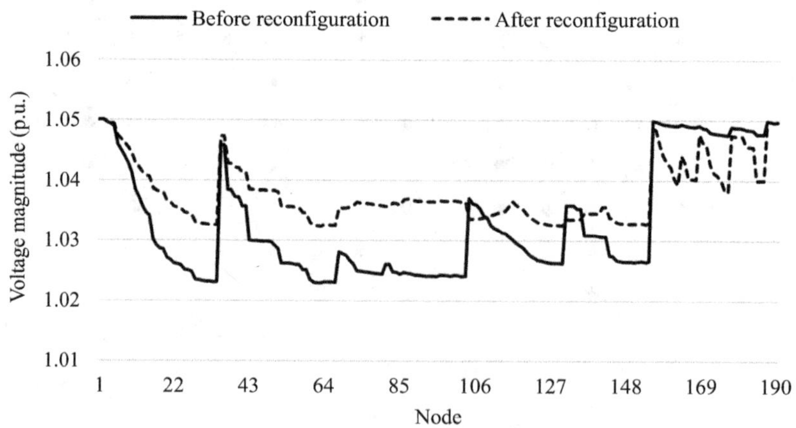

Figure 13.9 Nodal voltage profile of the Vietnamese electrical distribution system under 100% loading condition

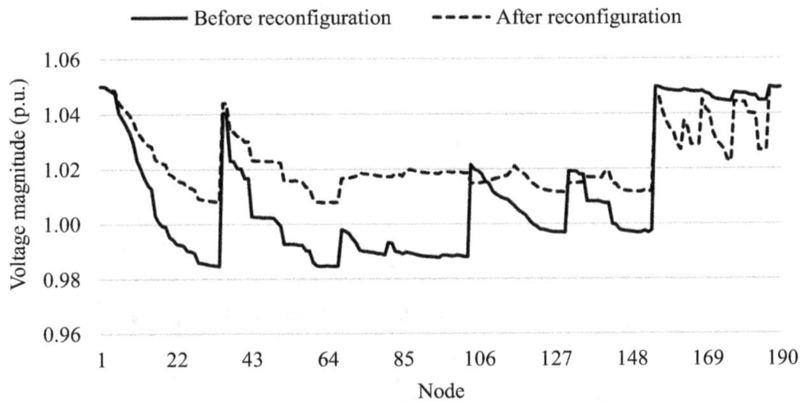

Figure 13.10 Nodal voltage profile of the Vietnamese electrical distribution grid under 200% loading condition

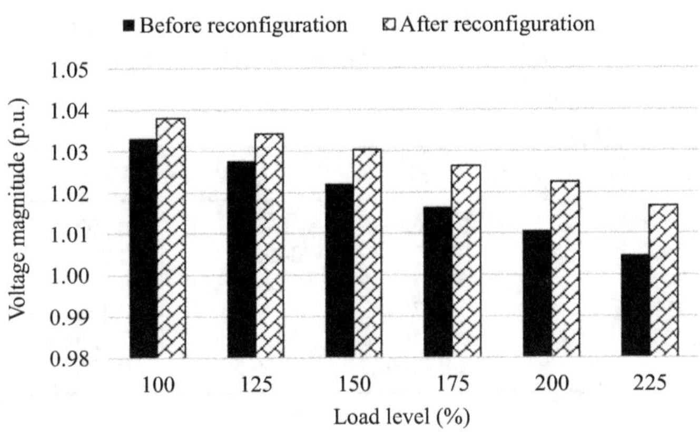

Figure 13.11 Average voltage results for the Vietnamese power distribution network

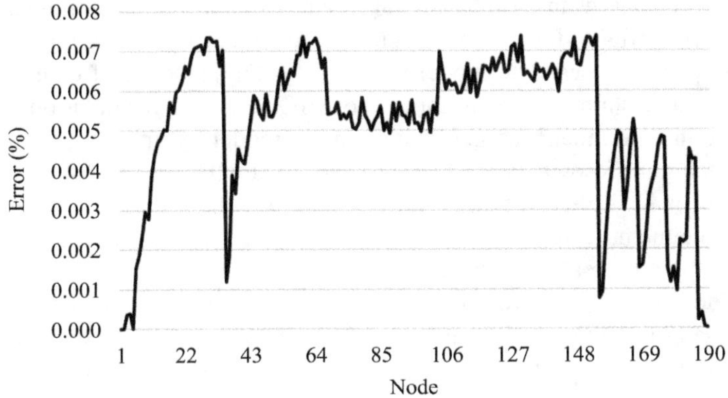

Figure 13.12 Voltage magnitude errors of the MD model for the Vietnamese power distribution grid under the 100% loading condition

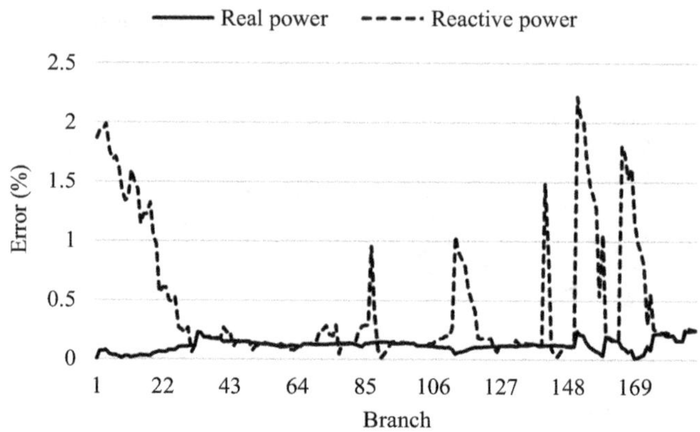

Figure 13.13 Branch power flow errors of MD formulation for Vietnamese power distribution system

Table 13.8 Comparative analysis of the overall loss power (kW) for the Vietnamese electrical distribution grid

Demand level	ACPF	MIQCP	Error (%)
100%	207.65	207.03	0.297
125%	346.77	345.93	0.241
150%	523.46	522.44	0.195
175%	738.75	737.57	0.159
200%	993.70	992.38	0.133
225%	1408.40	1406.50	0.135

13.6 Conclusion

This chapter proposed an MIQCP-based model to address the optimization issue of network reconfiguration in the electricity distribution grids with the goal of minimizing power losses. The model incorporates the presence of distributed generating units and shunt compensators. The MIQCP-based formulation is derived from an original nonlinear integer programming model by applying the linearized load flow technique known as the MD approach. The effectiveness of the suggested methodology was verified through two case studies: the 33-node distribution system and a Vietnamese real-world 190-bus network. Simulation results confirmed that network reconfiguration significantly reduces power losses and enhances voltage profiles. Additionally, the errors in nodal voltage and total power loss between the MIQCP-based MD model and the exact ACPF method are negligible, demonstrating the correctness and calculation efficiency of the suggested approach.

Appendix A

See Tables A.1 and A.2 for data for the 33-node distribution network and Vietnamese 190-node network.

Table A.1 Data for the 33-node distribution network (load at receive bus)

Sending bus	Receiving bus	R (Ω)	X (Ω)	P (kW)	Q (kVAr)
1	2	0.0922	0.047	100	60
2	3	0.493	0.2511	90	40
3	4	0.366	0.1864	120	80
4	5	0.3811	0.1941	60	30
5	6	0.819	0.707	60	20
6	7	0.1872	0.6188	200	100
7	8	0.7114	0.2351	200	100
8	9	1.03	0.74	60	20
9	10	1.044	0.74	60	20
10	11	0.1966	0.066	45	30
11	12	0.3744	0.1238	60	35
12	13	1.468	1.155	60	35
13	14	0.5416	0.7129	120	80
14	15	0.591	0.526	60	10
15	16	0.7463	0.545	60	20
16	17	1.289	1.721	60	20
17	18	0.732	0.574	90	40
2	19	0.164	0.1565	90	40
19	20	1.5042	1.3554	90	40
20	21	0.4095	0.4784	90	40
21	22	0.7089	0.9373	90	40
3	23	0.4512	0.3083	90	50
23	24	0.898	0.7091	420	200
24	25	0.896	0.7011	420	200
6	26	0.203	0.1034	60	25
26	27	0.2842	0.1447	60	25
27	28	1.059	0.9337	60	20
28	29	0.8042	0.7006	120	70
29	30	0.5075	0.2585	200	600
30	31	0.9744	0.963	150	70
31	32	0.3105	0.3619	210	100
32	33	0.341	0.5302	60	40
33	21	8	2	2	
34	12	22	2	2	
35	25	29	0.5	0.5	

Table A.2 Data for the Vietnamese 190-node network (load at receive bus)

Sending bus	Receiving bus	Length (m)	P (kW)	Q (kVAr)
1	3	100	405	196.1
3	4	100	129.6	62.8
4	5	100	291.6	141.2
5	6	900	0	0
6	7	300	81	39.2
7	8	300	453.6	219.6
8	9	400	145.8	70.6
9	10	500	145.8	70.6
10	11	900	0	0
11	12	400	0	0
12	13	600	0	0
13	14	500	60.75	29.4
14	15	200	0	0
15	16	1800	0	0
16	17	460	0	0
17	18	340	0	0
18	19	180	145.8	70.6
19	20	2300	145.8	70.6
20	21	650	60.75	29.4
21	22	1100	0	0
22	23	250	0	0
23	24	300	81	39.2
24	25	1500	0	0
25	26	180	0	0
26	27	650	259.2	125.5
27	28	3500	145.8	70.6
28	29	400	60.75	29.4
29	30	500	60.75	29.4
30	31	450	202.5	98.1
31	32	200	0	0
32	33	900	0	0
33	34	500	202.5	98.1
6	35	400	145.8	70.6
35	36	100	202.5	98.1
11	37	100	453.6	219.6
37	38	100	453.6	219.6
12	39	520	145.8	70.6
39	40	620	324	156.9
13	41	200	510.3	247.1
41	42	100	453.6	219.6
16	43	500	145.8	70.6
43	44	300	202.5	98.1
44	45	100	0	0
45	46	900	202.5	98.1
46	47	100	202.5	98.1
45	48	600	202.5	98.1

(continued)

Table A.2 Continued

Sending bus	Receiving bus	Length (m)	P (kW)	Q (kVAr)
48	49	500	202.5	98.1
17	50	300	259.2	125.5
18	51	400	81	39.2
22	52	800	0	0
52	53	800	0	0
53	54	900	202.5	98.1
52	55	400	81	39.2
23	56	800	145.8	70.6
56	57	300	81	39.2
25	58	200	81	39.2
58	59	2200	40.5	19.6
28	60	500	145.8	70.6
32	61	1700	129.6	62.8
61	62	700	60.75	29.4
62	63	1100	81	39.2
33	64	300	0	0
64	65	700	60.75	29.4
65	66	950	202.5	98.1
64	67	800	202.5	98.1
18	68	500	145.8	70.6
68	69	200	145.8	70.6
69	70	300	324	156.9
70	71	760	81	39.2
71	72	700	0	0
72	73	1500	0	0
73	74	500	0	0
74	75	200	129.6	62.8
75	76	400	0	0
76	77	400	60.75	29.4
77	78	550	324	156.9
78	79	600	0	0
79	80	3300	202.5	98.1
72	81	300	60.75	29.4
81	82	100	526.5	254.9
74	83	600	145.8	70.6
76	84	300	129.6	62.8
79	85	700	202.5	98.1
73	86	600	145.8	70.6
86	87	200	0	0
87	88	300	202.5	98.1
88	89	500	0	0
89	90	700	81	39.2
90	91	300	0	0
91	92	500	24.3	11.8
92	93	300	0	0

(continued)

Table A.2 Continued

Sending bus	Receiving bus	Length (m)	P (kW)	Q (kVAr)
93	94	620	60.75	29.4
94	95	1900	324	156.9
89	96	100	0	0
96	97	360	0	0
97	98	500	81.00	39.20
98	99	650	202.5	98.1
96	100	300	324	156.9
97	101	400	129.6	62.8
91	102	1000	324	156.9
93	103	1000	60.75	29.4
1	104	4700	145.8	70.6
15	105	1000	0	0
104	105	300	0	0
105	106	100	0	0
106	107	220	0	0
107	108	500	510.3	247.1
108	109	120	324	156.9
109	110	600	648	313.8
110	111	250	453.6	219.6
111	112	180	453.6	219.6
112	113	200	145.8	70.6
113	114	100	0	0
114	115	470	145.8	70.6
115	116	500	145.8	70.6
116	117	720	81	39.2
117	118	420	81	39.2
118	119	500	145.8	70.6
119	120	960	81	39.2
120	121	700	0	0
121	122	550	145.8	70.6
122	123	620	0	0
123	124	300	0	0
124	125	400	405	196.1
125	126	600	145.8	70.6
126	127	500	145.8	70.6
127	128	300	145.8	70.6
128	129	100	81	39.2
129	130	1800	202.5	98.1
106	131	700	0	0
131	132	400	81	39.2
131	133	400	145.8	70.6
107	134	100	453.6	219.6
134	135	100	129.6	62.8
114	136	500	0	0
136	137	400	81	39.2
137	138	150	145.8	70.6
136	139	100	453.6	219.6
139	140	300	1012.5	490.3

(continued)

Table A.2 Continued

Sending bus	Receiving bus	Length (m)	P (kW)	Q (kVAr)
115	141	300	145.8	70.6
141	142	300	145.8	70.6
121	143	100	81	39.2
143	144	900	202.5	98.1
123	145	200	145.8	70.6
145	146	800	0	0
146	147	400	202.5	98.1
147	148	300	202.5	98.1
148	149	380	145.8	70.6
149	150	840	0	0
150	151	420	259.2	125.5
146	152	300	145.8	70.6
150	153	600	202.5	98.1
124	154	800	202.5	98.1
2	155	400	0	0
155	156	150	145.8	70.6
156	157	800	0	0
157	158	550	0	0
158	159	370	0	0
159	160	280	60.8	29.4
160	161	630	0	0
116	162	1000	81	39.2
161	162	350	81	39.2
158	163	300	60.8	29.4
159	164	500	453.6	219.6
161	165	600	145.8	70.6
165	166	500	145.8	70.6
166	167	480	81	39.2
2	168	1200	0	0
157	168	1600	0	0
168	169	900	81	39.2
169	170	280	0	0
170	171	2200	202.5	98.1
171	172	1000	202.5	98.1
172	173	350	0	0
173	174	600	145.8	70.6
174	175	1200	81	39.2
87	176	1000	202.5	98.1
175	176	300	202.5	98.1
168	177	900	202.5	98.1
177	178	300	259.2	125.5
178	179	400	202.5	98.1
179	180	500	202.5	98.1
170	181	800	81	39.2
181	182	660	202.5	98.1
182	183	900	202.5	98.1

(continued)

Table A.2 Continued

Sending bus	Receiving bus	Length (m)	P (kW)	Q (kVAr)
174	184	100	0	0
184	185	400	129.6	62.8
184	186	400	129.6	62.8
2	187	970	129.6	62.8
187	188	620	0	0
155	189	1000	259.2	125.5
188	189	1100	259.2	125.5
188	190	300	145.8	70.6

References

[1] Civanlar, S., Grainger, J.J., Yin, H., and Lee, S.S.H. 'Distribution feeder reconfiguration for loss reduction'. *IEEE Transactions on Power Delivery,* 1988;3(3):1217–1223.

[2] Lavorato, M., Franco, J.F., Rider, M.J., and Romero, R. 'Imposing radiality Constraints in distribution system optimization problems'. *IEEE Transactions on Power Systems,* 2012;**27**(1):172–180.

[3] Rao, R.S., Ravindra, K., Satish, K., and Narasimham, S.V.L. 'Power loss Minimization in distribution system using network reconfiguration in the Presence of distributed generation'. *IEEE Transactions on Power Systems,* 2013;28(1):317–325.

[4] Dorostkar-Ghamsari, M.R., Fotuhi-Firuzabad, M., Lehtonen, M., and Safdarian, A. 'Value of Distribution network reconfiguration in presence of renewable energy resources'. *IEEE Transactions on Power Systems,*2016;31(3):1879–1888.

[5] Shen, F., Huang, S., Wu, Q., Repo, S., Xu, Y., and Ostergaard, J. 'Comprehensive Congestion management for distribution networks based on dynamic tariff, reconfiguration, and re-profiling product'. *IEEE Transactions Smart Grid,* 2019;10(5):4795–4805.

[6] Tian, Z., Wu, W., Zhang, B., and Bose, A. 'Mixed-integer second-order cone programing model for VAR optimisation and network reconfiguration in active distribution networks'. *IET Generation, Transmission & Distribution,* 2016;10(8):1938–1946.

[7] Lee, C., Liu, C., Mehrotra, S., and Bie, Z. 'Robust distribution network reconfiguration'. *IEEE Transactions Smart Grid,* 2015;6(2):836–842.

[8] Zhan, J., Liu, W., Chung, C.Y., and Yang, J. 'Switch opening and exchange method for stochastic distribution network reconfiguration'. *IEEE Transactions on Smart Grid,* 2020;11(4):2995–3007.

[9] Farahani, V., Vahidi, B., and Abyaneh, H.A. 'Reconfiguration and capacitor Placement simultaneously for energy loss reduction based on an improved reconfiguration method'. *IEEE Transactions on Power Systems,* 2012;27(2):587–595.

[10] Chen, S., Hu, W., and Chen, Z. 'Comprehensive Cost minimization in distribution Networks using segmented-time feeder reconfiguration and reactive power control of distributed generators'. *IEEE Transactions on Power Systems,* 2016;31(2):983–993.

[11] Wang, Y., Xu, Y., Li, J., He, J., and Wang, X. 'On the radiality constraints for Distribution system restoration and reconfiguration problems'. *IEEE Transactions on Power Systems,* 2020;35(4):3294–3296.

[12] Yang, Z., Xie, K., Yu, J., Zhong, H., Zhang, N., and Xia, Q.X. 'A general formulation of linear power flow models: Basic theory and error analysis'. *IEEE Transactions on Power Systems,* 2019;34(2):1315–1324.

[13] Yang, J., Zhang, N., Kang, C., and Xia, Q. 'A state-independent linear power flow Model with accurate estimation of voltage magnitude'. *IEEE Transactions on Power Systems,* 2017;32(5):3607–3617.

[14] Bai, L., Wang, J., Wang, C., Chen, C., and Li, F. 'Distribution locational marginal Pricing (DLMP) for congestion management and voltage support'. *IEEE Transactions on Power Systems,* 2018;33(4):4061–4073.

[15] Taylor, J.A. and Hover, F.S. 'Convex models of distribution system reconfiguration'. *IEEE Transactions on Power Systems,* 2012;27(3):1407–1413.

[16] Zimmerman, R.D., Murillo-Sanchez, C.E., and Thomas, R.J. 'MATPOWER: Steady-state operations, planning, and analysis tools for power systems research and education'. *IEEE Transactions on Power Systems,* 2011;26(1):12–19.

Index

www.ingramcontent.com/pod-product-compliance
Lightning Source LLC
LaVergne TN
LVHW011557200926
848484LV00001B/326
9781839539770